Wastewater Reuse and Management

Sanjay K. Sharma • Rashmi Sanghi
Editors

Wastewater Reuse and Management

Springer

Editors
Sanjay K. Sharma
Department of Chemistry
JECRC University
Jaipur, India

Rashmi Sanghi
R-2 Media Lab
Indian Institute of Technology
Kanpur, UP, India

ISBN 978-94-007-4941-2 ISBN 978-94-007-4942-9 (eBook)
DOI 10.1007/978-94-007-4942-9
Springer Dordrecht Heidelberg New York London

Library of Congress Control Number: 2012951337

Printed on acid-free paper

Springer is part of Springer Science+Business Media (www.springer.com)

This book is for the coming generations, to make them aware about the "cost" of water.

-Sanjay K. Sharma and Rashmi Sanghi

Preface

Anyone who can solve the problems of water will be worthy of two Nobel prizes - one for peace and one for science.

John F. Kennedy

When the well is dry, we learn the worth of water.

Benjamin Franklin

Water pollution and water treatment are strictly interrelated, meaning that it is not possible to explore one topic without the other. Researchers all over the world have been trying hard to address these issues while exploring new ways to fight water scarcity. Water reuse stands as one of the smartest options to fight water scarsity, although it has been overlooked by the research community so far.

However, as the increase in the world population is causing a rise in industrial developments and an overexploitation of our water resources, it is quite obvious that the volume of waste water is destined to rise accordingly. Thus, waste water management has become a crucial issue that needs to be addressed correctly.

This book presents the studies of some of the most prestigious international scientists and gathers them in three different sections: *Wastewater Management and Reuse*, *Wastewater Treatment Options*, and *Risk Assessment*. The result is an insightful analysis of waste water management, its treatments, and the processes that have been studied, optimized and developed so far to sustain our environment.

Wastewater Reuse and Management is a valuable resource to academic researchers, students, institutions, environmentalists, and anyone interested in environmental policies aimed at safeguarding both the quality and the quantity of water.

We would like to thank Springer for publishing this title, and we sincerely welcome feedback from our valuable readers and critics.

Reuse water, save water!

Sanjay K. Sharma
Rashmi Sanghi

Acknowledgments

The time has come to express our sincere gratitude to all our friends, supporters, and well wishers. We are heartily obliged for the support they have shown us while writing "***Wastewater Reuse and Management.***"

First of all, we would like to thank all the esteemed contributors of this book. Without their contribution, none of this would have not been possible.

Professor Sharma would like to start by expressing his sincere gratitude to his teachers, Dr. R.K. Bansal, Dr. R.V. Singh, Dr. R.K. Bhardwaj, and Dr. Saraswati Mittal, the *Gurus* behind all his academic achievements and publications.

He then acknowledges Ackmez Mudhoo, Dr. Nabuk Eddy, Dr. Dong Chen, Dr. V.K. Garg, and all his friends and colleagues at the Jaipur Engineering College and Research Centre (JECRC) for their active interest and moral support.

Finally, he praises his family. His parents, Dr. M.P. Sharma and Mrs. Parmeshwari Devi; his wife, Dr. Pratima Sharma; and all his family members for their never-ending encouragement, moral support and patience over the months spent writing this book. A special thank you goes to his children, Kunal and Kritika: valuable moments of their lives have been missed because of his busy schedule.

Professor Sanghi would like to thank her family, who supported her all along; her children, Surabhi and Udit, for their admirable patience and understanding that gave her the strength to face this challenging project; her husband, Dheeraj Sanghi, whose encouragement and optimism at every stage have been precious; and her family friend, Prof. Sudhir Jain, for her support, insights, and critical comments. Thank you also to her group of friends at IIT Kanpur for the relaxing walks they took together. Last but not least, a big thank you goes to Dr. Sanjay Sharma, editor of this book who "dragged" her into this amazing project.

We are beholden to many other people whose names we have not been able to mention here but whose guidance has been very valuable. Finally, we would like to thank our valuable readers and critics for encouraging us to do more and more research on this issue.

Save water! Think green!

Sanjay K. Sharma
Rashmi Sanghi

About the Editors

Prof. (Dr.) Sanjay K. Sharma is a very well-known author and editor of many books, research journals, and hundreds of articles from the last twenty years. His recently published books are "Green Corrosion Chemistry and Engineering" (from Wiley-VCH, Germany), "Green Chemistry for Environmental Sustainability," "Handbook on Applications of Ultrasound: Sonochemistry and Sustainability" (both from CRC Taylor & Francis Group, LLC, Florida, Boca Raton, USA) and "Handbook of Applied Biopolymer Technology: Synthesis, Degradation and Applications" (from Royal Society of Chemistry, UK). He has also been appointed as Series Editor by Springer's UK for their prestigious book series "Green Chemistry for Sustainability." His work in the field of Green Corrosion Inhibitors is very well recognized and praised by the international research community. Other than this, he is known as a person who is dedicated to educate people about environmental awareness, especially for rain water harvesting.

Dr. Sharma has 13 Books of Chemistry from National-International Publishers and over 48 research papers of National and International repute to his credit.

Dr. Sharma is also serving as editor-in-chief for four international research journals "RASAYAN Journal of Chemistry," "International Journal of Chemical, Environmental and Pharmaceutical Research," "International Journal of Water Treatment & Green Chemistry," and "Water: Research & Development." He is also a reviewer for many other international journals including the prestigious Green Chemistry Letters and Reviews.

Presently he is working as Professor and Head, Department of Chemistry at JECRC University, Jaipur (Rajasthan) India, where he is teaching engineering chemistry and environmental engineering courses to B. Tech. students and spectroscopy courses to PG students and pursuing his research interests. He is a member of American Chemical Society (USA), International Society for Environmental Information Sciences (ISEIS, Canada) and Green Chemistry Network (Royal Society of Chemists, UK) and is also life member of various international professional societies including International Society of Analytical Scientists, Indian Council of Chemists, International Congress of Chemistry and Environment, Indian Chemical Society, etc.

E-mail: drsanjay1973@gmail.com

Dr. Rashmi Sanghi is currently working as a research consultant at the Indian Institute of Technology Kanpur and guest faculty at the LNM Institute of Information Technology, Jaipur. After obtaining her D.Phil degree from Chemistry Department, University of Allahabad, India, in 1994, she has been working at the Indian Institute of Technology Kanpur India as a research scientist. She was a visiting scientist in the chemistry department of Rutgers University, Piscataway, NJ, USA in 1997 and worked with Professor Alan S. Goldman.

She is passionate about environmental green chemistry, and her major research interests are bioremediation, biopolymers and biosynthesis of nanomaterials using microbes and/or polysaccharides. Her research mainly focuses on the development of methods that can help in minimizing or eliminating the hazardous substances in the environment. Some of this work related to the green environmental chemistry is nothing but outstanding. She has traveled worldwide for various academic activities and/or professional talks. Her work on design and application of biopolymers in wastewater remediation shows great promise as evident by her patents "A method for preparing auto capped nano particles such as CdS in continuous flow columns" using fungus and "Poly(acryl amide) grafted Cassia grandis-silica hybrid: efficient metal ion adsorbent." She has over 90 international journal publications to her credit. She has published three books on green chemistry: "Green Chemistry: Environment Friendly Alternatives" (2003), "Green Chemistry and Sustainable Development" (2005) and "Green Chemistry for Environmental Remediation" (2011). She is a member of many academic societies and reviewer of many international journals. Rashmi Sanghi is the recipient of the CHEMRAWN VII Prize for Atmospheric and Green Chemistry for the year 2012.

E-mail: rsanghi@iitk.ac.in; rsanghi@gmail.com

Contents

Contributors

Tuty Emilia Agustina Chemical Engineering Department, Sriwijaya University, Palembang, Indonesia

M. Al Aukidy Department of Engineering, University of Ferrara, Ferrara, Italy

T.V. Arjunan Department of Mechanical Engineering, Coimbatore Institute of Engineering and Technology, Coimbatore, Tamilnadu, India

A. Bes-Piá Department of Chemical and Nuclear Engineering, Universidad Politécnica de Valencia, Valencia, Spain

Giuseppe Castaldelli Biology and Evolution Department, University of Ferrara, Ferrara, Italy

Mahesh Chandra Chattopadhyaya Environmental Chemistry Research Laboratory, University of Allahabad, Allahabad, U. P, India

Nicolò Colombani Earth Sciences Department, University of Ferrara, Ferrara, Italy

Patrice Monteiro de Aquim Chemical Engineering Department, Laboratory for Leather and Environmental Studies, Federal University of Rio Grande do Sul, Porto Alegre, RS, Brazil

A. Elkamel Department of Chemical Engineering, Petroleum Institute, Abu Dhabi, UAE

Department of Chemical Engineering, University of Waterloo, Waterloo, ON, Canada

M.V. Galiana-Aleixandre Department of Chemical and Nuclear Engineering, Universidad Politécnica de Valencia, Valencia, Spain

A. Galletti Department of Engineering, University of Ferrara, Ferrara, Italy

Ravindra Kumar Gautam Environmental Chemistry Research Laboratory, University of Allahabad, Allahabad, U. P., India

Guang Liu School of Earth and Space Science, University of Science and Technology of China (USTC), Hefei, Anhui, China

Advanced Lab for Ecological Safety and Human Health, Suzhou Institute of USTC, Suzhou, Jiangsu, China

Gui-Jian Liu School of Earth and Space Science, University of Science and Technology of China (USTC), Hefei, Anhui, China

Mariliz Gutterres Chemical Engineering Department, Laboratory for Leather and Environmental Studies, Federal University of Rio Grande do Sul, Porto Alegre, RS, Brazil

Jacqueline Xiao Wen Hay Chemical Engineering Discipline, School of Engineering, Monash University, Bandar Sunway, Selangor Darul Ehsan, Malaysia

Andrew F.A. Hoadley Department of Chemical Engineering, Monash University, Clayton Campus, Victoria, Australia

Keisuke Ikehata Pacific Advanced Civil Engineering, Inc., Fountain Valley, CA, USA

Andreas Kallioras School of Mining and Metallurgical Engineering, National Technical University of Athens, Athens, Greece

Giovanni Libralato Department of Environmental Sciences, Informatics and Statistics, University Cà Foscari Venice, Dorsoduro, Venice, Italy

ECSIN - European Center for the Sustainable Impact of Nanotechnology, Veneto Nanotech S.C.p.A, Rovigo, Italy

Pei Nie Lim Chemical Engineering Discipline, School of Engineering, Monash University, Bandar Sunway, Selangor Darul Ehsan, Malaysia

Su Lin Lim Chemical Engineering Discipline, School of Engineering, Monash University, Bandar Sunway, Selangor Darul Ehsan, Malaysia

Víctor M. Luna-Pabello Laboratorio de microbiología Experimental, Departamento de Biología, Facultad de Química, Universidad Nacional Autónoma de México, Ciudad Universitaria, Universidad Nacional Autónoma de México, México, D.F, Mexico

Akrama Mahmoud Laboratoire de Thermique Energétique et Procédés (EAD 1932), ENSGTI, Pau, France

Micòl Mastrocicco Earth Sciences Department, University of Ferrara, Ferrara, Italy

J.A. Mendoza-Roca Department of Chemical and Nuclear Engineering, Universidad Politécnica de Valencia, Valencia, Spain

Abdul Wahab Mohammad Scale-up and Downstream Processing Research Group, Department of Chemical and Process Engineering, Faculty of Engineering and Built Environment, Universiti Kebangsaan Malaysia, Bangi, Selangor Darul Ehsan, Malaysia

Jacek Namieśnik Department of Analytical Chemistry, Chemical Faculty, Gdańsk University of Technology, Gdańsk, Poland

Jérémy Olivier Laboratoire de Thermique Energétique et Procédés (EAD 1932), ENSGTI, Pau, France

Luis Alfredo Ortega-Clemente Boca del Río Institute of Technology (ITBoca), Division of Graduate Studies and Research, Veracruz, Mexico

Fotios-Konstantinos Pliakas Department of Civil Engineering, Democritus University of Thrace, Xanthi, Greece

Héctor M. Poggi-Varaldo CINVESTAV del IPN, México, Department Biotechnology and Bioengineering, Environmental Biotechnology R&D Group, CINVESTAV del IPN, México, D.F, Mexico

Randolf Rausch Deutsche Gesellschaft fuer Internationale Zusammenarbeit (GIZ) GmbH, Riyadh Office, Riyadh, Kingdom of Saudi Arabia

Justyna Rogowska Department of Analytical Chemistry, Chemical Faculty, Gdańsk University of Technology, Gdańsk, Poland

Y. Saif Department of Chemical Engineering, University of Waterloo, Waterloo, ON, Canada

Department of Chemical Engineering, Petroleum Institute, Abu Dhabi, UAE

K. Sampathkumar Department of Mechanical Engineering, Tamilnadu College of Engineering, Coimbatore, Tamilnadu, India

Christoph Schuth Institute of Applied Geosciences, Darmstadt Technical University, Darmstadt, Germany

P. Senthilkumar Department of Mechanical Engineering, KSR College of Engineering, Tiruchengode, Tamilnadu, India

Sanjay K. Sharma Department of Chemistry, JECRC University, Jaipur, India

Jean Vaxelaire Laboratoire de Thermique Energétique et Procédés (EAD 1932), ENSGTI, Pau, France

P. Verlicchi Department of Engineering, University of Ferrara, Ferrara, Italy

Ta Yeong Wu Chemical Engineering Discipline, School of Engineering, Monash University, Bandar Sunway, Selangor Darul Ehsan, Malaysia

Xue-Bin Yin School of Earth and Space Science, University of Science and Technology of China (USTC), Hefei, Anhui, China

Advanced Lab for Ecological Safety and Human Health, Suzhou Institute of USTC, Suzhou, Jiangsu, China

Zi-Xiang Chen School of Earth and Space Science, University of Science and Technology of China (USTC), Hefei, Anhui, China

Advanced Lab for Ecological Safety and Human Health, Suzhou Institute of USTC, Suzhou, Jiangsu, China

Amit Chanan Faculty of Engineering and Information Technology School of Civil and Environmental Engineering, University technology Sydney, NSW, Australia

Saravanamuth Vigneswaran Faculty of Engineering and Information Technology School of Civil and Environmental Engineering, University technology Sydney, NSW, Australia

Jaya Kandasamy Faculty of Engineering and Information Technology School of Civil and Environmental Engineering, University technology Sydney, NSW, Australia

Bruce Simmons School of Sciences and Health, University of Western Sydney, Hawkesbury Campusm, Richmond, NSW, Australia

About the Contributors

Tuty Emilia Agustina is currently Senior Lecturer in the Department of Chemical Engineering at Sriwijaya University, South Sumatera, Indonesia. Dr. Agustina completed her PhD in the area of Advanced Oxidation Processes (AOPs) from Curtin University of Technology, Western Australia. Her research areas are in environmental engineering and industrial wastewater treatment with special emphasis on wasterwater treatment by AOPs.

Mustafa Al Aukidy took a degree in Civil Engineering at the University of Baghdad (Iraq). Currently he is a Ph.D. student at the Department of Engineering at the University of Ferrara (Italy). His research fields regard the environmental pollution issues and in particular occurrence and removal of micropollutants from wastewater, risk environmental assessment and environmental policy.

T.V. Arjunan is Professor of Mechanical Engineering at the Coimbatore Institute of Engineering and Technology, Coimbatore, India. He received his B.E. and M.E. in Mechanical Engineering from Bharathiyar University and his Ph.D. in Mechanical Engineering from Anna University. His areas of interest are solar thermal application, heat transfer and alternative fuels.

Amparo Bes-Piá obtained a degree of Chemical Engineering in 1999 and Ph.D. in 2003 about textile wastewater treatment using physical-chemical and membrane treatments. Since 2001 lecturer at the Universitat Politècnica de València teaching wastewater treatment at the Industrial Engineering School. Research field in recycling of industrial wastewater, mainly by membrane processes. Several publications in scientific international journals and contributions to international congresses.

Giuseppe Castaldelli is a biologist who has spent most of his professional life studying nitrogen biogeochemistry, eutrophication and nitrate contamination of groundwater. He is now researcher in ecology at the Department of Biology and Evolution of the University of Ferrara (IT).

Amit P. Chanan is the General Manager Strategic Assets at the NSW State Water Corporation. He is an advocate for sustainable water resource management. Amit has led the path in changing the direction of Sydney's local councils from traditional drainage engineering entities to holistic water cycle managers. His knowledge and leadership were imperative in making Kogarah Council (where he previously worked as Director of Assets & Services) one of the leaders in integrated water cycle management in Australia, and the first in Sydney metropolitan to adopt decentralised wastewater recycling.

Mahesh Chandra Chattopadhyaya obtained Ph.D. degree from IIT Bombay. He joined University of Allahabad as lecturer in 1974 and thereafter he became Reader and Professor of Chemistry. He served the University as Head of the Department of Chemistry during the period 2008–2010. Besides teaching Inorganic and Analytical chemistry he also taught Environmental Chemistry at the University. Twenty-four students obtained doctoral degree under his supervision from the university and 6 are working. He has published more than 100 papers in national and international journals and delivered talks in several international conferences. He is currently the president of Indian Chemical Society.

Nicolò Colombani is a hydrogeologist specialized in contaminated groundwater systems and application of reactive transport models. He obtained his Ph.D. degree in 2011 and he is now working as a contract researcher at the Earth Sciences Department of the University of Ferrara (IT).

Patrice Monteiro de Aquim is head of leather group of Master's Material Degree at Feevale University. She graduated in Chemical Engineering at Pontifícia Universidade Católica do Rio Grande do Sul (2002) and obtained his Master in Chemical Engineering (2004) and Ph.D. in Chemical Engineering at Federal University of Rio Grande do Sul (2009) focusing mainly on issues related to leather, sustainable tanning and water reuse.

A. Elkamel is a Professor of Chemical Engineering at the University of Waterloo, Canada, and is currently on visit to The Petroleum Institute. He obtained his B.S. in Chemical and Petroleum Refining Engineering and B.S. in Mathematics from Colorado School of Mines, M.S. in Chemical Engineering from the University of Colorado-Boulder and Ph.D. in Chemical Engineering from Purdue University. His specific research interests are in computer-aided modelling, optimisation and simulation with applications to the petroleum and petrochemical industry. He has contributed more than 200 publications in refereed journals and international conference proceedings and serves on the editorial board of several journals, including the *International Journal of Process Systems Engineering*, *Engineering Optimisation*, and the *Int. J. Oil, Gas, Coal Technology*.

M.V. Galiana-Aleixandre obtained a chemical engineering degree in 1999 and Ph. D. in 2010. Since 2001 she works in the tannery Industrias del Curtido S.A. in Valencia (Spain). Currently, she is responsible for research and development in the company. She is also an external associate of the Chemical and Nuclear Engineering Department of the Universitat Politècnica de València since 2003.

A. Galletti took a degree in Environmental Engineering at the University of Ferrara (Italy) and a Ph.D. in Environmental Engineering. He worked for 5 years at the University of Ferrara in the Environmental and Sanitary Engineering group and his research fields refer to treatments of wastewaters (urban, hospital, industrial) and options for reuse for agricultural and industrial purposes. Now he is working at CH2MHILL as Process Engineer.

Ravindra Kumar Gautam did his post-graduation in Environmental Science in 2009 from University of Allahabad, India. Thereafter he worked for one year in National Environmental Engineering Research Institute (NEERI), Nagpur, India. He qualified for CSIR-UGC National Eligibility Test for Junior Research Fellowship. Currently he is engaged in doctoral work in the University of Allahabad, India.

Mariliz Gutterres is a Chemical Engineer (1984), M.Sc. Engineer (Brazil, 1996) and Doctor in Chemistry (2001). She is head of the Laboratory for Leather and Environment Studies at UFRGS and author of papers, publications and lectures in journals, congresses and conferences. She also participates in several committees and in the management of leather industry associations.

Jacqueline Xiao Wen Hay is currently a Ph.D. student at Monash University, Malaysia. She received her Bachelor of Engineering (Chemical Engineering) with Honors from Monash University in 2011. Her research area is on wastewater reuse as a substrate in biohydrogen production through photo-fermentation process.

Andrew F.A. Hoadley is a senior lecturer in the department of Chemical Engineering at Monash University, Australia. His research interests include designing processing plants for better environmental performance, upgrading of waste materials, industrial ecology and sustainability, dewatering, steam drying, and wastewater treatment and related to these areas. He graduated from Melbourne University Chemical Engineering in 1981 and completed his Ph.D. at the University of Cambridge (UK) in 1988.

Keisuke Ikehata received his doctorate in Civil and Environmental Engineering in 2003 from the University of Alberta in Edmonton, Alberta, Canada. Dr. Ikehata is a registered professional engineer in Alberta, Canada, and an Advanced Water R&D Manager at Pacific Advanced Civil Engineering, Inc. in Fountain Valley, California. He has more than 14 years of experience in basic and applied research in water and wastewater science and engineering. His areas of expertise include advanced wastewater treatment for water reuse, water chemistry and biology, environmental microbiology and biotechnology and xenobiotics toxicology.

Andreas Kallioras is a Lecturer at National Technical University of Athens (Greece), Guest Fellow at Environmental Research Center, UFZ, Leipzig (Germany) and Visiting Lecturer at Technical University of Darmstadt (Germany). His main focus is groundwater hydrology, arid and coastal hydrogeology and transboundary water management.

Jaya Kandasamy is an Associate Professor in the Centre for Technology in Water and Wastewater at the University of Technology Sydney. Prior to taking on the academic role at the university, Dr Kandasamy worked at the Sydney Water Corporation and NSW Department of Land and Water Conservation.

Giovanni Libralato is a postdoctoral researcher and a contract professor at the University Cà Foscari Venice (Italy) and a researcher at the European Center for the Sustainable Impact of Nanomaterials (ECSIN, Veneto Nanotech S.C.p.A.) in Rovigo (Italy). He earned a B.S., M.S. and Ph.D. in Environmental Sciences from University Cà Foscari Venice. His research interests include (eco)toxicity, toxicant effects on populations, factors modifying toxicity, aquatic ecotoxicology, inorganic water chemistry, wastewater treatment, soil remediation, recovery and reuse of by-products.

Pei Nie Lim is currently a Ph.D. student at Monash University, Malaysia. She received her Bachelor of Engineering (Chemical Engineering) with Honors from Monash University in 2010. Her research area is on organic solid waste management through vermicomposting process.

Su Lin Lim is currently a Ph.D. student at Monash University, Malaysia. She received her Bachelor of Engineering (Chemical Engineering) with Honors from Monash University in 2011. Her research area is on organic waste management through biological and vermicomposting processes.

Víctor M. Luna-Pabello is Titular Professor "B" Full Time at the National Autonomous University of Mexico (UNAM). Biologist, Master of Science (Biology) and Doctor of Sciences (Biology) in the Faculty of Sciences at UNAM with a post-doctoral stay in the National Institute for Applied Sciences in Toulouse, France.

Akrama Mahmoud is a senior research fellow in the Chemical Engineering Department at the University of Pau at the Laboratoire de Thermique Energétique et Procédés, Pau, France. His research interests include advanced/assisted liquid/solids separations technologies, purification processes involving mineral and vegetable suspensions, protein extraction and electrodeionization. He is a co-inventor of TAMD process. He received his Ph.D. in Chemical Engineering from Institute National Polytechnique de Lorraine-Ecole Nationale Supérieure des Industries Chimiques, Nancy, France.

Micòl Mastrocicco is a hydrogeologist specialized in lowland groundwater systems and groundwater quality assessment. She obtained her Ph.D. degree in 2003 and she is now lecturer in hydrogeology and applied hydrogeology at the Earth Sciences Department of the University of Ferrara (IT).

J.A. Mendoza-Roca obtained a degree in Industrial Engineering in 1995. He is Associate Professor in the Chemical and Nuclear Engineering Department of the Universitat Politècnica de València and received his Ph.D. in 2000. His research is devoted to membrane technologies and wastewater recycling. He has published numerous works and research papers about treatment and reuse of textile and tannery wastewaters.

Abdul Wahab Mohammad is currently a Professor in Membrane and Separations Technology in the Department of Chemical and Process Engineering, Universiti Kebangsaan, Malaysia. His research interests are on membrane separations, water and wastewater treatment and process economics. He was one of the recipients of the Prince Sultan International Prize for Water in 2008, and currently he is one of the Editorial Board Members in *Desalination*.

Jacek Namieśnik (born 1949) obtained his Ph.D. in 1978. Since 1996 he has become Full Professor. He is a renowned specialist in the field of analytical chemistry and environmental chemistry and author or co-author of more than 400 publications. He has supervised 39 Ph.D. works and co-supervised 5 Ph.D. dissertations.

Jéremy Olivier received his Ph.D. in Chemical Engineering from the University of Pau and Pays de l'Adour, France. He is presently an Associate Professor in Chemical Engineering at the University of Pau and Pays de l'Adour. His current field of research focuses on sewage sludge treatment (dewatering, electro-dewatering).

Luis Alfredo Ortega-Clemente is a Full Professor of Boca del Río Institute of Technology (ITBoca), of the Division of Graduate Studies and Research in Boca del Río City, Veracruz, Mexico. He obtained his title of Biochemical Engineer in the Veracruz Institute of Technology (ITVer) of Veracruz City, Mexico, Master of Science and Doctoral of Science in Biotechnology from CINVESTAV del IPN in México City, Mexico and a postdoctoral stay in UNAM in México City.

Fotios-Konstantinos Pliakas is Associate Professor at the Department of Civil Engineering, Democritus University of Thrace, Greece. His main focus is groundwater management, groundwater flow modeling, management of aquifer recharge, saline water intrusion, climate change and groundwater resources, transboundary rivers and aquifers.

Héctor M. Poggi-Varaldo is a Full Professor in the Department of Biotechnology and Bioengineering at the Centre for Higher Studies and Research (CINVESTAV del IPN) in Mexico City, Mexico. He holds a Sc.D. degree in Biotechnology from CINVESTAV del IPN, an M.Sc. Environ. Eng. from the National Autonomous University of Mexico (UNAM), and a B.Sc. Chem. Eng. from the National University of Uruguay (UROU).

Randolf Rausch is Technical Director of the GIZ-IS Project "Water Resources Studies" in the Kingdom of Saudi Arabia and Professor at Technical University of Darmstadt (Germany). He has been working in the Middle East since 2003. His main focus of interest is integrated water management and groundwater modeling.

Justyna Rogowska (born 1979) obtained her Ph.D. degree in Chemistry at the Gdańsk University of Technology in 2011. Her major research interests include environmental chemistry and ecotoxicology (with particular emphasis on the marine environment) as well as the legal aspects of environmental protection, in particular of marine environmental protection. She is co-Author of six publications (including four papers in journals on the ISI Master Journal List) and one book chapter.

Y. Saif is currently a Reach/Teaching associate at the Petroleum Institute, Abu-Dhabi, UAE. He received his B.S. in Chemical Engineering from King Abdul-Aziz University, M.S. in Chemical Engineering from Dayton University and Ph.D. in Chemical Engineering from the University of Waterloo. His research interests are in process systems engineering and optimization with applications to waste treatment and minimization and the oil and gas industry.

K. Sampathkumar is Associate Professor of Mechanical Engineering at the Tamilnadu College of Engineering, Coimbatore, India. He received his B.E. degree in Mechanical Engineering from Bharathiyar University and his M.S. and Ph.D. in Mechanical Engineering from Anna University. His areas of interest are solar desalination, energy policies, heat transfer and engineering education.

Christoph Schuth is a Chair in Hydrogeology at Technical University of Darmstadt (Germany) and is the Scientific Director WR in IWW Water Centre (Germany). His main focus is hydrogeology of arid environments, coastal hydrogeology, water treatment and transboundary water management.

P. Senthilkumar is Professor of Mechanical Engineering at the KSR College of Engineering, Tiruchengode, India. He received his M.E. in Mechanical Engineering from Bharathiyar University and his Ph.D. in Mechanical Engineering from Indian Institute of Technology Madras (IITM), Chennai. He has also received numerous awards and citation of excellence as an engineering educator.

Bruce Simmons is a Research Scientist at the University of Western Sydney (USW). In a career spanning 30 years Bruce has worked as an analytical chemist, designated NSW government researcher, divisional deputy manager, senior consultant, senior academic leader (Head of School and Dean of Faculty) and, since 2001, UWS academic and research scientist. His expertise includes Water Resources Planning and Management and Environmental Management Systems.

Jean Vaxelaire obtained his Ph.D. in Chemical Engineering from the "Institute National Polytechnique de Lorraine" in the school of Chemical Engineering (Ecole Nationale Supérieure des Industries Chimiques), Nancy, France, in 1994. He is presently an Associate Professor in Chemical Engineering at the University of Pau and Pays de l'Adour. His current field of research includes sewage sludge treatment (dewatering, drying and incineration).

P. Verlicchi took her degree in Chemical Engineering at the University of Bologna (Italy) and her Ph.D. in Environmental and Safety Chemical Engineering. She is currently an Assistant Professor in Environmental and Sanitary Engineering at the University of Ferrara (Italy) and her research fields refer to the treatments and management of wastewaters (urban, hospital, industrial) and options for reuse for agricultural and industrial purposes.

Saravanamuth Vigneswaran is a Professor in the Centre for Technology in Water and Wastewater at the University of Technology Sydney. He is the Director of the School of Civil and Environmental Engineering at UTS. He has published over 175 technical papers and authored two books (both through CRC press, USA). He has established research links with the leading laboratories in France, Korea, Thailand and the USA.

Ta Yeong Wu is currently a Senior Lecturer in the School of Engineering, Monash University. His current research interests include sustainable waste management, biohydrogen production and advanced oxidation process. He has been serving as an Associate Editor for *International Journal of Environmental Science and Technology* since 2011.

Xue-Bin Yin is an Associate Professor in the Department of Environmental Sciences and the Director of Advanced Lab for Selenium and Human Health, at University of Science and Technology of China (USTC). His research is related to the biogeochemistry of trace elements (such as, selenium, zinc, iodine and fluorine), with a special emphasis on their biological transformation process, health issue and utilization in agriculture and foods.

Zi-Xiang Chen received his dual Ph.D. from the University of Science and Technology of China and City University of Hong Kong in July of 2012. His current research interest lies in two areas: (1) Using Nitrogen and Oxygen isotopic compositions to identify nitrate sources and assess Nitrogen cycle process in surface water system; (2) Developing the smart method of determining the Nitrogen and Oxygen isotopic compositions.

LIU Gui-Jian received his Ph.D. from China University of Mining and Technology in 1999. Currently, he works in the Department of Earth and Space Sciences at the University of Science and Technology, China. He leads and manages a research team whose interest focuses primarily on: (1) environmental geochemistry (organic and inorganic); (2) fuel and environment; (3) trace element and human health; (4) quaternary period and global change; and (5) mineral environment and safety risk. Professor. LIU Gui-Jian is the author of over 100 scientific publications in connection with his research area.

LIU Guang received his Bachelor's degree in 2007 from the University of Science and Technology of China where he is presently a Ph.D. student. His research area focuses on: (1) using nitrogen and oxygen isotopic compositions to identify nitrate sources and assess nitrogen cycle process in lake water system; (2) using oxygen isotope compositions to identify phosphate sources and assess phosphorus cycle process in lake water system; and (3) developing new methods for nitrate and phosphate isotopic compositions detection.

Abbreviations

Nomenclature

Sets

AS	Set of air stripper units
C	Set of pollutants
MIX	Set of mixer nodes
O	Set of unit operations
P	Set of product streams
PERM	Set of final permeate product streams
PV	Set of pervaporation stages
SFPER	Set of final permeate streams in RO network
SFREJ	Set of final reject streams in RO network
SIN	Set of inlet wastewater streams
SPU	Set of pump stages
SPU	Set of pump stages in RO network
SRO	Set of RO stages
SROPER	Set of permeate streams from RO stage
SROREJ	Set of reject streams from RO stage
SSP	Set of splitter nodes
STU	Set of turbine stages
STU	Set of turbine units in RO network
UO	Set of utility units
USI	Set of inlet utility streams
USO	Set of utility-exit streams
VOC	Set of volatile organic compounds
WI	Set of inlet wastewater streams
WO	Set of wastewater-exit streams

Parameters/Variables

$f(x,y)$	Objective function
M	Large number
y	A binary variable
Y	Vector of binary variables
x	A continuous variable
X	Vector of continuous variables
$h(x)$	Vector of functions in equality constraints
$g(x)$	Vector of functions in inequality constraints
$\mathrm{NMd}_{\mathrm{SRO}}$	Number of parallel modules at every RO stage belongs toSRO
a_{MRO}	Cost of the single RO module
$\mathrm{PPu}_{\mathrm{pu}}$	Power consumption by a pump unit belong to *SPU*
$\mathrm{PTu}_{\mathrm{tu}}$	Power production by a turbine unit belong to *STU*
$a_{\mathrm{pu,f}}$	Fixed cost coefficient of a pump unit *pu*
$a_{\mathrm{tu,f}}$	Fixed cost coefficient of a turbine unit *tu*
$a_{\mathrm{pu,o}}$	Fixed cost coefficient for the operating cost of a pump unit *pu*
$a_{\mathrm{tu,o}}$	Fixed cost coefficient for the operating cost of a turbine unit *tu*
ΔP_{SPU}	Pressure difference across a pump unit *pu* belong to *SPU*
F_{SPU}	Flowrate through a pump unit *pu* belong to *SPU*
y_{SPU}	Binary variable defines the existence of pump unit *pu* belong to *SPU*
y_{STU}	Binary variable defines the existence of pump unit *tu* belong to *STU*
ΔP_{STU}	Pressure difference across a turbine unit *tu* belong to *STU*
F_{STU}	Flowrate through a turbine unit *tu* belong to *STU*
$\mathrm{Fp}_{\mathrm{SRO}}$	Permeate flowrate from a RO stage belong to *SRO*
ΔP_{MRO}	Pressure difference across a RO module
W	Water permeability coefficient
SA	Surface area of a RO stage belong to *SRO*
r_i	Inner radius of the hollow fiber membrane
r_o	Outer radius of the hollow fiber membrane
l	RO hollow fiber membrane length
l_s	RO hollow fiber membrane seal length
$xp_{\mathrm{c,SRO}}$	Component concentration of *c* in any permeate stream from RO stage
$xr_{\mathrm{c,SRO}}$	Component concentration of *c* in any reject stream from RO stage
$x_{\mathrm{c-avg,SRO}}$	Average component concentration of *c* in any RO stage
K_{c}	Component permeability coefficient of *c* belong to *C*
$\mathrm{Fp}_{\mathrm{SRO}}$	Permeate flowrate of a RO stage belong to *SRO*
y_{SRO}	Binary variable of a RO stage belong to *SRO*
F_{SRO}	Inlet feed to a RO stage belong to *SRO*
$\mathrm{NMd}_{\mathrm{SRO}}$	Number of parallel modules in RO stage belong to *SRO*
P_{SRO}	Inlet feed pressure to a RO stage belong to *SRO*
$\mathrm{Fr}_{\mathrm{SRO}}$	Reject stream from a RO stage belong to *SRO*

F_{SSP}	Inlet feed to a splitter node belong to *SSP*
$F_{SSP,MIX}$	Stream assignment from a splitter node belong to *SSP* to a mixer node belong to *MIX*
F_{MIX}	Exit stream from a mixer node belong to *MIX*
$y_{SSP,MIX}$	Binary variable defines the stream match between a splitter and a mixer belong to *SSP* and *MIX*
P_{MIX}	Pressure of a stream from a mixer node belong to *MIX*
$P_{SSP,MIX}$	Pressure of a stream $F_{SSP,MIX}$
F_{SPER}	Flowrate of a final permeate product stream belong to SPER
$x_{c,SPER}$	Concentration of a pollutant c in the final permeate stream belong toSPER
F_{in}	Inlet wastewater stream belong to SIN
F_{in-pu}	Stream assignment from inlet node *in* to a pump unit *pu*
$F_{in-frej}$	Stream assignment from the inlet node *in* to the final exit node *frej*
$F_{in-fper}$	Stream assignment from the inlet node *in* to the final exit node *fper*
$F_{rorej-pu}$	Stream assignment from the RO reject stream *rorej* to a pump node *pu*
$F_{roper-pu}$	Stream assignment from the RO permeate stream *roper* to a pump node *pu*
F_{pu}	Pump *pu* feed stream
F_{pu-ro}	Stream assignment from a pump node *pu* to RO stage *ro*
F_{roper}	Permeate stream from RO stage *ro*
F_{rorej}	Reject stream from RO stage *ro*
$F_{roper-fper}$	Permeate stream assignment from RO stage to the final permeate node *fper*
F_{fper}	Permeate stream at the final permeate product node *fper*
$F_{rorej-ro}$	Reject stream assignment from RO stage to another RO stage
$F_{rorej-tu}$	Reject stream assignment from RO reject stream to a turbine node *tu*
$F_{rorej-frej}$	Stream assignment from a RO reject stream to the final reject node *frej*
F_{tu}	Inlet feed stream to a turbine unit *tu*
F_{tu-ro}	Stream assignment from a turbine unit *tu* to RO stage
$F_{tu-fper}$	Stream assignment from a turbine unit *tu* to the final permeate node *fper*
$F_{tu-frej}$	Stream assignment from a turbine unit *tu* to the final reject node *frej*
a	Specific surface area of packing
a_{Ph}	Specific hydraulic surface area of packing
$SurA_{PV}$	Pervaporation stage surface area
C_L	A parameter related to the packing material
$Cost_{Capital}$	Capital cost of the units
$Cost_{operating}$	Operating cost of the treatment
C_P	A parameter related to the packing material

C_S	A parameter related to the packing material
C_V	A parameter related to the packing material
D_L	Diffusivity of a solute in the liquid phase
$\mathrm{D_{AS}}$	Diameter of the air stripper tower
D_V	Diffusivity of a solute in the gas phase
F_A	Air flow rate
$\dot{F}_A$	Molar air flow rate
F_{MIX}	Exit flow stream from mixer MIX
$F_{\mathrm{SSP,MIX}}$	Inlet flow from splitter SSP to mixer MIX
F_{VOC}	Molar flowrate of a VOC in a pervaporation stage
$\dot{F}_W$	Molar wastewater flow rate
F_W	Wastewater flow rate
g	Gravity acceleration
h_L	Specific liquid holdup
hn	Henry's constant
$H_{\mathrm{C,OL}}$	Height of the transfer unit for component c
H_{AS}	Height of the air-stripping column
k_{VOC}	Mass transfer coefficient of a VOC in the concentration polarization layer
K_{VOC}	Overall mass transfer coefficient of VOC in the pervaporation stage
K_{water}	Overall mass transfer coefficient of water in a pervaporation stage
l_m	Thickness of the PV membrane
Lt_m	Length of a PV membrane
MW	Molecular weight
$N_{\mathrm{C,OL}}$	Number of the transfer unit in the air-stripping column for component c
P_{MIX}	Pressure of mixer MIX
$\mathrm{Pm_{VOC}}$	Permeability coefficient of a VOC in the membrane
$\mathrm{Pm_{water}}$	Permeability coefficient of water in the membrane
P_{Perm}	Permeate pressure in a pervaporation stage PV
$\mathrm{PPu_{Air\ blower}}$	Power consumption by a vacuum pump
$\mathrm{PPu_{pump}}$	Power consumption by a pump
$\mathrm{PPu_{Vaccum\ pump}}$	Power consumption by a vacuum pump
P_{SSP}	Pressure of splitter SSP
P_{AS}	Pressure in the column
$\mathrm{Re_L}$	Reynolds number of the liquid phase
$\mathrm{Re_V}$	Reynolds number of the gas phase
u_{L}	Superficial velocity of the liquid phase
$u_{\mathrm{L,S}}$	Superficial velocity of the liquid phase at the loading point
u_{V}	Superficial velocity of the gas phase
$u_{\mathrm{V,S}}$	Superficial velocity of the gas phase at the loading point
x	Mole fraction of a solute in the liquid phase
y_{PV}	Binary variable for pervaporation stage PV

y_{AS}	Binary variable for the air stripper tower belong to AS
$va_{VOC,Perm}$	Average molar concentration of a VOC in the permeate side of a pervaporation
va	Mole concentration of a solute in the gas phase
ρ_M	Molar density of wastewater mixture in a pervaporation stage
ρ_L	Density of the liquid phase
ρ_V	Density of the gas phase
μ_L	Viscosity of the liquid phase
μ_V	Viscosity of the gas phase
$\in$	Void fraction
ψ_S	Resistance coefficient in the liquid and gas critical velocities equations
ψ_L	Resistance coefficient of trickle packing
β_L	Mass transfer coefficient in the liquid phase
β_v	Mass transfer coefficient in the gas phase
α_{pu}	Fractional constant for the fixed cost part of a pump unit
α_{tu}	Fractional constant for the fixed cost part of a turbine unit
π_{MRO}	Osmotic pressure in a RO module
γ	RO module constant
η	RO module constant
μ	Viscosity of water

Abbreviations

AS	Air stripper
ASB	Air stripper box
DB	Distribution box
DCM	Dichloromethane
ED	Electrodialysis
EDC	Ethylene dichloride
EDI	Electordeionization
GAMS	General Algebraic Modeling System
GDB	Generalized Disjunctive Programming
GS	Gas separation
HB	Heat integration box
MCP	Monochlorophenol
MEB	Mass exchange box
MF	Microfiltration
MILP	Mixed integer linear program
MINLP	Mixed integer nonlinear program
NF	Nanofiltration
NLP	Nonlinear program
PB	Pump box
PV	Pervaporation

PVB	Pervaporation box
RO	Reverse osmosis
ROB	Reverse osmosis box
RON	Reverse osmosis network
SEN	State equipment network
STN	State Task Network
TAC	Total annualized cost
TB	Turbine box
TCE	Trichloroethylene
TCP	Trichlorophenol
UF	Ultrafiltration
VOCs	Volatile organic compounds

Notation

CWs	Constructed wetlands
HRT	Hydraulic retention time
HSSF-CW	Horizontal subsurface flow constructed wetlands
OCPs	Organochlorine pesticides
PCBs	Polychlorinated biphenyls
SF-CWS	Surface-flow CWs
SSF-CWs	Subsurface flow CWs
VSSF-CW	Vertical subsurface flow constructed wetland

Part I
Wastewater Management and Reuse

Chapter 1
Wastewater Management Journey – From Indus Valley Civilisation to the Twenty-First Century

Amit P. Chanan, Saravanamuth Vigneswaran, Jaya Kandasamy, and Bruce Simmons

1.1 Introduction

Throughout the human existence on this planet, water and wastewater infrastructure never attracted as much public attention as it did during the second half of the nineteenth century. Following extensive debate on appropriate solutions for problems associated with hygiene and public health, options were chosen from several competing alternatives, while others foreclosed. Sociocultural impacts of these late nineteenth century decisions have lasted until the present day [1]. The shape of our cities and centralised approach embedded within sanitary engineers mindset today owe its existence to the options selected nearly 200 years ago.

The current selection of water and wastewater management options, particularly in the developing world, has the opportunity to learn from 200 years of experience. It is critical that these lessons influence the design of future wastewater management systems because undoubtedly the decisions being made today will impact on the future generations.

This chapter highlights the development of wastewater management through human evolution and settlement history.

A.P. Chanan (✉)
State Water Corporation, Sydney

S. Vigneswaran • J. Kandasamy
Centre for Technology in Water and Wastewater, UTS, Sydney

B. Simmons
University of Western Sydney, Richmond

S.K. Sharma and R. Sanghi (eds.), *Wastewater Reuse and Management*,
DOI 10.1007/978-94-007-4942-9_1,

1.2 Water Supply and Sanitation in Early Civilisations

Perhaps one of the earliest recorded instances of the awareness of water management and the role it plays in human prosperity comes from the ancient Indian text the *Atharva Veda*. It is estimated that the *Atharva Veda* was compiled between 10,000 and 8,000 years Before Christ (B.C.), making it over 10,000 years old [2]. Atharva Veda III.13.5 assigns the king with the responsibility '*to build dams to control the wildly flowing waters so they can be utilised*' by the community. The need for water resource development to provide supply security has grown from there and has become a significant component of human endeavour.

Given the documented awareness of water management in the *Vedas* dating back to over 10,000 years, it is not surprising that the Indus Valley Civilisation in northwest India (now in Pakistan) that flourished 4,000 years ago has shown a sophisticated appreciation of water-related health needs [3]. According to Jansen [4], their sophisticated planning of the structural features of water supply and effluent disposal systems was particularly impressive. '*Fresh water was supplied by a network of wells, sunken cylindrical shafts several meters deep built of wedge-shaped, standard-size bricks. The waste water and other sewage of almost every house was channelled into the drain running along the street outside*' ([4], p.177).

Jansen [4] further states that the inner-urban water supply and effluent disposal systems of the Indus Valley Civilisation are major accomplishments, being the first in the history of mankind to create such waterworks. The homogeneous distribution of wells for water supply and drains and/or soak pits for wastewater disposal throughout the urban area appeared to be independent of social parameters such as house size and consequent social status.

Although not widely known and acknowledged as such, the network of effluent drains built of brick masonry along the streets of Mohenjo-Daro (Indus Valley Civilisation) is an astonishing feat of civil engineering achieved over 4,000 years ago. These drains are evidence that the birth of sanitary engineering that is often attributed to the public health revolution of the nineteenth century Britain, in fact, actually happened thousands of years earlier on the Indian subcontinent. Picture 1.1 below shows a typical brick-lined drain of the Indus Valley Civilisation.

According to Jansen [4], key attributes of the sanitary drains of the Indus Valley Civilisation included the following:

- They typically ran along past the houses on one side of the streets, about 50–60 cm below the surface.
- The U-shape in cross section with the sides and bottom of the drains was built of bricks set in clay mortar.
- Various coverings (loose bricks, flagstones or wooden boards) were used for the open top.
- Covers could be removed to clean the drains as needed.
- A typical drain cross section ranged between 17–25 cm broad and 15–50 cm high, giving a capacity range of between 260 and 1,200 cm^2.
- Drains sloped at a gradient of about 2 cm per metre, with curved structures sited such that frictional loss was cut to a minimum.

Picture 1.1 Over 4,000 years old sanitary drainage system, Indus Valley Civilisation (Source: www.fotopedia.com/items/flickr-4557770060, Image by Benny Lin)

- Where a drain had to traverse a longer distance or where several drains met, a cesspool was installed to avoid clogging by allowing solids to settle. The suspended matter gradually formed a deposit, which could be removed via steps leading down into the pit. These cesspools were likely to be covered over with wooden planks.

This sanitary and water engineering superiority of the Indian subcontinent remained unsurpassed for more than 2,000 years until the coming of the Romans and the rise of civil engineering and architecture [4].

1.3 Greco–Roman Influence on Water Management

According to Chanan and Simmons [5], the Greek and Roman civilisations achieved the next level of urban water management. Until then, cities and towns were largely constructed adjacent to rivers and streams to ensure a continuum of water supply and waste transport. The security of the Roman political system, in particular, allowed for the development of water transport systems from remote

Picture 1.2 Roman aqueduct (Source: http://images.cdn.fotopedia.com/flickr-2354917044-image.jpg)

sources. With the formation of new centres and the expansion of existing ones, new supplies had to be identified, captured, transported and stored (Landels 1998; cited in [5]). Picture 1.2 below shows an aqueduct built during the Roman Empire at Nimes, France. Similar aqueducts were built across Europe to transport water over long distances.

According to Cotruvo and Cotruvo [6], between 343 B.C. and 225 A.D., Roman engineers constructed aqueducts to transport and supply water, and by the time of the collapse of the Roman Empire, over 500 km of aqueducts had been constructed. Water management and technology however declined in Europe after the fall of the Roman Empire. Political and military insecurity and insufficient state wealth meant that the building of large-scale water supply and sewage schemes were not attempted in Europe, and dependence on decentralised options continued until the eighteenth century.

1.4 Water Management and the Industrial Revolution

The invention of the microscope by Anton van Leeuwenhoek in the seventeenth century was instrumental step in advancing water treatment technology. However, it took another 200 years before the connection between microscopic organisms in water and public health was made [7]. According to Cotruvo and Cotruvo [6], John Snow was the first to make the connection between the 1854 cholera epidemic in London that killed over 30,000 people and drinking water contaminated by sewerage, with the discovery of a leaking sewer next to a well. Prior to Snow's discovery, it was thought that bad air and smell was the cause of cholera [7]. In 1883, Heinrich Robert Koch showed that microorganisms in drinking water caused cholera and typhoid outbreaks and within years also revealed that these diseases could be prevented by filtering contaminated water [6].

The Industrial Revolution saw significant growth in population, scientific knowledge and technology. The size of our cities grew rapidly with the Industrial Revolution, and the shape of cities changed into high-density urban centres. This exponential growth in the size of the cities and the resulting crowded dwellings caused an increase in pollution of the local water sources and associated public health issues. According to Hegger [1], it took a considerable part of the nineteenth century before governments intervened to improve hygienic conditions, which was necessary to combat public health problems such as the major cholera epidemics. It was in fact the mass outbreak of epidemics in the nineteenth century that triggered government involvement in the water industry [7].

The large quantities of water utilised and the sewage generated, as well as the increased storm water from the vast urban areas, needed a new management approach. Technical advancements resulting from the industrial revolution enabled water infrastructure construction to a size that was previously unimaginable.

By the end of the nineteenth century, all major northern European cities had built or were building new systems to distribute water and evacuate liquid wastes [8]. It was this critical infrastructure construction role that saw the status of engineer elevated as '*midwife to the Enlightenment and Republican vision of a new age of reason*' ([9], p.9).

1.5 Water Engineering in Twentieth Century

The capital and imperialist might of Europe during nineteenth and first half of twentieth century saw the European water technologies along with their negative and positive effects dispersed throughout the world, including Australia [10]. According to Chanan and Simmons [5], the twentieth century could best be described as the century of applying big engineering solutions to water management. During this period, extensive networks of canals, dams and reservoirs were designed and constructed worldwide. Since 1950, the number of large dams (15 m or taller) has grown from 5,000 to over 45,000 today. More than 85% of those were built in the 35-year period between 1965 and 2000 [11].

According to Pinkham [12], a conventional wisdom emerged within the twentieth century water utilities that the centralised approach was the best water management strategy for most communities. Large centralised systems were perceived to be most reliable, easiest to manage and least costly per capita. Townships on the urban fringe or regional communities without a sewage system that relied on on-site systems for wastewater disposal were viewed as temporary measures until such time as local growth allows these areas to be serviced by centralised systems [12].

The centralised water management system built during the twentieth century is based on sourcing large quantities of water from one location (often a different

Table 1.1 Status of Australian water infrastructure [16]

Infrastructure category	Rating	Explanation
Water	B–	Minor changes required in one or more of the areas to enable infrastructure to be fit for its current and anticipated purpose
Wastewater	C+	Major changes required in one or more of the areas to enable infrastructure to be fit for its current and anticipated purpose
Storm water	C–	Major changes required in one or more of the areas to enable infrastructure to be fit for its current and anticipated purpose

hydrological catchment), adding a number of nutrients to it during its once-only use and finally disposing the waste stream at another point location (Livingston 2007). Mouritz [10] described it as the '*Big Pipe In–Big Pipe Out*' strategy, commenting on the size of pipes needed for bringing drinking water into the city and for removing wastewater and storm water from the city. Big pipe engineering became the standard water management technique [13].

Newman [13] further states that the economy of scale often associated with the centralised approach to water management changed once the form of our cities changed. The century old solution of big pipes now means a lot of big pipes. Thomas and McLeod [14] estimated that for a typical modern Australian city, up to 85% of the capital investment principally by government agencies goes towards the provision of the pipes for water, wastewater and storm water services and less than 15% to 20% is actually spent in water treatment. Mouritz [10] and Chanan et al. [15] called for a more balanced investment approach, thereby advancing treatment to meet the goals of sustainability instead of investing in transporting water over longer and longer distances.

Centralised systems built during the first half of the last century now face a huge backlog of deferred maintenance. In the United States of America, for instance, the gap between what is currently being spent on maintenance and upgrades and what is actually needed amounts to billions of dollars per year [12]. The situation in Australia is similar (see Table 1.1, with an estimated underinvestment of $3 billion in water infrastructure [16]). Other ways of providing urban water services must therefore be given a serious consideration.

By the end of the twentieth century, society reached a turning point, and the flaws with the big pipe solution became widely acknowledged. A number of authors such as Winneberger [17], Niemczynowicz [18], Chu and Simpson [19], Mouritz [10], Anderson [20] and Davis [21] called for a re-examination of the traditional public health focus of managing our water systems.

Perhaps the most notable criticism of the twentieth century centralised water management came from Winneberger [17], p.3) who stated: '*for over 100 years since that initial widespread use in London, the water flushing toilet always has used substantial amounts of pure drinking water to transport relatively minuscule quantities of body waste through a network of sewer pipes and rivers, lakes or oceans......... Mixed with the body waste is wastewater from baths, sinks and other water using household appliances... Recent advances in sanitation engineering technology have raised serious questions about such methods.*'

Such a system not only results in receiving waters becoming more polluted, it also causes water sources to become scarce and nutrients such as nitrogen and phosphorus to be diluted and removed from where they could potentially be reused [22]. The loss of valuable nutrients has been a well-documented concern ever since the early days of sewage systems. A German chemist, Justus Von Liebig, argued that even in an era of increasing industrialisation and urbanisation, natural cycles of plant nutrients should not be broken down and that it was necessary that humans gave back their nutrients to agriculture [1].

1.6 Water Management of the Future

Sustainable water management of the twenty-first century has to be significantly different from the traditional twentieth century water management paradigm, which focused on meeting the demand for water by augmenting supply and disposing wastewater to prevent the spread of disease. Sustainable urban water systems need to focus on achieving a 'closed loop' through initiatives such as water recycling and reuse. As highlighted in Fig. 1.1, such an approach not only reduces the amount of water imported into the urban area, it also treats storm water runoff and sewage as resource and cuts down on their wasteful discharge into waterways [23].

Niemczynowicz [18] called for a drastic change from the traditional approach to water management, including current wastewater treatment technologies be complemented and ultimately replaced by economically efficient and environmentally sound technologies. He proposed the following key principles to guide this drastic new approach:

- An integrated systems approach to solving water problems that considers both structural and non-structural elements, in contrast to just narrow-minded technological approaches
- Multidisciplinary cooperation to solve complex problems, in contrast to engineers making the best augmentation decision
- Small-scale options, in contrast to the twentieth century technological monumentalism

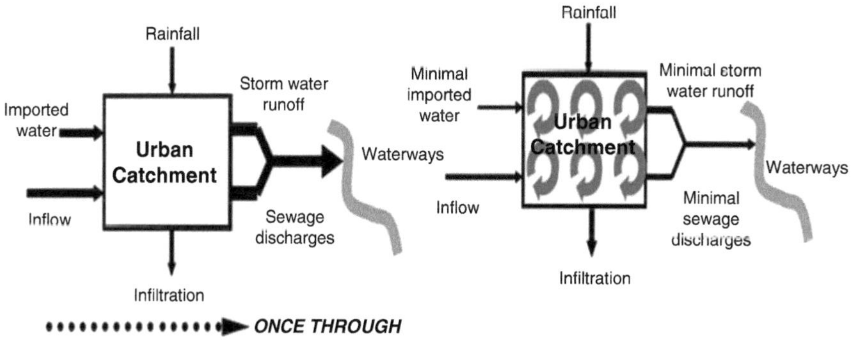

Fig. 1.1 Traditional versus sustainable urban water cycle management [23]

- Source control, in contrast to 'end-of-pipe' approaches to wastewater and storm water treatment
- Local disposal and reuse, in contrast to twentieth century approach of transporting in big pipes and wastefulness

Similarly, Newman [13] also argued that the large-scale regional water utilities framework is an ineffective mode of providing the specialised local level knowledge needed for integrated water cycle management. Detailed knowledge of local natural processes is a necessary prerequisite for water-sensitive urban design and implementing appropriate water recycling initiatives. He states, '*Such intimate knowledge of local soils, slopes, creeks, wetlands, as well as knowledge of the urban aspects of nature, i.e. open space, community gardens, street trees are ideally suited to the role of a local environmental scientist working in a local authority*'.

1.6.1 A Soft Path for the Future

The utilities sector is commonly described as institutions engaged in the provision of utility services such as electricity, natural gas, water supply and sewage removal. Despite the variation in services provided, there are some unique economic commonalities within this sector irrespective of the services. These commonalities originate from the fact that utilities require significant investment in infrastructure to be able to meet their customer demands. Over the years, market-driven transformation in one utility area has invariably followed into the other. Water utilities in particular have closely observed the developments in the electricity utilities and have typically followed the trends originating from that area.

Lovins et al. [24] highlight a fundamental change that is emerging in the electricity sector, which they described as a *shift in the scale of electricity supply from doctrinaire gigantism to the right size for the job*. Similar to the twentieth century journey of the water sector, the electricity sector also followed an identical trend – as the demand for electricity increased in growing cities, the generation plants were augmented, and the vertically integrated utilities' costs declined [24]. By late twentieth century, this trend however reversed, and the option of indefinite augmentation no longer remained a sensible option.

What is particularly noteworthy for the water sector though is the emerging role of decentralised options in the electricity sector, as highlighted by Lovins et al. [24], p.3):

> In 1976, the concept of largely "distributed" or decentralized electricity production was heretical; in the 1990s, it became important; by 2000, it was the subject of cover stories in such leading publications as *The Wall Street Journal*, *The Economist*, and *The New York Times*; and by 2002, it was emerging as the marketplace winner. This change is exactly the sort of "inflection point" described by Andrew Grove of Intel in his 1996 book *Only the Paranoid Survive: How to Exploit the Crisis Points That Challenge Every Company and Career*. Grove describes an inflection point as a pivotal, wrenching transformation that sorts businesses between the quick and the dead. If properly understood and exploited, an inflection point is the key to making businesses survive and prosper.

Table 1.2 Water management paradigm shift [12]

The old paradigm	The emerging paradigm
Human waste is a nuisance	Human waste is a resource
Storm water is a nuisance	Storm water is a resource
Build to demand	Manage demand
Demand is a matter of quantity	Demand is multifaceted
One use (throughput)	Reuse and reclamation
Grey infrastructure	Green infrastructure
Bigger/centralised is better	Small/decentralised is possible, often desirable
Limit complexity: employ standard solutions	Allow diverse local solutions
Integration of different arms of water cycle by accident	Physical and institutional integration to manage different arms of urban water cycle
Collaboration = public relations	Collaboration = engagement

Wolff and Gleick [25], Pinkham [12], Fane [26] and Vigneswaran et al. [27] all described water management of the future as following a '*soft path*', a term borrowed from the works of Amory Lovins (Rocky Mountain Institute) in the electricity sector in late 1970s. The *soft path* approach to water management emphasises the optimisation of end-use efficiency, small-scale management systems and the incorporation of fit-for-purpose water use [15]. A key characteristic of this new way of water management would be the use of diverse, locally appropriate and commonly decentralised infrastructure. Local reuse schemes for wastewater, and where appropriate for storm water, form an important supply option in this 'soft path' approach [26]. Table 1.2 below provides a summary of the shift in the values concerning the urban water cycle that is essential to make the transition into the proposed '*soft path of water management*' for the future [12].

A number of authors have argued on the merits of decentralised wastewater infrastructure; these include Anderson [20], Wilderer and Schreff [28], Tchobanoglous [29], Pinkham [12], Fane [26], Stenkes [30] and Livingstone [31]. Advantages of decentralised wastewater management with wastewater treated close to source and greater opportunities for reuse are now well documented in the literature.

1.7 Role for Water Recycling and Reuse

Water recycling involves treating wastewater from sewage system to suitable levels so that the treated water can be safely reused for desired applications [32]. According to Marks [33], water recycling is recognised by policy makers and the water supply industry as being an important strategy in the management of urban water supplies. Using reclaimed water offers the potential for exploiting a new resource that can be substituted for existing finite resource, depending on its intended use [34]. This concept is called 'source substitution', and is not new. According to Chanan and Ghetti [32], it has been over 50 years since the United

Nations Economic and Social Council resolved that, '*No higher quality water, unless there is a surplus of it, should be used for a purpose that can tolerate a lower grade*'.

It is necessary to note that water is used for various purposes in our cities, with varying quality and quantity requirements. Many urban, commercial and industrial uses can be met with water of less than potable water quality. After use, the wastewater generated also varies in quality and quantity. According to Hermanowicz and Asano [35], careful linking of wastewater source with appropriate water demand at a local level while giving due consideration to quality and quantity could provide a sustainable and economic source of water for our cities.

Wolff and Gleick [25] and Fane [26] described this concept of matching quality with intended use as '*water quality cascading*'. It recommends that only the minimum required quality should be provided for a given end-use. According to Fane [26], water quality cascade involves matching the quality of water to its use, and water is reuse as its quality decreases. '*The most common example of domestic water quality cascading would be greywater diversion from showering to toilet flushing, or washing machine diversion to garden washing*' ([26], p.45).

Yet another form of the source substitution and water quality cascading concepts is the proposed '*Close loop water cycle concept*', which provides an ideal blueprint for planners of future cities and towns [36]. As outlined in Fig. 1.2 below, at the scale of a household, a neighbourhood, an institution or a community, water would have to be managed in a closed loop. Every drop of water that would come into the loop would be managed at least twice before being sent out of the loop.

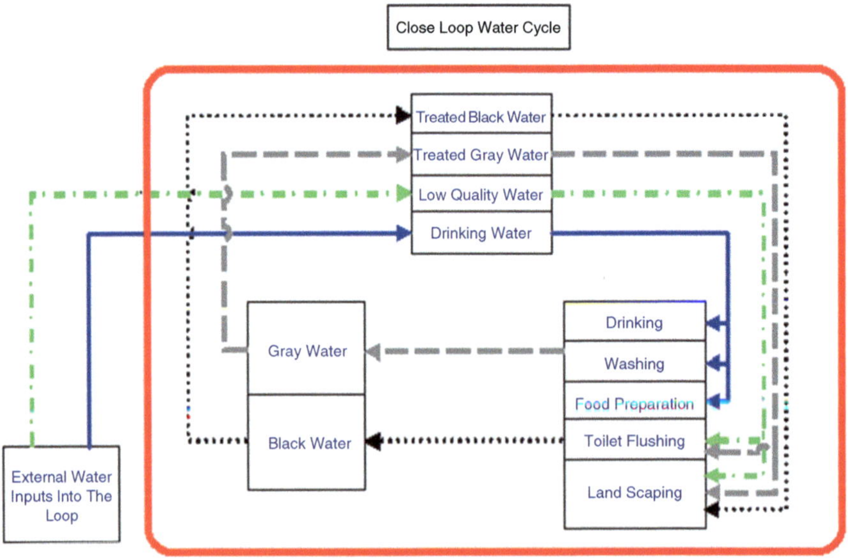

Fig. 1.2 Close loop water cycle concept [36]

Opponents of water recycling suggest that recycled water must be treated to drinking water standards and pumped back over long distances from coastal sewage treatment plants to water storages. Such claims are rather incongruous and can only be substantiated if reuse schemes were to be implemented as traditional 'centralised' system, with extensive conveyance and plumbing networks, high capital, maintenance and energy costs [15]. However, if reuse is implemented as decentralised schemes providing 'fit for purpose' water where it is needed, they are easy to build, operate and involve low energy costs. In order to explain this distinction better, the following section provides an overview of the types of water reuse and their respective applications.

1.7.1 Types of Water Reuse

According to Kermane [37], wastewater reuse may be planned or unplanned, with unplanned reuse described as a situation where downstream water supply is drawn from a source to which wastewater from upstream communities has been discharged. Most commonly used example to describe such unplanned reuse is London's water supply, with several towns upstream of London discharging their wastewater into the River Thames. Water industry folklore states that London's water has on average gone through seven kidneys before reaching a tap in London.

Planned reuse on the other hand is not a result of the geographic destiny along a river; it is in fact deliberate use of treated wastewater, and it can be direct or indirect, and it can be further classified as centralised or decentralised [37]. Figure 1.3 above provides a summary of the types of planned water reuse schemes,

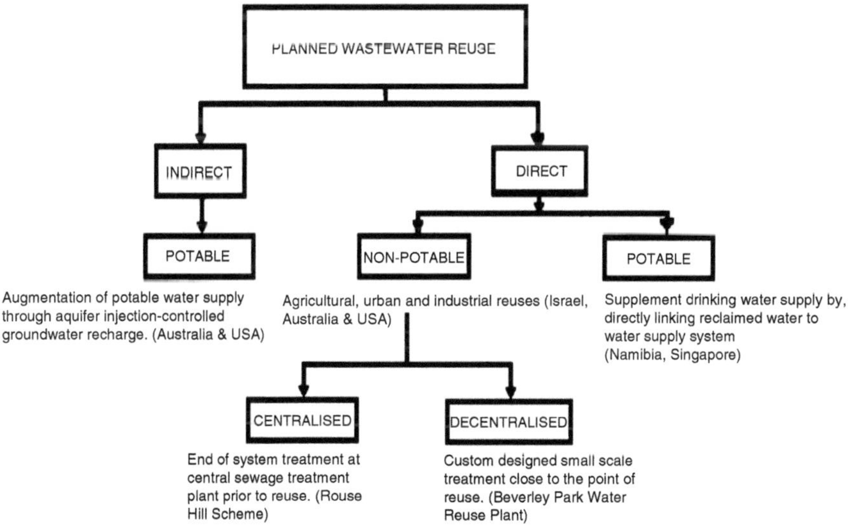

Fig. 1.3 Types of planned wastewater reuse (Modified from Kermane [37])

along with known examples of each. IWA WaterWiki [38] defines direct reuse as the use of recycled water delivered directly into a water distribution or into a water treatment plant. Indirect reuse on the other hand is the use of recycled water delivered into a river, reservoir or groundwater aquifer from which water supply is drawn at a point downstream [38].

Analysis of a number of water reuse projects in Europe concluded that a major benefit of water recycling is the production of alternative water resources 'near the point of use' [39]. Given that only 15% of water used in urban areas is for potable purposes, the potential for non-potable water recycling in urban areas is extremely high. Concepts of source substitution [34], water quality cascading [25] or closed-loop water cycle [36] all call for matching quality with intended use.

1.7.2 *Decentralised Reuse Systems*

According to Clark ([40]; cited in [26]), decentralised water infrastructures minimise the distance water, and wastewater must be transported, and the size and cost of the pipe networks. Asano [41] and Gikas and Tchobanoglous [42] have further categorised the decentralised water reuse schemes into three main types, as illustrated in Fig. 1.4 below.

- *Upstream type*: The upstream type of system is typically used to treat wastewater generated at the outskirts of a centralised collection system, where opportunities for reusing the reclaimed water exist due to large suburban parks, golf course or median strip irrigation [41]. Typically, this type of reuse scheme is associated with new housing developments and new

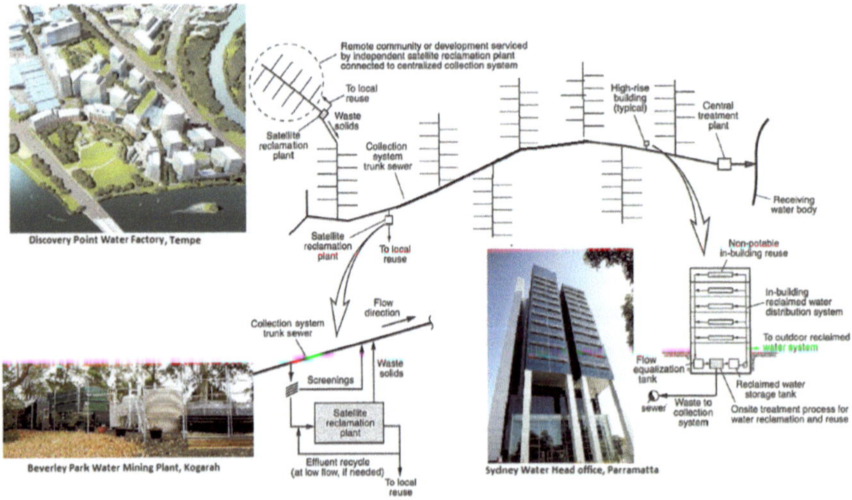

Fig. 1.4 Types of decentralised reuse scheme (Modified from Gikas and Tachobanoglous [42])

commercial centres on the outskirts of the city (Fig. 1.4). An example of this type of reuse is currently being implemented in South Sydney at Discovery Point Development in Tempe.

- *Extraction type*: Wastewater to be recycled in the extraction type of systems is literally mined from the collection system en route to the central treatment plant [42]. The Beverley Park Water Reuse project that won the International Water Association's Project Innovation Award for 2008 is in fact an extraction-type scheme as per this classification, and it is therefore commonly referred to as a *water mining* scheme. Typical applications of extraction-type satellite systems include park or green-belt irrigation, water reuse in commercial centres and cooling tower applications.
- *Interception type*: Asano [41] described the interception type systems as those in which wastewater is intercepted before reaching the collection system. Intercepted wastewater is in fact diverted to a satellite (decentralised) system for treatment and reused locally. Typical applications for this type of treatment systems are for water reuse in high-rise commercial buildings. A good example of interception type water reuse scheme is located at the Sydney Water Corporation's head-office building in Parramatta, Australia.

1.8 Conclusion

As discussed in this chapter, the journey of wastewater infrastructure development has been an interesting one. The sociopolitical situation of the day influenced its design and implementation. The soft path for water management has now emerged as the best pathway for future of wastewater management, with decentralised options including reuse as an integral component of this approach.

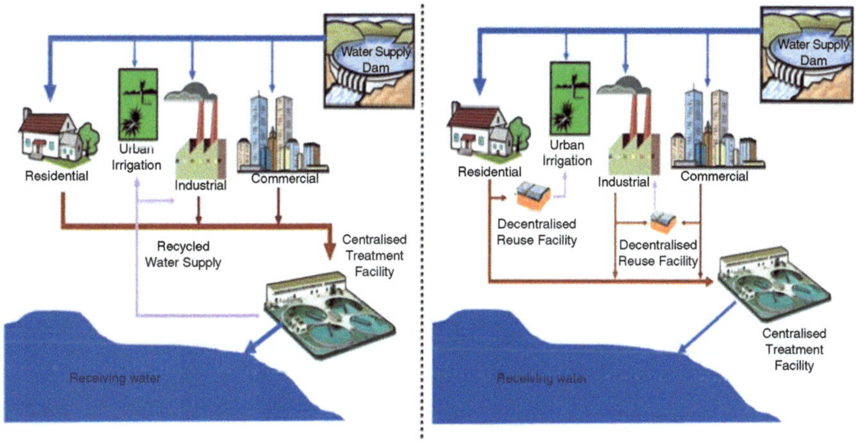

Fig. 1.5 Sewer mining versus centralised reuse [43]

There are several advantages associated with moving away from conventional centralised sewer management systems and adopting innovative concepts such as water mining. Some of these advantages are highlighted by Chanan and Kandasamy [43], and three most notable ones include:

- Decentralised treatment and local reuse reduces the need for inefficient transport of water over long distances from centralised treatment facilities (as outlined in Fig. 1.5). This approach also provides flexibility of extracting water practically anywhere and to treat it closer to the site of its intended use [44].
- Strategically designed network of decentralised treatment facilities reduces the impact associated with unforeseen events such as natural or man-made disasters and is therefore better in terms of disaster recovery when compared to a centralised sewage system.
- Decentralised wastewater treatment schemes provide opportunities for local community to be involved in managing their local water cycle, thereby raising community awareness.

References

1. Hegger D (2007) Greening sanitary systems: an end-user perspective. In: PhD thesis, Wageningen University, Wageningen, Netherlands
2. Siddharth B (1999) The celestial key to the vedas: discovering the origins of the world's oldest civilization. Inner Traditions International, Vermont
3. Edwards M (2000) Indus civilisations. National Geographic 197(6)
4. Jansen M (1989) Water supply and sewage disposal at Mohenjo-Daro, world archaeology. Archaeol Public Health 21(2):177–192
5. Chanan A, Simmons B (2002) Dams to demand management: a journey through development of urban water supply. In: Proceedings of river symposium 2002, Brisbane, Sept 2002
6. Cotruvo JA, Cotruvo JA Jr (2003) Nontraditional approaches for providing potable water in small systems: part 1. Am Water Works Assoc J 95(3):69–76
7. Fazakerley VW (2005) Critical issues for the future of the Australian urban water supply industry. In: PhD thesis, Curtin University, Perth
8. Reid D (1991) Paris sewers and sewermen. IBT Global, London
9. Cosgrove D (1990) An elemental division: water control and engineered landscape. In: Cosgrove D, Petts G (eds) Water engineering and landscape. Belhaven Press, London
10. Mouritz M (1996) Sustainable urban water systems: policy and professional praxis. In: PhD thesis, Murdoch University, Perth
11. The World Commission on Dams (2000) Dams and development: a new framework for decision making, United Nations
12. Pinkham R (2004) 21st century water systems: scenarios, visions, and drivers, Rocky Mountain Institute, Snowmass, Colorado, Feb 2004
13. Newman P (2007) Sustainability and the urban water system, case study, Institute for Sustainability and Technology Policy, Murdoch University
14. Thomas J, McLeod P (1992) Australian research priorities in the urban water services and utilities area. Division of Water Resources No. 7 CSIRO
15. Chanan A, Kandasamy J, Vigneswarans V, Sharma D (2009) A gradualist approach to address Australia's urban water challenge. Desalination 249(3):1012–1016

16. Engineers Australia (2005) Australian infrastructure report card, Barton, Australian Capital Territory, Aug 2005
17. Winneberger J (1974) Manual of gray water treatment practice. Ann Arbor Science Publishers, Ann Arbor
18. Niemczynowicz J (1992) Water management and urban development: a call for realistic alternatives for the future. Impact Sci Soc 42(2):133–147
19. Chu C, Simpson R (1994) Ecological public health: from vision to practice. A joint publication of Centre for Health Promotion, University of Toronto and Institute of Applied Environmental Research, Griffith University
20. Anderson J (1996) The potential for water recycling in Australia – expanding our horizons. Desalination 106:151–156
21. Davis C (2008) Fourth generation water systems, a paper for the local government and Shires Association of NSW, University of Technology Sydney, Feb 2008
22. Livingston D, Stenekes N, Colebatch H, Ashbolt N, Waite TD (2004) Water recycling and decentralised management: the policy and organisational challenges for innovative approaches. In: Proceedings of WSUD 2004 – cities as catchments conference, Adelaide, South Australia, 21–25 Nov 2004
23. Stenekes N, Livingston D, Colebatch HK, Waite TD, Ashbolt NJ (2004) Sustainable water management in Australia: an institutional analysis. In: Proceedings of good water governance for people and nature: water roles for law, Institutions & finance conference, Dundee, Scotland, August 29–September 2, 2004
24. Lovins A, Datta E, Feiler T, Rabago K, Swisher J, Lehman A, Wieker K (2002) Small is profitable: the hidden economic benefits of making electrical resources the right size. Rocky Mountain Institute, Snowmass
25. Wolff G, Gleick P (2002) The soft path for water. In: The world's water 2002–2003: the biennial report on freshwater resources, Oakland, California, pp 1–32
26. Fane S (2005) Planning for sustainable urban water: system approaches and distributed strategies. In: PhD thesis, institute for sustainable futures, University of Technology Sydney
27. Vigneswaran S, Davis C, Kandasamy J, Chanan A (2009) Urban wastewater treatment: past, present and future, in water and wastewater treatment technologies. In: Vigneswaran S (ed) Encyclopedia of life support systems (EOLSS), UNESCO Initiative. Eolss Publishers, Oxford
28. Wilderer P, Schreff D (2000) Decentralized and centralized wastewater management: a challenge for technology developers. Water Sci Technol 41(1):1–8
29. Tchobanoglous G, Burton FL, Stensel HD (2003) Wastewater engineering: treatment, disposal, and reuse, 4th edn. McGraw Hill Inc., New York, 1819 pp
30. Stenekes N (2006) Sustainability and participation in the governing of water use: the case of water recycling. In: PhD thesis, School of Civil and Environmental Engineering, University of NSW, Sydney
31. Livingston DJ, Ashbolt NJ, Colebatch HK (2006) Institutional barriers to decentralised systems, water. J Aust Water Assoc 33(3):75–77
32. Chanan A, Ghetti I (2006) Beverley Park water reclamation project: pioneering water reclamation in Sydney. In: Proceedings of Institute of Public Works Engineering Australia, NSW divisional annual conference 2006, Sydney, 21–24 May 2006
33. Marks JS (2006) Taking the Public seriously: the case of potable and non potable reuse. Desalination 187:137–147
34. US EPA (2004) Guidelines for water reuse. US EPA Municipal Support Division, Washington, DC, 2004
35. Hermanowicz SW, Asano T (1999) Abel Wolman's "the metabolism of cities" revisited: a case for water recycling and reuse. Water Sci Technol 40(4–5):29–36
36. Hakim (2002) Water demand and pollution control: key to securing and safeguarding the water supplies for the MENA in the 21st century. WHO Regional Centre for Environmental Health Activities, Amman, Jordan

37. Keremane G (2007) Urban wastewater reuse for agriculture: governance paradigms and institutional arrangements in Australia and India. Ph.D. thesis, University of South Australia, Adelaide
38. IWA WaterWiki (2011) Water recycling glossary. By Victoria Beddow. IWA Publishing. http://iwawaterwiki.org/xwiki/bin/view/Articles/Water+Recycling+Glossary. Accessed 25 Sept 2011
39. Lazarova V, Rougé P, Sturny V, Arcangeli J (2006) Evaluation of economic viability and benefits of urban water reuse and its contribution to sustainable development. Water Pract Technol 1(2):11. doi:10.2166/WPT.2006025
40. Clark R (1993) Is big really best? In: Proceedings of ecological sustainable water management – a threat or an opportunity, Sydney Water Board Seminar, Sydney, 15 November 1993
41. Asano T (2007) Water reuse: issues, technologies and applications. Metcalf and Eddy Inc. McGraw-Hill, New York, p 1570
42. Gikas P, Tchobanoglous G (2009) The role of satellite and decentralized strategies in water resources management. J Environ Manag 95(1):144–152
43. Chanan A, Kandasamy J (2009) Water mining: planning and implementation issues for a successful project. In: Vigneswaran V (ed) Waste water treatment technologies, encyclopedia of life support systems. UNESCO Initiative. Eolss Publishers, Oxford
44. Khan S (2007) Water recycling in Australia – a blog site http://waterrecycling.blogspot.com/2007/03/sewer-mining.html. Sunday, 4 Mar 2007
45. Butler R, MacCormick T (1996) Opportunities for decentralized treatment, sewer mining and effluent re-use. Desalination J 106:273–283
46. Livingston DJ (2008) Institutions and decentralised urban water management. In: PhD thesis, School of Civil and Environmental Engineering, University of New South Wales, Sydney, Apr 2008
47. Niemczynowicz J (1999) Urban hydrology and water management – present and future challenges. Urban Water 1:1–14
48. Tsagarakis KP (2005) Recycled water valuation as a corollary of the 2000/60/EC water framework directive. Agric Water Manag 72(1):1–14

Chapter 2
Integration of Membrane Processes for Optimal Wastewater Management

Y. Saif and A. Elkamel

2.1 Introduction

Depletion of water resources and wastewater generation from industrial and domestic sectors bring into existence issues with water resources conservation and wastewater management. Pollution occurrence and health concerns by direct discharge of wastewater streams into the environment are other facets of the problems which call for responsible actions. Strict regulations have been forced worldwide to set boundaries on the limits of discharging wastewater streams into the environment. Compliance with these regulations is nowadays mandatory practice. Associated with these activities, large amount of money is being spent annually to deal with these problems. Therefore, optimal wastewater management is a necessary step toward money saving and meeting environmental regulations.

The challenges within the area of wastewater management demand innovative development of processes and chemicals that effectively reduce water consumption and wastewater generation. In the chemical industry, several processes and chemical agents are jointly incorporated within a system (e.g., chemical plant) to transform raw chemicals into desired products. Such a system normally has different configurations in the conceptual design phase. In other words, the combination of the processes with the chemical agents is not unique for the system which takes the transformation task of feed to products. Besides, the operational conditions of the processes and chemical agents have direct impact on the overall performance of the system. Consequently, decisions about the system optimal configuration and its operation have significant impact on the system economy and the waste generation.

Y. Saif • A. Elkamel (✉)
Department of Chemical Engineering, University of Waterloo, Waterloo, ON, Canada

Department of Chemical Engineering, Petroleum Institute, Abu Dhabi, UAE
e-mail: ysaif@pi.ac.ae; aelkamel@uwaterloo.ca

S.K. Sharma and R. Sanghi (eds.), *Wastewater Reuse and Management*,
DOI 10.1007/978-94-007-4942-9_2,

Recent trends toward effective wastewater management focus on minimizing water intake through process integration [1]. The tools seek manipulation of the process streams with their chemical constituents in the design phase through mixing and splitting. Undesirable chemical compounds can be kept within the plant through recycling and reuse. Sometimes, restrictions on the process water quality prevent recycling or reuse of specific streams with undesirable water qualities. Regeneration of these streams through wastewater conditioning processes allows maximum utilization of wastewater streams within a given system. In 1996, roughly 40% of total industrial wastewater was recycled with considerable variation between the industrial sectors [2]. In addition, decentralization of wastewater treatment has resulted in pollutant source reduction replacing "end-of-pipe" treatment approaches [3].

During the last 35 years, continuous improvements of membrane processes have been achieved. Improvements in selectivity flux and operating practice of membrane processes have enabled them to become competitive with more conventional processes. In 1998, the annual sales of membrane systems reached 19 billion USD with 10% annual growth in the water and wastewater treatment sectors [4]. Clearly, membrane technologies have become appealing for water and wastewater treatments from the point of view of pollution prevention, efficient operation, and cost reduction.

2.2 Wastewater Contaminants and Their Impacts

Every water and wastewater problem varies from location to location in terms of its constituents. Water and wastewater are usually characterized in terms of their organic, inorganic, and biological contaminants. The organic part can contain carbohydrates, oil and grease, surfactants, and priority pollutants, e.g., benzene, ethylbenzene, toluene. Inorganic compounds can include nonmetals such as arsenic and selenium and/or metals (e.g., chromium, lead, silver). In addition, microorganisms and pathogens are often present in water and wastewater [5].

Public health and environmental concern have led to strict regulations for drinking water quality and wastewater discharge. Exposure to the previously mentioned chemicals and biological constituents can be very harmful to all life forms. Dissolved oxygen depletion is a serious threat to aquatic life through decomposition processes and the potential formation of toxic gases such as hydrogen sulfide. Discharge of high loads of nutrients, such as nitrogen and phosphorus, leads to excessive and destructive growth of algae. By law, effective water and wastewater treatment is mandatory to achieve and maintain health and safety standards and minimize pollution.

2.3 Membrane Processes: Classifications and Some Applications

Membranes are selective thin layer materials that can be used to separate different species. The separation can be accomplished when a driving force is applied across a membrane. Due to membrane selectivity, water and wastewater streams can be separated into lean and concentrated products. Separation requires the application of driving forces such as pressure, electrical potential, or chemical activity gradients across the membrane. The use of synthetic membranes with appropriate structure and properties can allow very efficient separation, often with substantial energy savings over more traditional separation techniques [6].

There are several ways to classify membranes. Based on their constituents, membranes are classified as being organic such as polymer membranes or inorganic in the case of metal and ceramic membranes. Membranes are also classified in terms of their geometry. These include flat sheet or cylindrical (tubular or hollow fiber) configurations. Another classification can be attributed to the operating driving forces such as pressure, electrical, or concentration gradients. In addition, membranes can be classified according to whether their structure is porous, dense, or composite. The varieties of the membrane structures and their operations give them very wide applications. Table 2.1 shows some practical applications of membranes processes in different industries [7].

Table 2.1 Practical industrial applications of membrane processes

Industrial sector	Membrane processes	Industrial sector	Membrane processes
Drinking water	NF, UF, RO	*Biotechnology*	
		Enzyme purification	UF
		Concentration of fermentation broth	MF
		SCP harvesting	MF,UF
		Membrane reactor	UF
		Marine biotechnology	MF,UF
Demineralized water	RO, ED, EDI	*Energy*	Proton exchange membrane
		Fuel cell	
Wastewater treatment	MF, NF, RO, ED	*Medical*	UF
	MF, UF	Control release	
Food industry		*Chemical industry*	
Dairy	UF,RO,ED	Hydrogen recovery	GS
Meat	UF,RO	CO_2 separation	GS
Fruit and vegetables	RO	Ethanol dehydration	PV
Grain milling	UF	Organic recovery	PV
Sugar	UF, RO, ED, MF, NF	Chlor-alkali process	Membrane electrolysis

2.4 Chapter Content

The design of integrated membrane system for wastewater treatment represents a process synthesis problem. The task is to separate pollutants present in different feed streams into lean and concentrated streams using the optimum combination of units, operating conditions, and distribution of process streams from an assortment of alternatives. Superstructure optimization is a valuable tool which serves the purpose of synthesizing an optimal flow sheet for the treatment of wastewater streams through membrane processes. Sections 2.5 and 2.6 provides an overview of process synthesis through superstructure optimization. A generic representation of integrated membrane systems is given to assist the construction of the superstructures and the formulation of the optimization models.

Section 2.7 discusses energy-integrated RO networks. Within the network, several stages of pumps, turbines, and RO trains are assembled in a superstructure framework to treat wastewater streams with multiple pollutants. The RO superstructure eventually gives all the combination alternatives of the abovementioned unit operations to achieve the treatment targets. An MINLP model is accordingly derived based on the given representation to describe the RO network. The model solution gives the optimum flow sheet and the operating conditions of the unit operations.

Volatile organic compounds (VOCs) constitute an important class of priority pollutants listed by the environmental protection agencies of most countries. An integrated membrane network is illustrated by presenting the air stripper/pervaporation system for the treatment of wastewater streams contaminated by VOCs. Section 2.7.1 presents the superstructure representation of an integrated air stripper/pervaporation system. Besides, a description of the MINLP model is provided for the integrated air stripper/pervaporation system. Section 2.8 gives two case studies for the energy-integrated RO networks and the air stripper/pervaporation systems.

2.5 Superstructure Optimization

The scope of process synthesis has evolved over years starting from the development of unit operations through optimization of chemical process flow sheet. Due to the high concern over environmental degradation and sustainable development, process plant integration becomes a popular subject. The common approaches for optimizing the process synthesis problems are the use of hierarchical decomposition, superstructures, and targeting techniques [8]. Despite its considerable value, hierarchical techniques cannot evaluate process alternatives simultaneously. On the other hand, the superstructure approach can handle wide range of practical synthesis problems. Targeting techniques apply physical knowledge to understand features of a feasible design. Nevertheless, the superstructure approach is the most favorable for process synthesis problems [9].

Superstructure optimization has been applied to design single units or to retrofit existing units. Examples are the optimization of a distillation column and hybrid units (e.g., distillation column coupled with pervaporation unit). An MINLP model was developed to analyze the optimal number of trays, feed location, and recycle streams for a distillation column [10]. Optimal synthesis of a reactive distillation column was addressed in terms of the optimal tray numbers, stream assignment between trays, and the determination of the reactive and separation zones [11]. The separation of azeotropic mixtures requires intensive energy consumption and may require introducing separating agents to alter the concentration of chemical species. An MINLP model for the optimal integration of a distillation column and a pervaporation unit to separate propylene/propane mixture was presented to reduce energy consumption by a single distillation unit [12]. These examples showed that superstructure optimization was successful in improving the design of individual units.

The optimization of a process flow sheet is a rather complicated issue due to the large number of alternatives [46]. The state space approach for process synthesis representation gives a framework for processes which involve mass and/or heat exchange (e.g., heat exchange networks, energy-integrated distillation networks, mass exchange networks) [44]. This representation provides a large number of alternative process layouts and different modeling relations to assess the interaction between units and network streams. Interesting designs of integrated distillation trains have been identified due to the richness of network alternatives considered. This approach has also been applied to reverse osmosis (RO) networks, pervaporation (PV) networks, and integrated mass exchange-RO networks for waste treatment and reduction [13–15].

State-task networks (STN) and state-equipment networks (SEN) have been proposed as two different concepts to represent superstructures for process synthesis problems [16]. The STN approach determines the quantitative (intensive and extensive) and qualitative properties of the streams and the tasks to be performed by the units. In a second step, the optimization routine assigns predefined equipment to carry out the tasks in the network. On the other hand, the first step of the SEN method determines all possible states and equipment in the network; this is followed by the optimization step to identify the different tasks that all equipment must complete. Generalized disjunctive programming (GDP) is applied to formulate these networks.

Unit operations can be viewed as sets of mass and heat exchanger units where the mass and/or heat flow are limited by driving forces between concentrated (or hot) streams and other diluted (or cold) streams [17]. A general modular framework for process synthesis has been proposed for the purposes of process intensification rather than optimization of conventional process units [18]. The representation of the unconventional unit operations within a superstructure is included as mass/heat and pure heat modules that allow heat and mass transfer between phases. The solution of a superstructure, which allows for extensive connectivity between modules, determines the module duties (e.g., distillation column, absorber, reactive distillation), optimal stream connectivity within the network, and the optimal operation of the modules. The proposed methodology has been applied to design distillation networks for azeotropic mixtures [19], combined reaction-separation systems [19], and heat-integrated distillation sequences [20].

Another modeling approach has been presented to address integrated reaction-separation networks [21–23]. The superstructure given in this case is flexible to represent reaction networks, separation process configurations, and reaction-separation systems. The flexibility in the problem representation is based on the description of generic units, namely, reaction-mass exchanger units and separation task units, which form the building blocks of the superstructure. The modeling approach was applied to several applications in chemical engineering which feature combined reaction-separation processes involving multiphases. Also, it was applied to wastewater treatment and natural gas sweetening applications [21, 22].

2.6 Integrated Membrane Superstructure

Chemical processing plants are a sequence of unit operations linked in a series and/or parallel way with possible recycle streams. The unit operations perform certain tasks on feed streams to alter their states in order to reach the final product states. A design engineer, however, will be given input-output information with the ultimate objective to map the best route between the input-output information while minimizing or maximizing given criteria. The design task can be visualized as a screening process among an assortment of different layouts of a treatment plant. The superstructure framework for process synthesis provides tools for building a family of process designs in a large representation. Thereafter, the mathematical formulation follows closely upon what has been framed in the superstructure representation. It is therefore important to formulate a good representation in order to avoid complicated mathematical formulations.

An integrated membrane process synthesis problem for the treatment of water or wastewater streams can be formulated from the following sets:

$$\mathrm{WI} = \{\mathrm{wi}_1, \ldots \ldots, \mathrm{wi}_n\}$$

is the set of wastewater streams entering the network in order to be processed by the system. The set has elements which represent different wastewater streams with different properties (e.g., flowrate, temperature, pressure, concentration).

$$C = \{1, \ldots \ldots, c\}$$

is the set of chemical compounds present in every stream in the system. These chemical compounds may have zero concentration in some of the streams.

$$\mathrm{USI} = \{\mathrm{usi}_1, \ldots \ldots, \mathrm{usi}_n\}$$

is the set of fresh utility streams entering the superstructure in order to assist the unit operations in the separation tasks. Examples of these streams are chemical agents, air, steam, etc.

$$O = \{o_1, \ldots \ldots, o_n\}$$

is the set of unit operations given in the superstructure. Every element represents a type of unit operation which may have several number of units arranged in parallel.

$$\mathrm{UO} = \{\mathrm{uo}_1, \ldots \ldots, \mathrm{uo}_n\}$$

is the set of utility unit operations provided in the superstructure to assist the unit operations in the separation tasks. Examples of these utility unit operations are pumps, turbines, heat exchangers, etc. It is worth pointing out that every element may have several utility units arranged in parallel.

$$\mathrm{USO} = \{\mathrm{uso}_1, \ldots \ldots, \mathrm{uso}_n\}$$

is the set of the utility streams exiting the network. These utility streams are assumed to have chemical properties unsuitable for the separation tasks.

$$\mathrm{WO} = \{\mathrm{wo}_1, \ldots \ldots, \mathrm{wo}_n\}$$

is the set of wastewater streams exiting the treatment network. It can be stated that these streams are with conditions satisfying the discharge limitations.

$$P = \{p_1, \ldots \ldots, p_n\}$$

is the set of product streams with target properties.

The design problem then requires a superstructure that embeds many design alternatives to achieve separation targets. The state space approach for process synthesis provides tools to assemble the superstructure of a process network [44]. Figure 2.1 represents a generic superstructure of a process synthesis problem. A distribution box (DB) accommodates several streams either entering or leaving the treatment network. Incoming flow streams (e.g., wastewater streams, WI) are split by the splitter nodes. Utility streams, *USI*, can also be split by the splitter nodes. The utility and the operation units, *O* and UO, receive single or multiple feed streams, and consequently, the state variables of the inlet streams change to other values satisfying the unit operation models if these units exist. The exit streams from the utility and unit operations loop back to the DB mixer nodes for further processing or leave the treatment network. The results of the separation work finally give the target product streams, *P*, and exhausted wastewater and utility streams, WO and USO respectively.

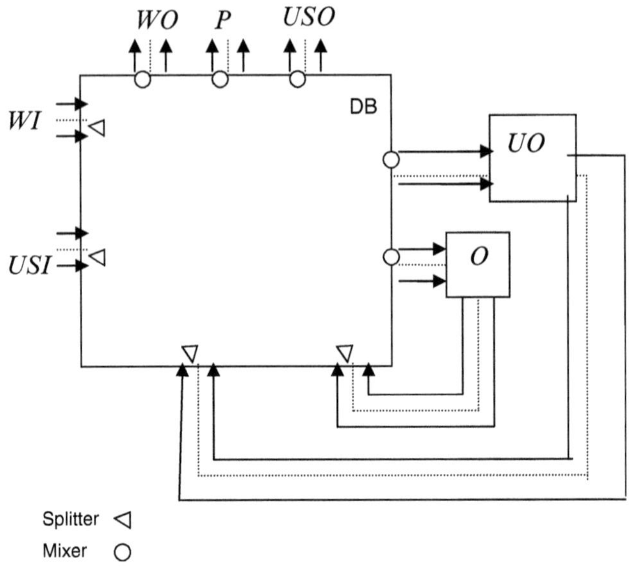

Fig. 2.1 Superstructure of integrated membrane systems

According to the superstructure representation and the unit operation models, the mathematical programming formulation can be derived in terms of a combination of parameters, continuous variables, discrete variables (e.g., binary and/or integer variables), equality and inequality equations, and an objective function to be maximized or minimized. The process synthesis optimization problem may be formulated as

$$
\begin{aligned}
&Min/Max \quad f(x,y) \\
&s.t. \\
&\qquad h(x) = 0 \\
&\qquad g(x) + M\ y \leq 0 \\
&\qquad x \in X, y \in Y
\end{aligned}
\tag{2.1}
$$

where x is the vector of the continuous variables which may represent flow, concentration, and pressure, etc., and y is the vector of binary variables which may represent the existence of units and/or streams. The equality constraints may include balances in the network (e.g., mass balance, component balance, energy balance, unit models). The inequality logical constraints show relations between the continuous variables and the binary variables (i.e., the existence of a unit requires that the value of the binary variable be set to 1). An objective function $f(x, y)$ may be defined to minimize or maximize a design criterion (e.g., total annualized cost, utility cost, unit surface area).

The given representation of integrated membrane superstructure takes into account parallel/series arrangements and recycle streams among the units that may exist in a treatment network. Such arrangement of units is observable in chemical process layouts. However, it includes no prior knowledge about the proper arrangement of the units within the superstructure since this issue is left to be determined by the optimizer. Further analysis of the relation between the superstructure representation and the associated MINLP models can be found elsewhere [24]. Next section gives the energy-integrated RO network description where one can apply the previously mentioned modeling tools.

2.7 Energy-Integrated RO Network

Reverse osmosis (RO) has been an effective technology for water and wastewater treatment. It is a pressure-driven process in which the membrane acts as a semipermeable barrier to allow primarily water to pass through as a purified permeate stream but retain pollutants in a concentrated stream. Due to their extremely small pore size, RO membranes have the capability of retaining molecules and ions. In addition, RO systems are modular and compact and consume only moderate energy during operation. These and other advantages have made RO systems strongly competitive against other separation processes.

RO networks are nonisobaric systems in which pumps deliver kinetic energy to wastewater streams, turbines extract energy from reject streams, and RO stages carry out the separation. The state space approach has been shown to give an adequate superstructure for RO networks [13]. Further extensions to RO networks have been proposed, such as the use of hybrid RO-mass exchange networks which combine RO units with other separation units [14]. RO and hybrid RO networks are formulated as nonconvex MINLP. Due to the high nonconvexity of the RO network [13], a genetic algorithm was applied to optimize the RO network considering different RO stages in different case studies [25]. However, the execution time for the algorithm was found to be prohibitively long when as many as ten unit operations were considered in the superstructure.

Optimal RO network and RO module dimensions were presented as an MINLP model [26]. The choice between module types (e.g., tubular, hollow fiber) was made on the basis of decision variables to determine optimum membrane dimensions and the surface area of the module. The performance model for RO module took care of pressure losses due to friction and flow in the module manifolds. In addition, the effect of concentration polarization was included to better estimate the osmotic pressure. Their analysis yielded the optimum hollow fiber module dimensions and series arrangement for the RO network.

Optimal RO membrane cleaning schedules and replacement were formulated as an MINLP problem [27]. The model in this case considers degradation of membrane performance due to irreversible and reversible fouling. Therefore, the optimal design determines the required membrane cleaning and replacement over a long time horizon. There was no attempt to determine the optimum RO network layout in this study. Instead, a circuit with two RO treatment trains in parallel was considered to produce permeate at a specified rate. Also, the fouling mechanism was not explicitly stated or related to the operating conditions.

Desalination by RO networks is a very well-established process. The design of RO networks normally does not exceed two RO stages in series in industrial practice. The effect of product quality having to meet multiple specifications (e.g., salt content in the final product, final product flow rate) on the RO network was addressed in a sensitivity analysis [28]. Their results show that the need to achieve very high product quality (e.g., low salt concentration in the product streams) may require further treatment of the RO permeate from one stage in a subsequent RO stage. In addition, the RO networks do not follow the common RO industrial layouts in terms of the RO stage number and the stream distributions within the network.

An RO design is determined on the basis of a superstructure which embeds all possible alternatives of a potential treatment network for water and wastewater streams. The superstructure contains several units of pumps, turbines, and RO stages. Following the concepts presented in Sect. 2.2, one should derive the optimization model based on the superstructure representation. In the following section, the superstructure of energy-integrated RO network is explained in details.

2.7.1 Superstructure

RO network is assumed to have three different types of unit operations. Pumps are necessary to raise the pressure of different wastewater streams. RO stages separate single feeds to different concentrated and diluted streams. Every RO stage is assumed to have parallel RO modules operating under the same conditions. Turbines serve essentially as units to recover kinetic energy from high-pressure streams. The RO network also has the ability to receive multiple wastewater streams with multiple pollutants. With the view of a superstructure, one should allow all possible connections between the unit operations, the unit-operations exit streams, and the inlet wastewater streams coming to the network. Figure 2.2 depicts the proposed superstructure for RO treatment network.

The superstructure is split into two parts, a distribution box (DB) where mixing/splitting of the streams occur, and another part has different boxes containing different unit operations which alter their feed stream conditions into desired exit conditions. The set of inlet wastewater streams represents different wastewater streams that need to be treated in the network. Every inlet wastewater stream is distributed over all the stages present in the unit-operation boxes, the set of final permeate nodes, and the set of final reject streams. Within every unit-operation box (e.g., turbine box (TB), pump box (PB), reverse osmosis box (ROB)), different stages of similar units may exist, and they operate under different conditions.

Every stage in every box receives a single feed stream. In any RO stage, the feed stream is split into a concentrated and a dilute stream if that stage exits in the

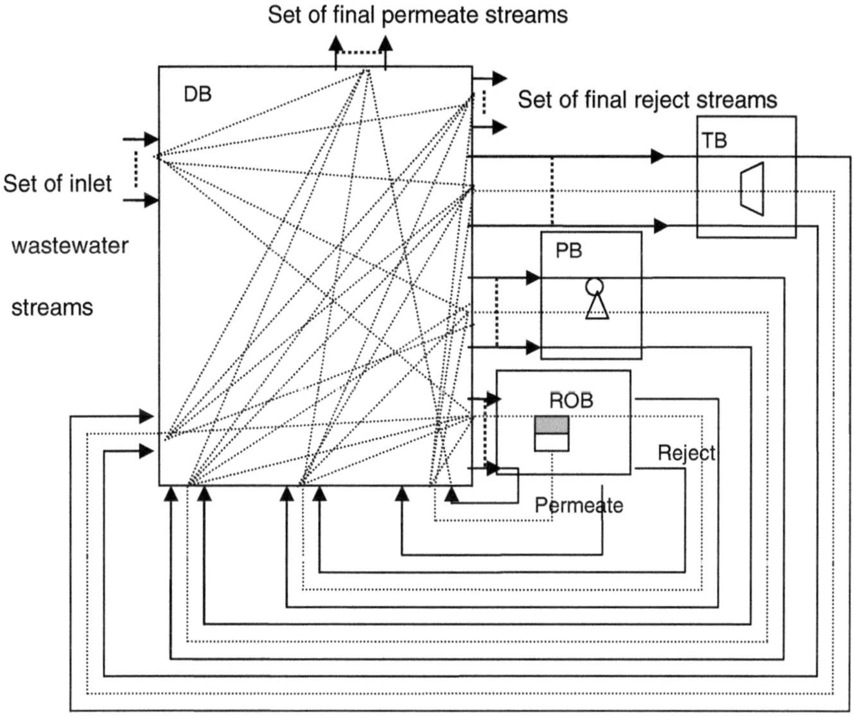

Fig. 2.2 RO superstructure representation

network. For every pump stage, the feed stream is pressurized. The turbine box is composed of different turbines which decrease their inlet feed pressure. All the exit streams from the unit operations are looped back to the DB for further recycle and bypass between the unit-operation stages, the set of final permeate streams, and the set of final reject streams. The stream distributions in the DB can be stated briefly as every incoming stream to the DB is distributed over all the exit streams in the DB. The given representation of RO superstructure gives all the possible alternatives for a potential treatment network.

2.7.2 MINLP Model Formulation

The superstructure is composed of several unit operations (e.g., turbines, pumps, RO stages) which represent the total cost of the network. Mixers and splitters nodes provide mixing and splitting for the streams within the network. The objective function of the mathematical programming model can be defined as to minimize the total annual cost (TAC) of the unit operations as

$$\text{Min} \qquad \text{TAC} = \sum^{\text{SRO}} a_{\text{MRO}}\,\text{NMd}_{\text{SRO}} + \sum^{\text{SPU}} a_{\text{pu,f}}\,\text{PPu}_{\text{pu}}^{\alpha_{\text{Pu}}} + \sum^{\text{SPU}} a_{\text{pu,o}}\,\text{PPu}_{\text{pu}} + \sum^{\text{STU}} a_{\text{tu,f}}\,\text{PTu}_{\text{tu}}^{\alpha_{\text{tu}}} - \sum^{\text{STU}} a_{\text{Ttu,o}}\,\text{PTu}_{\text{tu}} \quad (2.2)$$

The total cost of the network is assumed to depend linearly on the number of modules (NMd_{SRO}) at every RO stage (SRO) through the fixed/operating cost of the single RO module (a_{MRO}). Pump and turbine fixed costs which are attributed to the power produced/recovered (PPu_{pu}, PTu_{tu}) at every pump or turbine stage belong to the sets (SPU, STU) raised to a fractional constant (α_{pu}, α_{tu}). $a_{\text{pu,f}}$ and $a_{\text{tu,f}}$ give the fixed cost coefficients for the pump and turbine stages, respectively. The pump operation cost and the turbine operation value are assumed to be a linear function with respect to the unit's power through the constants $a_{\text{Pu,o}}$ and $a_{\text{Tu,o}}$.

The power produced by any pump is the pressure difference across the unit ΔP_{SPU} multiplied by the total flow through the unit F_{SPU}, Eq. 2.3. A binary variable y_{SPU} defines the existence of any pump if the pressure difference across the unit has a nonzero value, Eq. 2.4:

$$\text{PPu}_{\text{SPU}} = F_{\text{SPU}}\,\Delta P_{\text{SPU}} \qquad \forall \text{SPU} \quad (2.3)$$

$$\Delta P_{\text{SPU}} \le \Delta P_{\text{SPU}}^{\text{UP}}\,y_{\text{SPU}} \qquad \forall \text{SPU} \quad (2.4)$$

For the turbine stages, the power recovered by any turbine is given by Eq. 2.5, and the existence of any turbine within the network is related to a binary variable y_{STU}, Eq. 2.6:

$$\text{PPu}_{\text{STU}} = F_{\text{STU}}(-\Delta P_{\text{STU}}) \qquad \forall \text{STU} \quad (2.5)$$

$$-\Delta P_{\text{STU}} \le \Delta P_{\text{STU}}^{\text{UP}}\,y_{\text{STU}} \qquad \forall \text{STU} \quad (2.6)$$

The permeate production Fp_{SRO} from an RO stage is described by a shortcut model [29]. This model relates the permeate flow with the pressure drop across the module ΔP_{MRO}, osmotic pressure of the feed stream π_{MRO}, and the total number of parallel modules present in the stage, Eqs. 2.7, 2.8, and 2.9:

$$\text{Fp}_{\text{SRO}} = \text{NMd}_{\text{SRO}}\,W\,\text{SA}\,\gamma\,(\Delta P_{\text{MRO}} - \pi_{\text{MRO}}) \qquad \forall \text{SRO} \quad (2.7)$$

$$\gamma = \frac{\eta}{1 + 16\,W\,\mu\,r_o\,l\,l_s\eta/r_i^4} \quad (2.8)$$

$$n = \frac{\tanh\left[(16\,W\mu\,r_o/r_i^2)^{1/2}(l/r_i)\right]}{\left[(16\,W\mu\,r_o/r_i^2)^{1/2}(l/r_i)\right]} \quad (2.9)$$

where W is the water permeability coefficient, SA is the RO module surface area, and γ is a parameter related to the RO module dimension and water properties. The component concentration in any permeate stream ($xp_{c,SRO}$) is related to the component average concentration at the reject side of the RO module ($X_{c-avg,SRO}$), the solute permeability coefficient (K_c), the osmotic pressure (π_{MRO}), the pressure drop in RO module (ΔP_{MRO}), water permeability, and the geometrical parameter of the RO module, Eq. 2.10:

$$xp_{c,SRO} = \frac{K_c \, x_{c-avg,SRO}}{W \gamma \left(\Delta P_{MRO} - \pi_{MRO}\right)} \qquad \forall c, SRO \qquad (2.10)$$

The total number of modules present in every RO stage is an integer variable. To simplify the RO balance Eq. 2.7, the integrality of the parallel module, NMd_{SRO}, is relaxed to give estimate for the RO stage surface area. This assumption is reasonable to reduce the combinatory of the mathematical program.

The osmotic pressure π_{MRO} at every stage is approximated by a rule of thumb for dilute solutions [30] as

$$\pi_{MRO} = OS \sum^{c} x_{c-avg,SRO} \qquad (2.11)$$

where OS represents a proportionality constant between the osmotic pressure and the total concentration of the solute species in that stage.

Every RO stage may exist if the stage binary variable is true. The stage binary variable is related to the permeate production from the RO stage as given by Eq. 2.12:

$$Fp_{SRO} \leq Fp^{UP}_{SRO} \, y_{SRO} \qquad \forall SRO \qquad (2.12)$$

The RO module may require bounds on the operation variables to improve the system productivity. Equation 2.13 enforces the inlet feed to any RO stage to be bounded between upper and lower limits. Also, the inlet feed pressure to any RO stage may not exceed an upper value as it is described by Eq. 2.14:

$$F^{LO}_{MRO} \, NMd_{SRO} \leq F_{SRO} \leq F^{UP}_{MRO} \, NMd_{SRO} \qquad \forall SRO \qquad (2.13)$$

$$P_{SRO} \leq P^{UP}_{SRO} \qquad \forall SRO \qquad (2.14)$$

Conservation of the total and component streams is described by Eqs. 2.15 and 2.16 over every RO stage:

$$F_{SRO} = Fp_{SRO} + Fr_{SRO} \qquad \forall SRO \qquad (2.15)$$

$$F_{SRO} \, x_{c,SRO} = Fp_{SRO} \, xp_{c,SRO} + Fr_{SRO} \, xr_{c,SRO} \qquad \forall c, SRO \qquad (2.16)$$

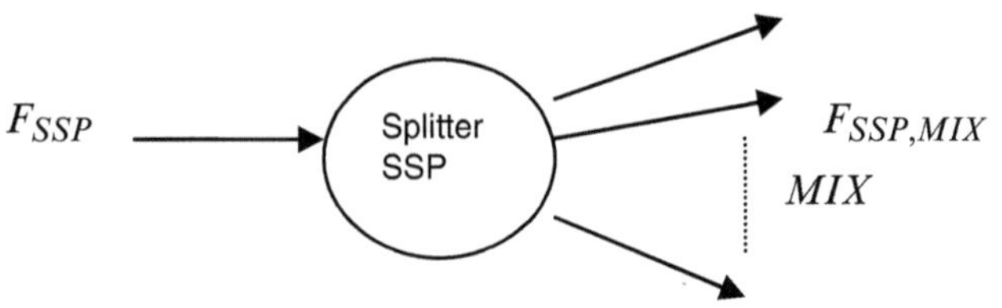

Fig. 2.3 Inlet and exit streams conditions in a splitter unit

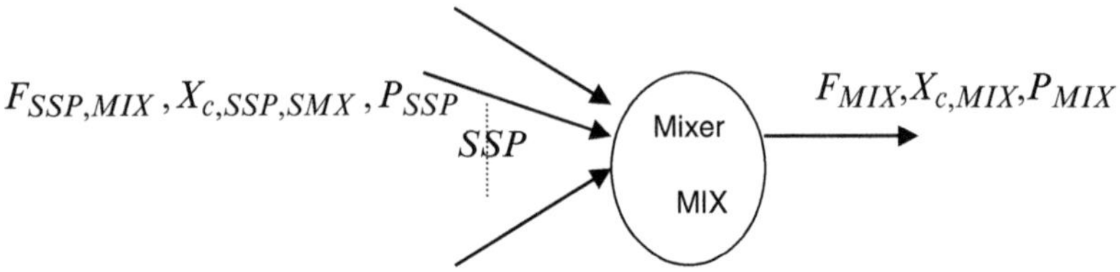

Fig. 2.4 Inlet and exit conditions for a mixer unit

In the DB, there are several splitting nodes for the incoming streams to the DB and mixing nodes before the unit operation stages and the final exit permeate and reject streams. Figure 2.3 shows a splitter node where a single feed stream (F_{SSP}) is split to several exit streams ($F_{SSP,\ MIX}$). Equation 2.17 gives total material balance over the splitter node:

$$F_{SSP} = \sum^{MIX} F_{SSP,MIX} \qquad \forall SSP \tag{2.17}$$

A mixing node is represented by Fig. 2.4 where several streams ($F_{SSP,\ MIX}$) are mixed through the mixer unit to yield a single stream. Total and component balances over the mixer node are given by Eqs. 2.18 and 2.19:

$$F_{MIX} = \sum^{SSP} F_{SSP,MIX} \qquad \forall MIX \tag{2.18}$$

$$F_{MIX}\, X_{c,MIX} = \sum^{MIX} F_{SSP,MIXX}\, X_{c,SSP,MIX} \qquad \forall c,MIX \tag{2.19}$$

Further, it is assumed here that mixing is not allowed between streams which have different pressure values. This condition is modeled through a binary variable $y_{SSP,MIX}$ which forces the flow of a stream $F_{SSP,MIX}$ to vanish if the stream pressure does not match the mixer-exit stream pressure, Eqs. 2.20, 2.21, and 2.22:

$$P_{MIX} - P_{SSP,MIX} \le M(1 - y_{SSP,MIX}) \qquad \forall SSP,MIX \tag{2.20}$$

$$P_{\mathrm{MIX}} - P_{\mathrm{SSP,MIX}} \geq -M\left(1 - y_{\mathrm{SSP,MIX}}\right) \qquad \forall \mathrm{SSP,MIX} \tag{2.21}$$

$$F_{\mathrm{SSP,MIX}} \leq F^{\mathrm{UP}}_{\mathrm{SSP,MIX}} \; y_{\mathrm{SSP,MIX}} \qquad \forall \mathrm{SSP,MIX} \tag{2.22}$$

Other constraints are imposed on the final exit streams from the network. Equations 2.23 and 2.24 give limitation on the final permeate streams (F_{fper}) to have minimum flow and the concentration of the components in these streams ($x_{\mathrm{c,fper}}$) not to exceed an upper value, respectively:

$$F_{\mathrm{fper}} \geq F^{\mathrm{LO}}_{\mathrm{fper}} \qquad \forall \; \mathrm{FPER} \tag{2.23}$$

$$x_{\mathrm{c,fper}} \leq X^{\mathrm{UP}}_{\mathrm{c,fper}} \qquad \forall \mathrm{c,FPER} \tag{2.24}$$

The set of Eqs. 2.2, 2.3, 2.4, 2.5, 2.6, 2.7, 2.8, 2.9, 2.10, 2.11, 2.12, 2.13, 2.14, 2.15, 2.16, 2.17, 2.18, 2.19, 2.20, 2.21, 2.22, 2.23, and 2.24 define a nonconvex MINLP for the RO treatment network. The mathematical program describes the unit existences and stream assignments within the network. It is worth pointing out that the nonconvexity of the RO mathematical program exhibits multiple optimal solutions. A designer has to solve the model from different starting points to explore the search space for improved solutions. In the RO case study, comments regarding the nonconvexity are given.

2.8 Integrated Air Stripper-Pervaporation Network

Groundwater and industrial wastewater streams are commonly contaminated with VOCs. VOCs are considered to be priority pollutants according to the US Environmental Protection Agency (EPA) due to their known or suspected carcinogenicity or toxicity [5]. The Henry's law constants for these compounds provide useful information concerning the behavior of VOCs in water as well as the applicability of a potential treatment technology. Air stripping has been applied for the treatment of wastewater contaminated with VOCs [31, 32]. However, this technology is inefficient if the VOCs have low Henry law constants. Consequently, the air stripper column height would have to be excessively high to treat such wastewater streams. Pervaporation is another option for the treatment of VOC which exhibits broader separation flexibility. In general, hybrid systems provide flexibility and enhanced performance over a single technology. In this regard, air stripper unit can remove VOCs with high volatility, while pervaporation complements the system by removing the low volatile compounds [33].

Pervaporation-based hybrid systems have been shown to be efficient over single technologies to achieve hard separation targets. Besides, these systems require less energy to operate and consequently reduced operational cost. Pervaporation synergistic effects basically overcome limiting conditions that may exist within chemical mixtures to enhance system throughput (i.e., altering azeotropic mixture composition, shifting reaction equilibrium toward products). Examples of such systems are hybrid pervaporation-distillation, pervaporation-reactor, and pervaporation-reactor-distillation. A review of pervaporation hybrid systems can be found elsewhere [34].

A simulation-based optimization study for hybrid air stripper-gas permeation addressed the recovery of VOC [35]. The air stream from the air stripper is continuously purified from the VOC in a closed loop configuration. Hybrid air stripping-pervaporation is presented to investigate the benefit of integrating an air stripper column with pervaporation unit [33]. The possibility of several water withdrawals from the air stripper column to the pervaporation unit was considered to determine if pervaporation has impacts on the air stripper performance. The degree of freedom of the mathematical model is reduced to allow the simulation of the system. Nonetheless, the simulation approach does not allow economical trade-off among the problem system of equations and variables.

In this section, an integrated air stripper-pervaporation system is revisited through superstructure optimization. Large alternative designs are embedded in a superstructure combined with utility units. The mathematical model is formulated as a nonconvex mixed integer nonlinear program (MINLP) which seeks an optimal treatment flow sheet for water streams with multiple VOCs. The approach is therefore rich in exploring a potential integrated flow sheet for the wastewater treatment.

2.9 Air Stripper-Pervaporation Superstructure

Hybrid air stripper-pervaporation superstructure is assisted by several utility units and air stream. The air stripper unit is coupled with an air blower to pressurize the air through the packed column. Another utility unit is a feed pump which raises the wastewater stream to the top of the packed column. The pervaporator requires a feed pump that pressurizes the wastewater feed to the required inlet pressure by the pervaporation unit. A vacuum pump is linked with the pervaporation permeate side. Figure 2.5 depicts the stream distribution and the unit operations in the proposed superstructure.

The air stripper box (ASB) is assumed to have several air stripper stages in the network. Every air stripper tower is linked with air blower and water pump units. The pervaporation box (PVB) is composed of several pervaporation stages, while every stage has a feed pump and a vacuum pump. Every pervaporation stage has several parallel pervaporation modules operating under the same operating conditions. The wastewater stream is first distributed over the unit operation nodes and the final reject stream nodes. Also, the air stream feed is linked only with the air stripper units. Within the network, the wastewater streams are linked

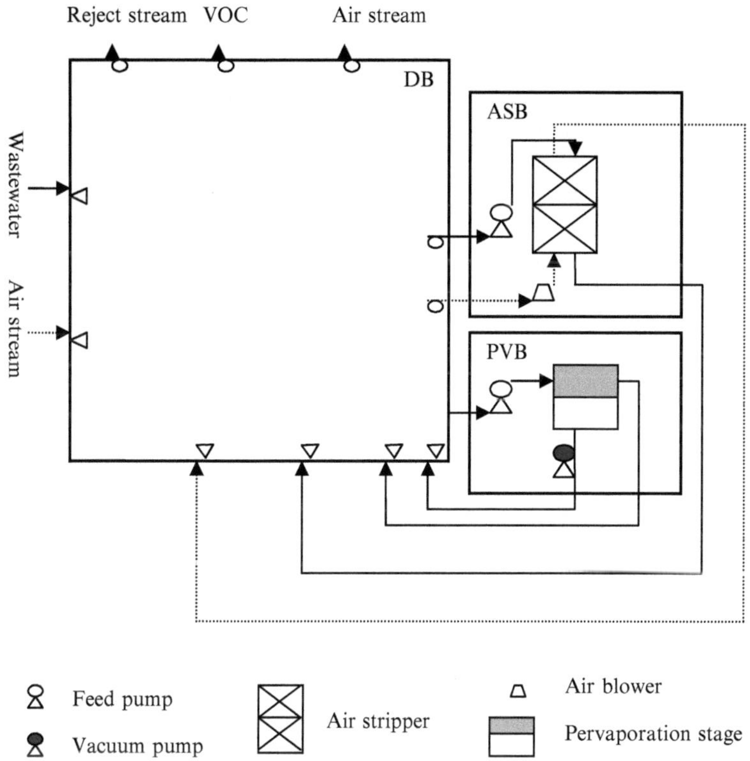

Fig. 2.5 Integrated air stripper-pervaporation superstructure

directly between the unit operations and the final reject streams. The permeate streams from all pervaporation units are combined to produce the final VOC stream.

It is worth pointing out the following remarks about the previous superstructure representation:

- The air stripper stage can represent a section of an air stripper unit or a stand-alone unit. Therefore, the decision to withdraw or inject streams from or to the unit can be left to the optimization algorithm. Coupling the pervaporation stages with the air stripper units will allow optional integration between the units and within sections of the air stripper units.
- Parallel/series arrangement between the units is given in the superstructure representation without any postulated design layout. Therefore, simultaneous evaluation of different layouts is embedded within the superstructure.
- The compact representation of the utility units within the unit operation boxes reduces the mathematical programming complexity and thus provides more emphasis on the integration between the unit operations.
- The superstructure representation is flexible to include other unit operations. Therefore, other integrated pervaporation systems can be modeled easily through the given representation by providing their appropriate mathematical models.

2.10 MINLP Formulation

The mathematical programming model describes basic material and component balances through all the mixer and splitter nodes in the superstructure. The mixing between the network streams requires that these streams have equivalent pressure values. A threshold of VOC concentrations in the reject streams is also predetermined to comply with given discharge regulations. In the case of VOC recovery, the previous inequality constraints can be reformulated over the permeate and the air streams by defining minimum recovery fractions. The air stripping model covers the operation of the unit up to the loading point. This condition is more practical and common with industrial practice [36]. The pervaporation model takes into account the effect of concentration polarization and pressure drop within the pervaporation module. The objective function is defined so as to minimize the total annualized cost of the unit operations,

$$\mathrm{TAC}_{\mathrm{treatment}} = \mathrm{Cost}_{\mathrm{capital}}\,(\mathrm{amortiz.}) + \mathrm{Cost}_{\mathrm{operating}} \tag{2.25}$$

The DB streams have several states that may allow prior elimination of mixing streams with different properties. For example, the pervaporation permeate streams can be collected directly to the final permeate stream set. The air streams within the network are only allowed to mix between each other (e.g., gas phase) and allowed to pass the network to the final air exit stream set. On the other hand, the wastewater streams (e.g., liquid phase) are allowed to mix at any mixing point that involves wastewater liquid stream. Therefore, every stream in the DB is characterized by flow rate, composition, pressure, and state. Mixing is only allowed between streams which have similar states (e.g., air, liquid wastewater, permeate streams). For every mixer, material and component balances hold:

$$F_{\mathrm{MIX}} = \sum^{\mathrm{SSP}} F_{\mathrm{SSP,MIX}} \qquad \forall \mathrm{MIX} \tag{2.26}$$

$$F_{\mathrm{MIX}}\, x_{\mathrm{C,MIX}} = \sum^{\mathrm{SSP}} F_{\mathrm{SSP,MIX}}\, x_{\mathrm{C,SSP}} \qquad \forall \mathrm{C,MIX} \tag{2.27}$$

In addition, mixing is only allowed between streams with equivalent pressure through definition of binary variables as

$$P_{\mathrm{MIX}} - P_{\mathrm{SSP}} \leq \mathrm{UP}\ \ y_{\mathrm{SSP,MIX}} \qquad \forall \mathrm{MIX,SSP} \tag{2.28}$$

$$P_{\mathrm{MIX}} - P_{\mathrm{SSP}} \geq \mathrm{LO}\ \ Y_{\mathrm{SP,MIX}} \qquad \forall \mathrm{MIX,SP} \tag{2.29}$$

The splitter nodes only require material balances as

$$F_{\mathrm{SSP}} = \sum^{\mathrm{MIX}} F_{\mathrm{SSP,MIX}} \qquad \forall \mathrm{SSP} \tag{2.30}$$

The demand constraints are set of inequalities which enforce discharge restrictions on the final reject streams as

$$x_{c,SFREJ} \leq x^{UP}_{c,SFREJ} \qquad \forall C,SFREJ \tag{2.31}$$

The mathematical model for the air-stripping unit describes the mass transport of VOCs from the wastewater stream to the air stream and the pressure drop along the column height [37–39]. The height of air stripper can be calculated from a material balance for every component present in the tower. This results with the following set of equations:

$$H_T = H_{C,OL} N_{C,OL} \tag{2.32}$$

$$H_{C,OL} = H_{C,L} + \frac{\dot{F}_W}{\dot{F}_A H'} H_{C,v} \tag{2.33}$$

$$N_{C,OL} = \frac{1}{CO_1}\left(\frac{CO_1 x_{in} + CO_2}{CO_1 x_{out} + CO_2}\right) \tag{2.34}$$

$$CO_1 = 1 - \frac{\dot{F}_W}{\dot{F}_A H'}; \quad CO_2 = -\frac{va_{out}}{H'} + \frac{\dot{F}_W}{\dot{F}_A H'} X_{in}; \quad H' = \frac{hn}{P_T} \tag{2.35}$$

Correlation of the mass transfer coefficients in the liquid and gas phases is given as

$$H_{C,L} = \frac{1}{C_L}\left(\frac{\mu_L}{\rho_L g}\right)^{1/6}\left(\frac{4 \in}{D_L u}\right)^{1/2}\left(\frac{u_L}{a}\right)^{2/3}\left(\frac{a}{a_{Ph}}\right) \tag{2.36}$$

$$H_{C,L} = \frac{u_L}{\beta_L a_{Ph}} \tag{2.37}$$

$$\begin{aligned} H_{C,v} &= \frac{1}{C_V}(\in - h_L)^{1/2}\left(\frac{4 \in}{a}\right)^{1/2}\left(\frac{1}{a}\right)^{3/2}\left(\frac{\rho_V u_V}{a\,\mu_V}\right)^{-3/4}\left(\frac{\mu_V}{\rho_V D_V}\right)^{-1/3}\left(\frac{u_V a}{D_v a_{Ph}}\right) \\ H_{C,v} &= \frac{u_v}{\beta_v a_{Ph}} \end{aligned} \tag{2.38}$$

The correlations for the specific liquid holdup h_L and the effective interfacial area a_{Ph}/a in the preloading region are given as

$$h_L = \left(12\,\frac{\mu_L}{g\,\rho_L}\,u_L\,a^2\right)^{1/3}; \quad u_V \leq u_{V,S} \tag{2.39}$$

$$\frac{a_{\text{Ph}}}{a} = 1.5\,(4\ \in)^{-0.5}\left(\frac{4\ \in u_L\rho_L}{a\,\mu_L}\right)^{-0.2}\left(\frac{u_L^2 a}{4\ \in\ g}\right)^{-0.45}\left(\frac{4\ \in u_L^2\ \rho_L}{a\ \sigma_L}\right)^{0.75} \tag{2.40}$$

The calculation for the air and the water velocities at the loading point is given as

$$u_{V,S} = \left(\frac{g\ \rho_L}{\psi_S\ \rho_V}\right)^{0.5}\left(\frac{\in}{a^{1/6}} - a^{0.5}\left(12\frac{\mu_L}{g\,\rho_L}\ u_{L,S}\right)^{1/3}\right)\left(12\frac{\mu_L}{g\,\rho_L}\ u_{L,S}\right)^{1/6} \tag{2.41}$$

$$\psi_S = \frac{g}{C_S^2}\left(\frac{L}{V}\ \left(\frac{\rho_V}{\rho_L}\right)^{0.5}\ \left(\frac{\mu_L}{\mu_V}\right)^{0.4}\right)^{-2\,n_S} \tag{2.42}$$

$$u_{L,S} = \frac{\rho_V}{\rho_L}\frac{L}{V}\ u_{V,S} \tag{2.43}$$

For

$$\frac{L}{V}\ \left(\frac{\rho_V}{\rho_L}\right)^{1/2} < 0.4\ :\ n_S = -0.326\ ;\ C_{S1} = 2.959$$
$$\frac{L}{V}\ \left(\frac{\rho_V}{\rho_L}\right)^{1/2} \geq 0.4\ :\ n_S = -0.723\ ;\ C_{S2} = 0.695\,C_{S1}\left(\frac{\mu_L}{\mu_V}\right)^{0.1588}$$

The pressure drop equations are given as

$$\frac{\Delta P}{H_T} = \psi_L\frac{a}{(\in - h_L)^3}\frac{F_V^2}{2}\frac{1}{K} \tag{2.44}$$

$$\begin{aligned} \psi_L &= C_P\left(\frac{64}{\text{Re}_V} + \frac{1.8}{\text{Re}_V^{0.08}}\right)\ \left(\frac{\in - h_L}{\in}\right)^{1.5}\ \text{EXP}\left(\frac{\text{Re}_L}{200}\right) \\ \text{Re}_L &= \frac{u_L\rho_L}{a\,\mu_L} \\ \text{Re}_v &= \frac{u_v\rho_v}{a\,\mu_v} \end{aligned} \tag{2.45}$$

The existence of an air stripper is a relation between the column height and a binary variable as

$$\begin{aligned} H_{\text{AS}} &\leq H_{\text{AS}}^{\text{UP}}\ y_{\text{AS}} \qquad \forall\ \text{AS} \\ H_{\text{AS}} &\geq H_{\text{AS}}^{\text{LO}}\ y_{\text{AS}} \qquad \forall\ \text{AS} \end{aligned} \tag{2.46}$$

The pervaporation model takes into account concentration polarization and the pressure drop in the spiral wound module [40]. For every VOC, the molar flow in every pervaporation stage is given as

$$F_{\mathrm{VOC}} = \mathrm{Sur}A_{\mathrm{PV}}\rho_{\mathrm{M}}\,K_{\mathrm{VOC}}\left(X_{\mathrm{VOC,ST}} - \frac{va_{\mathrm{VOC,Perm}}\,P_{\mathrm{Perm}}}{hn}\right) \tag{2.47}$$

The overall mass transfer coefficient of a VOC and water can be estimated as

$$\begin{aligned} \frac{1}{K_{\mathrm{VOC}}} &= \frac{l_m}{\mathrm{Pm}_{\mathrm{VOC}}} + \frac{1}{k_{\mathrm{VOC}}} \\ \frac{1}{K_{\mathrm{Water}}} &= \frac{l_m}{\mathrm{Pm}_{\mathrm{water}}} \end{aligned} \tag{2.48}$$

The mass transfer coefficient in the concentration polarization layer and the pressure drop is given as

$$k_{\mathrm{VOC}} = 0.5\,D_L\left(\frac{1.14u_L\,\rho_L}{\mu_L}\right)^{0.46}\left(\frac{\mu_L}{\rho_L\,D_L}\right)^{0.33} \tag{2.49}$$

$$\frac{\Delta P}{\mathrm{Lt_m}} = 2.54\,u_L^{1.551} \tag{2.50}$$

The existence of a pervaporation stage is relation between the pervaporation binary variable and its surface area as

$$\begin{aligned} \mathrm{Sur}A_{\mathrm{PV}} &\le \mathrm{Sur}A_{\mathrm{PV}}^{\mathrm{UP}}\,y_{\mathrm{PV}} \qquad \forall \mathrm{PV} \\ \mathrm{Sur}A_{\mathrm{PV}} &\ge \mathrm{Sur}A_{\mathrm{PV}}^{\mathrm{LO}}\,y_{\mathrm{PV}} \qquad \forall \mathrm{PV} \end{aligned} \tag{2.51}$$

2.11 Case Studies

In this section, we analyze two case studies for integrated membrane systems design. The first case study deals with the treatment of wastewater streams generated from pulp and paper plant. The other case handles the issue of VOC's recovery from ground water stream.

2.11.1 Pulp and Paper Wastewater Treatment

This case study presents the optimization of two wastewater stream treatment cases with two organic pollutants, mono-chloro phenol (MCP) and tri-chloro phenol (TCP), from the pulp and paper industry by DuPont B-9 RO modules. Input data for the optimization problem is given in Table 2.2. The superstructure includes

Table 2.2 Input data for the pulp and paper wastewater treatment network

Stream 1 flow rate, kg/s	6
Stream 2 flow rate, kg/s	25
Feed composition of solute 1 in stream 1	26.00×10^{-6}
Feed composition of solute 1 in stream 2	12.00×10^{-6}
Feed composition of solute 2 in stream 1	3.00×10^{-6}
Feed composition of solute 2 in stream 2	4.00×10^{-6}
Minimum final permeate flow rate 1, kg/s	4.5
Minimum final permeate flow rate 2, kg/s	9
Maximum final permeate composition of solute 1 in permeate 1	8.8×10^{-6}
Maximum final permeate composition of solute 2 in permeate 1	1.4×10^{-6}
Maximum final permeate composition of solute 1 in permeate 2	8.8×10^{-6}
Maximum final permeate composition of solute 2 in permeate 2	1.4×10^{-6}
Minimum flow rate per module, kg/s	0.21
Maximum flow rate per module, kg/s	0.46
Maximum feed pressure, Pa	28.58×10^{5}
Pressure drop per module, Pa	0.405×10^{5}
Pure water permeability (W), kg/s·N	1.2×10^{-10}
Solute transport parameter (K_{c1}), kg/m^2·s	2.43×10^{-4}
Solute transport parameter (K_{c2}), kg/m^2·s	2.78×10^{-4}

three RO stages, three pumps, and three turbine units. The solution strategy aims at generation of several starting points for the solver in the General Algebraic Modeling Systems GAMS 20.5 (GAMS), [41, 42, 45]. The best solution (152,406 $/year) is identified after the execution of the solution algorithm within 24 CPU seconds. Figure 2.6 depicts the RO treatment and the operating conditions in the network. A general global optimization algorithm is also useful if one attempts to verify the global optimality of RO networks [24].

The RO treatment network shows partial treatment of the network-feed streams. Interesting result from the case study shows that partial mixing of the feed-wastewater streams is beneficial to attain reduced treatment cost. In fact, the first wastewater stream (6 kg/s) is partially mixed with the other stream to reduce the concentration of the pollutant prior to the RO treatment unit. This observation, within the current case study conditions, opposes the current proposal of distributed treatment network. However, routing pollutants between different streams can be observed to meet the final product stream requirements

2.11.2 Groundwater Treatment

This system includes two volatile VOCs (TCE, DCM) and a semi-volatile component (EDC) present in a groundwater stream. Input data for the case study are given in Table 2.3, and the cost in the objective function is given in Table 2.4. The superstructure involves two air stripper units and two pervaporation stages. The optimal solution of the treatment network (376,862 $/year) features one air

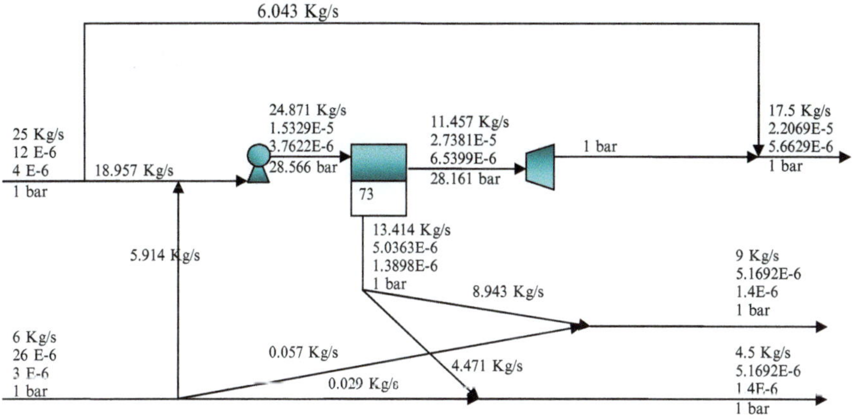

Fig. 2.6 Optimal design of the RO for the pulp and paper wastewater treatment

Table 2.3 Input date for the groundwater stream case study

$F_{\text{Water}} = 1.582(\text{kmol/s})$	$a = 86.8(\text{m}^2/\text{m}^3)$
$X_{\text{TCE}} = 1{,}100 \times 10^{-6}$	$C_L = 1.08$
$X_{\text{DCM}} = 46 \times 10^{-6}$	$C_P = 0.35$
$X_{\text{EDC}} = 7.1 \times 10^{-6}$	$C_S = 2.959$
Recovery = 99%	$C_V = 0.322$
$hn\ (\text{TCE}) = \frac{T(^{\circ}\text{K})}{0.2194} 10^{\left(5.874 - \frac{1{,}871}{T(^{\circ}\text{K})}\right)}\ (\text{bar})$	$hn\ (\text{DCM}) = \frac{T(^{\circ}\text{K})}{0.2194} 10^{\left(4.561 - \frac{1644}{T(^{\circ}\text{K})}\right)}\ (\text{bar})$
$hn\ (\text{EDC}) \frac{T(^{\circ}\text{K})}{0.2194} 10^{\left(4.434 - \frac{1{,}705}{T(^{\circ}\text{K})}\right)}\ (\text{bar})$	$\epsilon = 0.947\ (\text{m}^3/\text{m}^3)$
$l_m = 5 \times 10^{-6}\ (\text{m})$	$Lt_m = 1\ (\text{m})$
$\text{Pm}_{\text{TCE}} = 2.822 \times 10^{-8}\ (\text{m}^2/\text{s})$	$\text{Pm}_{\text{DCM}} = 5.59 \times 10^{-9}\ (\text{m}^2/\text{s})$
$\text{Pm}_{\text{EDC}} = 3.83 \times 10^{-9}\ (\text{m}^2/\text{s})$	$\text{Pm}_{\text{water}} 2.48 \times 10^{-13}\ (\text{m}^2/\text{s})$

stripper and one pervaporation stage, Fig. 2.7. This flow sheet follows a series arrangement where the pervaporation stage acts as a clean-up step for the air stripper effluent to meet the discharge requirements [43].

Within the operating temperature range of 25–55 °C, the treatment network has similar structure. The changes of the pervaporation stage membrane surface area and air stripper height over this temperature range are given in Fig. 2.8. An increase in temperature leads to a reduction in required pervaporation membrane surface area, while the air stripper column height remains almost constant at 6 m height. This is due to the increase of the removal efficiency in the air stripper column with

Table 2.4 Cost data for the air stripper-pervaporation units

Air stripper data	Pervaporation data
Capital cost	*Capital cost*
Packing material: 50 mm NORPAC®	Membrane: PDMS
Tower and packing cost $= 33{,}000\, D_{AS}^{0.57} H_{AS} + 1{,}105\, D_{AS}^{2} H_{AS}$	Membrane $= 720\,\mathrm{Sur}A$
Feed pump $= 12.265\left(\mathrm{PPu_{pump}}\right)^{0.53}$	Feed pump $= 46.6E3\ \left(\mathrm{PPu_{pump}}\right)^{0.53}$
Air blower $= 14.2E3\ \left(\mathrm{PPu_{Air\,blower}}\right)^{0.82}$	Vacuum pump $= 16.6E3\left(\mathrm{PPu_{Vacuum\,pump}}\right)^{0.53}$
$\mathrm{PPu_{pump}} = \mathrm{MW}\, F_{\mathrm{pump}} H_{AS} / g$	$\mathrm{PPu_{pump}} = F_{\mathrm{pump}} \Delta P_{\mathrm{pump}} / \rho_M$
$\mathrm{PPu_{Air\,blower}} = \left(\left(\frac{P_{\mathrm{out}}}{P_{\mathrm{in}}} \right)^{0.283} - 1 \right) \left(\frac{F_{\mathrm{air}} R T}{0.1415} \right)$	$\mathrm{PPu_{Vaccum\,pump}} = F_{\mathrm{pump}} \Delta P_{\mathrm{Vaccum\,pump}} / \rho_M$
Operational cost	*Operational cost*
Tower operating cost = 10% of the fixed cost	Membrane replacement cost $= 169\,\mathrm{Sur}A$
Power cost $= 0.05\ \$/\mathrm{Kwh}$	
Days of operation = 300 Day/year	

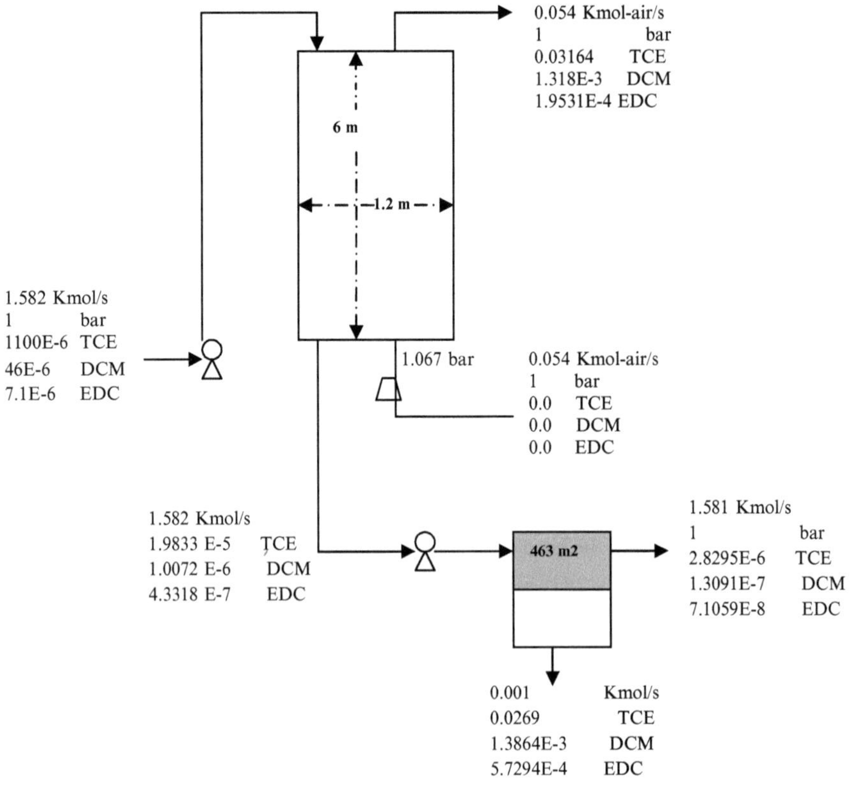

Fig. 2.7 Optimal network design of multicomponent VOC's treatment

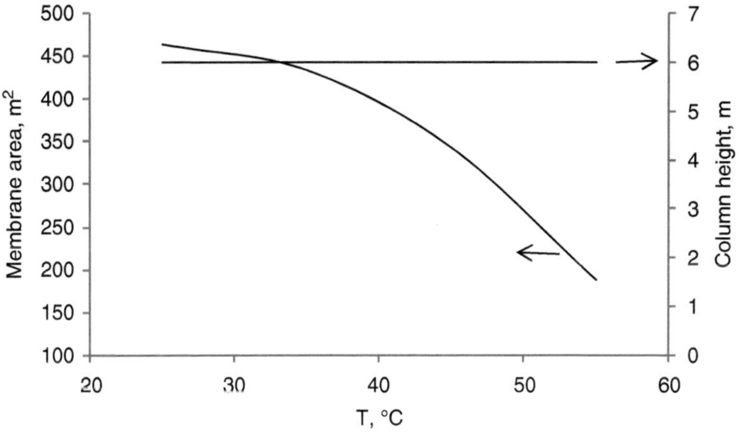

Fig. 2.8 Effect of the operating temperature on the membrane area and column height

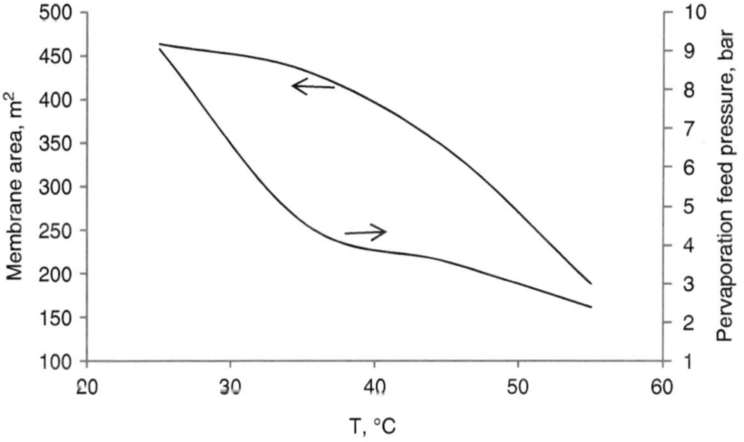

Fig. 2.9 Effect of temperature on the optimal pervaporation membrane area and feed pressure

increasing operating temperature. In addition, the required pervaporation feed pressure is reduced with increase of the operating temperature (Fig. 2.9). In general, approximately 40% reduction of the treatment cost can be achieved by increasing the feed temperature to 55 °C.

The aforementioned analysis shows the significant effect that temperature has on the separation efficiency and thus the treatment cost. However, the cost to generate this thermal energy has not been considered in the optimization routine. Further analysis of the proposed approach would be beneficial to consider a heat integrated hybrid air stripper-pervaporation system. Henry law constant shows that a reduction in its value imposes separation limitations on the air stripper efficiency. The presence of less volatile compounds may force the selection of pervaporation as a treatment option alone. Thus, the optional treatment is problem-specific (i.e., the network depends on the type of VOCs). A direct conclusion of the series arrangement of air stripper-pervaporation units may not be obvious since the optimal stream assignments within the network can only be determined from the solver optimal solution. Another important issue is how to treat the VOCs that end up in the gas streams discharged from the air stripper units. This will entail additional costs, and a designer should analyze it in the design phase.

2.12 Conclusion

Superstructure optimization has been considered in this chapter as an effective tool to study integrated membrane synthesis problems. The approach postulates rich design alternatives in the early design phase to analyze them simultaneously. The selection criterion of a potential design is based on an economical objective function. According to the given superstructure, an MINLP model is formulated to describe the network. The model optimal solution provides the stream distribution through the treatment units and the units optimal operational conditions. Integrated membrane systems were considered through two case studies for the treatment of wastewater streams.

References

1. Dunn RF, El-Halwagi M (2003) Process integration technology review: background and applications in the chemical process industry. J Chem Technol Biotechnol 78:1011–1021
2. Schaefer K, Exall K, Marsalek J (2004) Water reuse and recycling in Canada: a status and need assessment. Can Water Resour J 3:195–208
3. Kuo WJ, Smith R (1997) Effluent treatment system design. Chem Eng Sci 3:4273–4290
4. Starthmann H (2001) Membrane separation processes, Current relevance and future opportunities. AIChE J 47:1077–1087
5. Eddy Metcalf (1991) Wastewater engineering treatment, disposal, and reuse. Irwin/McGraw-Hill, New York
6. Baker RW (2004) Membrane technology and applications, 2nd edn. Wiley, New York
7. Wenten IG (2002) Recent development in membrane science and its industrial applications. Songklanakarin J Sci Technol 24:1009–1024
8. Grossmann IE, Daichendt MM (1996) New trends in optimization-based approaches to process synthesis. Compt Chem Eng 6:665–683
9. Westerberg AW (2004) A retrospective on design and process synthesis. Compt Chem Eng 28:447–458
10. Viswanathan J, Grossmann IE (1993) Optimal feed locations and number of trays for distillation columns with multiple feeds. Ind Eng Chem Res 32:2942–2949
11. Ciric AM, Gu D (1994) Synthesis of nonequilibrium reactive distillation processes by MINLP optimization. AIChE J 9:1479–1487
12. Kookos IK (2003) Optimal design of membrane/distillation column hybrid processes. Ind Eng Chem Res 8:1731–1738

13. El-Halwagi MM (1992) Synthesis of reverse osmosis networks for waste reduction. AIChE J 38:1185–1198
14. El-Halwagi MM (1993) Optimal design of membrane-hybrid systems for waste reduction. Sep Sci Technol 28:283–307
15. Srinivas BK, El-Halwagi MM (1993) Optimal design of pervaporation systems for waste reduction. Compt Chem Eng 17:957–970
16. Yeomans H, Grossmann IE (1999) A systematic modeling framework of superstructure optimization in process synthesis. Compt Chem Eng 6:709–731
17. El-Halwagi MM, Manousiouthakis V (1989) Synthesis of mass exchange networks. AIChE J 8:1233–1244
18. Papalexandri KP, Pistikopoulos EN (1996) Generalized modular representation framework for process synthesis. AIChE J 4:1010–1032
19. Ismail SR, Pistikopoulos EN, Papalexandri KP (1999) Modular representation synthesis framework for homogeneous azeotropic separation. AIChE J 45:1701–1720
20. Proios P, Pistikopoulos EN (2006) Hybrid generalized modular/collocation framework for distillation column synthesis. AIChE J 3:1038–1056
21. Linke P, Kokossis AC (2004) Advanced process systems design technology for pollution prevention and waste treatment. Adv Environ Res 8(2):229–245
22. Linke P, Kokossis AC (2003) Attainable reaction and separation processes from a superstructure-based method. AICHE 49(6):1451–1470
23. Mehta VL, Kokossis AC (2000) Nonisothermal synthesis of homogeneous and multiphase reactor networks. AIChE J 11:2256–2273
24. Saif Y, Elkamel A, Pritzker M (2009) Superstructure optimization for the synthesis of chemical process flowsheets: application to optimal hybrid membrane systems. Eng Optim 41:327–350
25. Vyhmeister E, Saavedra A, Cubillos FA (2004) Optimal synthesis of reverse osmosis systems using genetic algorithms. In: European symposium on computer-aided process engineering-14, Lappeenranta
26. Maskan F, Wiley DE, Johnston LPM, Clements DJ (2000) Optimal design of reverse osmosis module networks. AIChE J 46:946–954
27. Lu YY, Hu YD, Xu DM, Wu LY (2006) Optimum design of reverse osmosis seawater desalination system considering membrane cleaning and replacing. J Membr Sci 282:7–13
28. Lu YY, Hu YD, Zhang XL, Wu LY, Liu QZ (2007) Optimum design of reverse osmosis system under different feed concentration and product specification. J Membr Sci 287:219–229
29. Evangellsta F (1985) A short cut method for the design of reverse osmosis desalination plants. Ind Eng Chem Process Des Dev 24:211–223
30. Weber W (1972) Physiochemical processes for water quality. Wiley, New York
31. Adams JQ, Clark RM (1991) Evaluating the costs of packed-tower aeration and GAC for controlling selected organics. AWWA 1:49–57
32. Dzombak DA, Roy SB, Fang H (1993) Air-stripper design and costing computer program. AWWA 10:63–72
33. Shah MR, Noble RD, Clough DE (2004) Pervaporation-air-stripping hybrid process for removal of VOCs from groundwater. J Membr Sci 241:257–263
34. Suk DE, Matsuura T (2006) Membrane-based hybrid processes: a review. Sep Sci Technol 41:595–626
35. Wijmans JG, Kamaruddin HD, Segelke SV, Wessling M, Baker RW (1997) Removal of dissolved VOCs from water with an air stripper/membrane vapor separation system. Sep Sci Technol 14:2267–2287
36. Hines AL, Maddox RN (1985) Mass transfer, fundamentals and applications. Prentice Hall, New York
37. Billet R, Schultes M (1991) Modeling of pressure drop in packed columns. Chem Eng Technol 14:89–95

38. Billet R, Schultes M (1993) Predicting mass transfer in packed columns. Chem Eng Technol 16:1–9
39. Billet R, Schultes M (1999) Prediction of mass transfer columns with dumped and arranged packing, updated summary of the calculation method of Billets and Schultes. Trans IChemE 77:498–504
40. Hickey PJ, Gooding CH (1994) Modeling spiral wound modules for the pervaporative removal of volatile organic compounds from water. J Membr Sci 88:47–68
41. Saif Y, Elkamel A, Pritzker M (2008) Optimal design of RO network for wastewater treatment and minimization. Chem Eng Proc 47:2163–2174
42. Saif Y, Elkamel A, Pritzker M (2008) Global optimization of reverse osmosis network for wastewater treatment and minimization. Ind Eng Chem Res 47:3060–3070
43. Elkamel A, Saif Y, Pritzker M (2009) Optimal design of a hybrid air stripping/pervapouration system for removal of multicomponent VOCs from groundwater. Int J Proc Sys Eng 1:46–65
44. Bagajewicz MJ, Manousiouthakis V (1992) Mass/heat-exchange network representation of distillation networks. AIChE J 11:1769–1800
45. Brooke A, Kendrik D, Meeraus A (1992) GAMS user's guide. Boyd & Fraser Publishing Co, Danvers
46. Daichendt MM, Grossmann IE (1996) Integration of hierarchical decomposition and mathematical programming for the synthesis of process network. Compt Chem Eng 22:147–175

Chapter 3
Recent Advances in the Reuse of Wastewaters for Promoting Sustainable Development

Ta Yeong Wu, Abdul Wahab Mohammad, Su Lin Lim, Pei Nie Lim, and Jacqueline Xiao Wen Hay

3.1 Introduction

The sustainability of human activities (predominantly production and consumption) is a growing concern among businesses, customers, governments, international bodies, and nongovernmental organizations [1]. Of late, the importance of sustainability is being emphasized globally so that the needs of the present generation can be met without compromising the ability of future generations to meet their own needs.

Water scarcity has become a global issue and is not only a problem limited to arid zones. Continuous population growth, increasing standards of living, climate changes, industrialization, agriculture, and urbanization are triggering the decrease in water resources worldwide [2]. According to United Nations predictions, between two and seven billion people will face water shortages by the year 2050 [3]. Additionally, it is estimated that 60% of the world population will reside in urban areas by 2030 [4], with the combined effect of increasing population and greater water demand [5], thus creating larger amount of wastewaters in concentrated areas [6]. Therefore, sustainability of critical water resources, particularly in urban centers, will rely upon effective and sustainable wastewater management.

Applications of a sustainability framework to wastewater treatment and management are very recent and limited, although a somewhat larger number of cases in water resource managements have been studied from the sustainability perspective [7].

T.Y. Wu (✉) • S.L. Lim • P.N. Lim • J.X.W. Hay
Chemical Engineering Discipline, School of Engineering, Monash University, Jalan Lagoon Selatan, Bandar Sunway 46150, Selangor Darul Ehsan, Malaysia
e-mail: wu.ta.yeong@monash.edu; tayeong@hotmail.com

A.W. Mohammad
Scale-up and Downstream Processing Research Group, Department of Chemical and Process Engineering, Faculty of Engineering and Built Environment, Universiti Kebangsaan Malaysia, Bangi, Selangor Darul Ehsan 43600, Malaysia

S.K. Sharma and R. Sanghi (eds.), *Wastewater Reuse and Management*,
DOI 10.1007/978-94-007-4942-9_3,

Generally, a sustainable wastewater management system could be based on the 5-Rs policy: reduction, replacement, reuse, recovery, and recycling [8]. Within this context, it is proposed that a wastewater management system based on the promotion of cleaner production and environmentally sound biotechnologies could be included as a part of global wastewater management in order to attain sustainable development [8, 9].

Reused wastewater could be regarded as a new water resource, especially when it is used in regions where freshwater is scarce. According to Gurel et al. [10], the reuse of wastewater provides financial, environmental, economical, and social benefits, such as the following: (1) an increase in available water resources; (2) a more rational allocation of freshwater resources and their conservation; (3) a potential reduction in pollutants discharged into freshwater; (4) a rational use of the nutrient content in the treated wastewater; and (5) a guarantee of a regular water supply, especially in water-scarce areas (e.g., farmers would have access to water for restricted irrigation, even in times of drought).

Wastewater treatment for reuse may have a lower cost than developing new water supply sources, particularly for low-quality reuse in toilet flushing and similar non-potable urban uses. However, reuse of untreated wastewaters may lead to adverse effects associated with the potential public health and environmental risks [10]. For example, the potential health risks and environmental impacts resulting from wastewater reuse for irrigation have been well documented [11]. Untreated wastewater usually contains a variety of different pathogens that are capable of surviving for extended periods of time in soil or on crop surfaces, which, in turn, enter the food chain. Infections due to helminths, bacteria, viruses, and protozoa constitute the health risks of regions using wastewater irrigation [10]. Health and environmental aspects are particularly sensitive issues and important prerequisites; wastewater effluent must not be used or be accepted as a replacement for conventional or possibly other nonconventional water sources unless it is adequately treated and safely applied [12].

While reuse of wastewaters has substantial sustainable merits, the potential risks of applications should not be taken lightly. These risks can be minimized by proper treatment, disinfection, and control of the effluent to meet the related guidelines and standards [10]. In developed countries, public institutions usually determine water quality objectives by considering health risks and encouraging appropriate wastewater treatment to achieve these goals. In fact, there are integrated programs for planned reuse of wastewater in most developed countries [12]. In order to reduce the impact of wastewater reuse, many countries have adopted standards and guidelines, especially based on Californian (USA) and WHO guidelines [10]. In general, the standards and guidelines available reflect the level of wastewater reuse as well as the state of wealth in that country (Lallana et al. 2001).

Recently, there have been many discussions on the progress of wastewater reuse globally to protect the environment and achieve sustainable resources. Instead of using disposal and end-of-pipe treatment for wastewater management, cleaner production, such as reuse, recycling, and other green technologies, is encouraged. Based on the current trends of wastewater reuse, it can be concluded that the number of countries investigating and implementing wastewater reuse has increased over the past decade. In fact, wastewater reuse is growing steadily not only in water-deficient areas but also in highly populated countries in temperate regions. The suitability of wastewater reuse, especially in arid and semiarid areas, is now universally recognized. Thus, this chapter primarily addresses some of the recent advances in wastewater reuse, together with the recommended pretreatment processing of raw wastewater before it can be reused as a water source. In addition, it is proposed that wastewater could be sustainably reused and/or transformed into low- (or preferably high-) value-added products, such as irrigation and industrial waters, fermentation substrates in the production of various metabolites, fertilizers, and animal feeds.

3.2 Reuse of Wastewater as Irrigation Water

Wastewater reuse for irrigation is widely implemented in many countries around the world. It is used mainly for agricultural irrigation as well as landscape irrigation [13–15]. Both treated and untreated wastewaters are reused for irrigation. For example, farmers in Pakistan depend on wastewater for irrigation due to its nutrient content and high water costs [16]. Wastewater from washing machines and the kitchen sink has also been reused to irrigate lawns above and below ground, respectively [17]. However, the reuse of untreated wastewater for irrigation may pose a risk to human health, environment soil quality, and crop productivity and quality [16, 18]. For example, it was reported that soil, crop leaves, and fruits were contaminated by fecal coliforms when untreated wastewater was reused. Contamination has also been reported in countries of the Mediterranean region, such as Turkey and Morocco, as well as other countries, such as Ghana and Mexico. In Mexico, farmers irrigate their land with diluted, untreated, or partially treated wastewater [12, 18, 19]. Thus, it is important to have wastewater guidelines and standards for wastewater reuse for irrigation. Table 3.1 shows the guidelines developed by the WHO and various countries regarding microbiological parameters. As is evident in this table, different standards are set for unrestricted and restricted irrigation and for different crop types.

Additionally, Fig. 3.1 and Table 3.1 demonstrate the heavy metal concentration limits and other water quality parameters, respectively for use in irrigation. Other water quality parameters used in determining irrigation water quality such as 5-days biochemical oxygen demand (BOD_5), chemical oxygen demand (COD), and dissolved oxygen (DO) content reflect the biological treatment efficiency, while others, such as boron, sodium adsorption ratio (SAR), and chloride, are set for crop protection. Table 3.1 and Fig. 3.1 illustrate that each country has set different limits according to their national needs.

Most countries are moving toward improving their wastewater treatment system to produce high-quality effluent for irrigation. The need for wastewater reuse principally stems from the widespread water scarcity problems around the world

Table 3.1 Microbiological standards and guidelines for agricultural reuse and water quality parameters for irrigation of wastewater

Countries/ organizations	Category[a] and reuse conditions	Coliforms		Parasitological and irrigation water quality parameters
		Total	Fecal	
Australia	Food crops in direct contact with water, e.g., sprays	–	<10/100 mL median value with four of five samples containing less than 40/100 mL	$BOD_5 = 40$ kg/ha/day
	Food crops not in direct contact with water (e.g., flood or furrow) or which will be sold to consumers cooked or processed	<1,000/100 mL median value with four of five samples containing less than 10,000/100 mL	–	Sodium adsorption ratio (SAR) <6
California-USA	Agricultural reuse – food crops	2.2/100 mL (7-days median) 23/100 mL (not to exceed in more than one sample in 30 days) 240/100 mL (max. one sample)	–	–
	Agricultural reuse – nonfood crops	23/100 mL (7-days median) 240/100 mL (not to exceed in more than one sample in 30 days)	–	–

Cyprus	(i) All crops – irrigation of leafy vegetables, bulbs, and corms eaten uncooked is not allowed	–	5/100 mL (these values must not be exceeded in 80 % of samples per month)	BOD_5 = 10 mg/L
			15/100 mL (maximum value allowed)	TSS = 10 mg/L
	(ii) Amenity areas of unlimited access and vegetables eaten cooked – potatoes, beetroots, *Colocasia* species	–	50/100 mL (these values must not be exceeded in 80 % of samples per month)	BOD_5 = 10 mg/L
			100/100 mL (maximum value allowed)	
	(iii) Crops for human consumption. Amenity areas of limited access	–	200/100 mL (these values must not be exceeded in 80 % of samples per month)	BOD_5 = 20 mg/L
			1,000/100 mL (maximum value allowed)	
	(iv) Fodder crops	–	1,000/100 mL (these values must not be exceeded in 80 % of samples per month)	BOD_5 = 20 mg/L
			5,000/100 mL (maximum value allowed)	
	(v) Industrial crops	–	3,000/100 mL (these values must not be exceeded in 80 % of samples per month)	BOD_5 = 50 mg/L
			10,000/100 mL (maximum value allowed)	
Israel	(C) Deciduous fruits, conserved vegetables, cooked and peeled vegetables, green belts, football fields, and golf courses	250/100 mL		BOD_5 = 35 mg/L DO = 0.5 mg/L TSS = 30 mg/L

(continued)

Table 3.1 (continued)

Countries/ organizations	Category[a] and reuse conditions	Coliforms		Parasitological and irrigation water quality parameters
		Total	Fecal	
	(D) Unrestricted crops, including vegetables eaten uncooked (raw), parks, and lawns	12/100 mL (80 %)		BOD_5 = 15 mg/L DO = 0.5 mg/L TSS = 15 mg/L
Italy	For reuse of wastewater	*E. coli* = 10/100 mL (for 80 % of samples, 100 maximum value)		BOD_5 = 20 mg/L COD = 100 mg/L TSS = 10 mg/L TN = 15 mg/L SAR = 10 Sulfates (SO_4^-) = 500 mg/L Chlorides = 250 mg/L
	Restricted irrigation	20/100 mL (mean value of 7 consecutive sampling days)	–	–
Jordan	(A) Cooked vegetables, parks, playgrounds, and sides of roads within city limits	*E. coli* = 100/100 mL		Helminth $\leq$ 1 egg/L BOD_5 = 30 mg/L COD = 100 mg/L DO > 2 mg/L TSS = 50 mg/L TN = 45 mg/L NO_3^- = 30 mg/L
	(B) Fruit trees, sides of roads outside city limits, and landscape	*E. coli* = 1,000/100 mL		Helminth $\leq$ 1 egg/L BOD_5 = 200 mg/L COD = 500 mg/L TSS = 150 mg/L TN = 70 mg/L NO_3^- = 45 mg/L

	(C) Field crops, industrial crops, and forest trees	–		Helminth $\leq$ 1 egg/L BOD_5 = 300 mg/L COD = 500 mg/L TSS = 150 mg/L TN = 70 mg/L NO_3^- = 45 mg/L
Morocco	For cultures intended for raw consumption	–	1,000/100 mL	Pathogenic parasites, eggs, cysts of parasites, larvae of Ankylostomides, and Fluococercaires of *Schistosoma haematobium* should be absent TSS = 100 mg/L NO_3^- = 30 mg/L SO_4^- = 250 mg/L Chlorides =105 mg/L
		Salmonella = absence in 5 L *Vibrio cholerae* = absence in 450 mL		
Palestine	For restricted irrigation	1,000/100 mL	–	BOD_5 = 45 mg/L COD = 150 mg/L DO > 0.5 mg/L TSS = 40 mg/L NO_3^- = 50 mg/L SO_4^- = 1 mg/L
	For unrestricted irrigation	1,000/100 mL	–	BOD_5 = 60 mg/L COD = 200 mg/L DO > 0.5 mg/L TSS = 50 mg/L NO_3^- = 50 mg/L SO_4^- = 500 mg/L

(continued)

Table 3.1 (continued)

Countries/ organizations	Category[a] and reuse conditions	Coliforms		Parasitological and irrigation water quality parameters
		Total	Fecal	
Tunisia	–	–		Helminth < 1 egg/L BOD_5 = 30 mg/L TSS = 30 mg/L
Turkey	Class I (very good for irrigation)	–	0–2/100 mL	BOD_5 = 0–25 mg/L TSS = 20 mg/L NO_3^- = 0–5 mg/L SAR < 10 SO_4^- = 0–192 mg/L Chlorides = 0–142 mg/L
	Class II (good for irrigation)	–	2/100 mL–20/100 mL	BOD_5 = 25–50 mg/L TSS = 30 mg/L NO_3^- = 5–10 mg/L SAR = 10–18 SO_4^- = 192–336 mg/L Chlorides = 142–249 mg/L
	Class III (usable for irrigation)	–	20/100 mL–100/100 mL	BOD_5 = 50–100 mg/L TSS = 45 mg/L NO_3^- = 10–30 mg/L SAR = 18–26 SO_4^- = 336–575 mg/L Chlorides = 249–426 mg/L
	Class IV (usable with caution for irrigation)	–	100/100 mL–1,000/100 mL	BOD_5 = 100–200 mg/L TSS = 60 mg/L NO_3^- = 30–50 mg/L SAR > 26 SO_4^- = 575–960 mg/L Chlorides = 426–710 mg/L

	Class V (unusable for irrigation)	–	>1,000/100 mL	$BOD_5 > 200$ mg/L TSS > 100 mg/L $NO_3^- > 50$ mg/L $SO_4^- > 960$ mg/L Chlorides > 710 mg/L
WHO	(A) Unrestricted irrigation: irrigation of crops likely to be eaten uncooked, sports fields, public parks	–	≤1,000/100 mL (geometric mean no. per 100 mL)	Intestinal nematodes ≤1 (arithmetic mean no. of egg/L) when children under 15 are exposed, treatment to <0.1 egg/L
	(B) Restricted irrigation: irrigation of cereal crops, industrial crops, fodder crops, pasture, and trees	–	≤ 105/100 mL	Intestinal nematodes ≤ 1 (arithmetic mean no. of egg/L) when children under 15 are exposed, treatment to <0.1 egg/L
	(C) Localized irrigation of crops in category B. if exposures of workers and the public do not occur	–	Not applicable	Not applicable

Adapted from Gurel et al. [10]

[a]Categories A, B, C, D, Class I, Class II, i, ii, etc. given in Table 3.1 are the original categories given in the standards or guidelines of the countries

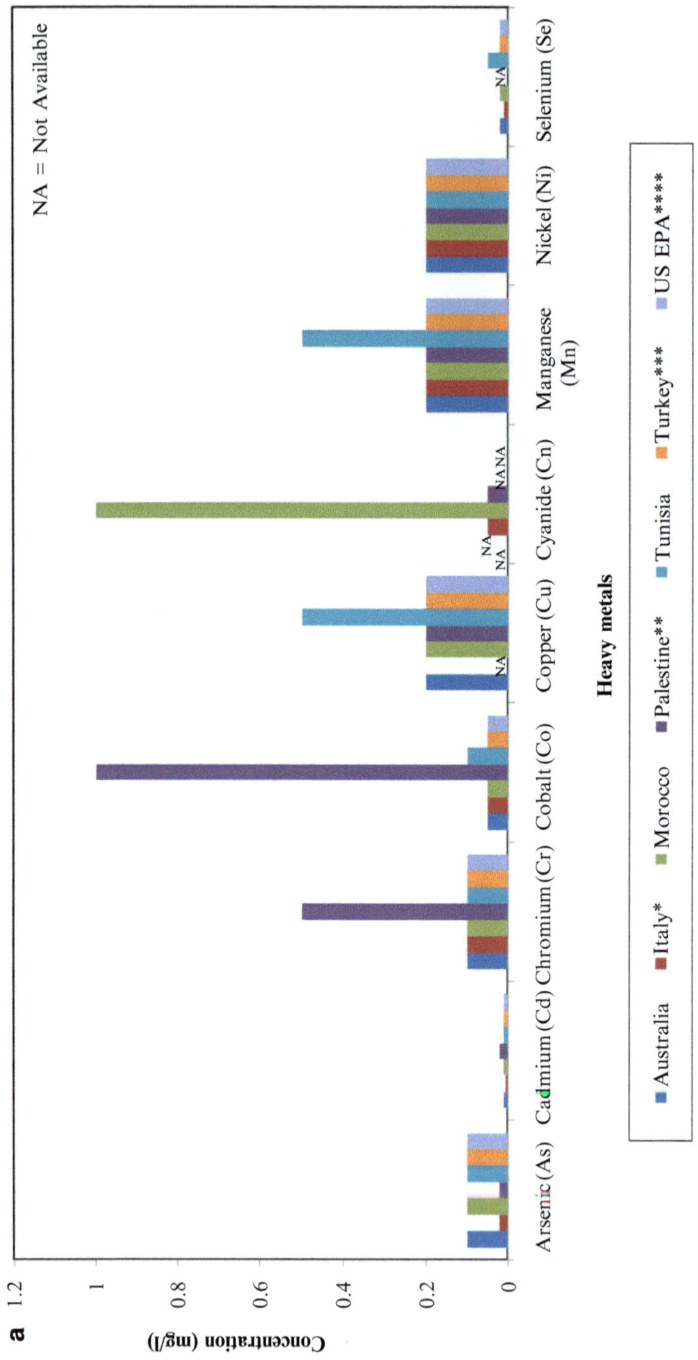
a
Concentration (mg/l)
1.2
1
0.8
0.6
0.4
0.2
0
NA = Not Available
Arsenic (As)
Cadmium (Cd)
Chromium (Cr)
Cobalt (Co)
Copper (Cu)
Cyanide (Cn)
Manganese (Mn)
Nickel (Ni)
Selenium (Se)
NA
Heavy metals
Australia
Italy*
Morocco
Palestine**
Tunisia
Turkey***
US EPA****

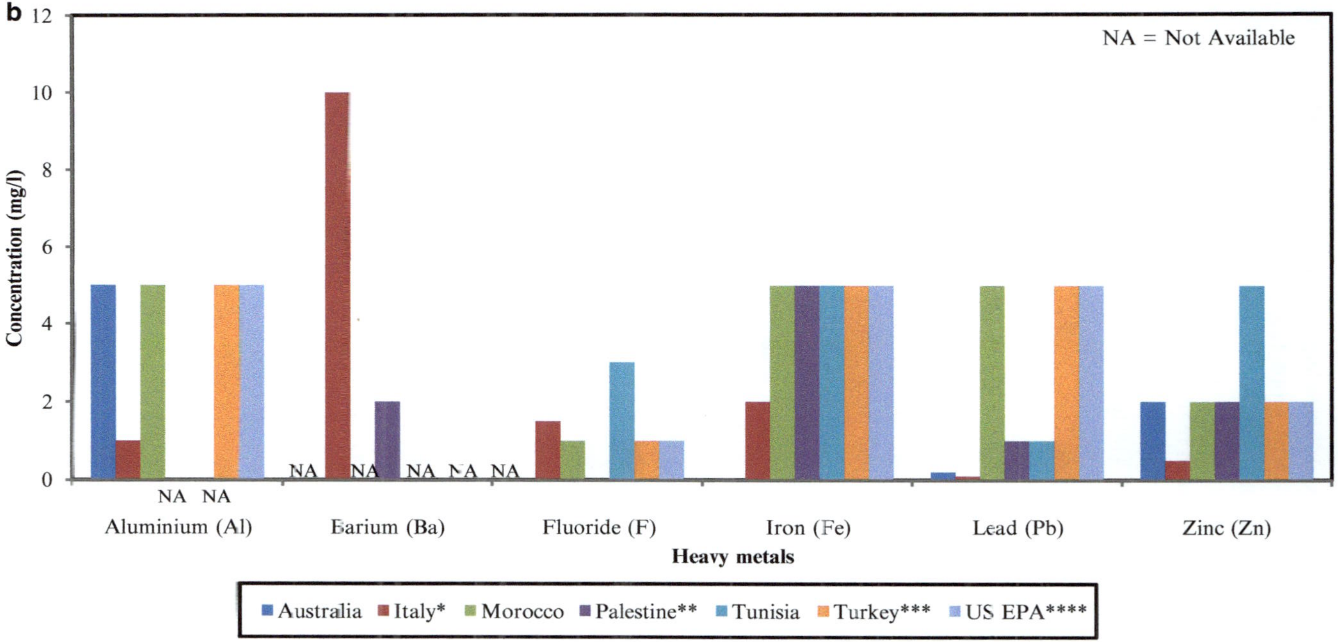

Fig. 3.1 Concentration of (**a**) arsenic, cadmium, chromium, cobalt, copper, cyanide, manganese, nickel, and selenium and (**b**) aluminum, barium, fluoride, iron, lead, and zinc in water for use in irrigation^{+} (Adapted from [10]). $^{+}$ The lowest concentration limit for heavy metal is mercury (Hg) = 0.001 mg/L in Italy, Morocco, Palestine, and Tunisia. *The reuse regulations in Italy are not only for irrigation but also for general reuse activities, **for restricted irrigation, ***limits for continuous irrigation under all soil conditions, and ****for long-term use

resulting from population growth [20–24]. Ongoing climate changes will also increase water losses from lakes, reservoirs, and dams, which will increase the demand for water [25]. Irrigation contributes largely to the world's food including vegetables, fruits, and other high-value crops [25]. In many areas, agricultural development is impossible without irrigation. Irrigation is crucial for increasing crop production, especially in areas with limited water resources or arid regions, such as countries in the Mediterranean region [12, 26–28].

Approximately 70–80% of water in most Mediterranean countries is used for landscape and agricultural irrigation. Treated wastewater provides a low-cost water source for irrigation in agriculture and reduces the dependence on freshwater [26]. In addition, treated wastewater contains significant amounts of nitrogen and phosphorus that are viable resources for irrigation [17, 29–31]. The available nutrients enhance soil nutrition, and consequently, less fertilizer, such as ammonium nitrate phosphate, is required for agriculture [13, 14].

Among all the Mediterranean countries, Israel is the pioneer in the development of wastewater reclamation. This practice was later followed by Cyprus, Jordan, and Tunisia [26]. In Israel, 65% of the total municipal sewage production is used for wastewater irrigation [32]. It was found that reclaimed wastewater yielded no significant difference as compared to freshwater when the former was used to irrigate olive trees [33]. Although salt loads were increased on the roots, no effect on olive growth and productivity was observed when reclaimed wastewater was used. Further research is needed to determine the long-term effect of irrigation with reclaimed wastewater [33]. Similarly, Hassanli et al. [34] also reported no significant difference between the use of treated water and freshwater for the irrigation of corn. However, a study performed by Akponikpè et al. [18] indicated that treated wastewater improved eggplant yield by an average of 40% compared to irrigation with freshwater.

In Cyprus, wastewater is reused for restricted irrigation, such as public amenity areas and golf courses. The cost of wastewater tertiary treatment is less than the cost of dam water or desalinated water [35]. Similar to most of the Mediterranean countries, Jordan is facing serious water shortage because of its increasing population. It is estimated that approximately 66% of water is used for irrigation. The high costs of other water resources are not feasible; thus, reuse of reclaimed wastewater for irrigation is encouraged [36]. Tunisia is one of the few countries with a national policy for wastewater reuse [37]. Approximately 30–43% of its treated wastewater is used for agricultural and landscape irrigation. Reclaimed water is used for irrigation of golf courses, public gardens, fodder crops, cereals, citrus, vineyards, and, in particular, olive trees [20, 29].

Other Mediterranean countries, such as Palestine, use treated wastewater irrigation of plants, olive trees, and fruit trees [38]. In Spain, effluent from treatment facilities is reused mainly for irrigation [23]. Illueca-Muñoz et al. [28] stated that further treatment of secondary level effluent was needed so that it could be reused for agricultural irrigation. A comparison of several tertiary treatments revealed that ultrafiltration was the most efficient with 50 and 100% removal of COD and total coliform, respectively. In Italy, the pilot-scale vertical flow constructed wetlands proved to be efficient in removing contaminants from municipal wastewater. However, pretreatment systems were needed to obtain high removal efficiencies to fulfill the Italian guidelines for irrigation reuse [39]. Generally, most treatment plants in Greece produce effluents that are suitable for reuse purpose in agricultural and landscape irrigation. The treatment processes used are activated sludge, disinfection (UV irradiation and chlorination), and sand filtration [21]. Additionally, more than 88% of the treated effluent is discharged to nearby farmlands. Thus, a lower cost is involved in irrigation [26].

Apart from Mediterranean countries, India also reuses treated wastewater for irrigation. For example, effluents from stabilization ponds are used to irrigate a wide range of crops, including those of sugarcane and coconut plantations. City wastewaters that oxidize in open channels are also used for crop irrigation as the heavy metals and nitrates are within the acceptable levels [40]. In Delhi city, the sewage treatment plant consists of many different treatment technologies: the activated sludge process; extended aeration; high-rate aeration; physical, chemical, and biological treatment (BIOFORE); and the trickling filter and oxidation pond. Effluent from the activated sludge process and the trickling filter is usually used for restricted and unrestricted irrigation. However, only effluent from an oxidation pond meets the guidelines for wastewater irrigation [41].

In central Mexico, a pilot-scale wetland was designed to treat the domestic sewage discharged. The constructed wetland reduces the amount of pollutants to an acceptable level for irrigation water. The low cost of this treatment system also makes it feasible in central Mexico as well as in other rural areas [42]. In Taiwan, wastewater from industrial parks is treated using activated sludge, chemical coagulation, Fenton oxidation, and ion exchange. The treatment processes manage to improve the wastewater quality, thus making it possible to be reused for agricultural irrigation. In addition, the preliminary cost of this treatment system is found to be reasonable [43]. In Australia, dairy processor wastewater has been treated in aerobic bioreactors before being reused for agricultural irrigation [44].

Membrane technologies, such as reverse osmosis and aerobic membrane bioreactors, have also been tested to treat wastewater for landscape and agricultural irrigation [45]. Similarly, You et al. [46] employed combined membrane technologies to treat textile-dyeing wastewater. Removal efficiencies for COD, BOD, total suspended solids (TSS), and true color were 88, 100, 100, and 97%, respectively. The effluent quality was proven to be suitable for reuse purpose in irrigation.

Wastewater treatment technologies, such as the membrane bioreactor (MBR) process and field treatment system (FTS), have been proven to be effective in treating industrial and municipal wastewater. Although both treatment processes produced effluent suitable for irrigation in gardening and agriculture, the FTR process was neither stable nor predictable [19]. Hence, the MBR process was preferred. You & Wu [45] also used membrane technology to treat high-cellulose-content wastewater, in which case the COD and BOD contents were reduced greatly while the TSS were completely removed from the wastewater. In addition, Lonigro et al. [47] showed that membrane treatment technologies were successful in removing parasites (*Giardia* cysts and *Cryptosporidium* oocysts) from wastewater for irrigation. Generally, the MBR process is not designed as an add-on technology to a secondary level treatment process but rather as a replacement technology to produce higher-quality effluents [48].

In general, most secondary-treated effluents can be reused for irrigation. However, for unrestricted irrigation, tertiary treatment is needed to protect human health and the environment [48, 49]. For example, disinfection technology using ozone or peracetic acid is used in Italy for indirect agricultural irrigation [48]. Gómez-López et al. [49] found that chlorination is still one of the best disinfection techniques when cost and environmental factors are considered.

Arid regions around the world are now facing increasing water scarcity partly due to population increase and climate change. In addition, the need for continuous food production requires adequate long-term water supplies for agricultural irrigation. Wastewater reuse is a feasible solution to the water shortage problems while maintaining global food production rates. Table 3.2 shows the wastewaters used and treatment technologies employed to treat wastewater for reuse as irrigation water.

3.3 Reuse of Wastewater as Industrial Water

The concept of wastewater reuse has been introduced and applied in certain industries. For example, dairy industry wastewater has been reused and recycled within a plant. The clean wastewater stream from boiler blowdown, clean-in-place washes, and pasteurized water, which has a high amount of suspended solids, can be treated and reused in a plant [44]. Sarkar et al. [51] also reused treated dairy wastewater in the dairy industry. The wastewater treatment processes they employed for treating raw dairy wastewater were coagulation, adsorption, and membrane separation. The quality of treated wastewater was found to be comparable to that of the process water used in the dairy industry; thus, the treated wastewater was suitable for reuse purposes.

The textile industry is an intensive industry with high water consumption and discharge. Wastewater discharge by this industry comes from bleaching, dyeing, printing, and stiffening processes. Textile wastewater has a high recalcitrant chemical oxygen demand, a large amount of suspended solids, a broadly fluctuating pH, and is highly colored. Thus, many treatment methods have been attempted to treat textile wastewater for reuse purposes within the same industry [52–54]. Water input to textile industries must be maintained at a high quality; hence, textile wastewater is usually treated in a chemical-physical plant or an active sludge biochemical plant. In addition, further treatments are sometimes necessary to improve the treated wastewater quality so that it can be reused for high-quality textile production and recycled for dyeing, washing, bleaching, and printing processes [52, 55, 56].

Textile wastewater usually undergoes the tertiary treatment level. The most common treatment processes used are membrane technologies, such as microfiltration,

Table 3.2 Summary of studies done on wastewater treatment for reuse purpose as irrigation water

Type of wastewater	Treatment processes	Main results/conclusions	Reference
Dairy processor wastewater	Aerobic bioreactors	Treated wastewater was reused for irrigation	[44]
Domestic and industrial wastewater	Aerobic-biological treatment	Treated wastewater characteristics were all within Tunisian limits. Treated effluent increased concentration of N, P, and K in the olive leaves	[29]
Domestic wastewater	Screening, oxidation, sedimentation, biological digestion	Reclaimed wastewater showed no significant difference when freshwater was used in olive irrigation. Reclaimed wastewater had no effect on olive tree growth and productivity although it increased the salt loads to the roots	[33]
Domestic sewage	Sedimentation terraces, stabilization pond, subsurface flow wetland, vertical flow wetland	Average COD removal = 84.9% TSS reduction = 95.9% Total nitrogen removal = 62.3%	[42]
High cellulose content wastewater	AMBR, RO	Removal efficiencies: COD = 96–99% BOD = 99–100% No TSS were detected in the treated effluent	[45]
Industrial park wastewater	Activated sludge, chemical coagulation, Fenton oxidation, ion exchange	COD removal >81% Quality of treated effluent met the standards for agricultural irrigation	[43]
Municipal wastewater	Secondary treatment level, membrane filtration	Membrane filtration system was successful in removing parasites (*Giardia* Cysts and *Cryptosporidium* Oocysts) from wastewater for irrigation.	[47]
	Vertical flow constructed wetlands	Removal efficiencies: COD, BOD, N, and K > 86% Na and Mg < 47% Water produced was not in high quality due to evaporation losses	[39]
Municipal and industry wastewater	MBR	Removal efficiencies: *E. coli* = 99.99% Total coliforms = 98.52% Arsenic = 82% Cadmium = 82%	[19]

(continued)

Table 3.2 (continued)

Type of wastewater	Treatment processes	Main results/conclusions	Reference
		Chromium = 97% Copper = 93% Lead = 99% Boron and manganese were not removed	
Sewage wastewater	Oxidation pond	Removal efficiencies: COD, BOD_5, turbidity, TKN > 80% Integrated efficiency > 98% Treated effluent from oxidation pond could be reused for unrestricted irrigation	[41]
Textile dyeing wastewater	AOMBR, RO	Removal efficiencies: COD = 88% BOD = 100% TSS = 100% True color = 97% Treated effluent from AOMBR/RO met the water reclamation criteria for irrigation	[46]
	Fenton oxidation, MBR	Removal ratio: TOC = 88.2% Color = 91.3% The quality of final effluent met the reuse criteria for landscape irrigation	[50]
Other wastewaters	No treatment was applied	Untreated wastewater was widely used in Pakistan for agriculture irrigation	[16]
	Activated sludge, extended aeration, WSP, direct filtration, UV disinfection	Reclaimed water was used for landscape and agricultural irrigation	[15]
	WSP: anaerobic, facultative, and maturation ponds	The effluent from this waste treatment plant was used for irrigation in Jordan Valley	[36]
	Secondary treatment level: activated sludge, ultrafiltration	Removal efficiencies: COD = 50% FC/100 mL = 100% Turbidity = 94%	[28]

Physiochemical process, DualSand, microfiltration, cartridge filter, RO DualSand, electrodialysis reversal	Removal efficiencies: COD = 71.4–98.1% BOD_5 = 89.6–98.3% TSS = 70–98.55 The treated effluent was reused in irrigation	[23]
Secondary treatment level, chlorination, UV treatment	The tertiary treatment process, chlorination, and UV treatment had the lowest environmental impact	[14]
Primary sedimentation, oxidation pond	Effluents were pumped to irrigate growing trees and feed crops, e.g., alfalfa hay, sudan grass, and winter grains. Trace amounts of heavy metals could affect the soil quality	[31]
Macrophyte ponds	The treated wastewater improved eggplant yield by 40% as compared to irrigation by freshwater	[18]
Activated sludge, sand filtration, disinfection by UV radiation	COD values were within the criteria set for agricultural irrigation. Both BOD_5 and TSS values were lower than the limit values for unrestricted irrigation. The treated effluent also showed no sign of coliforms	[21]

ultrafiltration, nanofiltration, and reverse osmosis, in which the technologies are very efficient at reducing the pollutants in the textile wastewater [20, 52, 56, 57]. Research conducted by Lu et al. [53] showed that the average removal efficiencies for COD, color, and turbidity using membrane technologies were approximately 93, 94.5, and 92.9%, respectively. Similarly, Alcaina-Miranda et al. [57] also reported high removals of COD, color, and turbidity from textile wastewater using a combination of the ultrafiltration and nanofiltration treatment processes. The high cellulose content of textile wastewater can also be treated with traditional anaerobic/aerobic activated sludge processes and membrane filtration [45]. You et al. [46] showed that the combination of the anaerobic-oxic membrane bioreactor (AOMBR) and reverse osmosis (RO) processes yielded 88 and 100% removal efficiencies for COD and BOD, respectively. Their study showed that membrane treatment produced treated textile wastewater suitable for reuse applications [45]. Although membrane technology is very costly, the cost is expected to decrease in the future. In addition, this technology can address the water quantity and quality issues [56, 58].

Apart from membrane technologies, granular-activated carbon adsorption was applied to textile wastewater. The effluent quality from this treatment method is lower but is of acceptable quality [56]. Advanced treatment methods including the Fenton process, polyaluminum chloride (PAC) coagulation, and ion exchange were also attempted to treat textile wastewater. The removal efficiency for COD after the Fenton process was 36%, and after PAC coagulation, it was 18%. In addition, the estimated cost for this treatment system was deemed reasonable in light of the expected increased water supply cost in the future [59]. Electrochemical treatment was also very effective in removing color and COD as well as chloride and sulfate ions from textile industry wastewater [55]. Research has also been conducted on removing Reactive Black 5 dye from textile wastewater through ozonation and membrane bioreactors [46, 60].

Generally, industrial wastewaters are reused within the same industry. For example, in France, treated industrial wastewaters are extensively used for cooling water, wash water, and process water [26, 48]. In addition, industrial uses represent 5% of freshwater use in Spain. This amount can be replaced by treated wastewater [14]. All reclaimed wastewaters are reused in industries except for cooling purposes in the food industry, which requires secondary treatment. There is also certain process water in industries that requires tertiary treatment, such as industrial applications in Belgium. In Belgium, wastewater is disinfected through ozone or peracetic acid. Quaternary wastewater treatment is also applicable for certain process water in industry which requires ultrapure water quality [48].

Wastewater reuse in industry is also realized through industrial water reuse synergies, which involves capturing, recovering, and reusing the wastewater from one industry to another industry operating in close proximity. This regional synergy has been performed in Kalundborg (Denmark), Kwinana (Western Australia), and, more recently, Port Melbourne (Australia) [61, 62]. Among the industries included in Port Melbourne were food, plasterboard production, automotive manufacturing, metal component manufacture, commercial fats and proteins production, newspaper printing, and precast concrete manufacture. From the study of these industries and the wastewater they produce, there are at least five possible synergy reuse options that could be considered as technically feasible. In all the wastewater treatment methods considered, membrane bioreactors are the central component. One option is recovering wastewater from food production (Kraft) using membrane bioreactors and reverse osmosis for subsequent reuse in plasterboard production (Boral). From an economic point of view, the water treatment options involve a substantial cost. However, resource scarcity and increasing water prices might offset the costly wastewater treatment technologies in the near future [62].

Other treated wastewaters, which have been attempted to be reused in the same industry, include the following: (1) process water from a dry-mill ethanol plant, which was treated through ozonation/catalytic ozonation to reduce the COD of thin stillage [63]; (2) poultry processing wastewater, which was treated using a ultra-filtration (UF) membrane process [58]; (3) rare earth manufacture process waste-water, which was treated using a membrane combination treatment method [64]; and (4) laundry wastewater, which was treated by a series of wastewater treatments [65]. The laundry wastewater treatment system consisted of a physicochemical pretreatment (coagulation, flocculation, dissolved air flotation (DAF)), sand filtration, ozonation, granular activated carbon (GAC) filtration, and cross-flow UF. The COD, TSS, and turbidity removal efficiencies were up to 80, 98, and 99%, respectively. The results showed that the produced effluent is suitable for reuse in the industry [65].

In India, treated sewage wastewater was reused in the cooling towers of a thermal power plant. Generally, turbidity is an important criterion for wastewater reuse in industries. Turbidity ranging from 1 to 30 NTU can be used in boilers and cooling towers [41]. In Jordan, collected wastewaters are characterized by their strong and high salinity attributes. However, the effluent can usually be treated using conventional activated sludge processes, extended aeration, and sand filtration. The treated effluent can then be reused as cooling water, boiler feedwater, and even process water [36]. A study presented by Adewumi et al. [17] showed that it is highly plausible that treated wastewater in South Africa could be reused for industrial processes.

In short, wastewater can be reused and recycled within the same industry after treatment. The reuse of treated wastewater helps combat water scarcity, reduce freshwater consumption, and reduce operational costs in the long run. Table 3.3 presents the wastewaters that have been attempted to be used as industrial waters after undergoing treatment processes.

3.4 Reuse of Wastewater as Fermentative Substrate

Because high-quality freshwater can actually be produced from wastewater, an extra effort for harvesting other resources, such as nutrients from wastewater, should also be considered in order to make overall wastewater management more sustainable. Indeed, wastewater treatment plants are categorized as point sources of

Table 3.3 Summary of studies done on wastewater treatment for reuse purpose as industrial water

Type of wastewater	Treatment processes	Main results/conclusions	Reference
Dairy industry wastewater	Coagulation, adsorption, membrane separation	Removal efficiencies: COD = 98% Conductivity = 96% Color and odor were completely removed after the treatment processes	[51]
High cellulose content wastewater	AMBR, RO	Removal efficiencies: COD = 96–99% BOD = 99–100% No TSS were detected in the treated effluent	[45]
Industrial wastewater	Microfiltration, reverse osmosis, artificial wetland treatment	Microfiltration and reverse osmosis treatment produced effluent with low total dissolved solids (50 mg/L)	[61]
	Membrane reactor, reverse osmosis, aerobic activated sludge process, anoxic stage, UV disinfection, upflow anaerobic sludge blanket	All reuse options and technologies seemed to be technically feasible, though more research was needed on the input water quality	[62]
Laundry wastewater	Physicochemical pretreatment (coagulation, flocculation, dissolved air flotation), sand filtration, ozonation, GAC filtration, cross-flow UF	Removal efficiencies: COD > 80% TSS = 98% Turbidity = 99% Nonionic surfactants = 87% Anionic surfactants = 93%	[65]
Poultry processing plants wastewater	Ultrafiltration	Ultrafiltration was an efficient method to recycle poultry processing wastewater and a solution to water resource scarcity	[58]
Rare earth manufacture process wastewater	Integrated air stripping pretreatment and low-pressure reverse osmosis	Removal efficiencies: Ammonia = 98% Fluoride ions = 99%	[64]

		COD > 80% BOD_5 > 85% TKN > 60% Fecal coliform and fecal streptococcus levels > log 5.5 Integrated efficiency > 97% Treated effluent was used in cooling towers in thermal power plant	
Textile industry wastewater	Activated sludge treatment, granular activated carbon adsorption	Removal efficiencies: COD = 42–50% TOC = 32–46% TSS = 68–74% Abs. 426 nm = 36–51% Abs. 550 nm = 41–45% Abs. 660 nm = 36–44%	[56]
	Electroflocculation	Removal efficiencies: COD = 80–100% Color = 70–90% Reduction of salt (sulfates and chlorides) content	[55]
	Fenton process, polyaluminum chloride coagulation, ion exchange	After Fenton process, the removal efficiencies are the following: COD = 36% TSS = 61% SAC_{436} = 35% SAC_{525} = 49% SAC_{620} = 67% After polyaluminum chloride coagulation, optimum removal efficiencies are the following: COD = 18% TSS = 83% SAC_{436} = 93% SAC_{525} = 32% SAC_{620} = 58%	[59]

(continued)

Table 3.3 (continued)

Type of wastewater	Treatment processes	Main results/conclusions	Reference
	Combined membrane processes	Anaerobic-oxic MBR and RO treatment removal efficiencies: COD = 88% TSS = 100% BOD = 97%	[46]
	Ultrafiltration, nanofiltration	The combination of ultrafiltration and nanofiltration treatment achieved high removals of COD, color, and turbidity	[57]
	Activated sludge treatment, nanofiltration	Yield = 9% Removal efficiencies: COD = 62% TDS = 62% The hardness of the treated water was within factory constraints	[20]
	Biological treatment, nanofiltration	COD removal efficiency > 91%	[52]
	Ozonation using semi-batch bubbling reactor	Ozonation process was able to remove Reactive Black 5 dye from wastewater	[60]
	Two-stage anaerobic-aerobic treatment, biological aerated filter process, membrane technology	Average removal efficiencies: COD = 93% Color = 94.5% Turbidity = 92.9% No TSS were found in the treated effluent	[53]
Other wastewater	Conventional activated sludge, extended aeration ponds, sand filtration	Projected effluent quality: COD = 200 mg/L TDS = 1,220 mg/L TSS = 30 mg/L NH_3-N = 15 mg/L P = 15 mg/L MBAS = 20	[36]

both nitrogen and phosphorous [6], which could be reused as nutrients in the fermentation medium. It was reported that certain wastewaters could be reused as fermentation media to produce biohydrogen, polyhydroxyalkanoates, polyhydroxybutyrate, enzymes, solvents, single-cell protein, and organic acid. Table 3.4 displays the various products or metabolites produced in bioprocesses by reusing wastewaters as substrates.

3.4.1 Biohydrogen Production

Biological hydrogen production attracts much attention as hydrogen can be used for clean, efficient, and renewable energy, which can be considered the most environmentally friendly route to produce future renewable energy. It was estimated by the US Department of Energy Hydrogen Program that the contribution of hydrogen to the total energy market will increase to 8–10% by 2025 [74]. There is a variety of technologies for biological hydrogen production; however, only photofermentation and dark fermentation will be emphasized here. Dark fermentation uses fermentative bacteria anaerobically to breakdown carbohydrate-rich substrates into biohydrogen and other products such as acids and alcohols. The photofermentation process uses photosynthetic bacteria and requires the input of light energy and various substrates, in particular organic compounds, to produce biohydrogen.

3.4.1.1 Photofermentation

Palm oil extraction produces approximately 50 million tons of palm oil mill effluent (POME) annually in Malaysia [70]. POME is rich in organic carbon with a biological oxygen demand (BOD) of 30,000 mg/L, a chemical oxygen demand (COD) of 50,000 mg/L, 6,000 mg/L of oil and grease, 59,350 mg/L of suspended solids, and 750 mg/L of total nitrogen [79]. Biohydrogen production by *Rhodopseudomonas palustris* PBUM001 from POME was performed by Jamil et al. [79], and the maximum biohydrogen production and COD reduction were achieved in 100% (v/v) POME concentration due to higher content of organic acids compared to the diluted POME.

Disposal of olive mill wastewater (OMWW) is an important environmental problem that has to be solved urgently as it contains a high organic matter content, with COD and BOD values reaching up to 200 and 100 g/L, respectively [78]. It was shown by Eroglu et al. [76] that OMWW could be utilized as a sole substrate source to produce biohydrogen by *Rhodobacter sphaeroides* O.U.001. The maximum amount of biohydrogen was obtained using 2% diluted OMWW together with valuable by-products, such as carotenoid and polyhydroxybutyrate. Biohydrogen could not be produced in concentrated media (containing more than 4% OMWW) due to the presence of high amounts of inhibitory substances whose relatively dark color reduced light penetration [102]. Other research has investigated the effect of

Table 3.4 Summary of studies done on wastewater treatment for reuse purpose as fermentative metabolite

Fermentative metabolites	Type of wastewater	Microorganism	Conditions	Main results/conclusions	Reference
Biohydrogen (dark fermentation)	Apple processing wastewater	–	Undiluted	COD removal efficiency = 6–7%	[66]
				H_2 production = 0.7–0.9 L H_2/L waste	
	Brewery wastewater	–	Optimum conditions:	Predicted production rate = 53.6 mL/h	[67]
			Waste concentration = 6.05 g/L	Predicted H_2 production = 149.6 mL H_2/g COD	
			Initial pH = 5.95	Experimental production rate = 50.5 mL/h	
			Temperature = 35.9 °C	Experimental H_2 production = 147.35 mL H_2/g COD	
	Cheese whey wastewater	Mixed microbial	Agitation speed = 200 rpm pH = 5.5 ± 0.05	H_2 production = 22 mmol/g COD	[68]
	Cassava starch processing wastewater	Enriched cultures of PK, SW, and PR	Temperature = 60 °C	PK H_2 production = 287 mL H_2/g starch	[69]
				SW H_2 production = 264 mL H_2/g starch	
				PR H_2 production = 132 mL H_2/g starch	
	POME	*Clostridium butyricum* EB6	Optimum conditions:	Production rate = 298 mL H_2/g carbohydrate	[70]
			COD concentration = 94 g COD/L	H_2 production = 849.5 mL H_2/h	
			Initial pH = 6.05		
			Temperature = 36 °C		

	Phenol-containing wastewater	*Clostridium butyricum*	Agitation speed = 150 rpm Temperature = 37 °C	COD removal efficiency = 30–40% (phenol < 600 mg/L) COD removal efficiency = 10% (phenol 1,200 mg/L) H_2 production = 1.4 mol H_2/mol glucose (200–400 mg/L phenol)	[71]
	Potato processing wastewater	–	Diluted 50% with water	COD removal efficiency = 6–7% H_2 production = 2.1–2.8 L H_2/L waste	[66]
	Probiotic wastewater	Mixed anaerobic consortia	Optimum conditions: Substrate concentration = 5 g/L Initial pH = 5.5 Temperature = 37 °C	H_2 production = 1.8 mol H_2/mol carbohydrate	[72]
	Starch wastewater	–	Optimum conditions: Starch concentration = 5 g/L Initial pH = 6.5 Temperature = 37 °C	COD removal efficiency = 64% H_2 production = 186 mL/g starch	[73]
Biohydrogen (photofermentation)	Brewery wastewater	*Rhodobacter sphaeroides* O.U.001	Light intensity = 9 klux Temperature = 28 °C Inoculum size = 30% v/v pH = 7–7.2	H_2 production = 0.22 L H_2/L waste	[74]
	Dairy wastewater	*Rhodobacter sphaeroides* O.U.001	Light intensity = 9 klux Temperature = 28 °C Inoculum size = 30% v/v pH = 7–7.2	COD removal efficiency = 21% H_2 production = 7.6 L H_2/L waste	[75]

(continued)

Table 3.4 (continued)

Fermentative metabolites	Type of wastewater	Microorganism	Conditions	Main results/conclusions	Reference
	Olive mill wastewater	*Rhodobacter sphaeroides* O.U.001	Light intensity = 200 W/m^2	COD reduction = from 1,100 to 720 mg/L	[76]
			Agitation speed = 300 rpm	BOD reduction = from 475 to 200 mg/L	
			Inoculum size = 10% v/v	Total recoverable phenol = from 2.32 to 0.93 mg/L	
			Waste concentration = 2%	H_2 production = 13.9 L H_2/L waste	
				Carotenoid production = 40 mg/L waste	
				PHB production = 60 mg/L waste	
		Rhodobacter sphaeroides O.U.001	Light intensity = 200 W/m^2	COD removal efficiency = 33% (CL)	[77]
			Temperature = 32 °C	31% (12 L/12D diurnal cycles)	
			Waste concentration = 2%	H_2 production = 50 mL H_2/L waste	
				Carotenoid production = 30 mg/L waste	
				PHB production = 61 mg/L waste	
		Rhodobacter sphaeroides O.U.001	Light intensity = 150 W	COD removal efficiency = 31%	[78]
			Waste concentration = 4%	Color removal efficiency = 65%	
				Phenol reduction = 81%	
	POME	*Rhodopseudomonaspalustris* PBUM001	Optimum conditions:	Predicted COD removal efficiency = 31.7%	[79]
			Light intensity = 4.0 klux	Predicted H_2 production = 1.05 mL H_2/mL waste	

			Agitation speed = 250 rpm	Experimental COD removal efficiency = 30.54%	
			Initial pH = 6 Inoculum size = 10% v/v Waste concentration = 100% v/v	Experimental H_2 production = 0.66 mL H_2/mL waste	
	Succinate wastewater	ZX-5	Light intensity = 4 klux Agitation speed = 400 rpm Temperature = 30 °C pH = 7	COD removal efficiency = 87.5% Production rate = 55 mL H_2/mL h H_2 production = 25 L H_2/L waste	[80]
	Tofu wastewater	*Rhodobacter sphaeroides* RV	Cell growth: Light intensity = 2 klux Temperature = 30 °C Biohydrogen production: Light intensity = 8.5 klux Temperature = 30 °C	Without dilution: TOC removal efficiency = 40% H_2 production = 1.9 L/L waste With 50% dilution: TOC removal efficiency = 66% H_2 production = 4.32 L/L waste	[81]
		Rhodobacter sphaeroides 49419, *Rhodobacter sphaeroides* AR-3	Light intensity = 6–7 klux Temperature = 30 °C Inoculum size = 10% v/v	*R. sphaeroides* 49419: COD removal efficiency = 55% Production rate = 6.7 mL/L h H_2 production = 0.81 L/L waste *R. sphaeroides* AR-3: COD removal efficiency = 45% Production rate = 14.2 mL/L h H_2 production = 1.73 L/L waste	[82]
Cellulase	POME	*Trichoderma reesei* RUT C-30	Temperature = 30^0 C Agitation speed = 150 rpm pH = 4.6	Maximum filter paper cellulase activity = 0.996 U/mL Maximum carboxyl methyl cellulase activity = 2.89 U/mL	[83]
		Isolate SO1	Temperature = 55^0 C Inoculum size = 5% v/v Agitation speed = 200 rpm	Maximum cellulase production = 12.11 U/mL	[84]

(continued)

Table 3.4 (continued)

Fermentative metabolites	Type of wastewater	Microorganism	Conditions	Main results/conclusions	Reference
			pH = 7	Shaking condition: % dry weight reduction = 69.7% Oil reduction in bottom layer = 85.3% Static condition: % dry weight reduction = 18.1% Oil reduction in bottom layer = 36.7%	
Citric acid	POME	*Aspergillus*	Temperature = 30 °C pH = 2–3 Inoculum size = 2% v/v Agitation speed = 150 rpm	Maximum citric acid concentration = 0.28 g/L COD reduction = 72%	[85]
		Aspergillus niger A103	Optimum condition: Substrate concentration = 2% w/w Wheat and glucose concentration = 4% w/w	Maximum citric acid concentration = 5.2 g/L COD reduction = from 99.6 to 10.6 g/L	[86]
Solvent	Brewery wastewater	*E. coli* KO11	Agitation speed = 120 rpm pH = 6	Maximum concentration ethanol = 12.18 g/L (1.4 vol%)	[87]
	POME	*Clostridium acetobutylicum* NCIMB 13357	Temperature = 35 °C Inoculum size = 10% v/v Agitation speed = 250 rpm pH = 5.8	Maximum concentration butanol = 1.6 g/L Maximum concentration acetone = 0.05 g/L	Masngut et al. (2007)
Itaconic acid	POME	*Aspergillus terreus* IMI 282743	Temperature = 35 °C Inoculum size = 5% v/v Agitation speed = 300 rpm	Maximum itaconic acid concentration = 0.079 g/L	[88]

			Incubation: Concentrations = 10^6 cell/mL. Temperature = 27 °C	activity = 877.9 ± 329.4 U/L Reduction of phenolic compounds = 70.2% COD reduction = 27.0%	
		Candida cylindracea	Incubation: Agitation speed = 180 rpm Temperature = 28 °C	Maximum lipase activity = 9.23 IU/mL	[90]
	POME	*Clostridium aurantibutyricum* ATCC 17777	Temperature = 37 °C Agitation speed = 200 rpm Inoculum size = 10% v/v pH = 6.8	Maximum lipase activity = 0.4 U/mL	[91]
Organic acid	Soybean wastewater	*Rhodobacter sphaeroides* Z03	Temperature = 12–25 °C Without light, oxygen supply, and nutrient	COD reduction = 96.2% (below 300 mg/L) Butyric acid concentration = 85.3 mg/L	[92]
PHA	Fermented sugar molasses wastewater	Mixed cultures	Acidogenic fermentation: pH = 6 Enrichment of GAOs: Temperature = 30 °C pH < 7.7 Accumulation of PHA: Temperature = 30 °C pH = 7.7	PHA molecular weight = (3.5–4.3) × 10^5 g/mol	[93]
	POME	*Rhodobacter sphaeroides*	Light intensity = 24 klux Temperature = 30 °C pH = 7	First stage: More than 70% conversion of organic acids (formic, acetic, and propionic acid). Second stage: PHA yield = 0.50 g/g, in the absence of formic acid (when pH is at 7)	[94]

(continued)

Table 3.4 (continued)

Fermentative metabolites	Type of wastewater	Microorganism	Conditions	Main results/conclusions	Reference
		Alcaligenes latus	Agitation speed = 150 rpm for 72 h, then 50 rpm. pH = 7	The molecular weight = 11,822 g/mol PHA yield = 160 mg/L	[95]
	Soybean milk wastewater	*Alcaligenes latus*	Agitation speed = 150 rpm for 72 h, then 50 rpm. pH = 7	The molecular weight = 1,120 g/mol PHA yield = 17 mg/L	[95]
	Starchy wastewater	*Alcaligenes latus*	Temperature = 55 °C	First stage: Propionic acid = 10–30% Butyric acid = 5–40% Second stage: PHA yield = 0.5 g/g TOC	[96]
PHB	Biodiesel wastewater	Mixed microbial consortia	Reactor substrate addition introduced at the beginning of the cycle. Consisted of 10% glycerol (v/v) and 90% tap water (v/v)	PHB molecular weights = 200–300 kDa PHB yield = 0.1 g/g methanol	[97]
		Cupriavidus necator DSM 545	Temperature = 34 °C pH = 6.8	Productivity = 1.1 g/L h Max PHB concentration = 38.1 g/L	[98]
	Sugar refinery wastewater	*Rhodobacter sphaeroides* O.U.001	Light intensity = 200 W/m^2 Temperature = 36 °C Waste concentration = 30% pH = 7	Anaerobic condition: PHB yield = 0.3 g/L (52.4% w/w) Aerobic condition: PHB yield = 0.5 g/L (70.4% w/w)	[99]
Protease	POME	*Aspergillus terreus* IMI 282743	Temperature = 37 °C Agitation speed = 250 rpm Inoculum size = 5% v/v	Maximum protease activity = 129 U/mL	[100]
Single-cell protein	Latex rubber sheet wastewater	DK 6	Light intensity = 3 klux Temperature = 30 °C	COD reduction = 90% BOD reduction = 90%	[101]

illumination on biohydrogen production using OMWW under different regimes of illumination, continuous light (CL), or 12 h light/12 h dark (12 h/12 days) diurnal cycles [77]. Clay treatment prior to biohydrogen production from OMWW was also carried out by Eroglu et al. [78]. The amount of biohydrogen produced doubled (31.5 L H_2/L OMWW) by using the wastewater after pretreatment as compared to biohydrogen produced using raw OMWW (16 L H_2/L OMWW) [78]. This is mainly due to the depletion of elemental nitrogen (an excess amount of nitrogen would repress the synthesis of nitrogenase and decrease biohydrogen production) after pretreatment, which leads to an enhancement of the nitrogenase enzyme and biohydrogen production [103].

Tofu wastewater (TWW), which is a carbohydrate- and protein-rich effluent, is considered as a serious environmental issue due to its high COD value. The main components of tofu wastewater include reductive sugar, sucrose, starch, and volatiles acids, with COD approximately 10,000–15,000 mg/L [82]. Zhu et al. [81] utilized TWW as a substrate in biohydrogen production with a wild type of anoxygenic phototrophic bacterium (*R. sphaeroides* RV). This study showed that through dilution of wastewater, the inhibitory factors were also diluted, resulting in an increase in biohydrogen production activity and the TOC removal ratio [81]. Biohydrogen generation from TWW was also studied by Zheng et al. [82] using a wild-type strain of *R. sphaeroides* 49419 and glutamine autotrophic mutant *R. sphaeroides* AR-3. The AR-3 mutant generated biohydrogen more efficiently in the medium with a higher NH_4^+ concentration because the inhibition of NH_4^+ to nitrogenase of AR-3 was released. TWW that normally contains 3–4 mmol/L of NH_4^+ is not suitable for the wild-type strain of *R. sphaeroides* because the presence of NH_4^+ at a concentration of 2 mmol/L or higher inhibits biohydrogen production due to its suppression on the synthesis of nitrogenase.

Brewery wastewater (BWW), with a high concentration of organic substances (an average COD of 0.8–2.5 kg/hL), as a substrate in biohydrogen production by *R. sphaeroides* O.U.001 was studied by Seifert et al. [74]. The best substrate yield was obtained after the BWW was filtered to a concentration of approximately 10% (v/v). At higher concentration (20%, v/v), the color of the medium became darker and reduced light penetration, which decreased the biohydrogen production. The potential of using dairy wastewater (DWW) as a substrate for producing biohydrogen was investigated by Seifert et al. [75]. Waste from dairy usually has a high concentration of organic substances with a COD value varying between 5 and 50 g O_2/L. Total inhibition of the process was observed at a concentration of 60% (v/v), containing 1.7 mM nitrogen in the NH_4^+. Such a high concentration of nitrogen suppresses the nitrogenase activity and significantly inhibits the biohydrogen production process [75].

Tao et al. [80] studied the characteristic of a new photosynthetic bacterial strain, ZX-5, for biohydrogen production and its application in succinate wastewater treatment. Through the photofermentation process, 25 L H_2/L WW was produced with an average production rate of 55 mL H_2/mL·h. In addition, 87.5% removal efficiency was achieved, and no succinate was detected in the final fermentation effluent.

3.4.1.2 Dark Fermentation

Clostridium butyricum EB6 was used for generating biohydrogen from POME through the fermentation process by Chong et al. [70]. Throughout the experiments, the biogas produced contained biohydrogen (60–75%) and carbon dioxide (25–40%). The accuracy of the optimum conditions was checked and verified with experimental results [70]. Cheese whey wastewater (CWW), which contains approximately 5% lactose, is suitable to be reused as a substrate in biohydrogen production. The potential of using CWW as a fermentative substrate in biohydrogen production was investigated by Azbar et al. [68]. They used mixed microbial communities under anaerobic and thermophilic conditions.

O-Thong et al. [69] studied the potential of reusing cassava starch processing wastewater (CSPW) for generating biohydrogen using thermophilic mixed cultures. CSPW is a carbohydrate-rich wastewater which contains high BOD, COD, and total solids (TS). The enriched cultures for Klong Pai Poo hot spring (PK), Ron hot spring (SW), and Romani hot spring (PR) could degrade starch in CSPW completely to biohydrogen, with by-products including acetic acid, lactic acid, and ethanol. Further research on starch wastewater was performed by Wei et al. [73] to produce biohydrogen.

Phenol, a toxic substance, is often found in industrial effluent. Tai et al. [71] utilized phenol-containing wastewater to fermentatively produce biohydrogen using *Clostridium butyricum.* Ethanol and acetate were the primary metabolites, and its yield increased with increasing phenol concentration until phenol >1,000–1,200 mg/L [71]. Biohydrogen production using mixed anaerobic consortia and probiotic wastewater as a substrate was reported by Sivaramakrishna et al. [72]. The biohydrogen production increased remarkably with an increase in the concentration of the substrate when the concentration was less than 5 g/L. When the concentration is more than 5 g/L, the biohydrogen yield decreased gradually as the concentration increased [72].

The food processing industry produces approximately 1.4 billion liters of wastewater annually in the United States alone. Food processing wastewater has high BOD and COD values, indicating that it is suitable to be reused as substrates for anaerobic treatment processes [158]. In a study by Ginkel et al. [66], apple processing wastewater and potato processing wastewater were utilized. They found that high-purity biohydrogen gas (60%) was produced throughout the experiments. Brewery wastewater was also reused as a substrate in biohydrogen production using a mixed culture from anaerobic sludge [67].

3.4.2 Polyhydroxyalkanoates (PHA) and Polyhydroxybutyrate (PHB)

PHA is a sustainable alternative to replace conventional plastics due to its biodegradability and capability of being produced from renewable resources. PHA can be synthesized naturally by many types of bacteria without producing any harmful by-products. Two common PHA monomers are hydroxybutyrate (HB) and hydroxyvalerate (HV) [104].

A two-stage process for the production of PHA from POME was proposed by Hassan et al. [94]. In the first stage, anaerobic treatment of POME was carried out to obtain organic acids, mainly acetic and propionic acid, which would then be converted to PHA by a phototrophic bacterium in the second stage. Results showed that at higher pH, acetic acid was the predominant acid formed while formic acid was the predominant acid formed at low pH. The PHA yield decreased gradually with increasing formic acid concentration [94]. Another study that utilized wastewaters from POME and the soybean milk industries for PHA production was carried out by Thammawong et al. [95]. It was observed that the maximum molecular weight from POME and soybean is obtained at 24 and 96 h, respectively, of oxygen limiting time and gradually decreases as a function of time [95].

The production of PHA from starchy wastewater in a two-step process involving microbial acidogenesis and acid polymerization was studied by Jian Yu [96]. The starchy organic wastewater was first digested in a thermophilic upflow anaerobic sludge blanket (UASB). Then, it was utilized and polymerized into PHA in the second stage. The composition of PHA was controlled mainly by the content of propionic acid in the effluent [96].

The growth of the biodiesel industry generates crude glycerol, which is produced at a rate of approximately 1 kg per 12.6 L of biodiesel [97]. It has been recognized as a potential feedstock for producing PHA as it is an organic-rich substrate. PHA synthesis on biodiesel wastewater using mixed microbial consortia was investigated by Dobroth et al. [97]. They found that the enriched mixed microbial consortia that produced exclusively PHB was utilizing the methanol fraction. The produced PHB exhibited thermal properties that were comparable with commercial PHB [97]. Waste glycerol was also reused for PHB production by *Cupriavidus necator* DSM 545 [98].

PHA can also be produced from fermented sugar molasses as demonstrated by Bengtsson et al. [93] using a mixed culture enriched in glycogen-accumulating organisms. PHA produced from waste by an open mixed culture is versatile with respect to thermal properties as the glass transition temperature, melting temperature, and melting enthalpy are controlled in broad ranges by the monomer composition. PHB can also be produced as a by-product during biohydrogen production from sugar refinery wastewater as shown by Yigit et al. [99].

3.4.3 Cellulase

Rashid et al. [83] investigated the production of cellulase by filamentous fungi using POME. *Trichoderma reesei* RUT C-30 was identified as a potential strain for cellulolytic enzyme production as compared to another strain, as it produced the highest filter paper cellulase and carboxyl methyl cellulase activities of 0.996 and 2.89 U/mL, respectively [83].

POME was also used by Laohaprapanon et al. [84] to produce cellulase and xylanase. The results indicated that the reduction of the dry weight and of oil and grease in the bottom layer of the POME is higher under a shaking condition compared to a static condition.

3.4.4 Lipase

Candida rugosa, Candida cylindracea, and *Yarrowia lipolytica* were the microorganisms used by Goncalves et al. [89] to produce lipase from OMWW. All strains showed that they were able to grow in the OMWW without dilution while consuming reducing sugars and reducing the COD of wastewater. However, *C. cylindracea* was the best strain to produce lipase while reducing phenolic compounds from the wastewater.

D'Annibale et al. [90] also utilized OMWW to produce microbial lipase. *C. cylindracea* gave the highest lipase activity of all the typologies of OMWW used. The lipase production from OMWW was markedly affected by the type of nitrogen source and was induced by the addition of olive oil.

Somrutai et al. [91] observed that *Clostridium aurantibutyricum* ATCC 17777 was able to produce lipase in the model medium for POME, in which a high rate of oil hydrolysis (46%) was seen at a pH of 6.8.

3.4.5 Protease

Prefiltered POME was used as the substrate for a wild-type strain, *Aspergillus terreus* IMI 282743, for protease production [100]. The process was conducted in two stages with the POME first being subjected to depth and surface filtration to remove the coarse solids found in the wastewater. Next, the prefiltered POME was concentrated twofold to form retentate. This experiment showed that retentate should be diluted prior to fermentation for better protease activity. A maximum protease activity of 129 U/mL was achieved in the medium containing 75% retentate of prefiltered POME.

A study of the effect of pulp wastewater on soil enzyme activities was conducted by Yan and Pan [105]. Five soil enzyme activities, acid phosphatase, alkaline phosphatase, urease, protease, and b-glucosidase, increased during short-term irrigation of a soil (25 days). That paper also showed that enzyme activities increased with increasing value of COD by the irrigated pulp wastewater. A higher value of COD has a higher supply of readily available substances for microorganisms and soil enzymes [105].

3.4.6 Organic Acid

The major by-products after the treatment of soybean wastewater by *Rhodobacter sphaeroides* Z08 were C2–C5 organic acids, predominantly butyric acids. In addition, Z08 had a cell yield of 0.28 g/g and a protein content of 52%, indicating that the pollutant could be degraded and bacterial proteins could be synthesized at the same time. No alcohol or C1 acid was detected. The wastewater COD treatment time was seen to drop almost linearly in the first 8 days, then drastically decreased for the next 2 days, and finally stabilized after 10 days [92].

Citric acid [85, 86] can also be produced using POME as the substrate under aerobic conditions [8]. Statistical optimization was carried out by Alam et al. [86] for converting POME into citric acid. A higher removal of COD (82%) with the production of citric acid (5.2 g/L) was achieved using the optimal conditions of a substrate concentration of 2% (w/w), wheat flour and glucose concentration of 2% (w/w), and the absence of ammonium nitrate. A liquid-state fermentation process of POME was also conducted by Jamal et al. [85] for the production of citric acid. *Aspergillus* A103 showed the highest concentration of citric acid (0.28 g/L), followed by A1020, A-22106, and others.

3.4.7 Single-Cell Protein

Latex rubber sheet wastewater, which contains both organic and inorganic matters as well as chemicals (formic acid, ammonia), can be reused for producing single-cell protein [101]. Chemical analysis of the cultures after the treatment of the enriched wastewaters showed that the protein content in the mixed culture (66.7%) was slightly higher than the protein content of the pure isolate DK 6 of *Rhodopseudomonas sp.* (65.2%). This observation indicates that single-cell protein (SCP) could be a potential by-product during the treatment of latex rubber sheet wastewater.

3.4.8 Solvent (Ethanol, Butanol)

The production of ethanol from brewery wastewater was investigated by Rao et al. [87]. The experiment was conducted using two types of medium: raw wastewater with no addition of enzyme and wastewater with the addition of commercial enzyme. The combination of a-amylase and pectinase appeared to enhance the consumption of both starch and maltotetraose in brewery wastewater, resulting in an increase of the ethanol yield by an average of 84% [87].

Work conducted by Masngut et al. [106] using *C. acetobutylicum* NCIMB 13357 indicated that POME is a viable fermentation medium for solvent production.

The experiment also compared the performance between an oscillatory flow reactor (OFR) and a stirred tank reactor (STR). The total solvent production in an OFR is 1.8 times higher than that of a STR, resulting in a total of 1.6 g/L of solvents. A high butanol concentration (1.6 g/L) was produced with no ethanol production and low acetone production (0.05 g/L) (Masngut et al. [106]).

3.5 Reuse of Wastewater as Animal Feed

Wastewaters have been previously attempted for reuse as animal feed. Some of the notable examples are the following: (1) treated and untreated wastewaters as well as stabilization pond effluent were reused for aquaculture in Africa and in Central, South, and Southeast Asia [30, 40, 107]; (2) treated water was reused for cultivation of fodder [30, 37]; (3) paper mill effluents were reused to provide protein supplement for animals [108]; and (4) wastewater was used for phosphorus recovery [109, 110].

One of the most common uses of wastewater as animal feed is for aquaculture [30]. Both untreated and treated wastewaters are widely used for aquaculture. Untreated wastewater as fish feed is common in Indonesia, Sri Lanka, Nigeria, India, Cambodia, and in particular, Vietnam and Thailand [41, 107, 111]. In Indonesia, diluted wastewater is used to culture carp, whereas in Sri Lanka, wastewaters from surface drains are used to culture tilapia. In addition, untreated domestic and industrial wastewaters are also used for aquaculture in Vietnam. The study also shows that there was no significant contamination of As, Cd, and Pb in the fishpond. However, it was noted that not much information was known about the accumulation of toxic elements in fish from the untreated wastewater system [111]. In some Asian countries, the reuse of untreated wastewater for aquaculture risks contaminating fish with pathogens and constitutes a health hazard [112]. According to the World Health Organization guidelines for the microbiological quality of wastewater, the fecal coliform (FC) level for aquaculture should not exceed 10^3MPN/100 mL [41, 113]. Thus, it is imperative that wastewater should be treated prior to reuse in aquaculture. Past findings also showed that treating wastewater prior to reuse could improve the efficiency of an untreated wastewater-fed system [107, 112].

Other countries, such as Malaysia, Hong Kong, Taiwan, Singapore, Israel, Hungary, and Germany, also culture fish and prawns in sewage-fed ponds [40]. Mara et al. [113] designed an efficient way to obtain minimal wastewater treatment and maximal fish production that is safe for human consumption. The treatment processes suggested by Mara et al. [113] involved an anaerobic and facultative pond; because the wastewater treatment system is very cost-effective, it is thus suitable to be applied in developing countries.

Primary-treated sewage effluent is rich in nutrients that are beneficial for the production of fish. [40, 114–116]. This is evident in a study by Dasgupta et al. [114], where similar gross fish production resulted from treated sewage effluent and a fertilizer-based system. In addition, treated sewage effluent can also enhance the productivity of water [116]. Sewage water was also used to raise common carp and *Sarotherodon galilaeus* in Nigeria [107]. China has been using sewage in aquaculture for more than 2,000 years. Generally, sewage effluent is reused in fishponds and lakes [40]. Sewage effluent can also be used for phosphorus recovery as demonstrated by Saktaywin et al. [109]. Phosphorus could be recovered by incorporating conventional anaerobic/oxic process with ozonation and crystallization [109].

Apart from sewage effluent, wastewater from aquaculture systems can also be reused for the production of fish. Wastewater can be reused for aquaculture after treatment through a membrane biological reactor (MBR). The MBR process is a combination of the activated sludge treatment and membrane filtration. The treated aquaculture wastewater is reused in the aquaculture system, while the activated sludge can be reused as protein feed for animals. This treatment technology gave high removal efficiency for TSS and $cBOD_5$, achieving 99.97–100%. Although using the MBR process involves a significant capital investment, the treated effluent usually reconciles well with the standard set for the reuse purpose. Furthermore, the wide application of this technology in the future may reduce the operational cost considerably [117, 118].

The olive oil industry produces approximately 4,401 liquid waste per 100-kg treated olives [119, 120]. Olive mill wastewaters were used to produce single-cell protein (SCP) through treatment with yeast [121]. The treatment process produced 3.35 g/L protein. In addition, the BOD and COD of olive mill wastewater were reduced by 40–50 and 50%, respectively [121]. Laconi et al. [120] also showed that the chemical-biological treatment process of olive mill wastewater produced a microbial biomass that could be used as animal feed. The wastewater went through an alkaline-oxidative treatment to reduce its polyphenolic content. After that, the wastewater was used as a culture medium for growing numerous microbial strains. The biomass produced was found to have significant amounts of vitamins A and E, calcium, potassium, iron, nicotinic acid, and unsaturated fatty acid. However, the final BOD did not meet the legal requirements; therefore, more research is required to obtain the optimum quantity of fungi or probiotic microorganisms involved in the treatment process. Another type of chemical-biological treatment process involves the use of a Fenton-like chemistry process and an aquaculture system as a final biological stage. This integrated system could remove the organic and phenolic components [122]. In addition, olive mill wastewater can also be reused for the production of animal feeds through solid fermentation [123]. Olive production is extensive in the Mediterranean region; thus, olive mill wastewaters play an important role as nutrient sources in animal feed [119].

Wastewater treatment processes such as sedimentation, single-stage aeration, and centrifugation are normally used to obtain wastewater sludge/solids that can be reused as an animal feed supplement. To obtain the solid constituents from dairy wastewater, a trickle-filter-type facility [124], membrane technology [125], and microbial conversion and flocculation [126] have been attempted. The sludge/solids are usually used to feed animals, such as swine, sheep, and pigs. The solids and precipitates from the wastewater treatment are used along with conventional animal feed, such as barley meal and soybean meal [124, 126]. Apart from dairy wastewater, process water from a corn dry milling plant could be centrifuged and evaporated to obtain distiller's grain which can be used as a high-quality and nutrient-rich livestock feed [63].

Treated palm oil mill effluent (POME) can also be used as animal feed [8, 127]. POME is generated abundantly in the palm oil industry, especially in Malaysia. It was reported that around 30 million tons of POME were generated in 2004 [128]. Wu et al. [127] reported that POME treatment with ultrafiltration membrane can recover and concentrate proteins as well as carbohydrates efficiently. The proteins and carbohydrates were retained together with suspended solids that could be reused as animal feed. POME is a dietary supplement for growing-finishing pigs, small ruminants, and poultry farming [8]. Babu et al. [129] also used POME to culture four species of fish. The culturing of silver carp, rohu, and mrigal using POME gave the maximum individual growth as 700, 550, and 600 g, respectively. Another fish culture, catla, had an average growth of 35.3 g [129].

Poultry processing wastewater [130], pineapple cannery effluent [131], and salad oil manufacturing wastewater [108] could also be converted into a valuable protein supplement or single-cell protein (SCP) for animal feeds. Generally, the treatment processes for producing SCP are enzymatic hydrolysis and/or fermentation using yeasts [130, 131]. For example, Zheng et al. [108] converted salad oil manufacturing wastewater by yeast isolate to obtained yeast protein. Apart from biomass production for animal feed, the wastewater treatment efficiency can be improved by adding inorganic nitrogen to obtain an initial range of N:C from 1:6 to 1:8.

Wastewaters are also used for phosphorus recovery. Phosphorus is an essential nutrient for all organisms and is therefore added to animal feed. Supplementing animal feed with phytases reduces the phosphate needed for animal feed as well as phosphorus content in animal manure [110, 132]. Watanabe et al. [110] used phosphohydrolysis to remove phosphorus from high-concentration organic wastes. Table 3.5 shows the potential of reusing treated or non-treated wastewater as animal feed.

3.6 Reuse of Wastewater as Fertilizer

The transformation of wastewater into fertilizer for use in agricultural land was initially thought to be impractical because of the risk of contamination with bacteria, viruses, or even parasites, which could affect human health or lead to serious environmental issues [136]. Several researchers reported that human urine, blackwater, or fecal matters are common raw materials to produce agricultural fertilizer after the treatment of wastewater through a separating system [136–138]. Tidåker et al. [138] found that after the urine separation system of household wastewater, the production of fertilizer fulfilled the requirements of fertilizer with a high nutrient value and without many hazardous compounds. The study also showed that there were several environmental benefits of using human urine as

Table 3.5 Summary of studies done on wastewater treatment for reuse purpose as animal feed

Type of wastewater	Treatment processes	Main results/conclusions	Reference
Aquaculture wastewater	Membrane biological reactors: activated sludge treatment and membrane filtration	Removal efficiencies: $cBOD_5$ = 99.8% TSS = 99.65–99.98% TVS = 99.96–99.99% TN = 91.8–95.5% TP = 65.2%–96.1%	[118]
	Membrane biological reactors	Removal efficiencies: $cBOD_5$ = 100% TSS = 99.98% TVS = 99.99% TN = 97.5% TP = 99.9%	[117]
Dairy wastewater	Trickle-filter-type facility	Animal feed containing up to 5% of dairy wastewater solids could be reused without any major health problems	[124]
	Immobilized lactic acid bacteria (*Lactobacillus plantarum*), chemical treatment	Precipitated materials: Dry matter = 10% (23% protein and 68% fat) COD removed by carboxy methyl cellulose = 65–78% COD removed by chitosan = 49–82%	[126]
	Membrane technology	Retentate obtained from concentrated dairy could be reused as animal feed.	[125]
Domestic and industrial wastewater	Untreated	It seemed that untreated wastewater did not contaminate the fish species with As, Cd, and Pb	[111]

(continued)

Table 3.5 (continued)

Type of wastewater	Treatment processes	Main results/conclusions	Reference
Olive mill wastewater	Treatment with yeast (*Candida krusei, Saccharomyces chevalierie, S. rouxii*)	COD removal from olive mill wastewater was 40–50%. *S. rouxii* showed the highest protein content of 3.35 g/L and had a yield of 0.45 g of biomass/g of glucose	[121]
	Alkaline-oxidative treatment, fermentation (yeast and fungi)	BOD removal efficiency = 90% Wet biomass obtained = 150–160 g/L of wastewater Partially dehydrated biomass content: Protein = 13 g% Row fiber = 6 g%	[120]
Phenolic protein precipitates (trub) from beer production	Drum dryer	Trub and spent yeast from beer production could be reused in dried protein feed preparations	[133]
Pineapple cannery effluent	Fermentation with yeast (*Candida utilis*)	Yeast single-cell protein content: Crude protein = 55.3% True protein = 52.2% COD removal = 90–95% Yeast single-cell protein was suitable to be removed as cattle feed	[131]
Salad oil manufacturing wastewater	Yeast isolate (*Candida utilis*)	Oil uptake rate = 0.96 kg oil/kg biomass/day Highest specific growth rate = 0.25/h Salad oil manufacturing wastewater could be converted by yeast isolate to yeast protein as animal feed	[108]
Sewage effluent	Anaerobic/oxic phosphorus removal process, ozonation, crystallization	Solubilization degree = 30% The treatment process had the potential to reduce sludge production and recover phosphorus in usable forms	[109]

	Anaerobic ponds, facultative ponds, stocking ponds	Higher water quality was found in the ponds located further from the input of sewage effluent. The growth, survival, and physiological health of fish were dependent on the water quality	[115]
	Primary treatment	Sewage-fed fish production system yielded similar results as fertilizer-based system. The use of treated sewage effluent also enhanced the water productivity by 64% higher than fertilizer-based system	[114]
Tequila industry wastewater	Centrifugation	From digestibility and feeding trials, fermented solids could be used as sheep feed	[134]
Wine distillation wastewater	Sedimentation, single-stage aeration	Good pollution removal efficiency and high production of biosludge which was suitable to be reused as animal feed	[135]

fertilizer production if the system was well designed. The advantages were the reduction of greenhouse gases, reduction in emission of nitrogen and phosphorus, and the reduction in the use of primary energy [138]. Karak and Bhattacharyya [137] later reconfirmed that human urine treated with Mg^{2+} can be reused as important slow-release phosphatic fertilizer because human urine contains high contents of readily available nutrients, such as nitrogen, phosphorus, and potassium.

Wu et al. [8] reported the potential of reusing palm oil mill effluent (POME) as a cheap organic fertilizer. POME could offer an alternative to fully or partially replace chemical synthesis fertilizer because POME contains a high concentration of carbohydrates, proteins, nitrogenous compounds, lipids, and minerals. However, raw POME might cause clogging and waterlogging of the soil which would lead to the killing of vegetation. Thus, it is suggested to use POME in a small amount at one time or that it goes through the composting process with the addition of shredded empty fruit bunches and poultry dropping as nutrient supplements [8].

According to Li et al. [139] and Luth et al. [140], swine wastewater could be reused in vermifiltration to produce earthworms and vermicompost. Soils treated with vermicomposts usually show significant increases in microbial biomass nitrogen, dehydrogenase activity, and orthophosphates in comparison with soils treated with inorganic fertilizers, where the former may influence plant growth indirectly [141, 142]. The earthworm population increased in just a few weeks when the feed was sufficient [139, 140]. Although previous studies showed evidence of nitrous oxide emissions and methane sinks induced by earthworm populations, Luth et al. [140] found a decrease in emissions of ammonia and nitrous oxide and a sink of methane in treatments with earthworms. They suggested that earthworm abundance could be used as a bioindicator of low-energy input, low greenhouse gas, and ammonia output in systems using fresh slurry with water recycling. Lansing et al. [143] indicated that swine wastewater with cooking grease in the digestion process was able to produce liquid fertilizer and the captured methane in the process could become a source of renewable energy. It is important to reduce the phosphorus concentration of wastewater before it is discharged into the environment. The PO_4, Mg, and NH_4 in swine wastewater could be crystallized and form struvite (MAP: magnesium ammonium phosphate), which could be used as a slow-release fertilizer [144]. Szögi et al. [145] stated that the implementation of technologies for phosphorus recovery from liquid swine waste would resolve the distribution issues of excess manure phosphorus and this excess nutrient could be transported off farms for agricultural uses.

Numerous studies have identified the suitability of transforming olive mill wastewater into agricultural fertilizer mainly because olive mill wastewater contains high organic loads [146–148]. However, olive mill wastewater also contains phytotoxic and antibacterial phenolic substances, which resist biological degradation [147]. Parades et al. [147] and Cayuela et al. [146] adapted the treatment of composting in order to improve the chemical and physicochemical properties of olive mill wastewater before the wastewater could be reused as fertilizer on crops. The composting process is a suitable method to increase the stability of olive mill wastewater because it usually exhibits limited porosity, a high concentration of lignin, lipids, and phenols, which generates a particular composting process with high temperatures and a long thermophilic phase [146]. During the composting process, organic matter in olive mill wastewater is biodegraded through an exothermic aerobic reaction, and CO_2, H_2O, and mineral salts as well as stable and humified organic material are produced [147].

Poultry manure wastewater [149], municipal wastewater [150], industrial wastewater [151], and fish meal wastewater [152] are also suitable to be reused to produce fertilizer for improving the fertility of soil. Anaerobic digestion technology was implied for poultry manure wastewater, but the organic and nutrient load might not meet the standard of discharge [149]. For industrial wastewater, dilution is never a solution for pollution. El Diwani et al. [151] investigated the removal of nitrogen from industrial wastewater using bittern as a source of the magnesium ion for struvite formation. The recovered struvite could be reused as a slow-release fertilizer to support the growth of broad bean plants [151]. Kim et al. [152] suggested reusing biodegraded fish meal wastewater as a fertilizer because the wastewater from fish meal production contains protein and organic waste, which are capable of promoting plant growth. Seafood processing wastewaters do not contain known toxic or carcinogenic materials and are more suitable to be reused as fertilizer, unlike municipal and industrial effluents [152]. Table 3.6 shows the potential of reusing treated or non-treated wastewater as fertilizer.

3.7 Other Reuse of Wastewater

Due to the increase of population and high demands on water resources, it is important to secure water resources for future usage. Hence, wastewater should be reused in many different areas, including agricultural irrigation, landscape irrigation, industrial and non-potable urban uses, recreation or environmental uses, groundwater recharge, and indirect potable uses in order to reduce the demands on potable sources of freshwater. Reuse of wastewater is attractive in many communities because the cost of producing treated wastewater has been proven to be lower than the cost of producing drinking water [17].

Deniz et al. [23] stated that reused wastewater could be added to other known water sources, such as groundwater wells and desalination plants. Díaz-Cruz and Barceló [156] reported that several sources of water have been used for recharge purposes, including surface water, reclaimed wastewater, urban and rural storm water runoff, and treated drinking water. Groundwater recharge is known to be an important issue because it can extend the clean water chain and provide major benefits in terms of improving the way in which water is used as a renewable resource [156]. However, there are issues and constraints in this application, such as the availability of suitable sites, groundwater contamination, salt and mineral buildup, and the toxicological effects of organic chemicals [157].

Table 3.6 Summary of studies done on wastewater treatment for reuse purpose as fertilizer

Type of wastewater	Treatment processes	Main results/conclusions	Reference
Fish meal wastewater	Aerobic biodegradation process	The levels of amino acids in the final broth were also comparable to those in a commercial fertilizer	[152]
Household wastewater (urine)	Ammonia recovery: Urine passes through air striping and subsequently be adsorbed in sulfuric acid to form ammonium sulfate (liquid fertilizer). Phosphorus recovery: Crystallization (struvite was formed)	Human urine could be effectively used as an alternative of chemical fertilizers	[137]
	Urine-separating system	Environmental benefits: Decrease in greenhouse gases Decrease in emission of nitrogen and phosphorus to the water Decrease the use of primary energy	[138]
Household wastewater (urine, blackwater, fecal matter)	Urine: Crystallization (struvite was formed)	Fertilizer analysis (urine): N = 9 g/L P = 0.7 g/L pH = 9 (ammonia volatilization $< 10\%$)	[136]
	Blackwater: Anaerobic treatment	Blackwater: Mineral N (ammonium) on total N increased due to both mineralization of organic nitrogen and reduction of N-fixing carbon compounds	
	Fecal matter: Composting process	Fecal matter: It was normally used as soil conditioner instead of fertilizer due to the low N availability of compost	
Industrial wastewater	Crystallization (struvite was formed)	Fertilizer analysis: pH = 6.91 Struvite yield = 6.975 in g NH_4 = 96.95 mg/L Mg^{2+} = 30 mg/L Ca^{2+} = 70 mg/L	[151]

Municipal wastewater	Step 1: Composting process Step 2: Vermicomposting process	Fertilizer analysis (composting): Water-holding capacity = 51.2% pH = 8.4 EC = 4.49 dS/m TC (dry weight) = 28.1% TOC (dry weight) = 26.9% TN (dry weight) = 1.9% C/N ratio = 14.6 TP (dry weight) = 1.2% TK (dry weight) = 11.3 mg/g Fertilizer analysis (vermicomposting): Water-holding capacity = 73.9% pH = 7.8–8.3 EC = 5.58–7.17 dS/m TC (dry weight) = 24.2–26.5% TOC (dry weight) = 23.0–25.3% TN (dry weight) = 2.3–3.4% C/N ratio = 7.1–11.0% TP (dry weight) = 1.4–2.1%	[153]
	Crystallization (form of struvite)	Fertilizer analysis (key contaminants): Ni ≤ 1.29 mg/kg Cr ≤ 1.35 mg/kg Zn = 13 mg/kg Hg = 4.23 mg/kg As ≤ 0.125 mg/kg Total PCB (mg/g) = 0 Removal efficiencies: NH_4-N = 89.35% COD = 39.78% Mg = 76.31% PO_4-P = 95%	[154]

(continued)

Table 3.6 (continued)

Type of wastewater	Treatment processes	Main results/conclusions	Reference
	Phosphorus recovery	Phosphorus content was increased from 0.2 to 1.3 mg/g. Polonite® had the largest sorption capacity for phosphorus	[150]
	Anaerobic digestion process	Compared to no fertilizer application, applying inorganic fertilizer increased the global warming potential by 2%, while applying sewage sludge increased the global warming potential by a lesser extent	[155]
Olive mill wastewater	Composting process	The analysis of enzymatic activity dynamics in composting air-dried samples represented a rapid, simple method and could be used as a reliable index for the characterization of olive mill wastewater compost stability	[146]
	Composting process	Fertilizer analysis: OM = 56.4% pH = 7.84 C_{org} = 293.6 g/kg C_w = 0.89% N_T = 31.1 g/kg NH_4^+-N = 315 mg/kg NO_3^--N = 2,739 mg/kg C/N = 9.4 P = 8.7 g/kg K = 35.1 g/kg Ca = 72.9 g/kg Mg = 10.5 g/kg	[147]
	Composting process	Fertilizer analysis: pH = 7.08 TOC = 27.74% TKN = 3.31% C/N = 8.39	[148]

POME	Composting process	Fertilizer analysis (clarified POME): C = 57.87% H = 8.26% N = 2.91% O = 30.96% C/N = 19.89 Fertilizer analysis (decomposed POME): C = 48.94% H = 5.76% N = 8.05% O = 37.25% C/N = 6.08	[8]
Pretreated poultry manure wastewater	Anaerobic digestion system	Wastewater removal efficiencies: NH_4^+-N removal = 85.4% Total COD removal = 54% Residual color removal = 50%	[149]
Swine wastewater	Digestion process (captured methane as a source of renewable energy, production of liquid fertilizer)	Fertilizer analysis: TKN = 553–605 mg/g NH_4^+-N = 387–447 mg/g TP = 90.6–224 g/kg PO_4^- = 51.5–139 mg/L	[143]
	Vermifiltration process	Wastewater removal efficiencies: C removal = 85% N removal = 65% The collected solids from the sieve and the vermifilter could be reused as organic fertilizer	[139]
	Vermifiltration process	Earthworms had a clear effect, most indirect, on gaseous emissions during vermifiltration of pig fresh slurry. The cast could be exported for use as vermicompost	[140]

(continued)

Table 3.6 (continued)

Type of wastewater	Treatment processes	Main results/conclusions	Reference
	Crystallization (struvite was formed)	Fertilizer analysis: Struvite purity = 95% Rate of struvite recovery = 117 g/m^3	[144]
	Separation process (2 stages: extraction of phosphorus and conversion of extracted phosphorus into useful product)	After flocculation, nonwoven polypropylene and monofilament filter bag fabrics with mesh size ≤ 200 μm retained more than 99% suspended solids and total phosphorus Solids content in the dewatered cake also increased from 1.5% to more than 90% in 20 days	[145]

Abbreviations in Tables 3.1, 3.2, 3.3, 3.4, 3.5, and 3.6

12 L/12D = 12 h light/12 h dark; *Abs* absorbance, *AMBR* aerobic membrane bioreactor, *AOMBR* anaerobic-oxic membrane bioreactor, *BOD* biochemical oxygen demand, *cBOD* carbonaceous biochemical oxygen demand, *CL* continuous light, *COD* chemical oxygen demand, *DO* dissolved oxygen, *EC* electrical conductivity, *FC* fecal coliform, *GAC* granular activated carbon, *MBAS* methylene blue active substances, *MBR* membrane bioreactor, *OM* organic matter, *PCB* polychlorinated biphenyls *PHA* polyhydroxyalkanoates, *PHB* polyhydroxybutyrate, *PK* Klong Pai Poo hot spring, *POME* palm oil mill effluent, *PR* Romani hot spring, *RO* reverse osmosis, *SAC* spectral absorption coefficient, *SW* Ron hot spring, *TC* total carbon, *TDS* total dissolved solids, *TK* total potassium, *TKN* total Kjeldahl nitrogen, *TN* total nitrogen, *TOC* total organic carbon, *TP* total phosphorus, *TSS* total suspended solids, *TVS* total volatile solids, *WSP* waste stabilization pond

Studies done by You and Wu [45] showed that treated wastewater could be reused for toilet flushing, but the reused water showed satisfactory qualities: *E. coli* (ND/100 mL), color (not uncomfortable), odor (not uncomfortable), BOD (<10 mg/L), and pH (6.0–8.5). In addition, Adewumi et al. [17] also showed the potential of reusing treated wastewater for toilet flushing applications. Indirect potable use of wastewater includes the blending of water with public water supplies and surface water augmentation [157]. However, there were also issues or constraints in this application including public acceptance and public health concerns [157].

3.8 Conclusions

As the demand for human consumption, agriculture, and industries grows, the most easily reachable resources become insufficient. The solution is to find other water sources to balance both the supplies and demands. In this approach, sustainability through reuse of wastewater is primarily seen as a technical and managerial issue [7]. It is undeniable that a cleaner production is, in the long run, a better option for managing wastewaters as opposed to end-of-pipe processes. However, although the emphasis on sustainability for wastewater reuse is well intentioned, it may sometimes raise false expectations. The limits imposed by the economic and social frameworks may present limitations to sustainable practices that could be applied on the large-scale. Consequently, there is usually no economic incentive to develop waste-free processes. Wastewater reuse is therefore limited unless the activity is subsidized, externalities are factored in, health protection is maximized, reusability of wastewater is successfully designed for commercial reuse and, most importantly, the government takes the initiative in legislating for sustainable industrial development. Since the economic framework depends on growth, production, and consumption, the initiatives to promote wastewater reuse can only come from the authorities, institutions, and decision-makers since their subtle actions could accelerate the research and development needed for promoting sustainability in wastewater management.

References

1. Graedel TE, Klee RJ (2002) Getting serious about sustainability. Environ Sci Technol 36:523–529
2. Verstraete W, de Caveye PV, Diamantis V (2009) Maximum use of resources present in domestic "used water". Bioresour Technol 100:5537–5545
3. United Nations (2006) The 2nd UN world water development report: water, a shared responsibility. United Nations Educational, Scientific & Cultural Organisation (UNESCO), Paris and Berghahn Books, NY
4. United Nations (2008) World urbanization prospects: the 2007 revision. United Nations Department of Economic and Social Affairs/Population Division

5. Postel SL, Daily GC, Ehrlich PR (1996) Human appropriation of renewable fresh water. Science 271:785–788
6. Carey RO, Migliaccio KW (2009) Contribution of wastewater treatment plant effluents to nutrient dynamics in aquatic system: a review. Environ Manag 44:205–217
7. Isosaari P, Hermanowicz SW, Rubin Y (2010) Sustainable natural systems for treatment and disposal of food processing wastewater. Crit Rev Environ Sci Technol 40:662–697
8. Wu TY, Mohammad AW, Jahim JM, Anuar N (2009) A holistic approach to managing palm oil mill effluent (POME): biotechnological advances in the sustainable reuse of POME. Biotechnol Adv 27:40–52
9. Wu TY, Mohammad AW, Jahim JM, Anuar N (2010) Pollution control technologies for the treatment of palm oil mill effluent (POME) through end-of-pipe processes. J Environ Manag 91:1467–1490
10. Gurel M, Iskender G, Ovez S, Arslan-Alaton I, Tanik A, Orhon D (2007) A global overview of treated wastewater guidelines and standards for agricultural reuse. Fresenius Environ Bull 16:590–595
11. Angelakis AN, Bontoux L, Lazarova V (2003) Challenges and prospectives for water recycling and reuse in EU countries. Water Sci Technol Water Supply 3(4):59–68
12. Pedrero F, Kalavrouziotis I, Alarcón JJ, Koukoulakis P, Asano T (2010) Use of treated municipal wastewater in irrigated agriculture – review of some practices in Spain and Greece. Agric Water Manag 97:1233–1241
13. Fatta-Kassinos D, Kalavrouziotis IK, Koukoulakis PH, Vasquez MI (2010) The risks associated with wastewater reuse and xenobiotics in the agroecological environment. Sci Total Environ. doi:10.1016/jscitotenv.2010.03.036
14. Meneses M, Pasqualino JC, Castells F (2010) Environmental assessment of urban wastewater reuse: treatment alternatives and applications. Chemosphere 81:266–272
15. Sala L, Serra M (2004) Towards sustainability in water recycling. Water Sci Technol 50(2):1–7
16. Ensink JHJ, Mahmood T, Hoek WVD, Raschid-Sally L, Amerasinghe FP (2004) A nation-wide assessment of wastewater use in Pakistan: an obscure activity or a vitally important one? Water Policy 6:197–206
17. Adewumi JR, Ilemobade AA, Van Zyl JE (2010) Treated wastewater reuse in South Africa: overview, potential and challenges. Resour Conserv Recycl 55:221–231
18. Akponikpè PBI, Wima K, Yacouba H, Mermoud A (2011) Reuse of domestic wastewater treated in macrophyte ponds to irrigate tomato and eggplant in semi-arid West-Africa: benefits and risks. Agric Water Manag 98:834–840
19. Battilani A, Steiner M, Andersen M, Back SN, Lorenzen J, Schweitzer A, Dalsgaard A, Forslund A, Gola S, Klopmann W, Plauborg F, Andersen MN (2010) Decentralised water and wastewater treatment technologies to produce functional water for irrigation. Agric Water Manag 98:385–402
20. Amar NB, Kechaou N, Palmeri J, Deratani A, Sghaier A (2009) Comparison of tertiary treatment by nanofiltration and reverse osmosis for water reuse in denim textile industry. J Hazard Mater 170:111–117
21. Bakopoulou S, Emmanouil C, Kungolos A (2011) Assessment of wastewater effluent quality in Thessaly region, Greece, for determining its irrigation reuse potential. Ecotoxcol Environ Saf 74:188–194
22. Bedbabis S, Ferrara G, Rouina BB, Boukhris M (2010) The effect of waste water irrigation on the extra virgin olive oil quality from the Tunisian cultivar Chemlali. Sci Hortic 125:556–561
23. Deniz F, Sadhwani JJ, Veza JM (2010) New quality criteria in wastewater reuse: the case of Gran Canaria. Desalination 250:716–722
24. Soliz D, Glenn EP, Seaman R, Yoklic M, Nelson SG, Brown P (2011) Water consumption, irrigation efficiency and nutritional value of Atriplex lentiformis grown on reverse osmosis brine in a desert irrigation district. Agric Ecosyst Environ. doi:10.1016/j.agee.2011.01.012

25. Turral H, Svendsen M, Faures JM (2010) Investing in irrigation: reviewing the past and looking to the future. Agric Water Manag 97:551–560
26. Angelakis AN, Marecos Do Monte MHF, Bontoux L, Asano T (1999) The status of wastewater reuse practice in the Mediterranean basin: need for guidelines. Water Res 33:2201–2217
27. Capra A, Scicolone B (2007) Recycling of poor quality urban wastewater by drip irrigation systems. J Clean Prod 15:1529–1534
28. Illueca-Muñoz J, Mendoza-Roca JA, Iborra-Clar A, Bes-Piá A, Fajardo-Montanana V, Martínez-Francisco FJ, Bernácer-Bonora I (2008) Study of different alternatives of tertiary treatments for wastewater reclamation to optimize the water quality for irrigation reuse. Desalination 222:222–229
29. Bedbabis S, Ferrara G, Rouina BB, Boukhris M (2010) Effects of irrigation with treated wastewater on olive tree growth, yield and leaf mineral elements at short term. Sci Hortic 126:345–350
30. Qadir M, Wichelns D, Raschid-Sally L, McCornick PG, Dreschsel P, Bahri A, Minhas PS (2010) The challenges of wastewater irrigation in developing countries. Agric Water Manag 97:561–568
31. Xu J, Wu L, Chang AC, Zhang Y (2010) Impact of long-term reclaimed wastewater irrigation on agricultural soils: a preliminary assessment. J Hazard Mater 183:780–786
32. Capra A, Scicolone B (2004) Emitter and filter tests for wastewater reuse by drip irrigation. Agric Water Manag 68:135–149
33. Segal E, Dag A, Ben-Gal A, Zipori I, Erel R, Suryano S, Yermiyahu U (2011) Olive orchard irrigation with reclaimed wastewater: agronomic and environmental considerations. Agric Ecosyst Environ. doi:10.1016/j.agee.2011.01.009
34. Hassanli AM, Ebrahimizadeh MA, Beecham S (2009) The effects of irrigation methods with effluent and irrigation scheduling on water use efficiency and corn yields in an arid region. Agric Water Manag 96:93–99
35. Papaiacovou I (2001) Case study – wastewater reuse in Limassol as an alternative water source. Desalination 138:55–59
36. Ammary BY (2007) Wastewater reuse in Jordan: present status and future plans. Desalination 211:164–176
37. Gharsallaoui M, Benincasa C, Ayadi M, Perri E, Khlif M, Gabsi S (2011) Study on the impact of wastewater irrigation on the quality of oils obtained from olives harvested by hand and from the ground and extracted at different times after the harvesting. Sci Hortic 128:23–29
38. McNeill LS, Almasri MN, Mizyed N (2009) A sustainable approach for reusing treated wastewater in agricultural irrigation in the West Bank – Palestine. Desalination 248:315–321
39. Morari F, Giardini L (2009) Municipal wastewater treatment with vertical flow constructed wetlands for irrigation reuse. Ecol Eng 35:643–653
40. Hosetti BB, Frost S (1995) A review of the sustainable value of effluents and sludges from wastewater stabilization ponds. Ecol Eng 5:421–431
41. Jamwal P, Mittal AK (2010) Reuse of treated sewage in Delhi city: microbial evaluation of STPs and reuse options. Resour Conserv Recycl 54:211–221
42. Belmont MA, Cantellano E, Thompson S, Williamson M, Sànchez A, Metcalfe CD (2004) Treatment of domestic wastewater in a pilot-scale natural treatment system in central Mexico. Ecol Eng 23:299–311
43. Lin SH, Chan HY, Leu HG (2000) Treatment of wastewater effluent from an industrial park for agricultural irrigation. Desalination 128:257–267
44. Verheyen V, Cruickshank A, Wild K, Heaven MW, McGee R, Watkins M, Nash D (2011) Characterization of organic particulates present in milk factory process waters used for reuse along with aerobically digested effluent wastewater. Bioresour Technol 102:2118–2125
45. You SJ, Wu DC (2009) Potential for reuse of high cellulose containing wastewater after membrane bioreactor treatment. Desalination 249:721–728

46. You SJ, Tseng DH, Deng JY (2008) Using combined membrane processes for textile dyeing wastewater reclamation. Desalination 234:426–432
47. Lonigro A, Pollice A, Spinelli R, Berrilli F, Di Cave D, D'Orazi C, Cavallo P, Brandonisio O (2006) *Giardia* Cysts and *Cryptosporidium* Oocysts in Membrane-filtered municipal wastewater used for irrigation. Appl Environ Microbiol 72(12):7916–7918
48. Bixio D, Thoeye C, De Koning J, Joksimovic D, Savic D, Wintgens T, Melin T (2006) Wastewater reuse in Europe. Desalination 187:89–101
49. Gómez-López MD, Bayo J, García-Cascales MS, Angosto JM (2009) Decision support in disinfection technologies for treated water reuse. J Clean Prod 17:1504–1511
50. Feng F, Xu Z, Li X, You W, Zhen Y (2010) Advanced treatment of dyeing wastewater towards reuse by the combined Fenton oxidation and membrane bioreactor process. J Environ Sci 22(11):1657–1665
51. Sarkar B, Chakrabarti PP, Vijaykumar A, Kale V (2006) Wastewater treatment in dairy industries – possibility of reuse. Desalination 195:141–152
52. Bes-Piá A, Iborra-Clar A, Garcia-Figueruelo C, Barredo-Damas S, Alcaina-Miranda MI, Mendoza-Roca JA, Iborra-Clar MI (2009) Comparison of three NF membranes for the reuse of secondary textile effluents. Desalination 241:1–7
53. Lu X, Liu L, Liu R, Chen J (2010) Textile wastewater reuse as an alternative water source for dyeing and finishing processes: a case study. Desalination 258:229–232
54. Oller I, Malato S, Sánchez-Pérez JA (2010) Combination of advanced oxidation processes and biological treatments wastewater decontamination – a review. Sci Total Environ. doi:10/1016/jscitotenv.2010.08.061
55. Ciardelli G, Ranieri N (2001) The treatment and reuse of wastewater in the textile industry by means of ozonation and electroflocculation. Water Res 35(2):567–572
56. Rozzi A, Malpei F, Bonomo L, Bianchi R (1999) Textile wastewater reuse in Northern Italy (Como). Water Sci Techol 39(5):121–128
57. Alcaina-Miranda MI, Barredo-Damas S, Bes-Piá A, Iborra-Clar MI, Iborra-Clar A, Mendoza-Roca JA (2009) Nanofiltration as a final step towards textile wastewater reclamation. Desalination 240:290–297
58. Avula RY, Nelson HM, Singh RK (2009) Recycling of poultry process wastewater by ultrafiltration. Innov Food Sci Emerg 10:1–8
59. Üstün GE, Solmaz SKA, Birgül A (2007) Regeneration of industrial district wastewater using a combination of Fenton process and ion exchange – a case study. Resour Conserv Recycl 52:425–440
60. Colindres P, Yee-Madeira H, Reguera E (2010) Removal of Reactive Black 5 from aqueous solution by ozone for water reuse in textile dyeing processes. Desalination 258:154–158
61. Beers DV, Bossilkov A, Berkel RV (2008) A regional synergy approach to advance sustainable water use: a case study using Kwinana (Western Australia). Australas J Environ 15:149–158
62. Giurco D, Bossilkov A, Patterson J, Kazaglis A (2010) Developing industrial water reuse synergies in Port Melbourne: cost effectiveness, barriers and opportunities. J Clean Prod. doi:10.1016/j.jclepro.2010.07.001
63. Singh S, Fan M, Brown RC (2008) Ozone treatment of process water from a dry-mill ethanol plant. Bioresour Technol 99:1801–1805
64. Zhang L, Xu B, Gong J, Bu T (2010) Membrane combination technic on treatment and reuse of high ammonia and salts wastewater in rare earth manufacture process. J Rare Earth 28:501–503
65. Ciabatti I, Cesaro F, Faralli L, Fatarella E, Tognotti F (2009) Demonstration of a treatment system for purification and reuse of laundry wastewater. Desalination 245:451–459
66. Ginkel SWV, Oh SE, Logan BE (2005) Biohydrogen gas production from food processing and domestic wastewaters. Int J Hydrog Energy 30:1535–1542
67. Shi XY, Jin DA, Sun QY, Li WW (2010) Optimization of conditions for hydrogen production from brewery wastewater by anaerobic sludge using desirability function approach. Renew Energy 35:1493–1498

68. Azbar N, Dokgoz FTC, Keskin T, Korkmaz KS, Syed HM (2009) Continuous fermentative hydrogen production from cheese whey wastewater under thermophilic aerobic conditions. Int J Hydrog Energy 34:7441–7447
69. O-Thong S, Hniman A, Prasertsan P, Imai T (2011) Biohydrogen production from cassava starch processing wastewater by thermophilic mixed cultures. Int J Hydrog Energy 36:3409–3416
70. Chong ML, Rahman NAA, Rahim RA, Aziz SA, Shirai Y, Hassan MA (2009) Optimization of biohydrogen production by *Clostridium butyricum* E86 from palm oil mill effluent using response surface methodology. Int J Hydrog Energy 34:7475–7482
71. Tai J, Adav SS, Su A, Lee DJ (2010) Biological hydrogen production from phenol-containing wastewater using *Clostridium butyricum*. Int J Hydrog Energy 35:13345–13349
72. Sivaramakrishna D, Sreekanth D, Himabindu V, Anjaneyulu Y (2009) Biological hydrogen production from probiotic wastewater as substrate by selectively enriched anaerobic mixed microflora. Renew Energy 34:937–940
73. Wei J, Liu ZT, Zhang X (2010) Biohydrogen production from starch wastewater and application in fuel cell. Int J Hydrog Energy 35:2949–2952
74. Seifert K, Waligorska M, Laniecki M (2010) Brewery wastewater in photobiological hydrogen generation in presence of *Rhodobacter sphaeroides* O.U.001. Int J Hydrog Energy 35:4085–4091
75. Seifert K, Waligorska M, Laniecki M (2010) Hydrogen generation in photobiological process from dairy wastewater. Int J Hydrog Energy 35:9624–9629
76. Eroglu E, Gunduz U, Yucel M, Turker L, Eroglu I (2004) Photobiological hydrogen production by using olive mill wastewater as a sole substrate source. Int J Hydrog Energy 29:163–171
77. Eroglu E, Gunduz U, Yucel M, Eroglu I (2010) Photosynthetic bacterial growth and productivity under continuous illumination or diurnal cycles with olive mill wastewater as feedstock. Int J Hydrog Energy 35:5293–5300
78. Eroglu E, Eroglu I, Gunduz U, Yucel M (2008) Effect of clay pretreatment on photofermentative hydrogen production from olive mill wastewater. Biosour Technol 99:6799–6808
79. Jamil Z, Annuar MSM, Ibrahim S, Vikneswary S (2009) Optimization of phototrophic hydrogen production by *Rhodopseudomonas palustris* PBUM001 via statistical experimental design. Int J Hydrog Energy 34:7502–7512
80. Tao Y, He Y, Wu Y, Liu F, Li X, Zhong W, Zhou Z (2008) Characteristic of a new photosynthetic bacterial strain for hydrogen production and its application in wastewater treatment. Int J Hydrog Energy 33:963–973
81. Zhu H, Ueda S, Asada Y, Miyake J (2002) Hydrogen production as a novel process of wastewater treatment-studies on tofu wastewater with entrapped *R. sphaeroides* and mutagenesis. Int J Hydrog Energy 27:1349–1357
82. Zheng G, Kang Z, Qian Y, Wang L (2009) Enhanced biohydrogen generation from organic wastewater containing NH_4^+ by photothrophic bacteria *Rhodobacter sphaeroides* AR-3. Front Environ Sci Eng China 3(4):387–392
83. Rashid SS, Md Alam Z, Karim MIA, Salleh MH (2009) Management of palm oil mill effluent through production of cellulases by filamentous fungi. World J Microbiol Biotechnol 25:2219–2226
84. Laohaprapanon T, Prasertsan P, H-Kittikun A (2007) Screening of thermotolerant microorganisms and application for oil separation from palm oil mill wastewater. Songklanakarin J Sci Teachnol 29:801–808
85. Jamal P, Alam MZ, Ramlan M, Salleh M, Nadzir MM (2005) Screening of *Aspergillus* for citric acid production from palm oil mill effluent. Biotechnology 4:275–278
86. Alam MZ, Jamal P, Nadzir MM (2008) Bioconversion of palm oil mill effluent for citric acid production: statistical optimization of fermentation media and time by central composition design. World J Microbiol Biotechnol 24:1177–1185
87. Rao K, Chaudhari V, Varanasi S, Kim DS (2007) Enhanced ethanol fermentation of brewery wastewater using the genetically modified strain *E. coli* KO11. Appl Microbiol Biotechnol 74:50–60

88. Wu TY, Mohammad AW, Jahim JM, Anuar N (2005) Penghasilan asid itakonik oleh *Aspergillus terreus* IMI 282743 dalam pelbagai jenis medium penghasilan. Pascasidang Kolokium Jabatan Kejuruteraan Kimia dan Proses, Universiti Kebangsaan Malaysia, Malaysia pp. 147–151
89. Goncalves C, Lopes M, Ferreira JP, Belo I (2009) Biological treatment of olive mill wastewater by non-conventional yeasts. Biosour Technol 100:3759–3763
90. D'Annibale A, Sermanni GG, Federici F, Petruccioli M (2006) Olive-mill wastewaters: a promising substrate for microbial lipase production. Biosour Technol 97:1828–1833
91. Somrutai W, Takagi M, Yoshida T (1996) Acetone-butanol fermentation by *Clostridium aurantibutyricum* ATCC 17777 from a model medium for palm oil mill effluent. J Ferment Bioeng 81:543–547
92. He J, Zhang GM, Lu HF (2010) Treatment of soybean wastewater by a wild strain *Rhodobacter sphaeroides* and to produce protein under natural conditions. Front Environ Sci Eng China 4(3):334–339
93. Bengtsson S, Pisco AR, Johansson P, Lemos PC, Reis MAM (2010) Molecular weight and thermal properties of polyhydroxyalkanoates produced from fermented sugar molasses by open mixed cultures. J Biotechnol 147:172–179
94. Hassan MA, Shirai Y, Kusubayashi N, Karim MIA, Nakanishi K, Hashimoto K (1996) Effect of organic acid profiles during anaerobic treatment of palm oil mill effluent on the production of polyhydroxyalkanoates by *Rhodobacter sphaeroides*. J Ferment Bioeng 83(5):485–488
95. Thammawong C, Thongkhong K, Lamtassana K, Sharp A, Opaprakasit P (2008) Production and characterization of polyhydroxyalkanoates (PHAs) from inexpensive substrates by *Alcaligenes latus*. Adv Mater Res 55–57:893–896
96. Yu J (2001) Production of PHA from starchy wastewater via organic acids. J Biotechnol 86:105–112
97. Dobroth ZT, Hu S, Coats ER, McDonald AC (2011) Polyhydroxybutyrate synthesis on biodiesel wastewater using mixed microbial consortia. Biosour Techno 102:3352–3359
98. Cavalheiro JMBT, De Almeida MCMD, Grandfils C, Da Fonseca MMR (2009) Poly(3-hydroxybutyrate) production by Cupriavidus necator using waste glycerol. Process Biochem 44:509–515
99. Yigit DO, Gunduz U, Turker L, Yucel M, Eroglu I (1999) Identification of by-products in hydrogen producing bacteria; *Rhodobacter sphaeroides* O.U.001 grown in the waste water of a sugar refinery. J Biotechnol 70:125–131
100. Wu TY, Mohammad AW, Md Jahim J, Anuar N (2006) Investigations on protease production by a wild-type *Aspergillus terreus* strain using diluted retentate of pre-filtered palm oil mill effluent (POME) as substrate. Enzyme Microb Technol 39:1223–1229
101. Kantachote D, Torpee S, Umsakul K (2005) The potential use of anoxygenic phototrophic bacteria for treating latex rubber sheet wastewater. Electron J Biotechol 8(3):314–323
102. Basak N, Das D (2007) The prospect of purple non-sulfur (PNS) photosynthetic bacteria for hydrogen production: the present state of the art. World J Microbiol Biotechnol 23:31–42
103. Koku H, Eroglu I, Gunduz U, Yucel M, Turker L (2002) Aspects of the metabolism of hydrogen production by *Rhodobacter sphaeroides*. Int J Hydrog Energy 27:1315–1329
104. Bengtsson S, Werker A, Christensson M, Welander T (2008) Production of polyhydroxyalkanoates by activated sludge treating a paper mill wastewater. Bioresour Technol 99:509–516
105. Yan JL, Pan GX (2010) Effect of wastewater irrigation on soil enzyme activities and respiration from a managed wetland. Soil Sediment Contam Int J 19:204–216
106. Masngut N, Takriff MS, Mohammad AW, Kalil MS, Kadhum AAH (2007) Performance of oscillatory flow reactor and stirred tank reactor in solvent fermentation from palm oil mill effluent. J Teknol 47(F):45–54
107. Bunting SW (2004) Wastewater aquaculture: perpetuating vulnerability or opportunity to enhance poor livelihoods? Aquat Resour Cult Dev 1(1):51–75
108. Zheng S, Yang M, Yang Z (2005) Biomass production of yeast isolate from salad oil manufacturing wastewater. Bioresour Technol 96:1183–1187

109. Saktaywin W, Tsuno H, Nagare H, Soyama T, Weerapakkaroon J (2005) Advanced sewage treatment process with excess sludge reduction and phosphorus recovery. Water Res 39:902–910
110. Watanabe T, Ikeda H, Masaki K, Fujii T, Iefuji H (2009) Cloning and characterization of a novel phytase from wastewater treatment yeast *Hansenula fabianii* J640 and expression in *Pichia pastoris*. J Biosci Bioeng 108(3):225–230
111. Marcussen H, Holm PE, Ha LT, Dalsgaard A (2007) Food safety aspects of toxic element accumulation in fish from wastewater-fed ponds in Hanoi, Vietnam. Trop Med Int Health 12 (2):34–39
112. Bunting SW (2007) Confronting the realities of wastewater aquaculture in peri-urban Kolkata with bioeconomic modelling. Water Res 41:499–505
113. Mara DD, Edwards P, Clark D, Mills SW (1993) A rational approach to the design of wastewater-fed fishponds. Water Res 27(12):1797–1799
114. Dasgupta S, Pandey BK, Sarangi N, Mukhopadhyay PK (2008) Evaluation of water productivity and fish yield in sewage-fed vis-à-vis fertilized based carp culture. Bioresour Technol 99:3499–3506
115. Murkherjee S, Jana BB (2007) Water quality affects SDH activity, protein content and RNA: DNA ratios in fish (*Catla catla, Labeo rohita* and *Oreochromis mossambicus*) raised in ponds of a sewage-fed fish farm. Aquaculture 262:105–119
116. Verdegem MCJ, Bosma RH, Verreth JAJ (2006) Reducing water use for animal production through aquaculture. Water Resour Dev 22(1):101–113
117. Sharrer MJ, Rishel K, Summerfelt ST (2010) Evaluation of a membrane biological reactor for reclaiming water, alkalinity, salts, phosphorus, and protein contained in a high-strength aquacultural wastewater. Bioresour Technol 101:4322–4330
118. Sharrer MJ, Tal Y, Ferrier D, Hankins JA, Summerfelt ST (2007) Membrane biological reactor treatment of a saline backwash flow from a recirculating aquaculture system. Aquac Eng 36:159–176
119. Alcaide EM, Nefzaoui A (1996) Recycling of olive oil by-products: possibilities of utilization in animal nutrition. Int Biodeterior Biodegrad 38(3–4):227–235
120. Laconi S, Molle G, Cabiddu A, Pompei R (2007) Bioremediation of olive oil mill wastewater and production of microbial biomass. Biodegradation 18:559–566
121. Gharsallah N (1993) Production of single cell protein from olive mill waste water by yeasts. Environ Technol 14(4):391–395
122. Boniardi N, Nano G (1998) Olive mill wastewater as valuable source of animal feed and energy. In: Pelino M, Pellacani GC (eds). Proceedings of the 1st National Congress Valorisation and Recycling of Industrial Wastes, Italy. Publisher: Mucchi Editore, Modena, Italy, pp. 41–48
123. Roig A, Cayuela ML, Sànchez-Monedero MA (2006) An overview on olive mill wastes and their valorisation methods. Waste Manag 26:960–969
124. Belyea RL, Williams JE, Gieseke L, Clevenger TE, Brown JR, Tumbleson ME (1990) Evaluation of dairy wastewater solids as a feed ingredient. J Dairy Sci 73:1864–1871
125. Luo J, Ding L, Wan Y, Paullier P, Jaffrin MY (2010) Application of NF-RDM (nanofiltration rotating disk membrane) module under extreme hydraulic conditions for the treatment of dairy wastewater. Chem Eng J 163:307–316
126. Dyrset N, Selmer-Olsen E, Havrevoll Ø, Ratnaweera H, Storrø I, Birkeland SE (1998) Feed supplement recovered from dairy wastewater by biological and chemical pretreatment. J Chem Technol Biotechnol 73:175–182
127. Wu TY, Mohammad AW, Jahim JM, Anuar N (2007) Palm oil mill effluent (POME) treatment and bioresources recovery using ultrafiltration membrane: effect of pressure on membrane fouling. Biochem Eng J 35:309–317
128. Singh RP, Embrandiri A, Ibrahim MH, Esa N (2011) Management of biomass residues generated from palm oil mill: vermicomposting a sustainable option. Resour Conserv Recycl 55(4):423–434

129. Babu MK, Sukumaran N, Kumar MGV, Murugesan AG, Jose GS, Hussain MIZ (2001) Fish culture in palm oil mill effluent. Int J Oil Palm Res 2:47–49
130. Najafpour GD, Klasson KT, Ackerson MD, Clausen EC, Gaddy JL (1994) Biological conversion of poultry processing waste to single cell protein. Bioresour Technol 48:65–70
131. Nigam JN (1998) Single cell protein from pineapple cannery effluent. World J Microbiol Biotechnol 14:693–696
132. Wang Y, Gao X, Su Q, Wu W, An L (2007) Cloning, expression, and enzyme characterization of an acid heat-stable phytase from *Aspergillus fumigatus* WY-2. Curr Microbiol 55:65–70
133. Russ W, Meyer-Pittroff R (2003) The use of phenolic protein precipitates (trub) from beer production in animal feed. Monatsschr Brauwiss 56(5–6):84–88
134. Iñiguez-Covarrubias G, Franco-Gómez MdJ, López-Ortiz G (1996) Utilization of recovered solids from tequila industry vinasse as fodder feed. Bioresour Technol 55:151–158
135. Wetter C, Poeppinghaus K, Morais H, Dias M, Mendes B, Santana F, Oliveira JFS (1989) Production of protein for animal feed stuff using organic wastewaters from wine distilleries. Wat Sci Techol 21(12):1853–1856
136. Winker M, Vinnerås B, Muskolus A, Clemens J (2009) Fertiliser products from new sanitation systems: their potential values and risks. Bioresour Technol 100:4090–4096
137. Karak T, Bhattacharyya P (2011) Human urine as a source of alternative natural fertilizer in agriculture: a flight of fancy or an achievable reality. Resour Conserv Recycl 55:400–408
138. Tidåker P, Mattson B, Jönsson H (2007) Environmental impact of wheat production using human urine and mineral fertilisers – a scenario study. J Clean Prod 15:52–62
139. Li YS, Robin P, Cluzeau D, Bouché M, Qiu JP, Laplanche A, Hassouna M, Morand P, Dappelo C, Callarec J (2008) Vermifiltration as a stage in reuse of swine wastewater: monitoring methodology on an experimental farm. Ecol Eng 32:301–309
140. Luth Robin P, Germain P, Lecomte M, Landrain B, Li Y, Cluzeau D (2011) Earthworm effects on gaseous emissions during vermifiltration of pig fresh slurry. Bioresour Technol 102:3679–3686
141. Arancon NQ, Edwards CA, Bierman P, Metzger JD, Lucht C (2005) Effects of vermicomposts produced from cattle manure, food waste and paper waste on the growth and yield of peppers in the field. Pedobiologia 49:297–306
142. Sim EYS, Wu TY (2010) The potential reuse of biodegradable municipal solid wastes (MSW) as feedstocks in vermicomposting. J Sci Food Agric 90:2153–2162
143. Lansing S, Martin JF, Botero RB, Nogueira da Silva T, Dias da Silva E (2010) Wastewater transformations and fertilizer value when co-digesting differing ratios of swine manure and used cooking grease in low-cost digesters. Biomass Bioenerg 34:1711–1720
144. Suzuki K, Tanaka Y, Kuroda K, Hanajima D, Fukumoto Y, Yasuda T, Waki M (2007) Removal and recovery of phosphorus from swine wastewater by demonstration crystallization reactor and struvite accumulation device. Bioresour Technol 98:1573–1578
145. Szögi AA, Vanotti MB, Hunt PG (2006) Dewatering of phosphorus extracted from liquid swine waste. Bioresour Technol 97:183–190
146. Cayuela ML, Mondini C, Sánchez-Monedero MA, Roig A (2008) Chemical properties and hydrolytic enzyme activities for the characterisation of two-phase olive mill wastes composting. Bioresour Technol 99:4255–4262
147. Paredes C, Cegarra J, Bernal MP, Roig A (2005) Influence of olive mill wastewater in composting and impact of the compost on a Swiss chard crop and soil properties. Environ Int 31:305–312
148. Zenjari B, El Hajjouji H, Ait Baddi G, Bailly J-R, Revel J-C, Nejmeddine A, Hafidi M (2006) Eliminating toxic compounds by composting olive mill wastewater-straw mixtures. J Hazard Mater A138:433–437
149. Yetilmezsoy K, Sapci–Zengin Z (2009) Recovery of ammonium nitrogen from the effluent of UASB treating poultry manure wastewater by MAP precipitation as a slow release fertilizer. J Hazard Mater 166:260–269

150. Hylander LD, Kietlińska A, Renman G, Simán G (2006) Phosphorus retention in filter materials for wastewater treatment and its subsequent suitability for plant production. Bioresour Technol 97:914–921
151. El Diwani G, El Rafie Sh, El Ibiari NN, El-Aila HI (2007) Recovery of ammonia nitrogen from industrial wastewater treatment as struvite slow releasing fertilizer. Desalination 214:200–214
152. Kim JK, Kin JB, Cho KS, Hong YK (2007) Isolation and identification of microorganisms and their aerobic biodegradation of fish-meal wastewater for liquid- fertilization. Int Biodeterior Biodegrad 59:156–165
153. Hait S, Tare V (2011) Optimizing vermistabilization of waste activated sludge using vermicompost as bulking material. Waste Manag 31:502–511
154. Uysal A, Dilsad Yilmazel Y, Demirer GN (2010) The determination of fertiliser quality of the formed struvite from effluent of a sewage sludge anaerobic digester. J Hazard Mater 181:248–254
155. Gilbert P, Thornley P, Riche AB (2011) The influence of organic and inorganic fertiliser application rates on UK biomass crop sustainability. Biomass Bioenerg 35:1170–1181
156. Díaz-Cruz MS, Barceló D (2008) Trace organic chemicals contamination in ground water recharge. Chemosphere 72:333–342
157. Asano T, Burton FL, Leverenz HL, Tsuchihashi R, Tchobanoglous G (2007) Water reuse: issues, technologies, and applications. Metcalf & Eddy, New York
158. Speece RE (1996) Anaerobic biotechnology for industrial wastewater. Archae Press, Nashville

Chapter 4
Reducing the Pollution from Tanneries by Process Wastewater Reuse and Membrane Technologies

M.V. Galiana-Aleixandre, J.A. Mendoza-Roca, and A. Bes-Piá

4.1 Introduction

Tannery activity has been traditionally contaminant, producing strong smells, organic wastes, and a high water consumption. Nowadays, tannery activity has evolved, and people responsible are fully aware that although there has been a large progress against contamination, there is still a long way to walk.

Tannery industry uses salty hides as raw material. To process them, it is necessary to remove the salt content, producing wastewaters with high conductivity. This high conductivity together with the organic contamination of the own hides gives way to a wastewater difficult to treat because of the high content of contamination (salts, organic matter, nitrogen, dyes, fats, etc.) [1].

From the environmental point of view, there are different ways of avoiding or minimizing the pollution in wastewaters in the tannery industry. These are:

- Use of less contaminant raw materials. In this way, fresh hides can be processed avoiding the conservation salt.
- Replace of some chemical products by others with less pollution potential. This can be carried out according to the IPPC procedure.
- Reduction of water consumption.
- Reuse of wastewaters, both in the same and in different processes either directly or after pretreatment.
- End of pipe treatment. Finally, when all the other alternatives are exhausted, we have to treat the global wastewaters to minimize the pollution in the discharging effluent. In this way, either conventional treatments such as physical-chemical and biological processes or clean technologies like membrane technologies can be applied.

M.V. Galiana-Aleixandre • J.A. Mendoza-Roca (✉) • A. Bes-Piá
Department of Chemical and Nuclear Engineering, Universidad Politécnica de Valencia, Camino de Vera s/n. 46022, Valencia, Spain
e-mail: jamendoz@iqn.upv.es

S.K. Sharma and R. Sanghi (eds.), *Wastewater Reuse and Management*,
DOI 10.1007/978-94-007-4942-9_4,

The application of the membrane technologies allows us to reduce the wastewater pollution and to reuse the permeate streams. The development of these techniques makes possible the production of clean water at a moderate cost, providing flexible solutions adapted to any necessity [2].

4.2 Processes in the Tannery

The raw matter in the tannery is the animal hide. The characteristics of the both sides of the hide are very different. Thus, grain side and flesh side have to be distinguished. The hide surface that contains the hair and oil glands is known as the grain side, while the other side, named flesh side, is much thicker and softer. Hides can be also divided from an anatomical point of view into three layers, i.e., epidermis, dermis, and subcutaneous tissue.

Raw hides undergo several stages during the tanning process. Depending on the type of hide used and the aimed end product, these stages may vary widely. Hides have the ability to absorb tannic acid and other chemical substances that prevent them from degrading, make them resistant to wetting, and keep them supple and durable.

The three types of hides and skins most often used in leather manufacture are from cattle, sheep, and pigs. Tannery processes are slightly different depending on the hide origin. The content of this chapter will be focused on cattle hides. Thus, Fig. 4.1 shows the different operations in the tannery [3].

4.2.1 Curing

Animal hides are first "cured," a process which involves hide salting and/or drying once it has been stripped from the animal. Since this step needs to be performed almost immediately after the hide is removed from the animal, it often takes place inside the meat-packing industry or at a nearby factory. Hides can be cured in one of two ways:

- Wet salting is done by salting the hide and then piling many skins together so that they can form a moist bunch. They are then left to cure for one month, so that the salt can completely be absorbed into the skin.
- Brine curing is more common than wet salting, as it is considered a faster and easier method. During brine curing, hides are positioned carefully in vats and smothered with a mixture of salt and disinfectant. After 10–16 h, the skins are completely cured and ready to move on to the next stage.

Figure 4.2 shows the appearance of cured hides.

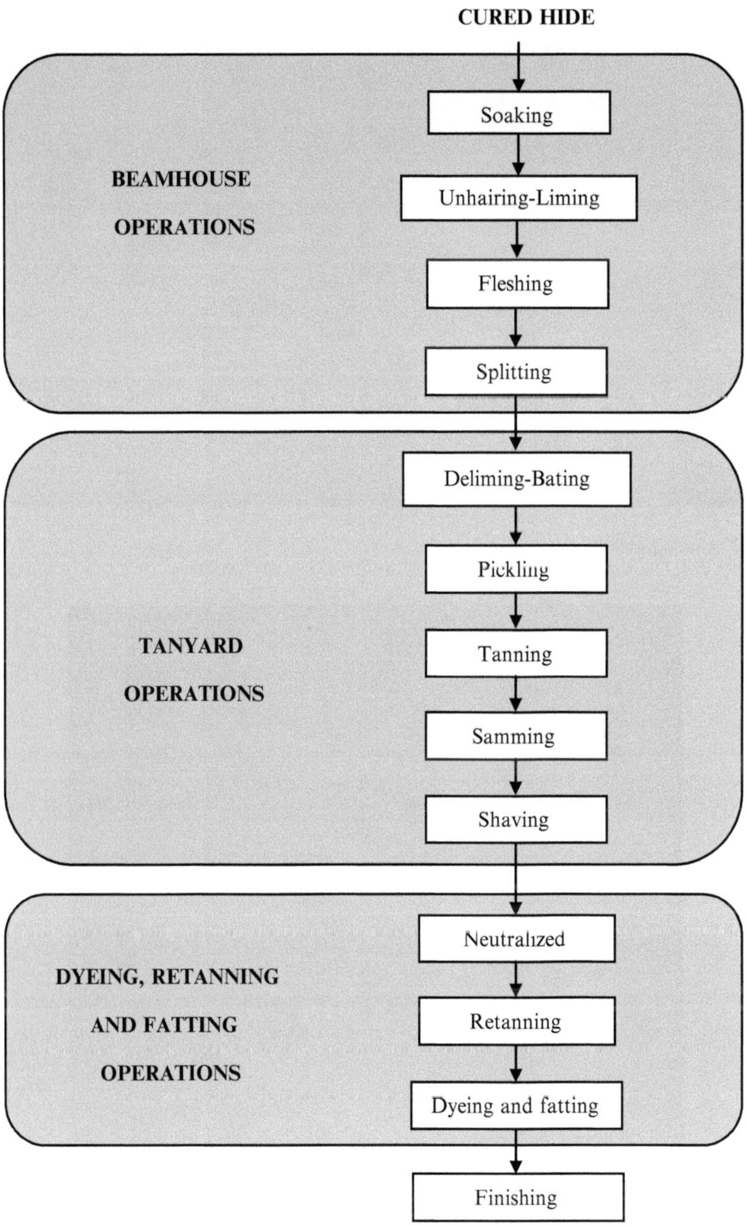

Fig. 4.1 Tanning process

Fig. 4.2 Cured hides

Fig. 4.3 Soaking operation

4.2.2 Beamhouse Operations

4.2.2.1 Soaking

Once the hides have been cured, they are then soaked in water solution containing surfactant, enzyme, and fungicide for hours or days (Fig. 4.3). Water helps removing dirt, debris, blood, and excess animal fats and restoring the moisture lost during curing.

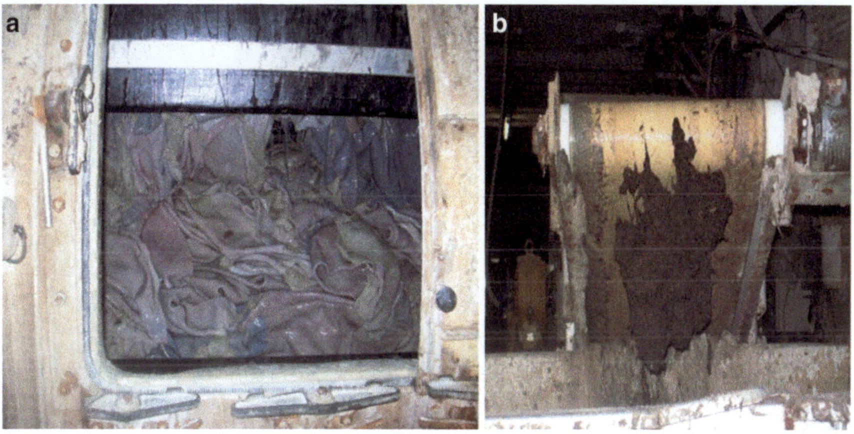

Fig. 4.4 Unhairing-liming (**a**) and hair recovery operations (**b**)

4.2.2.2 Unhairing-Liming

Following the soaking, the hides are unhaired to ensure that the grain is clean and that the hair follicles are free of hair toots. The chemical products used for it are mainly lime and sulfides. This process is carried out in drums as it is shown in Fig. 4.4a. Periodically, the bath is withdrawn from the drum so that the hair is not solubilized, what would produce a considerably increase in the COD of the process effluent. Thus, the bath is filtered for hair removal and recovery (Fig. 4.4b) and pumped into the drum again.

There are other options to carry out the unhairing that would reduce the wastewater pollution. It consists of replacing the sulfides by enzymatic products. However, the enzymatic process is more difficult to control.

4.2.2.3 Flesh Removal

After unhairing-liming, animal hides are moved through a machine which strips the flesh part from the hide. This process can be observed in Fig. 4.5.

4.2.2.4 Splitting

This operation can be performed after flesh removal or after the tanning process. The hide is divided into two layers, obtaining two hide types: grain hide and split hide. This operation can be observed in Fig. 4.6.

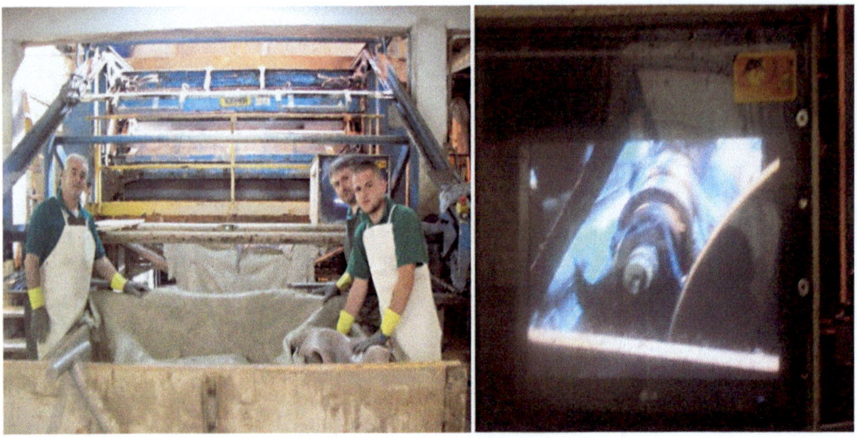

Fig. 4.5 Fleshing operation

Fig. 4.6 Splitting operation

4.2.3 Tanning Operations

4.2.3.1 Deliming-Bating

After the hair and debris have been removed from the skin, hides are mixed in a drum with a solution containing ammonium salts or carbon dioxide in order to remove the lime from the hide. After that, hides are treated with enzymes (bating operation) with the aim of imparting softness, stretch, and flexibility to the leather.

4.2.3.2 Pickling

The objective of the pickling is preparing the hide for the tanning by reducing its pH. This is carried out with sulfuric acid. A brine solution is added to avoid that the hide spoils.

4.2.3.3 Tanning

The tanning is the operation that converts the hide into leather. It means that a degradable stuff is transformed into a stable matter. Hides are often treated several times during the process of tanning. The type of tanning procedure used depends largely on the hide itself and the aimed resulting product.

Vegetable Tanning

Vegetable tanning consists of stringing hides on large frames, located inside large vats, and exposing them to tannin, a natural product found in the bark, wood, leaves, and fruits of chestnut, oak, and hemlock trees. Hides are transferred to many different bins during this step, each containing a stronger solution of tannin. Vegetable tanning prevents the hide from decay and shrinkage.

Hides which are tanned with a vegetable tanning agent solution produce stiff leathers, such as those used in luggage, furniture, leashes, belts, hats, and harnesses.

Mineral Tanning

Hides which are tanned with minerals are pickled first in an acid and salt mixture until pH 3. The chrome tanning materials (chromium-sulfate solution) are introduced, and the pH is raised. This process is much faster than vegetable tanning, and it usually takes 1 day. Next, the chrome-tanned leather is piled down, sammed, and shaved to the desired thickness.

Mineral or chrome tanning is performed on skins which will be used for softer, stretcher leathers, such as those found in purses, bags, briefcases, shoes, gloves, boots, jackets, pants, and sandals.

4.2.4 Dyeing, Retanning, and Fatting Operations

4.2.4.1 Neutralization and Retanning

Depending on the desired product, the hides undergo a retanning process, which also involves adding moisture back into the leather and adjusting the pH (neutralization). The retanning process offers leather the desired final feature.

Fig. 4.7 Finishing operation

4.2.4.2 Dyeing

Typical dyestuffs are aniline-based compounds that combine with the leather in order to form an insoluble compound. The type of dyeing will depend on the requirements of the final product.

4.2.4.3 Fatting

Fatliquoring is the process of introducing oil into the skin before the leather is dried to replace the natural oils lost in beamhouse and tanning processes. Fatliquoring is usually performed in a drum using an oil emulsion at temperatures of about 60–66 °C for 30–40 min.

After fatliquoring, the leather is wrung, set out, dried, and finished. The finishing process refers to all the steps that are carried out after drying.

4.2.5 Finishing Operations

The final step in the tanning process involves finishing the leather. This is done by covering the grain surface with a chemical compound and then brushing it. Light leathers are buffed and sandpapered to cover imperfections. Leathers which are buffed for long periods of time become suede.

Waxes, pigments, dyes, glazes, oils, and other solutions are also added to make the leather more appealing to the buyer. In Fig. 4.7 an example of a finishing operation can be observed.

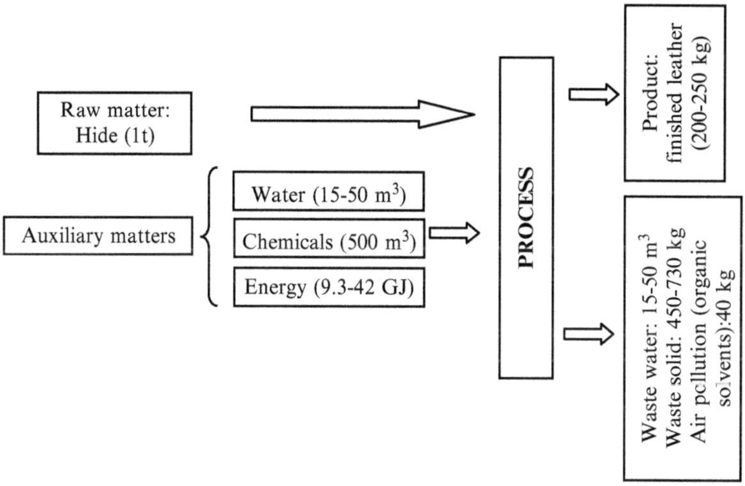

Fig. 4.8 Inputs and outputs in a tannery

Table 4.1 Global wastewater characteristics for a tannery processing 1 t of raw hide

Wastewater volume	15–50 m^3
Chemical oxygen demand (COD)	230–250 kg
Biological oxygen demand (BOD)	100 kg
Suspended solids (SS)	100 kg
Chrome	5–6 kg
Sulfide	10 kg

4.3 Main Environmental Concerns

Once the processes occurring in the tannery have been explained, the main environmental concerns, especially those related with wastewater, are described in the following sections.

Figure 4.8 illustrates of an input/output overview for a conventional (chrome tanning) process for bovine-salted hides per ton of raw hide treated *[Integrated Pollution Prevention and Control (IPPC), Reference Document on Best Available Techniques for the Tanning of Hides and Skins. February 2003]* [4].

It can be observed that the inputs (raw and auxiliary matters) and outputs (final product and waste) have been considered on the basis of 1 t of raw matter. The hide together with chemicals, water, and energy produces the final leather as product and residual air and water streams in addition to a considerable amount of solid wastes.

Focusing on wastewater, their characteristics can be observed in Table 4.1.

4.3.1 Residual Streams from the Processes

Figure 4.9 shows the main processes of the tannery illustrating the main chemicals used in the most polluting operations. Besides, the main characteristics of the wastewaters from these operations are also shown.

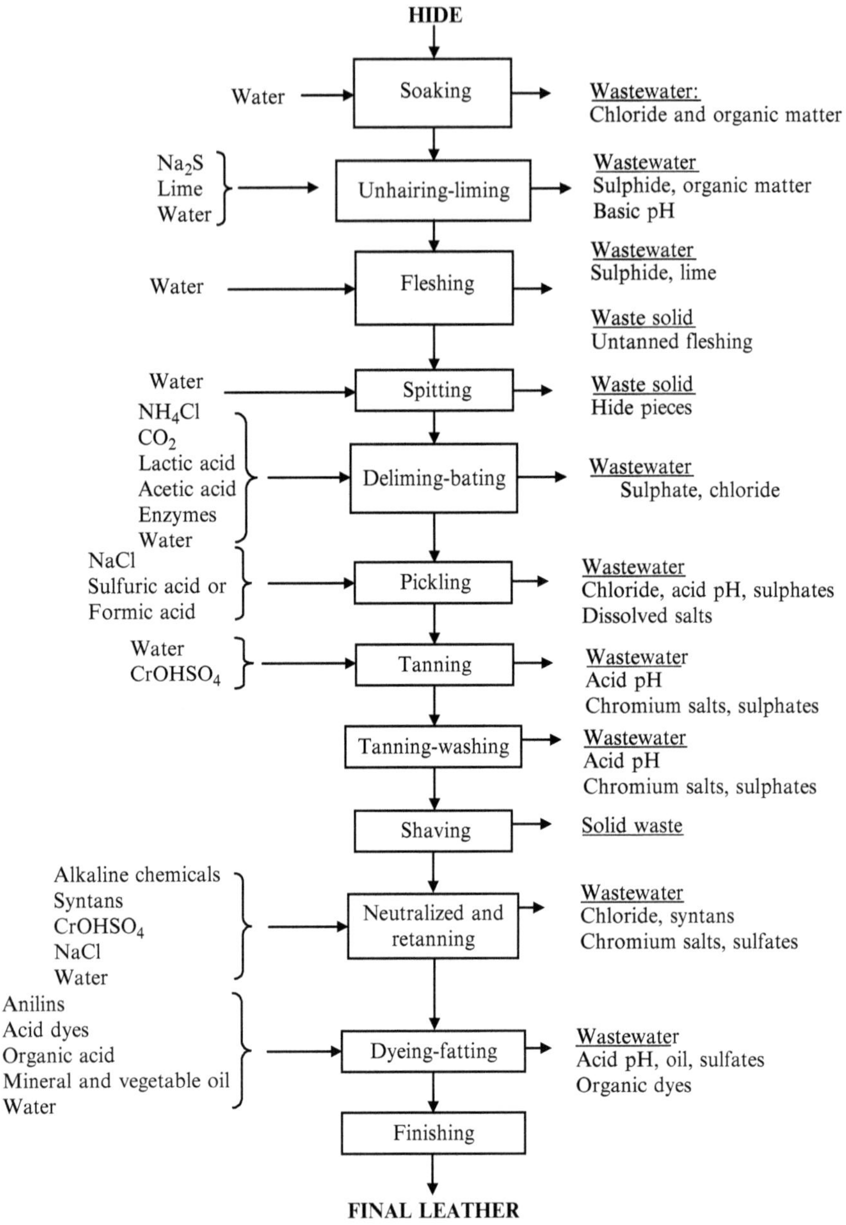

Fig. 4.9 Main processes of a conventional tannery with the main chemicals and wastewater characteristics

As it can be observed in the figure, most of the operations in a tannery require water consumption. This type of industry has always been characterized by high water consumption, but many years of application of the best available techniques together with the environmental legislation turned out into a freshwater reduction.

However, this entails the increase in the pollutant concentration (sulfate, chloride, or organic matter).

The main contributions to wastewater diminution come from the wet processing in the beamhouse, the tanning operations, and the post-tanning operations.

4.3.1.1 Wastewaters from Beamhouse Operation

Wastewaters from beamhouse come from the soaking and the unhairing-liming. Both are alkaline wastewater. *Soaking wastewater* is characterized by containing blood, globulins, surfactants, biocide, and salts, mainly sodium chloride. *Unhairing-liming* wastewater volume is lower than soaking wastewater volume. Its main characteristics are high concentrations of sulfide, organic matter, surfactants, and fats.

4.3.1.2 Wastewater from Tanning Operations

Wastewaters from tanning operations are divided in three groups: deliming-bating, pickling, and tanning. All of these are acidic wastewater. *Deliming-bating wastewater* is characterized by high concentrations of ammonium salts, acids, and enzymes. The main feature of *pickling wastewater* is its high conductivity due to the high salts concentration. Finally, t*anning wastewater* is characterized by high concentrations of trivalent chromium, sulfates, fats, and fungicides.

4.3.1.3 Post-tanning Operation

Wastewaters from post-tanning operations are divided into three groups: neutralizing and retanning, dyeing, and fatting. All of these are acidic wastewater.

Neutralizing and retanning wastewater is characterized by high concentration of salts (bicarbonates, formats, and bisulfites), trivalent chromium, and syntans. Concerning *dyeing wastewater*, it has to be commented that it is characterized by the presence of dyes and organic matter. At last, f*atting wastewater* contains oils and greases.

4.4 Reuse of Process Wastewater

In order to reduce the water and chemicals consumption and the wastewater pollution, the recycling and reuse of residual baths is of paramount importance. In this section, the reuse of different residual floats is described. In particular, there are three residuals baths which can be reused:

- Unhairing-liming wastewater as a part of soaking solution
- Pickling wastewater as a new pickling solution
- Tanning wastewater as a new tanning solution

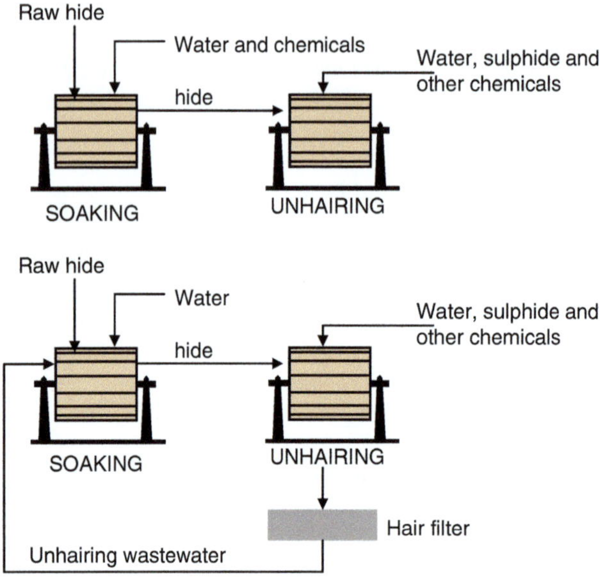

Fig. 4.10 Comparison of the traditional soaking processes (*top part* of the figure) and the proposed scheme (*bottom part* of the figure)

4.4.1 Unhairing-Liming Recycling

As explained in Sect. 4.2.2, unhairing-liming process is performed in a drum by mixing the hides with an alkaline solution that contains mainly lime (used to swallow the hide), sulfides for the hair removal, and surfactant. Immediately after the unhairing, the residual bath is taken out of the drum and driven to a grid to separate hair before its solubilization. At the end of the operation, a residual bath with a pH value of about 12 and with a high content of organic substances (proteins), lime, and sulfide is left. This wastewater is the effluent with the highest BOD and COD contribution to the total wastewater (70 and 55%, respectively). Thus, its reuse is very important.

Besides, the use of unhairing wastewater as a part of the soaking solution entails savings of water and chemicals, since the chemicals added in the soaking are also present in the unhairing wastewater. Sulfides recycled to the soaking drums will begin to react with the hide, what implies a saving in sulfide in the subsequent unhairing. However, this sulfide concentration will not be high enough to begin the hair elimination from the hide. Figure 4.10 compares the traditional process with the proposed scheme.

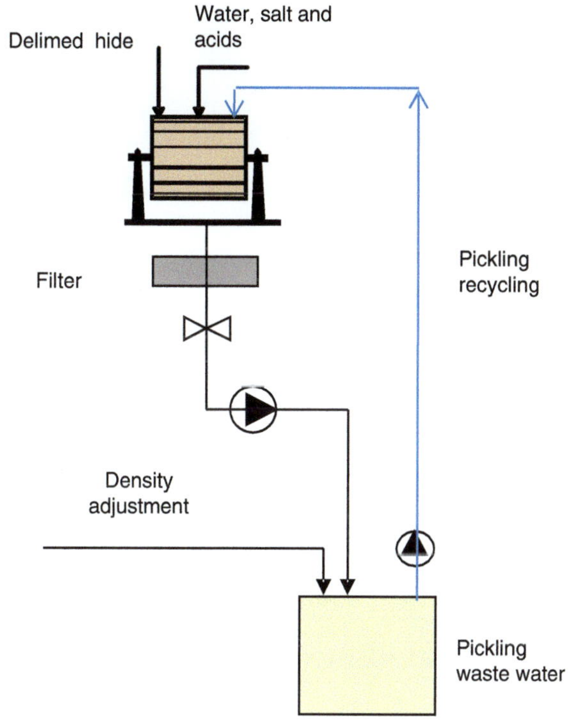

Fig. 4.11 Scheme of the pickling reuse

4.4.2 Pickling Wastewater Reuse

Pickling process is performed in a drum by mixing the hides after deliming and batting with salt (sodium chloride) and acids (formic an sulfuric) in order to obtain a pH near to 3. After pickling, the residual bath is taken out of the drum and driven to a grid to separate solids (little pieces of hide). This bath can be stored in a tank where density is adjusted. The initial density in the pickling process is 9 ºBé, while at the end of the process, the density is 5.5 ºBé. For this reason, it is necessary to adjust density before recycling pickling bath.

However, a great majority of experts did not fully support this view because it can have a negative effect on the quality of leather, especially aniline leather. But there are some tanneries that have been making this recycling many years without any problem. Figure. 4.11 shows schematically the pickling wastewater reuse.

4.4.3 Tanning Wastewater Reuse

Tanning process is performed in a drum by mixing the pickled hide with chromium salts in order to obtain a tanning hide (leather). Tanning process starts at pH 3 and

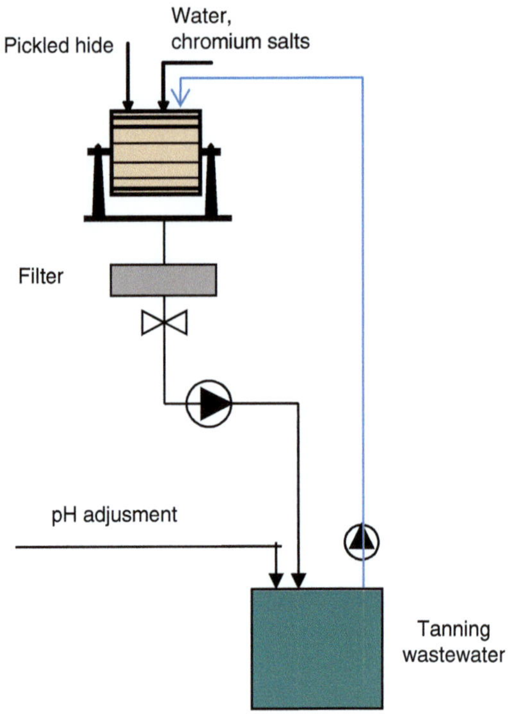

Fig. 4.12 Comparison of the traditional tanning processes (on the *left part* of the figure) and the proposed scheme (on the *right part* of the figure)

finish at pH around 4. This change of pH is necessary to fix chromium salts inside the hide. After tanning, the residual bath is taken out of the drum and driven to a grid to separate solids (little pieces of hide). This bath can be stored in a tank, where pH is adjusted before its recycling. Figure 4.12 compares the traditional tanning process with the proposed scheme.

4.4.4 Leather Quality After Residual Baths Reuse

Reuse of the tanning exhausted bath is nowadays very usual. However, the pickling wastewater reuse is not fully accepted as mentioned above, and the reuse of unhairing wastewater in the soaking is hardly carried out due to the risk of beginning the unhairing in the soaking drums. Thus, it is very important to be able to check the final leather quality when the process includes the reuse of these residual baths.

In this way, laboratory experiments with hide samples have been reported in the literature. Both mechanical properties and subjective characteristics are evaluated, and their values are compared with the standards.

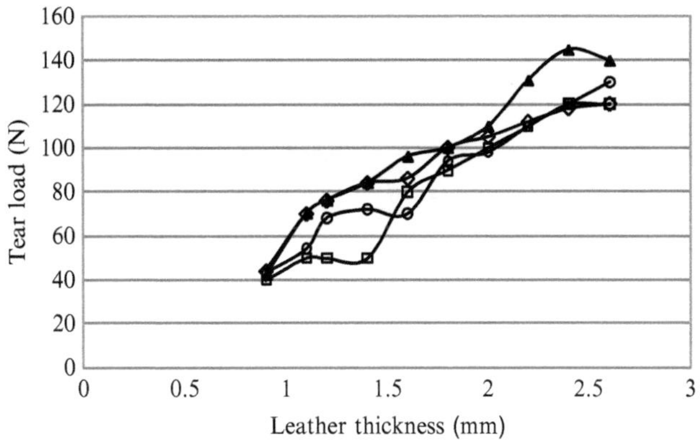

Fig. 4.13 Tear load test results (◇ unhairing recycling, ▲ tanning recycling, ○ pickling recycling, □ standard value)

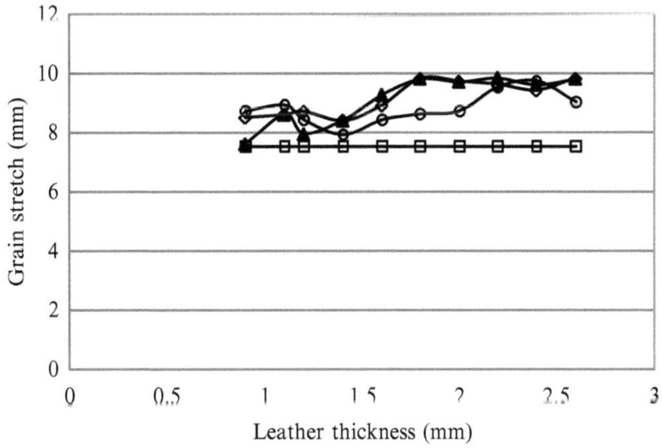

Fig. 4.14 Grain stretch test results (◇ unhairing recycling, ▲ tanning recycling, ○ pickling recycling, □ standard value)

Figures 4.13, 4.14, and 4.15 show the determined values for the mechanical properties (tear load, grain stretch, and tensile strength) of the leather obtained in three cases:

1. Soaking with a mixture of 50% of unhairing wastewater and 50% of freshwater
2. Pickling with 100% of pickling wastewater after adjusting the density at 9 ºBé
3. Tanning with 100% of tanning wastewater adjusting the pH at 3.0

The mechanical properties should be evaluated for hide pieces of different thickness as shown in figures. The procedures for the evaluation of the mechanical properties are described by ISO standards [5–7].

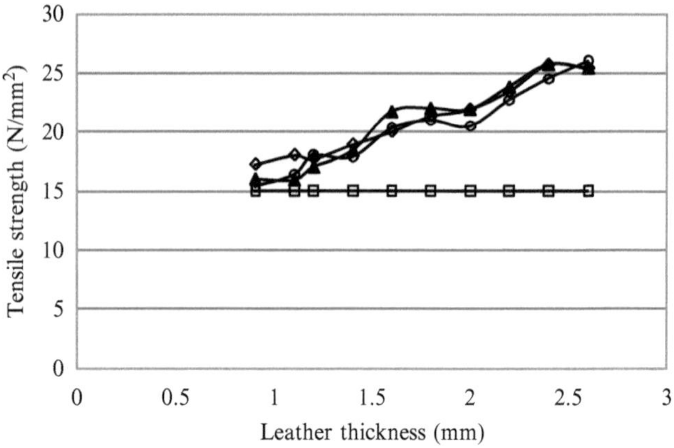

Fig. 4.15 Tensile strength test results (◇ unhairing recycling, ▲ tanning recycling, ○ pickling recycling, □ standard value)

Table 4.2 Comparison of pollution generated between conventional tannery processes and those including exhausted baths reuse

Parameter	Process	Soaking ($kg/t_{raw\ hide}$)	Unhairing-liming ($kg/t_{raw\ hide}$)	Deliming-batting ($kg/t_{raw\ hide}$)	Pickling ($kg/t_{raw\ hide}$)	Tanning ($kg/t_{raw\ hide}$)	Others ($kg/t_{raw\ hide}$)
COD	A	69.30	87.00	8.00	10.90	12.20	24.60
	B	106.80	0	8.00	0.10	0.10	24.60
SO_4^{-2}	A	2.60	7.30	1.50	16.50	13.40	23.00
	B	2.30	0	1.50	0.20	0.10	23.00
ST	A	76.2	101.70	13.70	264.80	207.30	195.20
	B	105.10	0	13.70	2.80	2.20	195.50
SO	A	53.40	56.20	7.60	99.50	93.00	84.00
	B	71.00	0	7.60	1.10	1.10	84.00
SI	A	22.8	45.5	6.1	165.3	114.3	111
	B	34.1	0	6.1	1.7	1.1	111
Cr III	A	0	0	0	0	10.0	0.001
	B	0	0	0	0	0.01	0.001
Water	A	5,772	3,108	3,108	2,220	2,220	4,440
	B	2,886	3,108	3,108	0	0	4,440

These figures show that all of the experiments yielded appropriate values of the mechanical properties. In this way, it can be stated that it is possible to reuse unhairing liquor in soaking process, pickling wastewater, and tanning liquor.

If the three liquors are reused, a reduction in the pollution generated in the wastewater would be achieved. Table 4.2 shows the comparison between the pollution generated without (process A) and with (process B) wastewater reuse for the main tannery processes.

It can be observed that it is possible to reduce around 34% of COD, 58% sulfates, 63% solids, and 99% chromium in wastewater by reusing the liquors. In the same way, the reduction in water consumption is estimated about 35%.

4.5 Application of Membranes to Process Wastewaters

In the two last decades of the twentieth century and the first decade of the current one, some authors have reported applications of membrane processes to effluents from some tannery operations. The application of membranes to the global wastewater after conventional treatments is out of the scope of this work.

Membrane processes are able to separate dissolved substances from the wastewater. They can be classified according the driving force responsible for the separation: concentration difference in both sides of the membranes, pressure difference, electric potential.... In tanneries, most of the applied membrane processes are those belonging to the pressure driven process (microfiltration (MF), ultrafiltration (UF), nanofiltration (NF), and reverse osmosis (RO)).

MF and UF membranes are microporous, and the main separation mechanism is sieving. Thus, solutes with a size higher than the membrane pore size will be separated. In this way, by means of UF, solutes with a molecular weight higher than 1,000 Da can be removed from the wastewater.

On the contrary, RO membranes are dense, and their separation mechanism is based on the solution-diffusion model. Since there are no pores on the active surface, this kind of membranes can separate monovalent ions.

Finally, NF is a technology classified between UF and RO. Its separation mechanism is based on both the solution-diffusion model and sieving. Then, multivalent ions can be separated by this technology.

4.5.1 Application of MF/UF

In the literature, the two most reported applications of the UF to tannery wastewaters are related with the unhairing effluent and with the pretreatment of the NF or RO, mainly applied to tanning wastewaters. The second application will be commented together with the NF/RO application.

Focusing on the first application, it has firstly commented that by means of UF membranes, proteins are separated from the inorganic chemicals used in the operation bath, i.e., lime and sulfides. In this way, the membrane permeate may be recycled to the unhairing drums with the subsequent savings both in water and in auxiliary matters.

Cortese and Drioli [8] carried out the UF of unhairing wastewater with organic membranes. These authors reported efficiency values for protein separation higher than 80%. Collivignarelli and Barducci [9] went deeper into the process and evaluated different alternatives for the management of the unhairing wastewater. They selected UF and reported data from a pilot plant working with a tubular membrane from polyamide at 45 °C.

The industrial application of the proposed UF depended on the solution of two big concerns: rapid membrane fouling, which would make the process economically unfeasible, and the management of the reject stream.

Table 4.3 Characteristics of the unhairing wastewaters

Parameter (units)	Values range
COD (mg/L)	18,000–22,000
TS (mg/L)	58,000–60,000
pH	12–12.2
Conductivity (mS/cm)	40–45
S^{2-} (mg/L)	6,500–7,000

Table 4.4 Characteristics of the permeate samples

Parameter (units)	Values range
COD (mg/L)	12,000–14,000
TS (mg/L)	50,000–52,000
S^{2-} (mg/L)	5,500–6,500
pH	12–12.2
Conductivity (mS/cm)	36–40

Table 4.3 shows the main characteristics of the unhairing wastewaters. These values were measured in wastewaters coming from a process where the unhairing bath was withdrawn periodically from the drum in order to separate the hair before its solubilization. If this is not performed, COD values will be considerably higher (near 40,000 mg/L).

From these values, it can be understood that fouling is a problem to be solved before the UF implementation. In this way, a possible solution can be based on an exhaustive wastewater pretreatment. However, this is very difficult to achieve, since pH cannot be lowered due to sulfide formation and most of the pollutants are dissolved, and conventional filtrations will hardly be efficient. Thus, the most successful results were obtained with unhairing effluents with lower pollution loads. In this way, the operation of an industrial plant was firstly reported for the residual bath from the scale removal stage for crocodile skin Aloy et al. [10, 11].

If effective pretreatments are difficult to implement, efficient membrane cleaning procedures have to be applied. Mendoza-Roca et al. [12] compared three cleaning types (with surfactants, enzymes, and sodium hypochlorite). In this chapter, the permeate reuse was also studied. For that, hide samples were processed using UF permeates as unhairing solutions.

Permeate characteristics can be observed in Table 4.4.

It has to be commented that COD of the permeate is produced in a high extent by the oxidation of inorganic compounds such as sulfides. This is the reason why it is not contradictory with the achievement of a removal of more than 80% of the proteins.

Finally, in the following paragraph, management of the membrane reject stream is going to be discussed. At the beginning of the studies, its application after sulfide removal to animal feeding was thought, what was further forbidden due to the discovered relation between animal proteins and the Creutz-feldt-Jakob disease. Nowadays, different alternatives are being considered including the external management after a further stage of concentration by evaporation.

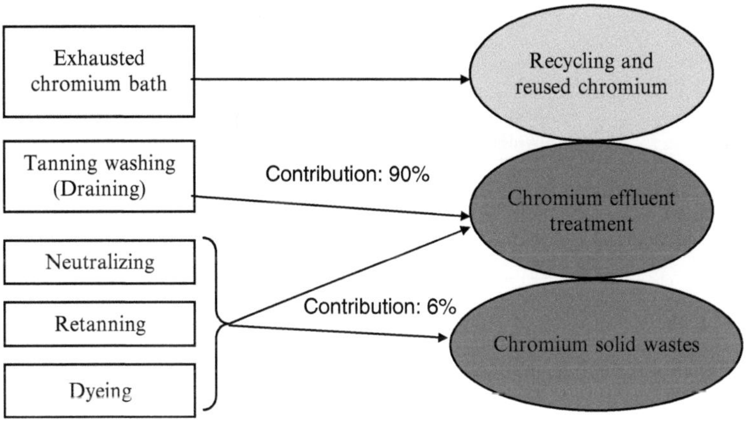

Fig. 4.16 Sources of chromium in the tannery

4.5.2 Application of NF

Once the reuse of the tanning residual bath has been evaluated (Sect. 4.4.3), it has to be stated that chromium concentration in wastewaters has to be further reduced. If this is not carried out, chromium will be separated together with the organic matter in the physic-chemical treatment of the global wastewater. This entails that the produced sludge could not be applied to agriculture.

Washing after tanning (tanning washing) still contains high chromium concentrations (more than 100 mg/L). The scheme of the contribution of the different operations to the final chromium concentration is shown in Fig. 4.16.

NF will divide the tanning washing effluent into two streams: a reject stream, where chromium sulfate is concentrated and can be recycled to the tanning drums, and a permeate stream that could be reused as washing solution in the tannery.

Taleb-Ahmed y Taha [13] studied the parameters influencing the chromium retention by NF membranes. These authors concluded that the chromium rejection mainly depended on the transmembrane pressure, the wastewater pH, and the chromium concentration in the membrane feed stream. The monovalent salts concentration did not influence significantly the chromium separation. Similar conclusions were obtained by Ortega and Lebrun [14].

Cassano et al. [15, 16] studied the application of an integrated membrane process to treat the tanning washing wastewaters. The process consisted of a combination of UF and NF. The UF stage was carried out with a 50,000 Da membrane at 1.7 bar and at a volume concentration factor of 6.44. At these operating conditions, 84 and 2.1% of suspended solids and chromium removal efficiencies were achieved, respectively. By means of the NF stage, a practically total chromium removal entailed a final concentrate with 10 g/L of chromium concentration. The operating transmembrane pressure was 16 bar.

NF has been applied for sulfate reuse in pickling wastewaters [17]. Although the direct reuse of the residual bath has already been discussed (Sect. 4.4.2), it has to be

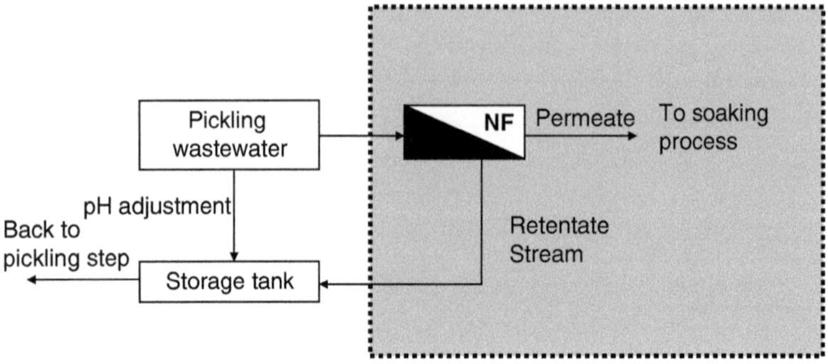

Fig. 4.17 Scheme of the application of NF to the excess pickling wastewater

mentioned that an excess of wastewater is produced in comparison with the water need for the process. This is due to the water loss from the hides in the pickling. This excess wastewater can be treated by NF membranes in order to obtain a concentrate rich in sulfates and a permeate that can be reused (Fig. 4.17).

4.5.3 Application of RO

For tannery wastewaters, RO constitutes an alternative to NF in the treatment of the tanning washing wastewaters. The advantage of the RO is the higher quality of permeate, and its main drawback is the higher-operating transmembrane pressure what entails a higher energy consumption. A combination between UF and RO for the tanning washing wastewaters was proposed by Scholz and Lucas [18].

4.6 Conclusion

Tanneries are industries with an enormous water and chemicals consumption. However, it is possible to recycle and reuse segregated residual streams from the most polluting processes. In this way, the recycle of unhairing wastewaters to the soaking and the direct reuse of the pickling and tanning reduce considerably the final pollutants discharge. Besides, membrane technologies can be applied to the unhairing and tanning washing wastewaters in order to minimize the sulfide and chromium concentrations in the final tannery effluent.

References

1. Salmerón J (1995) Generación y tratamiento de residuos en la industria del curtido de Valencia. Confederación Empresarial Valenciana, Valencia, Spain
2. Mulder M (1996) Basic principles of membrane technology. Kluwer Academic, Dordrecht
3. Adzet Adzet JM (1985) Química de tenería. AQEIC, Barcelona

4. European Commission (2003) Integrated Pollution Prevention and Control (IPPC). Best available techniques reference document on the tanning of hide and skins. http://www.prtr-es.es/. Accessed 09 Nov 2009
5. ISO 3372 (2002) Determination of tensile strength and elongation
6. ISO 3377:2 (2002) Physical and mechanical test- determination of tear load
7. ISO 3379 (2005) Determination of distension and strength of grain – ball burst test
8. Cortese B, Drioli D (1982) Ultrafiltration processes for pollution control and chemical reuse in the tanning industry. Desalination 34:131–139
9. Collivignarelli C, Barducci G (1984) Waste recovery from the tannery industry. Waste Manag Res 2:265–278
10. Aloy M, Donikian S (1992) Ultrafiltration de bains de tannerie. Industrie du Cuir 4:68–74
11. Aloy M, Vulliermet B (1998) Technologies membranaires pour le traitement des bains résiduaires de tannerie. Industrie du Cuir 2:44–50
12. Mendoza-Roca JA, Galiana-Aleixandre MV, Lora-García J, Bes-Piá A (2010) Purification of tannery effluents by ultrafiltration in view of permeate reuse. Sep Purif Technol 70:296–301
13. Taleb-Ahmed M, Taha S (2002) The influence of physic.chemistry on the retention of chromium ions during nanofiltration. Desalination 145:103–108
14. Ortega LM, Lebrun R (2005) Application of nanofiltration in the recovery of chromium (III) from tannery effluents. Sep Purif Technol 44:45–52
15. Cassano A, Della Pietra L, Drioli E (2007) Integrated membrane process for the recovery of chromium salts from tannery effluents. Ind Eng Chem Res 46(21):6825–6830
16. Cassano A, Drioli E, Molinari R (1997) Recovery and reuse of chemicals in unhairing, degreasing and chromium tanning processes by membrane. Desalination 133:251–261
17. Bes-Piá A, Cuartas-Uribe B, Mendoza-Roca JA, Galiana-Aleixandre MV, Iborra-Clar MI, Alcaina-Miranda MI (2008) Pickling wastewater reclamation by means of nanofiltration. Desalination 221:225–233
18. Scholz W, Lucas M (2003) Techno-economic evaluation of membrane filtration for the recovery and re-use of tanning chemicals. Water Res 37:1859–1867

Chapter 5
Wastewater Reuse Focused on Industrial Applications

Mariliz Gutterres and Patrice Monteiro de Aquim

5.1 Introduction

Water is essential to life. However, its indiscriminate use may lead to the shortage of its precious asset. The concept of water being a never-ending resource with a limitless renewable capacity belongs to the past. One way to optimize water use is by adopting reuse techniques, including recycling, where the same water is used once or twice for different purposes or for the same purpose.

Wastewater reuse is a rational practice that contributes to environmental protection. This technique is being implemented to reduce water consumption and the high costs of water treatment. Therefore, more than an environmental measure, water reuse has an economic impact. Wastewater reuse must be a safe, environmentally conscious, responsible, and accountable process, which meets the recycling requirements and the current legislation, as well as the directives of the World Health Organization, since irresponsible wastewater reuse may cause various problems.

Water scarcity is faced all around the world. It may be defined in various ways [1]. In short, scarcity occurs when there is less than 200 m^3 of freshwater available per person per year. Scarcity problems were detected in North Africa, Southern Africa, the Horn of Africa, and the Sahel. There are serious shortages in southwestern Asia, northern India, Pakistan, and northern China. Some Visegrad countries (Bulgaria, Romania, Poland, and Hungary), a large part of the southwestern USA, and Western Australia as well as Mexico and the Middle East are also facing scarcity problems. Among Visegrad countries over 70% of the total freshwater inflows from only one river, the Danube, which flows through several

M. Gutterres (✉) • P.M. de Aquim
Chemical Engineering Department, Laboratory for Leather and Environmental Studies, Federal University of Rio Grande do Sul, Rua Eng. Luiz Englert, s/n, Porto Alegre, RS 90040 040, Brazil
e-mail: mariliz@enq.ufrgs.br

S.K. Sharma and R. Sanghi (eds.), *Wastewater Reuse and Management*, DOI 10.1007/978-94-007-4942-9_5, countries. The concern over the future of water supply will be focused first in those areas where countries are using more water than they receive in rainfall. Wastewater reuse has become an essential component of water demand management (WDM) in many countries like Jordan, Kuwait, Israel, Spain, Australia, Namibia, Germany, United Kingdom, and the United States of America [2].

Wastewater reuse is the process of reusing treated or nontreated water for the same or another purpose. This reuse can be direct or indirect, planned or unplanned, according to the following classification:

- *Direct reuse* occurs when the water used for a given purpose can be reused without previous treatment to be applied to a less noble purpose. Some examples include wastewater use in industries or in irrigation.
- *Indirect wastewater reuse* occurs when wastewater used for domestic or industrial purposes is treated to be available for another use.
- *Water recycling* involves reusing *water* for the *same* application for which it was originally used. In this case, the water is treated.
  The types of wastewater reuse are classified in four main categories: urban, industrial, agricultural, and groundwater recharge, as it is described.
- *Urban reuse:* As the water supply requirements of large metropolis are growing, the option of reuse of wastewater for domestic purposes is increasingly being considered. There is potential for reuse at all system scales, from household level to the large irrigation schemes. The choice is conventionally technical and economic one, and this will be introduced to effectively measure once the community as a whole is involved in the working of reuse systems. Urban reuse systems provide reclaimed wastewater for various *non-potable purposes,* including irrigation of public parks, recreation centers, school yards and playgrounds, and landscaped areas surrounding public buildings and facilities; *irrigation of landscaped areas surrounding residences,* general wash down, and other maintenance activities; *irrigation of landscaped areas surrounding commercial, office, and industrial developments*; *commercial uses* such as vehicle washing facilities and laundry facilities and *ornamental landscape uses and decorative water features*, such as fountains, reflecting pools, and waterfalls; *dust control and concrete production* for construction projects; *fire protection* through reclaimed water fire hydrants; and *toilet and urinal flushing* in commercial and industrial buildings.
- *Industrial reuse:* The reuse of wastewater in industrial processes, activities, and operations has been widely explored. Besides preventing the waste of potable water, the reuse of wastewater for industrial purposes is advantageous for the companies for it avoids its waste. The industrial sector is responsible for a great part of the global freshwater consumption [3], reaching up to 20% of the available resources, and, at the same time, it is one of the main causes of their continual degradation. This occurs due to the rapid growth in the world population, which leads to the ever-increasing need to increase the production of consumer goods. To ensure safe wastewater reuse, minimum water requirements should be used for a given industrial process or operation. Industrial reuse has

increased substantially since the early 1990s; for many of the same reasons, urban reuse has gained popularity, including water shortages and increased populations, particularly in drought areas, and legislation regarding water conservation and environmental compliance. Industrial process/boiler-feed requirements and utility power plants are ideal facilities for reuse due to their large water requirements for cooling, ash sluicing, radioactive waste dilution, and flue gas scrubber requirements. Petroleum refineries, chemical plants, and metal working facilities [4] are among other industrial facilities benefiting from reclaimed water not only for cooling but for process needs as well.

- *Agricultural reuse:* It is the application of reuse wastewater in agricultural production and the cultivation of planted forests. Despite the great consumption of water in the industry, it is more significant in agriculture, which has the greatest potential for reuse. The agricultural use of wastewater corresponds up to 80% of the consumptive use, that is, during water use, a certain amount of water is collected from the sources, and, following the use, a smaller amount and/or a certain amount of lower quality water is returned, that is, part of the collected water is consumed during its use, in some countries, which should be taken into consideration by decision makers when wastewater reuse priorities are discussed [5]. Wastewater reuse may have various benefits because some wastewaters have concentrations of substances that can be used in agricultural production. However, some types of water may contain harmful contaminants. Thus, analysis should be performed to assess the viability of water reuse conditions in order to avoid public health problems. Properly treated effluents may have the following applications: in *cultures of noncommercially processed food,* for superficial irrigation of any food culture, including food eaten in its raw state; in *cultures of commercially processed food,* for the superficial irrigation of orchards and vineyards; and in *non-nourishing cultures,* of foraging, fibers, and grains and for *watering thirsty animals.*
- *Groundwater recharge:* Artificial recharge of aquifers with treated effluents can have many applications, including the increased availability and storage of water, salinity control in coastal aquifers, and land subsidence control. This practice may be relevant in some cities supplied by underground water where the natural recharge of aquifers has been reduced due to the increase in the number of waterproof areas. Managed aquifer recharge (MAR) encompasses several physical processes used to replenish and augment groundwater resources and is often an integral part of planned potable reuse [6]. MAR can include direct injection in the subsurface or infiltration via surface-spreading basins. The process of surface spreading has the added benefit of additional constituent removal and transformation in the basin and during percolation (referred to as soil aquifer treatment or SAT). Treatment in the basin is provided passively by volatilization and photodecomposition and during SAT from physical filtration, adsorption to soil particles, microbial biotransformation, and dilution with native groundwater [7]. Surface spreading using recharge basins is one of the oldest and most common methods for groundwater recharge [8].

5.1.1 Development of a Program for the Conservation and Reuse of Water

The implementation of reuse requires the establishment of quality guidelines, criteria and standards, and procedures that regulate the practice of wastewater reuse. It is important to establish a well-defined program that can be executed by a group of people committed to the success of the program. In the manual of conservation and reuse of water for the industry, the development stages of a program for the conservation and reuse of water are presented. The implementation of such a program requires the full understanding of the use of water (quantitative and qualitative) in all types of construction, external areas, and processes in order to identify the largest consumers and the best actions to be taken in the technology field, as well as the mechanisms of control to be incorporated to the system of water management adopted [9]. The constraints (limits) to the implementation of a program for wastewater conservation and reuse are obtained by the achievement of goals and by planning actions within the scope of a plan of continuous improvement [10]. Some suggestions regarding programs of implementation of reuse of water are shown in Table 5.1:

5.2 Wastewater Characterization for Reuse

The pollution potential of wastewater is directly proportional to the concentration of its pollutant components. The characterization of wastewaters makes it possible to establish the wastewater treatment system necessary to ensure wastewater reuse feasibility or to avoid degradation of receiving waters by discharge. Water reuse is considered an important solution to water supply issues in many countries. Accordingly, the main purpose of wastewater treatment is to meet the quality requirements for public or private users and to protect the consumer from pathogens and impurities in the water potentially deleterious to humans, animals, plants, and the environment as a whole.

Wastewater reuse applications require different water quality specifications, which require different purification grades, varying from simple processes to more advanced ones with selection of combined treatments technologies. The limiting factor for wastewater reuse [12] is many times the quality of the available wastewater associated to the treatment processes (technology) and potential hazards for secondary users. The resources employed for wastewater reuse are often available nearby (in urban use or industrial sites). The hydraulic characteristics should be observed and comply with the demands and distribution requirements. Water obtained from recycling or reuse systems [13] should fulfill the criteria of hygienic safety, aesthetics, and environmental tolerance as well as technical and economic feasibility. Specific parameters and analytical control methods are established to assess water quality and treatment efficiency. The most critical parameters are controlled according to risk

Table 5.1 Programs aimed to the implementation of reuse of water

Reference	Stages and relevance
FIESP/CIESP [9]	Preliminary technical assessment: document analysis, field survey
	Assessment of water demand: analysis of physical loss, analysis of waste, identification of the different levels of quality of water. Assessment of water supply: concessionaires, direct capture, pluvial waters, effluent reuse, underground waters
	Study of technical and economic viability: creation of the solution matrix, technical and economic analysis
	Technical details: technical specifications, technical details
	Management system: water consumption monitoring plan, training program for managers and users, maintenance routines, specific procedures
Paulino [10]	Establishment of goals and priorities
	Selection of water program manager or managers continuously trained and updated to operate and disseminate the program
	Planned allocation of initial investments expected to be reduced as savings are generated, ensuring the resources necessary for further investments
	Support of the top executive management during the elaboration of water management plans
	Optimization of the use of water, ensuring a better performance of the involved consumption activities
	Research, development, and innovation in industrial processes or other activities with observation of the quality standards and attempting to reduce costs
	Development and implementation of a management system that shall ensure the maintenance of good consumption indexes and the perfect performance and monitoring of hydraulic systems, equipments, and processes throughout time, contributing to the reduction and maintenance of costs during the useful life of the equipment
	Multiplication of the program for the conservation and reuse of water for all the users of the system
	Disclosure of the results obtained in order to encourage and enhance the commitment of the involved users
	An exclusion or premature assessment of each one of the above mentioned stages may compromise the efficiency of the initiatives taken
	In order to maintain the level of savings obtained, the management systems must include operational, institutional, educational, and legal actions
Hespanhol [5]	Define the criteria for the treatment of effluents for water reuse and for the proposal of technologies suitable to the regional or local climatic, technical, and cultural characteristics
	Establish criteria for the economic and financial assessment of reuse programs and projects
	Establish norms and programs aimed to provide information, environmental education, and public participation in water reuse programs and projects
	Establish a system for monitoring, assessing, and disseminating national, regional, and local programs for the conservation and reuse of water
	Establish a policy of reuse of water, defining objectives and goals, types of reuse, priority areas, and local and/or regional conditions for the implementation of the practice
	Propose institutional structures for the promotion and management of programs and projects of reuse at the national, regional, and local levels

(continued)

Table 5.1 (continued)

Reference	Stages and relevance
	Establish a legislation including directives, standards, and codes of practice
	Establish regulations, including assignments, responsibilities, incentives, and penalties
Bixio et al. [11]	May help water professionals interpret existing environmental regulations and formulate acceptable alternatives to address water resource management issues
	It may reassure local authorities and operators that they are following the most suitable wastewater recycling and reuse guidelines
	The purposed Aquarec manual on management practices aims at disseminating the knowledge acquired to enable the efficient implementation of water reuse across the European Union and Australia. The manual was designed to spread the knowledge on practices, or combination of practices, that have proved to be effective and easy to implement and operate
	The manual applies to a wide range of non potable water reuse initiatives and builds upon five major milestones
	Inventory of case practices across seven world regions
	Conventional literature review on management practices
	Inquiry to managers of medium-to large-scale facilities to determine where present-day practice lies in relation to what is seen as "best practice"
	International workshop of water reuse professionals to discuss specific data gaps
	Extensive peer review of the draft documents to include the insight and know-how of a larger number of professionals from water reuse utilities, manufacturers, consultants, and research institutions

assessment approaches of water destination. In order to minimize the analytical costs [14], suitable indicator substances or indicator effects should be identified, and a suitable monitoring strategy combined with a Hazard Analysis and Critical Control Points (HACCP) system should be developed.

5.2.1 Constituents in Wastewater

Physical properties and chemical and biological characteristics of waters and wastewaters are interrelated, for example, physical and chemical characteristics of water bodies affect the abundance, species composition, stability, productivity, and biological conditions of populations of aquatic organisms. Regarding physical properties, for instance, temperature may affect the gases and, in some cases, the dissolved inorganic solids. Another physical property – turbidity – is associated with suspended solid concentrations.

The important constituents that deserve consideration in wastewater treatment are included in the following categories [15]: suspended solids, biodegradable organic pollutants, pathogens, nutrients, priority pollutants, refractory organic pollutants, heavy metals, and dissolved inorganic solids. Secondary treatment standards for wastewater are concerned with the removal of biodegradable organic agents, total suspended solids, and pathogens. Most strict standards that have been recently developed concern the removal of nutrients, heavy metals, and priority pollutants. When wastewater is to be reused, standards generally include additional requirements for the removal of refractory organic pollutants, heavy metals, and sometimes dissolved inorganic solids.

5.2.1.1 Biological Constituents

The biological methods used for assessing water constituents include collection, counting, and identification of aquatic organisms; biomass measurements, measurements of metabolic activity rates, measurements of toxicity, bioconcentration, and bioaccumulations of pollutants; and processing and interpretation of biological data. One of the most important attributes of high-quality water is to be free of disease-causing organisms. Testing to detect and quantify individual pathogens are difficult to perform and time-consuming. Measurements of most probable number (MPN) of coliform organisms are used to assess the presence of pathogenic bacteria and the effectiveness of the disinfection process. Specific microorganisms (bacteria, protozoa, helminths, and viruses) are assessed in connection with the operations of the treatment plant and the analysis of wastewater for reuse. Microorganisms are also responsible for biological wastewater treatment. Algae play an important role in the eutrophication of lakes, being used in wastewater treatment in stabilization ponds. Material discharges can be biologically oxidized when utilized as substrate by bacterial flora, and the dissolved oxygen present in natural waters provides a source of oxygen for this process. The methods are used to evaluate the toxicity of treated wastewaters to aquatic species [15, 16].

5.2.1.2 Chemical Constituents

The chemical constituents in water are classified as inorganic and organic. Inorganic chemical constituents include nutrients, nonmetallic constituents, metals, and gases. Inorganic pollutants include inorganic acids, bases and salts, nutrients (nitrogen and phosphorus), sulfide, and inorganic species. The analyses to measure aggregate organic matter comprise a number of organic constituents that cannot be separately distinguished. Methods commonly used in laboratories to analyze high levels of organic matter (greater than 1 mg/L) include biochemical oxygen demand (BOD), chemical oxygen demand (COD), and total organic carbon (TOC). Individual organic compounds are determined to assess the presence of priority pollutants identified by the U.S. Environmental Protection Agency (USEPA), as well as a number of new emerging compounds are of concern. Both inorganic and organic pollutants are classified as priority pollutants according to their known or suspected carcinogenicity, mutagenicity, teratogenicity, or high acute toxicity. Toxic substances interfere directly, or through synergistic action, with the native aquatic population of receiving waters.

5.2.1.3 Physical Properties

The most important physical characteristic of water is its total solid content, which is detected due to the contributions of floating matter, settleable matter, colloidal matter, and matter in solution in water or wastewater. Several differentiations of solid indicators (volatile solids, fixed, suspended, and settleable) are used to assess the potential of reuse of wastewater and to determine the most suitable type of operations and processes for its treatment. Other important physical characteristics include particle size distribution, turbidity, color, taste, temperature, conductivity, specific gravity, and odor. Generally, a change of about 5 °C may significantly affect the balance and health of an aquatic environment. Examples of discharges [17] with high temperatures include condenser discharges and boiler blowdown, and colored effluents can be generated from dyeing operations, canneries, and from chemical and explosive manufacture. Substances such as phenolic compounds, solvents, petroleum products, and reduced sulfur compounds can impart taste and odor to water, although at minute concentrations. The types of wastewater discharges that cause turbidity are those from hydraulic mining operations, coal washeries, carbonization of coal, and petroleum refineries.

5.2.1.4 Microbiological Parameters and Pathogens Survival

Wastewater generally contains significantly high concentrations of pathogens. The wastewater treatment process tends to remove pathogens from the treated wastewater, thereby concentrating the pathogens in the sewage sludge. Like any other living organisms, pathogens thrive only under certain conditions. Each pathogen species has different tolerance to different conditions. The pathogen reduction requirements are therefore based on the need to reduce all pathogenic populations. Some of the factors that influence the survival of pathogens include pH, temperature, competition from other microorganisms, sunlight, contact with host organisms, proper nutrients, and moisture level [18].

Bacteria have a long history of use as indicators in microbiological testing. Due to the costs and the complexity of analyzing actual pathogens, wastewater professionals and regulators have relied for decades on traditional fecal indicators. However, there is scarce data available on how the removal of coliforms compares to the removal of protozoan pathogens, viruses, or other bacteria. Microorganism analyses can be performed in untreated wastewater and throughout the various phases of the treatment process. Differences in wastewater characteristics and treatment process will affect microorganism concentrations in wastewater treatment facilities. The increase of health hazards and concerns in relation to wastewater, the growing number of wastewater treatment facilities, and the need for obtaining additional water resources through wastewater reuse have demanded the urgent use of more accurate and sophisticated biological control tools for wastewater recycling [14, 19].

Escherichia coli is always present in high numbers in domestic wastewater, and its detection by culture methods is relatively inexpensive. Hence, *E. coli* has become the indicator of choice for fecal pollution of water. Its occurrence suggests the possible presence of pathogenic microorganisms. Yet it is not an unequivocal indicator of the presence of pathogens. While *E. coli* is a valuable warning indicator in potable water supplies, its value in domestic wastewater and biosolid applications is reduced because the source of water and sludge is always fecally polluted [20]. The scientific community was always aware that detection and quantification of *E. coli* is not sufficient to define the quality of a given wastewater treated, reclaimed, or discharged into the environment. Some pathogens are more resistant to conventional wastewater treatment (including chlorination), and its sources are not the feces of warm-blooded animals. Therefore, *E. coli* is an insufficient tool to reflect the quality changes due to wastewater treatment processes, conventional or advanced and extensive or intensive. In addition, it does not allow the control of wastewater disinfection [14].

Through routine monitoring of coliform bacteria, several alternative monitoring approaches have been suggested as indicators of the presence of potential pathogens. Types of waterborne pathogens and used indicators are shown in Table 5.2 (modified from WERF [19], Salgot [14]). More suitable indicators and new molecular biological methods are under development in order to establish the biological quality of different types of wastewater that enable a faster determination of specific microorganisms.

5.2.1.5 Contaminants of Emerging Concern

The difficult aspect of wastewater reuse is the removal of harmful contaminants to meet the water supply parameters, as required. Wastewaters contain contaminants of emerging concern (CECs), which include pharmaceutically active compounds, household chemicals, personal care products, disinfection by-products, pesticides, endocrine-disrupting chemicals and dioxins, furans, and dioxin-like polychlorinated biphenyls (PCBs), as well as other organic micropollutants and harmful substances. Organic compounds that tend to resist conventional methods of wastewater treatment are recognized as refractory organics. The long-term health effects of trace amounts of contaminants in water are not well-known.

Extensive lists of organic micropollutants have been discussed, for it is known that a significant number of trace chemical contaminants persist in the environment and some of these chemicals are known or suspected of deleterious implications to the environment, but only some of them were found to be associated to toxicity in humans. On the other hand, their determination is very expensive, and it is difficult to find an indicator for such a huge number of substances with different properties and different origins. Some typical compounds and their indicator substances are referred [14]: surfactants, mineral oil, pesticides (e.g., Diuron; 2,4-D), complex-forming substances (e.g., Ethylenediaminetetraacetic acid, EDTA), chloride solvents (if AOX > limit, e.g., TCE), aldehyde, aromatic organic solvents

Table 5.2 Descriptions and indicators for waterborne pathogens in wastewaters

Waterborne pathogens	Indicator	Examples	Description	Comments about indicator usage
Bacterial	Total coliform	*Escherichia, Klebsiella, Citrobacter, Enterobacter*	Facultatively anaerobic	Not widely used indicator
	Fecal coliform	*Escherichia, Klebsiella*	Facultatively anaerobic	Fecal coliform (FC) determination is the most used indicator, despite the problems and discussions *E. coli* determination is slowly substituting faecal coliform
	Enterococci	*Enterococcus faecalis, Enterococcus faecium*	Aerotolerant anaerobe	Subgroup of fecal streptococci and tend to be more persistent than fecal coliform, specially through wastewater treatment processes
	Other bacteria	*Clostridium perfringens, Staphylococcus aureus, Salmonella* spec., *Pseudomonas aeruginosa, Legionella pneumophila*	*Clostridium perfringens* is an opportunistic pathogen, produces enterotoxin, obligately anaerobic	Bacteria used by characterization of bathing waters, groundwater, etc.
Viruses	Enteroviruses	*Polioviruses, echoviruses, coxsackievirus*	Infect mammalian cells	Tend to be more infective and decay more slowly than bacterial pathogens An accepted indicator still does not exist
	Bacteriophage	Coliphages – viruses that infect *E. coli* and other coliform bacteria	Infect host by attaching to fertility fimbriae	The presence of survival coliphages is related to bacterial concentrations Similarities of certain coliphages and human enteric viruses
Protozoan Parasites	*Giardia intestinalis*	*Flagellated protozoan, phylum Mastigophora*	Cyst is infective form	Analytical tools not well developed until now
	Cryptosporidium parvum	*Coccidian protozoan, phylum Apicomplexa*	Oocyst is infective form, resistant to disinfection	Analytical tools not well developed until now
Helminths, nematodes and other helminths	Nematode eggs	*Ascaris, Trichuris, Ancylostoma*	On research better concentration method and viability.	Discouraging: many negative results in many countries. In some cases other helminths (i.e., *Taenia*) are important for risk related to animal health
Fungi, algal toxins	Unknown	–	–	Few cases detected

(e.g., benzene), PAHs (e.g., benzo(a)pyrene), phenols, pharmaceuticals (e.g., carbamazepine, X-ray, contrast media, sulfamethoxazole), and endocrine disruptors (E-Screen).

Though many of these contaminants are present at very low levels (ng/L), potential health concerns have given rise to numerous scientific investigations regarding the occurrence, concentration, transport, degradation, and aquatic fate. The fate and transport of a suite of 26 CECs and bulk organic matter [7] during a surface-spreading aquifer recharge operation were assessed. Individual CECs were identified along with bulk water quality parameters that could be used to indicate the performance for this particular surface-spreading aquifer recharge operation. Target of CECs included one herbicide, an X-ray contrast media, one plasticizer, one stimulant, one preservative, one surfactant, two chlorinated flame retardants, four personal care products, and 14 pharmaceuticals. Increased sediment toxicity has been strongly correlated with increased pesticide application rates within a watershed. Also, pharmaceutically active compounds (PhACs) have been detected [21] at biologically active concentrations within urban streams, and effects on nontarget organisms have been observed. A study has investigated the viability of using in situ near-real-time sensors to predict the concentrations of fecal coliforms, prometon (a residential herbicide), atrazine (an agricultural herbicide), and caffeine (a wastewater indicator). Such pollutants can be used as indicators of sources that contribute to what is often termed *urban stream syndrome*.

There are concerns [22, 23] about exposure to chemicals that may act as endocrine disruptors (environmental estrogens or estrogenic xenobiotics). Also, the potential for development of antibiotic resistance is of concern. Various health effects [24] in humans and wildlife species have been associated to endocrine-disrupting chemicals (EDCs), with increased vitellogenin (VTG) levels in male and juvenile female fish, altered sexual ratios, reproductive abnormalities, and neuroendocrine disruptions in some aquatic species, providing some evidence of this association. Research has also revealed a possible relation between EDCs and adverse human health effects, such as male testicular and prostate cancers and female breast cancers. Furthermore, EDCs can be biologically active in fish even at the low ng/L levels in ambient water. Three estrogens, estrone (E1), 17β-estradiol (E2), and 17α-ethinylestradiol (EE2), are generally considered to be the main EDCs of concern in wastewater effluents.

The term "dioxins" refers to the family of chlorinated dibenzo-*p*-dioxins (PCDDs), and "furans" designates the family of polychlorinated dibenzofurans (PCDFs). The coplanar polychlorinated biphenyls (PCBs) were found to have toxic effects similar to those of dioxins and are believed to operate by the same general mechanism [25, 26]. Dioxins, furans, and dioxin-like PCBs are ubiquitously distributed throughout the environment and can be detected in the air, water, soil, sediment, and biota. They enter the environment as complex mixtures resulting mainly from incineration, accidental fires, or spills involving PCBs. It is estimated that over 96% of dioxins in the environment have originated from emissions to air [27], which are then deposited on plant, soil, and water surfaces. Dioxins, furans, and dioxin-like PCBs have been detected in freshwater, wastewater, and drinking water samples, and therefore, exposure to dioxins can potentially occur through various sources of contaminated water, including wells, surface water, and in swimming areas [28]. Toxic equivalency factors (TEFs) for 2,3,7,8-substituted polychlorinate dibenzo-p-dioxins (PCDD) and dibenzofurans (PCDFs) and dioxin-like polychlorinated biphenyls (PCBs) congeners have been developed by the World Health Organization to simplify the risk assessment of complex mixtures. The WHO 2005 toxic equivalency factors (TEF) were used to calculate the toxic equivalents (TEQs) [29]. Dioxin guideline values in acceptable human intakes given by USEPA 1996 are 0.006 pg TEQ/kg bw/day and by WHO 1998 are 1–4 pg TEQ/kg bw/day [27]. In Australia, the estimated annual release of dioxins from water is 3.2 g/TEQ (less than 0.2% of the total annual release of 1,778 g/TEQ) [28].

5.2.1.6 Ecotoxicological Test

Due to the enormous number of potentially polluting substances contained in wastewaters, a chemical-specific approach is insufficient to provide the necessary information on water quality. Therefore, biological test systems with living cells or organisms give a global response to the pool of micropollutants present in the sample. Most toxicity studies are single-factor experiments with constant exposure concentrations. Efforts have been made in an attempt to predict joint toxicity of several parameters. The measurement of biochemical responses to chemical contaminants may help improve the assessment of biologically significant exposures to toxic chemicals and enhance the ability to assess the risk of effects on the health and survival of toxicant exposed populations. One of the purposes of (eco) toxicological tests is to determine the maximum concentration which does not affect a given relevant parameter of the test organisms ([23, 30, 31]; Hoeven 2008).

Toxicity tests have been included in regulatory requirements. USEPA has also included the need for effluent toxicity evaluation in the National Pollutant Discharge Elimination System (NPDES). Brazil requires the use of toxicity assays [32] in final effluents, and each state defines its own requirements and tests. The state of São Paulo, in 2000, and the state of Rio Grande do Sul, in 2006, have included ecotoxicity tests in enforcement actions. The maximum toxicity level of a final effluent allowed to be discharged is calculated based on the effluent dilution rate in the river and its ecotoxicity.

Toxicity tests on several trophic level using, for example, the following organisms are recommended: producer organisms – algae (e.g., *Scenedesmus subspicatus, chlorella vulgaris, Scenedesmus subspicatus*), primary consumers – microcrustaceans (e.g., *Daphnia magna, Cladocera*), secondary consumers – like fish (*Danio rerio, Brachydanio rerio*), and decomposers – bacteria (*Photobacterium phosphorium, Vibrio fischeri*). The crustacean *Daphnia magna* [23, 33–35] is the most commonly used organism in the study and control of water quality and is used in biological assays to determine the toxicity of wastewater. *Daphnia magna* is used as the sensor organism to measure the lethal concentration that causes the death of 50% (LD_{50}) of a group of test animals as a criterion to assess the toxicity of wastewater.

The phytotoxicity of the wastewater samples is also assessed by measuring the rate of seed germination and growth of plant species like the *monocotyl Sorghum saccharatum*, the *dicotyl garden cress Lepidium sativum, mustard Sinapis alba*, and *Lemna minor L.* [30, 33]. Due to their settled lifestyle, plants are constantly exposed to pollution. In addition, plants are the major energy source for nearly all higher eukaryotes, and, thus, they play an active role in transferring contaminants to higher trophic levels [31].

The CONSEMA Resolution No.129/2006 issued by the state of Rio Grande do Sul, Brazil, requires genotoxic through reverse mutation assay using *Salmonella* strains or Ames test. To determine the potential genotoxic effects of micropollutants in water [30], single-cell gel electrophoresis assay (Comet assay) was performed. The alkaline version of the Comet assay is a sensitive method that quantitatively measures DNA damage.

Biomarkers such as pigment content (chlorophyll and carotenoids) and enzyme activity (like peroxidase) are used as parameters for toxicity tests. A bioindicator of cell biomembrane injury, estimated by malondialdehyde (MDA) content, was tested for the first time in the ecotoxicological risk assessment.

5.2.1.7 Other Water Quality Parameters

Relevant metals in treatment, reuse, and disposal of treated effluents and biosolids are As, Cd, Cd, Ca, Cr, Co, Cu, Fe, Pb, Mg, Mn, Hg, Mo, Ni, K, Se, Na, W, V, and Zn, where Ca, Fe, Mg, K, and Na are the macronutrients essential for biological growth and Cr, Co, Cu, Mg, Mn, Mo, Ni, Se, W, V, and Zn are micronutrients. Heavy metal contamination of wastewaters and water bodies may originate from different local industries, commercial activities, transportation, dust, and other sources. The concentration of heavy metals in wastewater effluents can be critical, especially in industrial effluents. For example, cadmium, chromates, lead, and mercury are often present in industrial wastes. These are usually found in metal-plating wastes and should be removed for pretreatment rather than mixed with the municipal wastewater. Copper and zinc are heavy metals found in wastewater reuse systems, which may originate from the corrosion of pipes. Toxic concentrations of copper and zinc tend to build in wastewater reuse systems with low makeup flows. The toxicity of heavy metals strongly depends on water chemistry and is reduced at high alkalinities and hardness. The bioavailable fraction of heavy metals to plants, their uptake, and their accumulation depend on the properties of soil, on the plant, and other factors. Soil factors include, soil pH, organic matter content, cation exchange capacity, moisture, temperature, and evaporation. Major plant factors are the species and variety, plant parts used for consumption, plant age, and seasonal effects. Prolonged consumption of heavy metals [14, 15, 23, 36] may lead to the disruption of numerous biological and biochemical processes in the human body, and they can be found in the tissue of fresh vegetables. For some uses like crop irrigation, the determination of heavy metals such as the carcinogenic Pb and Hg is advised at regular time intervals.

Salinity is generally used to describe and measure seawater or some industrial wastes, and the parameter of total dissolved solids (TDS) concerns the amount of dissolved solids (typically various compounds of salts, minerals, and metals) in a given volume of water. Another way to determine the approximate TDS content is by measuring the conductivity of a water sample and converting the resistance (in microomhs) to TDS. Dissolved salts [36] causing salinity in wastewater exert an osmotic effect on plant growth. An increase in osmotic pressure of the soil solution increases the amount of energy the plant needs to expend to take up water from the soil. As a result, respiration is increased and the growth and yield of plants decline. However, it has been found that not all plant species are susceptible. Some areas were irrigated [37] by saline water so that the soil has been so loaded with salt that crop productivity has decreased, or only special saline resistant could grow. Salinity also affects the soil properties such as dispersion of particles, stability of aggregates, soil structure, and permeability. Hardness is a term used to express the properties of certain highly mineralized waters (high TDS concentrations). Discharges containing relatively high concentration of salinity [17] and/or hardness can impair the quality of lower riparian potable water supplies. Examples of wastes that may fall within this classification are oil-field brines, acids, alkalis, mine drainage, and brine from ion-exchange units. In reclaimed water that is chlorinated, chlorine residuals of less than 1 mg/L do not affect plant foliage, but chlorine residuals in excess of 5 mg/L can cause severe plant damage when sprayed directly on foliage [15].

The nutrients nitrogen and phosphorus are used for plant life and favor the fast growth of unwanted algae and aquatic plants. When they are discharged in excessive amounts on land, they can also lead the pollution of groundwater. Nitrogen and phosphorus removal is often required where reclaimed water is discharged to recreational and sensitive water bodies used for groundwater recharge or other reuse applications. Among the nitrogen compounds, ammonia and nitrites have poisonous effects on aquatic organisms. The nitrogen in reclaimed water [15, 38] for irrigation or crops can replace equal amounts of commercial fertilizer during the early to midseason crop-growing period. However, excessive nitrogen in the final part of the growing period may be detrimental to many crops, causing excessive vegetative growth, delayed or uneven maturity, or reduced crop quality. If alternate low-nitrogen water is available, a switch in water supplies or blending of reclaimed water with other water supplies can been used to keep nitrogen under control.

5.2.1.8 Water Quality Indices

Water quality indices (WQI) have been developed with the aim of providing summarized information on water quality. A water quality index is a unitless number that ascribes a quality value to an aggregate set of measured chemical, physical, and microbiological parameters [39]. Water quality indices usually consist of sub-index scores assigned to each parameter by comparing its measurement with a parameter specific rating curve, optionally weighted and combined into the final index [40]. Basically, a water quality index attempts to provide a mechanism for presenting a cumulatively derived numerical expression to define a certain level of water quality [41]. WQI summarizes large amounts of water quality data into simple terms (e.g., excellent, good, bad, etc.). The concept of WQI is based on the comparison of water quality parameters with the respective regulatory standards and gives a single value to the water quality of a source, which translates the list of constituents and their concentrations present into a sample (Khan et al. 2003, [42, 43]). Several regulatory agencies issued useful water quality criteria for the protection of beneficial uses [44]. The Wastewater Polishing Index (WWPI) defined by a weighted average of six parameters (suspended solids (SS), BOD_5, chemical oxygen demand (COD), ammonia, total phosphorus, and *Escherichia coli*), each one transformed into a sub-index scale from 0 to 100, was aimed to provide the fast assessment of the quality achieved by different polishing treatments for water discharged into surface water bodies and for reuse purposes [39].

5.2.2 Guidelines and Standards for Wastewater Reuse

In order to protect public health and the environment, regulatory approaches stipulate water quality standards in conjunction with requirements for treatment, sampling, and monitoring. Significant restraints to wastewater reuse include the potential risks to public health due to the chemicals and potential bacteria within the effluent and the potential for reduced sewer- or streamflows. Agricultural irrigation reuses, in general, require lower quality levels of treatment; on other extreme domestic reuse options, reuses need the highest treatment level. The requirement treatment levels for other reuse options lie between them. Depending on the water source and the treatment process, water reuse applications are divided in categories and their constraints are informed [4, 12, 22, 36, 45, 46]. These categories, their main constraints, and some characteristics of the systems for wastewater are shown in Table 5.3.

Cooling water is the predominant industrial reuse. The basic types [4] of cooling water systems that use reclaimed water are (1) *once-through cooling water systems* which involve a simple pass of cooling water through heat exchangers. There is no evaporation and, therefore, no consumption or concentration of the cooling water; (2) *recirculating evaporative cooling water systems* which use water to absorb process heat and then transfer the heat by evaporation. As the cooling water is recirculated, makeup water is required to replace water lost through evaporation. Water must also be periodically removed from the cooling water system to prevent a buildup of dissolved solids in the cooling water. Cooling towers and spray ponds are evaporative cooling systems that use reclaimed water.

The following hierarchy for acceptability of reuse (in decreasing order) is suggested [1]: (1) irrigating forests, (2) irrigating forage crops, (3) irrigating human food crops, (4) garden and parkland irrigation, (5) livestock watering, (6) cooling,

Table 5.3 Potential constraints and systems for the categories of waste water reuse

Categories and types of wastewater reuse	Potential constraints	Systems and considerations
Agricultural irrigation Reuses: crop irrigation, commercial nurseries	Surface and groundwater pollution if not properly managed Marketability of crops and public acceptance Effect of water quality, particularly salts on soil and crops Public health concern related to pathogens Nitrogen is not always beneficial to crops	Agricultural irrigation has been the largest reported reuse of wastewater. In addition to provide a low-cost water source, reported benefits are increase in crop yields: reduced dependence on chemical fertilizers and increased protection against frost damage Osmotic and specific-ion effects can reduce the yield potential of crops if high salinity water is used for irrigation High concentrations of nitrogen late in the season can adversely affect fruit quality, cause unnecessary vegetative growth, and delay maturity
Landscape irrigation Reuses: parks, school yards, playgrounds, landscaped areas around commercial, office, and industrial developments, landscaped areas around residences, freeway medians, greenbelts	Potential for human contact Reuse water must be treated to high-level need to avoid risk of spreading diseases Other concerns like odor, insects, and problems deriving from buildup of nutrients	Many landscape irrigation projects involve dual distribution systems, which consist of one distribution network for potable water and a separate pipeline to transport reclaimed water
Industrial recycling and reuse Reuses: cooling, boiler makeup, process water, heavy construction	Constituents in reclaimed wastewater related to scaling, corrosion, biological growth, and fouling Public health concerns, particularly aerosol transmission of pathogens in cooling water	Direct water reuse or water recycling if supplemental treatment is required to regenerate water quality Cooling water is the predominant industrial reuse. Basic types of cooling water systems are (1) once-through cooling water system and (2) evaporative recirculating cooling water systems (cooling towers and spray ponds)

	Biofilm growth in the recirculation system	Quality requirements for boiler makeup water depend on the pressures at which the boiler is operated. Generally, the higher the pressures, the higher the quality of water required
	Presence of microorganisms (pathogens or otherwise) with nutrients such as N and P in warm and well-aerated conditions creates ideal environments for biological growth	Reclaimed wastewater is ideal for many industrial purposes, which do not require water of high quality. Industrial process water uses vary greatly, and water quality requirements tend to be industry-specific. The effects on the final products, the plant equipment, and the health and safety of the workers have to be considered
		The potential risk of contaminants in wastewater to the system must be considered with the hydraulic and thermal loads
Ground water recharge	Organic chemicals in reclaimed wastewater and their toxicological effects	Groundwater replenishment by assimilation and storage of reclaimed water in groundwater aquifers
	Total dissolved solids, nitrates, and pathogens in reclaimed wastewater	Groundwater can either be recharged using injection wells or soil aquifer treatment (SAT)
		Groundwater recharge establishing hydraulic barriers against salt-water intrusion in coastal areas
		Use of treated wastewater for artificial recharge of groundwater has become a fast-growing way to treat and store effluent underground for subsequent recovery and unrestricted reuse
Recreational and environmental uses	Health concerns of bacteria and viruses	Reclaimed water has been applied to wetlands for a variety of reasons including: habitat creation, restoration and/or enhancement, provision for additional treatment prior to discharge to receiving water, and provision for a wet weather disposal alternative for reclaimed water
Reuses: lakes and ponds, marsh enhancement, streamflow augmentation, fisheries, snowmaking, water traps	Eutrophication due N and P in receiving water	
	Toxicity to aquatic life	

(continued)

Table 5.3 (continued)

Categories and types of wastewater reuse	Potential constraints	Systems and considerations
Non-potable urban uses Reuses: fire protection, air conditioning, toilet flushing	Public health concerns on pathogens transmitted by aerosols Effect of the quality of wastewater on scaling, corrosion, biological growth, and fouling Cross-connection	For fire flow-requirements, the volume of water to be stored at any given time, of pipe size and residence time, should be increased Reuse can be planned either within single households/building or on a larger-scale use through a reticulation system Enhancing potential for domestic water by separating gray water from "black water" toilet waste
Potable reuse	Constituents in reclaimed wastewater, especially trace organic chemicals and their toxicological effects Aesthetics and public acceptance Health concerns about pathogen transmission, particularly viruses	The opportunity for potable reuse occurs either by blending in water supply storage reservoirs or, in extreme conditions, by direct input of highly treated wastewater into the water distribution system

(7) industrial cleaning, (8) industrial process, (9) fisheries, (10) recreational water, and (11) public water supply: (a) gray water and (b) potable use.

According to the Royal Decree 1620/2007 of Spain, treated wastewater reuse is forbidden in the following situations: (a) for human consumption, except in situations of declared disasters; (b) for the specific uses of the food industry; (c) for use in hospital installations and other similar uses; (d) for the breeding of filtering mollusks in aquaculture; (e) for recreational use as swimming waters; (f) for use in fountains and ornamental waters in public spaces or inside public buildings; and (g) for any other use that the health authorities may deem to be a hazard to human health. The use in refrigerating towers and evaporation condensers is subject to very stringent requirements and forbidden in urban areas and in places with public or commercial activities [47].

The purple color has been adopted for reclaimed water lines [4]. There are standard methodologies for the implementation and management of water reclamation and reuse projects and several different water quality guidelines related to the different end uses. All guidelines were useful, but none has been entirely satisfactory because of the wide variability in field conditions. The guidelines were broken down into two broad schools of thought: (1) the so-called "better safe than sorry" approach, which is aimed to reduce the risk as much as possible – a representation of this school is the Californian Title 22 regulation. Several countries including Australia, Italy, Cyprus, and Israel adopted a similar approach; (2) a less restrictive approach designed to facilitate reuse, recognizing that regulations should be realistic and realizable in the context in which they are to be applied. A representation of this school is the World Health Organization's "guidelines for the safe use of wastewater and excreta in agriculture and aquaculture" of 1989. Several Mediterranean countries supported this approach, including a few European ones (France, Sicily, and Andalusia) [11].

Among the risks associated to reclaimed water use, the possible transmission of infectious diseases by pathogens is the most important concern. The difficulty of defining microbiological quality [48] of the media that receive reclaimed wastewater, particularly the agricultural systems where reclaimed water is used for irrigation, and the need to establish and conduct joint studies of all media implied in wastewater reuse systems have been reported. A WHO report issued in 1989 recommended the use of the epidemiological method for determining health risks associated with reuse practices. Wastewater reuse standards from Germany, China, the USA, Japan, Australia, and Queensland are reported [49], and although most of them are applied for the reclaimed municipal wastewater, these guidelines were aimed to be used as a basis for the establishment of the guideline of gray water recycling.

For many countries in Europe, either the guidelines of WHO that establish minimum qualities for direct reuse or the standards from USEPA form the basis for any decision or for granting permission to any kind of reuse. The first water quality criteria for reuse of wastewater in irrigation were set in 1933 by the California State Health Department. Standards or guidelines for other possible reuses such as groundwater recharge, industrial uses, etc., are not common, mainly because such types of reuses are not widespread [36]. The Council Directive 91/271/EEC concerning urban wastewater treatment has the objective to protect the environment from the adverse effects of urban wastewater discharges and discharges from certain industrial sectors and concerns its collection, treatment, and discharge so that the full implementation of the directive will contribute to obtain treated wastewaters for reuse. This indicates that treated wastewater and sludge from wastewater treatment shall be reused whenever appropriate. Disposal routes shall minimize the adverse effects to the environment.

The Royal Decree 1620/2007 establishes the legal regime for treated wastewater reuse and quality criteria based on maximum admissible values that take into account sanitary risks and other pollutants depending on the uses to which the regenerated water is destined. The Royal Decree establishes new microbiological and pollutant parameters that measure the quality of the regenerated water. The microbiological parameters are intestinal *nematodes*, *E. coli*, *Legionella spp.*, *Taenia saginata*, and *Taenia solium*. The chemical pollutants are suspended solids, turbidity, nitrates, total nitrogen, and total phosphorus. The reuse of wastewater for agricultural uses needs the application of tertiary treatments capable of obtaining the established quality limits. These maximum limits are the following: intestinal nematodes (1 egg/10 L), *Escherichia coli* (200 CFU/100 mL), suspended solids (20 mg/L), and turbidity (10 NTU) [50].

A guideline for gray water reuse for restricted and nonrestricted agricultural irrigation was released by the WHO. For restricted irrigation, the number of helminth eggs and the number of *E. coli* shall be lower than 1/1 L and 10^5/100 mL, respectively. For unrestricted irrigation, the number of helminth eggs and the number of *E. coli* shall be lower than 1/1 L and 10^3/100 mL, respectively [51]. The German Berliner Senate Office for Construction and Housing has established a gray water reuse guideline, in which parameters like BOD, oxygen concentration, total coliform, fecal coliform, and *Pseudomonas aeruginosa* are required [13].

In countries where domestic reuse is also widely practiced, such as Japan, there are standards for such reuse [36]. In Italy, specific legal requirements for reclaimed wastewaters to be reused for agricultural, industrial, or civic purposes refer to 55 parameters. However, regarding domestic wastewater reuse, most problems that arise are due to high concentrations of six parameters selected to define the Wastewater Polishing Index (WWPI): 20 mg BOD_5/L, 100 mg COD/L, 10 mg SS/L, 2 NH_4 mg/L, 2 mg Ptot/L, and *E. coli* (CFU/100 mL) 10 as 80th percentile 100 as the maximum value by physical or chemical treatments and 50 as 80th percentile 200 as the maximum value by lagooning or constructed wetlands [39]. Treated wastewater can be reused for toilet flushing, landscaping and irrigation, or for cooling water, depending on the quality provided for Taiwan EPA treated wastewater reuse criteria. For treated wastewater to be used for toilet flushing, the *E. coli* count, color, BOD, and pH are the parameters of concern, while an extra criterion of turbidity is taken into consideration for landscaping and irrigation water. It is pH, turbidity, total hardness, total alkalinity, chloride, total dissolved solids, and suspended solids that are of concern for treated wastewater to be reused as cooling water [52]. In Brazil, international reference standards are generally adopted regarding wastewater reuse or else technical guidelines provided by private agencies. Concerning rainwater harvesting systems, the project and dimensioning directives are prescribed in the *Norma Brasileira* – NBR, 15.527/2007 – which establishes the requirements for rainwater harvesting from roofs in urban areas for non-potable purposes. Thus, rainwater can be used for non-potable purposes after proper treatment.

In the framework of the international project "Integrated Concepts for Wastewater Reuse" (Aquarec), an independent task force composed by water and wastewater utility professionals, consultants, and researchers was assigned the task and produced a manual of management practices for wastewater reuse schemes. The project has identified over 200 wastewater reuse projects in Europe out of 3,300 water reclamation projects globally. The review considered seven geographical regions: (a) North and (b) Latin America, (c) Europe, (d) Mediterranean Region and Middle East, (e) sub-Saharan Africa, (f) Oceania, and (g) Japan. Japan has the largest number of reuse projects (over 1,800), followed by the USA (over 800), which is the largest producer, with a volume of reused water estimated at close to 6.5 Mm^3/day. Almost 100 sites were identified in the Mediterranean and Middle East area, with more than 50 in Latin America and 20 in sub-Saharan Africa [11, 53]. In 2004, it was estimated that 700 Mm^3/year of wastewater was reused in Europe, which is less than a fifth of the estimated potential for wastewater reuse [54].

5.3 Industrial Wastewater Management

5.3.1 Water Usage by Industry

The industry water is withdrawn from aquifers, surface water, or municipal drinking water network. Water is applied as a utility and is used for various purposes. The water cycle in industry is generally not closed. Used water not consumed in the process is usually discharged to water bodies usually after treatment. In spite of the greatest potential for polluting water in industry, there is a high potential for saving water by reuse or recycling in the process, greater than by agriculture or domestic usage. In Spain, for example [55], from the total volume of treated effluents reused, only 0.7% corresponds to industrial wastewater.

The industrial sector takes up an average 22% of the globally used water. Water usage intensity is significant to all businesses whose sustainability will depend on its availability, cost, and quality [56]. The relation between the industrial and agricultural activity indicates that the total water intensity (or water footprint) of an organization or region not only is limited to the operational water directly used in the processing of the final product but also includes the water used in all processes involved in the product supply chain [57].

In general, the quantity and quality of water required for the development of various consumer processes in an industry depends on the type of product/process and production capacity. Industrial water demand and wastewater [58] production are sector-specific. Industries may require large volumes of water for cooling (power plant, steel, mills, and distillation industries), processing (breweries, pulp, and paper mills), cleaning (textile mills, abattoirs), transporting products (beet and sugar mills), and flushing wastes. The treatment requirements for industrial water use are related to the water use application. In general, the water meets the following applications and requirements in industry (adapted from FIESP [59]) (Table 5.4).

The amount of water needed to meet the needs of various industrial activities is influenced by many factors, such as segment, production capacity, regional climatic conditions, availability of water, production method, age of facilities, operational practice, local culture, technology innovation, and research investments, among others. Proper water consumption indicators make it possible to assess water use efficiency in order to improve water-based processes, minimizing the impacts generated, either qualitative or quantitative. Also, the indicators make it possible a benchmarking between the industries of one given segment [59].

Industrial symbiosis is a known application of industrial ecology principles where the waste materials of one industry are used as inputs for another industry. It concerns the exchange of by-products, water, energy, and process wastes. The keys to industrial symbiosis are collaboration and the synergistic possibilities offered by geographic proximity. Localized industrial ecology in the form of industrial symbiosis could also have the broader benefit of linking to regional development [57, 60]. The concept of industrial ecology (IE) requires that an industrial system be viewed not in isolation from its surrounding systems but in concert with them.

Depending on the industrial process, the concentration and composition of the waste floss may vary significantly. In particular, industrial wastewater may have a wide variety of micro-contaminants, which add to the complexity of wastewater treatment [58]. A term largely used to evaluate the impact of industrial or other waste on a treatment work or stream is the "Population Equivalent." Population equivalent refers to the amount of oxygen-demanding substances whose oxygen consumption during biodegradation equals the average oxygen demand of the wastewater produced by one person, assuming that one person demands 54 g of BOD per 24 h. Water usage [61] can be also referred as prorated on a per-employee basis in service industries, on a per-hectare basis for zoos, parks, and gardens, on a per-unit-of-turnover basis for manufacturing industries, and on a per-unit-of-production basis for many industries. Water footprint is another indicator [62] that shows the impact of consumption on environment with water equivalent, which allows a detailed quantification of water use directly and indirectly by segments for the domestic and abroad final consumption. A framework for calculating the national water footprint (NWF) with input–output method on China was presented for the primary (agricultural), secondary (industrial), and tertiary (services) sectors.

In industrial wastewater discharges, many manufacturing facilities generate relatively constant flow rates during production, but the flow rates change markedly during cleanup and shutdown. Wastewater flow rates from industrial sources vary according to the type and size of the facility, the degree of water reuse, and the on-site

Table 5.4 Major uses and water quality requirements in industry

Water uses in industry	Water quality requirements
Human consumption: Drinking water used in sanitary environments, changing rooms, kitchens and refectories, water fountains, safety equipment, or in any other activity that involves direct human contact	Quality standards of water for human consumption
Product and reactant: water is a raw material which cannot be replaced. Examples: beverages, hydrolysis	Water quality may vary significantly, and the use of water of quality equal to or higher than water for human consumption can be accepted when the main purpose is the protection of the health of end consumers and/or the guarantee of end product quality
Use as a secondary fluid: Water can be used as a secondary fluid in several activities such as the preparation of chemical suspensions and solutions, e.g., intermediate compounds, vehicles, or in washing operations	If this water comes into contact with the end product, the end product standard will vary and this water shall have more restricted uses, depending on the process and on the desired end product
	If there is no contact between the water and the end product, the standards will be less restrictive than those required for water for human consumption, especially regarding disinfectant residual concentration
Use in energy generation: For this type of application, the water can be used by means of transformation of kinetic, potential, or thermal energy accumulated in the water in mechanical energy and then in electric energy	Depending on the process of water treatment, the water shall meet very different quality standards. Concerning the use of potential or kinetic energy, the water is used in its natural state, and it can be used in its raw form, as it has been collected from rivers, lakes, and reservoirs. Care must be taken to prevent that large materials, sweepings, damage the energy generation devices. Concerning the use of thermal energy, after heating and evaporation, water shall meet high quality standards in order to avoid problems in the steam generator or in the energy conversion device
Use as heating and/or cooling fluid: The water is used as a fluid to remove heat from reactive mixtures or devices that need cooling, or due to the operation conditions established, since high temperatures may compromise the system performance, as well cause equipment damage	The water used in steam form shall meet a very high quality standard. In turn, the use of water as cooling fluid requires a much less restricted quality standard for it concerns the protection and useful life of the equipment that shall come into contact with the water
Other uses: Use of water to fight fires, water green areas, or incorporated to several by-products generated in industrial processes, in the solid, liquid, or gaseous phase	Many applications require the observation of a greater number of parameters in order to minimize the risks to the process, product, or system where the water shall be used. The following tables contain data on water requirements for industrial applications

wastewater treatment methods. For industries that do not count on facilities to treat wastewater or reuse programs, it can be assumed that about 85–96% of the water used in the various operations and processes will become wastewater. For industries with facilities to recycle wastewater, separate estimates based on actual water consumption records must be made. Average domestic (sanitary) wastewater contributions from industrial facilities may vary from 30 to 95 L/capita.d [15].

For the water and wastewater management, which has the main purpose of ensuring water resource conservation, the principle of wastewater recycling or reuse dictates that wastewater of any marginal quality water can be used as long as proper treatment is provided to meet the water quality requirements for the intended purpose. Indeed, water quality required for recycling or reuse is different from the quality acceptable for discharge, either into a surface water body or municipal sewers, and depends on technical considerations.

Since water is used in many applications, its quality varies because of the introduction of several chemical or biological contaminants. The effective treatment of wastewater in order to meet water quality standards and to ensure public health protection is a critical element of water reuse systems. The water quality parameters that are used to evaluate reclaimed water are based on current practice in water and wastewater treatment and industrial process requirements. The changes in water quality needed to improve wastewater have become the basis for wastewater treatment. In practice, the treatment is carried out to meet the standards of regulatory agencies in order to ensure the protection of the aquatic environment and other beneficial uses.

5.3.2 *Water and Wastewater Management for Reuse*

Water demand management (WDM) aims to promote and balance the purposes of efficiency, equity, and sustainability in the management and allocation of water resources, satisfying the existing needs for water supplying with reduced consumption. WDM covers a wide range of technical, economic, educational, capacity-building, and policy measures that need to be applied by water resource planners, water supply agencies, and end users. WDM is normally viewed as a useful tool in achieving Integrated Water Resource Management (IWRM) [63].

Cleaner production in industrial processes concerns the concomitant operations of an industrial process at many levels to provide a more efficient use of natural resources (raw materials, energy, and water) and reduce the generation of waste and emissions in the source. Unlike end-of-pipe treatment systems, cleaner production in most industrial processes can be applied to different stages of the process, and a project implemented by stages according to a company's needs and possibilities. Good housekeeping measures arise after appropriate brainstorming and using common sense. Cleaner production and related concepts of eco-efficiency, pollution prevention, and industrial ecology are guiding posts for the business journey to sustainable development.

Industries that invest in water saving and waste minimization techniques have a better market image since people are increasingly concerned with the rational use of natural resources and environmental degradation. Such industries are more likely to be awarded environmental certifications.

In many production processes, water is used in once-through systems since it is relatively cheap and not fit for reuse again. End-of-pipe treatment in which all effluents are grouped together and the combined stream is submitted to the final treatment system is operated on a large scale by industrial treatment facilities.

One important issue concerns the relevance of using the same (high) water quality for various purposes once some water inputs in process streams can require water with inferior quality using wastewater streams. Wastewater stream segregations and decentralized treatment are often more effective and efficient since direct reuse can be introduced and lower amounts of water, higher concentration levels, and less variations in contaminants are observed. Making the water fit-for-reuse again requires investments in treatment (by recycling) and additional piping or reservoirs. Also, besides saving water, this system reduces the volume for the end-of-pipe wastewater treatment plant.

A careful analysis of the stages of an industrial process that uses water will often show savings that can be obtained without detriment to product quality. The elimination of leaks in water piping or unnecessary use of water, more efficient use of water through countercurrent washing tanks, and reuse of batch solutions that can be made up to required strength are typical methods that reduce the cost of water employed in a process and simultaneously reduce the cost of treatment of the resultant waste. Cooling and condensate waters should also be segregated from process waters for separate discharge, provided they are of a quality good enough to justify such discharge, and in an amount that justifies the higher cost of separate collection [17].

As more advanced technologies are applied for wastewater treatment and purification, the practical benefits of water reclamation and reuse are evident. The cost-effectiveness of industrial wastewater reuse depends on a variety of industry-specific factors that relate to the quantity and quality of water needed and the proximity of water resources to industrial users.

In wastewater management, some main terms are defined. First, the differentiation between treated water and reclaimed water should be taken into account. Treated water is the wastewater that has undergone a treatment process to meet the regulatory standards for discharge in water bodies, whereas reclaimed water is the water whose quality has to fulfill the requirement for the use for which it is intended, where an additional or complementary treatment can be applied.

A particular differentiation between reuse and recycling of water for industrial applications is usually employed. Reuse is when water is used for the same or another purpose without needing to be processed. Water recycling is defined as reusing water following any treatment (regeneration) to ensure that it is fit for purpose as reclaimed water (which may include output release).

Industrial treated wastewater can be reused for any other (external) purposes (e.g., irrigation or groundwater recharge) than consumption in industrial facilities.

When the use of wastewater in the plant involves reclaimed water from one industrial plant stream to be used for another destination stream or process without additional treatment to regenerate water quality, it is referred as water reuse and is also called "cascading."

In water recycling, reused cleaner water can be introduced to the process at a level that depends on the amount acceptable in the recycled water and the quality of the input water. Therefore, recycling systems are generally semi-open systems whose calibration depends on the balance of accumulation within the process [46].

A better understanding of the benefits of water and wastewater reductions on industrial sites may be gained with the consideration of the hydrological and hydrogeological characteristics of the region [64]. Once storm water is usually of better quality than industrial discharges, industrial water minimization programs can be focused on storm water usage practices.

The cost of wastewater reclamation is often cited as the key criterion for the evaluation of treatment train performance, as well as other technical and engineering criteria. When all reasonable efforts have been made to reduce waste flows and total waste loads, it is wise to consider the persistent waste treatment problem and the practical solutions of end-of-pipe treatment.

5.3.3 Wastewater Treatment Processes

Industries located inside the city boundaries usually discharge their wastewater to sewers. Moreover, most industrial wastes can be more effectively treated in combination with domestic sewage than alone when industrial wastes lack nutrients. Because of its heterogeneous composition, domestic sewage can provide nutrients such as carbon, nitrogen, and/or phosphorus essential for bacterial nutrition. Most manufacturing wastes are more amenable to biological treatment after dilution with domestic wastewater. However, large volumes of high-strength wastes must be considered in the sizing of municipal treatment plant. The discharge of industrial wastewater into publicly owned treatment plants may have serious consequences, such as release of toxic substances, interference with the treatment operation, sludge deposits, corrosion of pipes, and hazardous effects such as toxic gases or risk of explosion. These problems can readily be avoided by pretreatment of wastewater at the industrial site before it is discharged into public sewer system [17, 65, 66].

Complete separate wastewater treatment plants are commonly installed for industrial effluent treatment, once the locations of many industrial plants are inaccessible to municipal sewers. Other limitations are contaminant constraints and the reduced capacity in municipal treatment, which makes the joint treatment unfeasible. Some locations count on industrial effluent treatment that meets good quality parameters after treatment, but municipal sewage effluent is not properly treated and contaminates the receiving waters.

Water and wastewater treatment plants are designed on a unit operation concept where one operation is optimized to accomplish one task, although more than one problem may be obviated by one single operation. In general, each operation produces effects on other downstream treatment processes and trade-offs between increasing the efficiency of the process. The choice of the treatment depends on the quality and variability of the raw water source and the purpose of the treatment. The system operation must handle the variations in raw water quality.

In wastewater treatment plants, preliminary mechanical devices precede physicochemical and/or biological treatment and are used to remove sand, grit, fibers, and floating substances and materials in order to avoid obstructions and to enhance the performance of the subsequent treatment stages. Detention tanks also regulate the flow of wastes. Physicochemical treatment comprises sedimentation aid for the coagulation of non-settleable solids or flotation methods. Chemical treatment concerns the use of chemicals to bring about a chemical change in the waste constituents, either to permit their removal or to convert them into less objectionable forms. Biological treatment [17] is usually necessary to remove soluble and colloidal organic material from wastes. These include biological lagoons, standard and high rate or recirculating trickling filters, activated sludge, including both conventional and high-rate or high-load forms, and sand filters.

Advanced treatment methods remove more contaminants from wastewater after conventional physicochemical and/or biological treatments. Common advanced methods remove phosphorous from solutions, oxidize ammonia to nitrate (nitrification), convert nitrate to nitrogen (denitrification), and remove or inactivate pathogenic bacteria and viruses (disinfection). In water reclamation, the objectives may be expanded to include the removal of heavy metals, organic chemicals, inorganic salts, and the elimination of all pathogens [65].

The economic disposal of sludges from industrial waste treatment processes is often one of the most difficult problems encountered. The combination of operations for sludge treatment (conditioning, dewatering, drying, incineration, materials recovery) and the disposal techniques required in a particular situation will depend on many variables.

The ongoing development on advanced treatment systems enables wastewater regeneration requirements for segregated or integrated water recycling or reuse and the removal of difficult contaminants to meet environmental regulatory standards for discharge. Treatment systems such as sand filtration, ion exchange, adsorption, disinfection, advanced oxidation processes, membrane filtration, reverse osmosis, and advanced oxidation meet the requirements of recycling and reuse. All these systems basically lead to the segregation of depurated water flows that can be reused for any purpose or recycled in the industrial process.

The cost of the treatment, including the equipment and operation, needs evaluation both from the company's business perspective and also the broader environmental benefits. These two aspects are not always compatible, and a final decision on reuse will require public information and input. From a company's point of view, there are two costs that must be compared for decision on reuse or recycling. The first is the cost of the connection to obtain primary source water plus the cost to discharge it. The second involves reuse where the cost of the water and its discharge are negligible, but the treatment is expensive. The financial costs can be easily determined and provide a preliminary indication of the possible advantages of reuse. The company must also consider other financial implications that are more difficult to assess [46].

5.3.4 Mass Integration on Water System

The study of the concept of water reuse has effectively started in the beginning of the 1980s. It was an attempt to solve the problem of Wastewater Allocation Planning, which consists in the determination of water flow through an interconnected network in the processes where the consumption of freshwater is minimized, whereas the processes are supplied with water that meets the standards. Therefore, the idea of minimizing the amount of effluents generated during the process and performing the treatment at the point of effluent generation, called distribution treatment, has arisen. The chemical manufacturers have shown great interest in the implementation of water recycling/reuse policies where the pollutants are recovered in wastewater flow processes using typical wastewater separation processes. Moreover, the possibility of distributed generation has become widely accepted. These techniques are called mass flow integration processes and is based on [67] principles of chemical engineering combined with system analysis that use graphic or optimization tools. The first step in the use of mass flow integration techniques is the development of a total representation of mass allocation of the system from the point of view of the involved species.

Reuse occurs when the effluent is used directly in other operations as long as the level of contaminants does not interfere with the process [68]. An effluent outflow of water from a treatment can be mixed with the effluent wastewater from another treatment or with freshwater. Sometimes, reuse regeneration is needed, which requires the total or partial removal of contaminants from the wastewater effluent, to make it possible to use such wastewater flow in another operation. The wastewater effluent is regenerated by partial treatment to remove the contaminants that might prevent its reuse and is then reused in another operation. Here again, the wastewater effluent can be mixed with an effluent from another treatment process or with freshwater. In this case, the wastewater reused after regeneration cannot be used in the process where it has been previously used [68]. Like reuse, recycling involves the reuse of processed flows that carry pollutants. In the case of recycling, the wastewater flow is used in the same process where it has been generated. Typically, wastewater separation techniques are key elements in recycling systems to recover valuable substances such as water.

Process integration techniques for the synthesis of water network have been widely accepted as a promising tool to reduce freshwater and wastewater flow rates via in-plant water reuse/recycle and have been studied over the past decades

[68–70]. There are three key components in the understanding of the process integration methodology ([67, 71], Rossiter 1995): (1) Process Synthesis: Due to the great number of alternatives to the process, it is important that the synthesis technique "extracts" the optimal solution(s); (2) Process Analysis: The detailed characteristics of the process are forecasted using analysis techniques that include mathematical models, empirical correlations, and computer-assisted process simulation tools; and (3) Process Optimization: the leads to interaction between synthesis and analysis. The implementation of strategies aimed at the minimization of wastewater effluents and water requires a mathematical assessment of the process in order to make it possible the development of mathematical models capable of simulating the process. These models are used in approximations of mathematical optimization in the search for optimal solutions. The method(s) selected for the resolution of the optimization problem depends primarily on the characteristics of the objective function and on whether it is explicitly defined, on the nature of the restrictions and on the number of dependent and independent variables.

In recent years, much research has been conducted on this topic mainly based on two traditional methods: graphical method and mathematical programming.

5.3.4.1 Graphical Method

Pinch analysis (or bottleneck point) is the most widely used methodology. According to this method, a set of good process flows is transferred to a set of bad process flows. The design of the mass exchangers is performed using principles of thermodynamics based on heuristics and process knowledge. The main limitations of this method concern the number of contaminants that can be included in the problem and in the synthesis of the exchanger network system (arrangement of connections between the units) that is not automatic. Besides, each mass transfer coefficient is considered as a linear function of the composition, which does not always reflect the real case [72]. In the method of mass exchanger network synthesis [73], the reduction in wastewater volume is performed with the integration of wastewater flows that may originate in the process itself or from external sources. Thus, it uses the equipment involved in mass exchange.

Bai et al. [74] did the following back over about this methodology: "the graphical method was initiated by Wang and Smith [68]. They introduced several important concepts, such as a limiting water profile and limiting composite curve and pinch. Then, the water demand and supply of a system can be described on the concentration–mass load diagram, whether water reuse, regeneration reuse, or regeneration recycling is considered. For regeneration reuse, at a given post-regeneration concentration, the optimal regeneration concentration is taken as the pinch concentration of the system in the paper. Later, Mann and Liu [75] found that when considering water regeneration reuse, some systems had another pinch point, regeneration pinch, which is higher than the conventional pinch for freshwater targeting. Castro et al. [76] introduced a multiple-pinch-points concept and proposed a novel reuse network targeting/design methodology for single-contaminant systems. This method can be used to design and optimize regeneration reuse water systems." The above research on optimizing regeneration reuse systems resorts to graphical methods and, thus, applies only to single-contaminant systems [74]. Bandyopadhyay and Cormos [77] also used a graphical representation to address water management issues of integrated processes that involve regeneration and recycle through a single treatment unit.

5.3.4.2 Mathematical Programming

The application of this method requires the formulation of the problem as a mathematical programming technique. The network system is obtained by optimization. The optimization problem for obtaining the mass integration network uses mathematical models. The mathematical programming technique demands advanced knowledge because the engineer must be able to judge the selected solution. The mathematical programming for integration was used to assess the optimal conditions of use of wastewater from refineries and other multicomponent plants in the study of Savelski and Bagajewicza [78]. The approximation of mathematical programming to analyze the suitability of the zero discharge levels of liquid effluent discharges in different industries [79] has demonstrated that the cost of regeneration and the cost of freshwater, as well as the concentration of treatment discharge, is the decisive factor in zero discharge levels.

When mathematical programming is used, regeneration reuse, as one of the possibilities to reduce the freshwater consumption or total cost of a water system, is incorporated in the superstructure [80–85]. The targeting and design of regeneration reuse systems are performed by solving the corresponding mathematical models. Based on a sequential three-step programming method, Xu et al. [86] investigated the relationship between the freshwater flow rate required and the regeneration concentration of a regeneration reuse system. The study showed that the regeneration of water at pinch concentration does not always minimize the freshwater consumption and there are three possibilities for the minimum regeneration concentration: below, above, or at the pinch. Based on sequential optimization [87], a handy method has been proposed to construct the optimal water supply line for a regeneration recycling system, and the improved mass problem table to target for freshwater, regenerated water, and regeneration concentration was introduced. The methodology has been extended to regeneration reuse systems. A mathematical programming formulation for the water integration in eco-industrial parks considering streams with several pollutants was presented [88]. The formulation was based on a superstructure that allows wastewater reuse in the same plant, water exchange with different plants, and a shared set of interceptors that must be selected to determine the network configuration that satisfies process equipment and environmental constraints.

5.3.5 Wastewater Reuse in Some Industries

The suitability of reclaimed water for use in industrial processes depends on the particular use. The electronics industry requires high-quality water similar to that of distilled water. On the other hand, textiles, pulp and paper, tanning, and metal fabricating require water of intermediate quality. The water quality requirements [4] for the chemical industry vary greatly according to the production requirements. Water of a relatively low quality can usually be used in the manufacture of petroleum and coal products.

The application of advanced wastewater treatment processes is expressive and abundantly described in the literature for food and textile industries overall. However, water reuse or recycling is not a widely discussed theme nor are these processes applied in industrial plants. The economic advantages of water reuse have been demonstrated in industrial plants, even when the treatment facilities need to be improved. Water management-related aspects and studies or cases showing efforts on water reuse and recycling are here briefly introduced for some industrial sectors.

The *textile sector* has a high demand for water. Its biggest impact on the environment is related to primary water consumption (80–100 m^3/ton of finished textile) and wastewater discharge (115–175 kg of COD/ton of finished textile, a large range of organic chemicals, low biodegradability, color, salinity) [89]. The waters used in textile manufacturing must be nonstaining; hence, they must be low in turbidity, color, iron, and manganese. Hardness may cause curds to deposit on the textiles and damage some of the processes that use soap. Nitrates and nitrites may cause problems in dyeing [4]. Many studies were published on advanced techniques applied to remove wastewater contaminants from the textile industry. Water produced by ultrafiltration–nanofiltration treatment of aerobically pretreated effluents [90] had reasonable COD, color, and conductivity, meeting the water reuse requirements for special processes such as dyeing very light colors. A project (BATTLE Project) [89] designed an innovative water scheme, implementing effluent reuse and applying a methodology that integrates the different aspects involved in reuse (data collection, online characterization, streams segregation, final effluent treatability evaluation and impact control, reclaimed water reusability evaluation, cost analysis), based on the prototypal multicriteria methodology.

In general, most guidelines address the quality requirements for water to be used and reused in the *food industry* but do not include criteria for food process water treatment and reuse [91]. The microbiological quality of the water to be reused must be guaranteed and monitored at all times in the food industry, and, therefore, hazard analysis critical control point (HACCP) is obviously used, which is a systematic safety management tool [92]. The treatment of dairy process waters by nanofiltration (NF) or reverse osmosis (RO) operations was proposed to concentrate dairy matter and produce purified water for reuse in dairy plants [93]. Low-contaminated wastewater from the food industry has a high potential for reuse. Treated water stream can be reused for cleaning and cooling processes, boiler feedwater [94]. Using chiller shower water from the meat processing industry, water of drinking water quality was obtained by applying a process combination involving pretreatment (sedimentation with fat skimming, cartridge filtration, and UV disinfection), a membrane stage (NF), and a posttreatment stage (UV oxidation and UV disinfection) [95]. A simultaneous combination of ozone and a granular activated carbon (O_3/GAC) as a tertiary treatment of wastewater generated from the activity of various food-processing industries in order to reduce their organic load to a minimum is reported, where the water was suitable for reuse [96]. Rational water management strategies to identify operational flaws were assessed for improvements at operational issues and physical loss identification in a poultry processing plant. Some sectors with high water consumption and wastewater generation were identified, and a reduction by almost 31% in water consumption was achieved [97].

A case study was carried out in a water network of a *petroleum refinery* where a nonlinear program (NLP) model targeting the minimization of freshwater consumption and/or operating costs was developed based on the conservation equations of chemical species (contaminants) and mass (water). The regeneration processes used by this network are an API gravity separator followed by activated carbon adsorption, reverse osmosis, and a Chevron wastewater stripper [98]. The water balance has been used to map the aqueous streams in petrochemical and chemical units, being a useful tool for the management and use of water in the productive process. Several types of streams in a water balance, clarified and distilled water, steam at varying pressures, condensed water, and wastewaters, have been included [99]. A pilot plant was designed for studying the methods of biological purification (aerobic and anaerobic) wastewater from petrochemical industries, which are polluted with ammonia derivatives. Then, water could be reused after purification in a production cycle [100]. A methodology for water use minimization has been developed by the Clean Technology Network of Bahia over a 10-year period in joint cooperative programs with the chemical, petrochemical, and copper metallurgy industries located in the Industrial Complex in Camaçari – Bahia, Brazil. The methodology comprises a set of tools including reconciled aqueous stream balances, database of aqueous streams; large-scale training leading to the identification of water minimization alternatives in the processes, water reuse optimization approaches; geographical information systems, as well as consideration of the region's hydro and hydrogeological characteristics and assessment of the possibility of using storm and wastewaters for industrial use [64].

In *tanneries*, the stages of conditioning and preparation of the hide for tanning, as well as the transportation and fixation of tanning substances, are all carried out in aqueous media. Most of the steps of these processes are carried out with freshwater, which is used as a vehicle for the diffusion of chemical products and the extraction of undesirable materials from the hide. Processing waters of slimming/dehairing and tanning operations are often reused as input of the same respective operations in industry, most of all by slimming/dehairing operations in the case of Brazil. A study [101, 102] provided tools for the minimization of water consumption at tanneries by applying the reuse of water at some stages of the process. Water consumption and concentration of some contaminant parameters for each stage of beamhouse, tanning, and pos-tanning operations are given, and alternatives of water reuse were evaluated. A study was carried out on pilot and industrial scale to demonstrate the viability of reuse practices in the tannery industry, which has shown the feasibility of wastewater streams reuse in beamhouse steps of leather processing. Finally, a mathematical program for mass integration was applied to solve the problem of water reuse in the beamhouse steps. The state of implementation of clean technologies to avoid water pollution in tanneries, directions to improve water management, and alternatives available to tanneries aiming at their improvement has been described [103]. There are different methods available to treat the chromium in wastewater generated from chrome tanning process. Chemical precipitation, coagulation, solvent extraction and membrane process, ion exchange, and adsorption methods are some of the concepts available to recover the chromium from the effluent. Chrome recovery by using magnesium oxide is one of the popular chrome recovery methods. There is abundant literature on the precipitation of chromium with the addition of lime, sodium hydroxide, and sodium bicarbonate [104].

The *brewing industry* is a large consumer of groundwater for brewing, rinsing, and cooling purposes. From four different water streams (wastewater after biological treatment, bottle rinsing water, rinsing water of the brewing room, and rinsing water of the bright beer reservoir) filtered with four different nanofiltration membranes, the results for the biologically treated wastewater were very promising [105]. The reuse of bottle washing wastewater from the beverage industry after ultrafiltration and reverse osmosis membrane filtration treatment systems not only reduce the consumption of potable water but also help conserve energy [106].

Pulp and paper industry internally recycle water to a very high degree. Modern mills approach a recycling ratio of 100%, using only 16,000–17,000 gal of fresh-water per ton (67–71 L/kg) [107]. Many industries, for example, pulp or pharmaceutical, produce wastewater containing high amounts of *cellulose*. The potential target applications for reuse of treated wastewater from a high-cellulose-containing effluent from a pharmaceutical factory were investigated. The effluents from the aerobic membrane bioreactor (AMBR)/RO and anaerobic activated sludge plus aerobic (AOMBR)/RO processes met the Taiwan EPA criteria for both effluent and treated wastewater reuse when the hydraulic retention time (HRT) for the AOMBR/RO and AMBR/RO processes was equal to or longer than 12 and 8 h, respectively [52].

References

1. Howell JA (2004) Future of membranes and membrane reactors in green technologies and for water reuse. Desalination 162:1–11
2. IWA – International Water Association (2008) Water reuse an international survey of current practice, issues and needs. In: Jimenez B (ed) Asano T Scientific. IWA Publishing, ISBN 1-84339-089-2, Technical report N. 20
3. Guelli U, Souza SMA, Melo AR, Ulson de Souza AA (2006) Re-utilization conditions of wastewaters from textiles industries. Resour Conserv Recycl 1:1–13

4. USEPA (2011) Guidelines for water reuse chapter 2. Types of reuse. http://www.epa.gov/nrmrl/pubs/625r04108/625r04108chap2.pdf. Accessed 21 Mar 2011
5. Hespanhol I (2003) Potencial de reuso de água no Brasil: agricultura, indústria, municípios, recarga de aquíferos. Bahia Análise & Dados, Salvador v.13, n. especial. http://www.hidro.ufcg.edu.br/twiki/pub/ChuvaNet/ChuvaTrabalhosPublicados/PotencialdereusodeguanoBrasilagriculturaindstriamunicpiosrecargadeaqferos.pdf. Accessed Mar 2011
6. Asano T et al (2007) Water reuse: issues, technologies, and applications. McGraw Hill, New York
7. Laws BV, Dickenson ERV, Johnson TA, Snyder SA, Drewes JE (2011) Attenuation of contaminants of emerging concern during surface-spreading aquifer recharge. Sci Total Environ 409:1087–1094
8. Todd D (1980) Groundwater hydrology. Wiley, New York
9. FIESP/CIESP/ANA (2004) Conservação e reúso de água: Manual de orientações para o setor industrial. http://www.ana.gov.br/Destaque/docs/d179-breuso.pdf. Accessed 10 Mar 2011
10. Paulino JA (2010) Engenharia no dia a dia. http://engenharianodiaadia.blogspot.com/2010/11/implantacao-de-programas-de-conservacao.html. Accessed Mar 2011
11. Bixio D, Thoeye C, Wintgens T, Ravazzini A, Miska V, Muston M, Chikurel H, Aharoni A, Joksimovic D, Melin T (2008) Water reclamation and reuse: implemetation and management issues. Desalination 218:13–23
12. Urkiaga A, de las Fuentes L, Bis B, Chiru E, Balasz B, Hernández F (2008) Development of analysis tools for social, economic and ecological effects of water reuse. Desalination 218:81–91
13. Nolde E (1999) Greywater reuse systems for toilet flushing in multi-storey buildings – over ten years experience in Berlin. Urban Water 1:275–84
14. Salgot M, Huertas E, Weber S, Dott W, Hollender J (2006) Wastewater reuse and risk: definition of key objectives. Desalination 187:29–40
15. Metcalf & Eddy (2003) Wastewater engineering: treatment and reuse, 4th edn. McGraw-Hill, Boston. ISBN 0-07-041878-0
16. APHA/AWWHA/WEF (1998) Standard methods for the examination of water and wastewater, 20th edn. APHA American Public Health Association, Washington. ISBN 0-87553-235-7
17. NIIR Board (2004) Modern technology of waste management: pollution control, recycling, treatment & utilization. National Institute of Industrial Research, Delhi. ISBN 8178330849
18. USEPA (2003) Environmental regulations and technology control of pathogens and vector attraction in sewage sludge. USEPA, Cincinnati
19. WERF - Water Environment Research Foundation (2004) Reduction of pathogens, indicator bacteria, and alternative indicators by wastewater treatment and reclamation processes. IWA Publishing, London
20. Yates V (2011) Pathogens in reclaimed water. http://www.geoflow.com/wastewater/pathogens.htm. Accessed 16 Mar 2011
21. Henjum MB, Hozalski RM, Wennen CR, Arnold W, Novak PJ (2009) Correlations between in situ sensor measurements and trace organic pollutants in urban streams. J Environ Monit. doi:10.1039/b912544b
22. Asano T (2001) Water from (waste)water – the dependable water resource. Stockholm water symposium, Sweden
23. Colt J (2006) Water quality requirements for reuse systems. Aquac Eng 34:143–156
24. Cao Q, Yu Q, Des W (2010) Connell Fate simulation and risk assessment of endocrine disrupting chemicals in a reservoir receiving recycled wastewater. Sci Total Environ 408:6243–6250
25. Berg VM et al (1998) Toxic equivalency factors (TEFs) for PCBs, PCDDs, PCDFs for humans and wildlife. Environ Health Perspect 106:775–792
26. Berg VM, Peterson RE, Schrenk D (2000) Human risk assessment and TEFs. Food Addit Contam 17:347–358

27. Australian Department of Environment and Heritage (2005) National plan for addressing dioxins in Australia. National Dioxins Program. Australian Government
28. Rodriguez C, Cook A, Devine B, Van Buynder P, Lugg R, Linge K, Weinstein P (2008) Dioxins, furans and PCBs in recycled water for indirect potable reuse. Int J Environ Res Public Health 5(5):356–367
29. Berg VM, Birnbaum LS, Denison M, De Vito M, Farland W, Feeley M et al (2006) The 2005 World Health Organization reevaluation of human and mammalian toxic equivalency factors for dioxins and dioxin-like compounds. Toxicol Sci 93:223–41
30. Radic S, Stipanicev D, Cvjetko P, Mikelic IL, Rajcic MM, Sirac S, Pevalek-Kozlina B, Pavlica M (2010) Ecotoxicological assessment of industrial effluent using duckweed (*Lemna minor L.*) as a test organism. Ecotoxicology 19:216–222
31. Wang W, Freemark K (1995) The use of plants for environmental monitoring and assessment. Ecotox Environ Safe 30:289–301
32. Rodrigues ES, Umbuzeiro GA (2011) Integrating toxicity testing in the wastewater management of chemical storage terminals – a proposal based on a ten-year study. J Hazard Mater. doi:10.1016/j.jhazmat.2010.12.083
33. Bakopouloun S, Emmanouil C, Kungolos A (2011) Assessment of wastewater effluent quality in Thessaly region, Greece, for determining its irrigation reuse potential. Ecotoxicol Environ Saf 74:188–194
34. Navarro AV, González MCR, López ER, Aguilar RD, Marçal WS (1999) Evaluation of daphnia magna as an indicator of toxicity and treatment efficacy of textile waste waters. Environ Int 25(5):619–624
35. Peltier W, Weber CI (1985) Methods for measuring the acute toxicity of effluents to freshwater and marine organisms. EPA/600/4-850/013, p. 140. Cincinnati: U.S. Environmental Protection Agency
36. Vigneswaran SM, Sundaravadivel M. Wastewater recycle, reuse, and reclamation – recycle and reuse of domestic wastewater. http://www.eolss.net/ebooks/Sample%20Chapters/C07/E2-14-01.pdf. Accessed 16 Mar 2011
37. Mohsen MS, Jaber JO (2002) Potential of industrial wastewater reuse. Desalination 152:281–289
38. Volkman S (2003) Sustainable wastewater treatment and reuse in urban areas of the developing world. http://www.cee.mtu.edu/peacecorps. Accessed 16 Mar 2011
39. Verlicchi P, Masotti L, Galletti A (2011) Wastewater polishing index: a tool for a rapid quality assessment of reclaimed wastewater. Environ Monit Assess 173:267–277
40. Yagow G, Shanholtz V (1996) Procedures for indexing monthly NPS pollution loads from agricultural and urban fringe watersheds. In: Proceedings of watershed 96 conference, pp. 431–434. Accessed 13 July 2009
41. Bordalo AA, Teixeira R, Wiebe WJ (2006) A water quality index applied to an international shared river basin: the case of the Douro River. Environ Manag 38:910–920
42. Abbası SA (2002) Water quality indices', state of the art report (pp. 73). Scientific contribution No. INCOH/SAR-25/2002. Roorkee: INCOH, National Institute of Hydrology
43. Boyacioglu H (2010) Utilization of the water quality index method as a classification tool. Environ Monit Assess 167:115–124
44. Said A, Stevens DK, Sehlke G (2004) Environmental assessment an innovative index for evaluating water quality in streams. Environ Manag 34(3):406–414
45. Ammary BY (2007) Wastewater reuse in Jordan: present status and future plans. Desalination 211:164–176
46. Genon G (2011) Water recycling and on-site reuse. https://www.unido.org/foresight/rwp/dokums_pres/water_symposiun_genon_214.pdf. Accessed 16 Mar 2011
47. Iglesias R, Ortega E, Batanero G, Quintas L (2010) Water reuse in Spain: data overview and costs estimation of suitable treatment trains. Desalination 263:1–10
48. Campos C (1999) Indicadores de contaminación fecal en la reutilización de agua residual regenerada en suelos. In: PhD thesis, University of Barcelona

49. Li F, Wichmann K, Otterpohl R (2009) Review of the technological approaches for grey water treatment and reuse. Sci Total Environ 407:3439–3449
50. Deniz F, Sadhwani JJ, Veza JM (2010) New quality criteria in wastewater reuse: the case of Gran Canaria. Desalination 250:716–722
51. WHO (2006) WHO guidelines for the Safe use of wastewater, excreta and greywater, vol 4. WHO, Geneva. ISBN 92 4 154685 9
52. You SJ, Wu DC (2009) Potential for reuse of high cellulose containing wastewater after membrane bioreactor treatment. Desalination 249:721–728
53. Bixio D, Thoeye C, De Koning J, Joksimovic D, Savic D, Wintgens T, Melin T (2006) Wastewater reuse in Europe. Desalination 187:89–101
54. Angelakis AN, Durham B (2008) Water recycling and reuse in EUREAU countries: trends and challenges. Desalination 218:3–12
55. Cortacáns JA (2008) Water reuse planning and management: the point of view of a contractor-operator. Desalination 218:74–80
56. WBCSD - World Business Council for Sustainable Development (2006) Business in the world of water WBCSD water scenarios to 2025. Atar Roto Presse SA, Geneva
57. Aviso KB, Tan RR, Culaba AB, Cruz JB Jr (2011) Fuzzy input output model for optimizing eco-industrial supply chains under water footprint constraints. J Clean Prod 19:187–196
58. Helmer R, Hespanhol I (2004) Water pollution control. SPon Press, London. ISBN 0419229108
59. FIESP-Federação e Centro das Indústrias do Estado de São Paulo. Conservação e reúso de água Manual de Orientações para o Setor Industrial, vol. 1 http://www.fiesp.com.br/publicacoes/pdf/ambiente/reuso.pdf. Accessed 16 Mar 2011
60. Giurco S, Bossilkov A, Patterson J, Kazaglis A (2011) Developing industrial water reuse synergies in Port Melbourne: cost effectiveness, barriers and opportunities. J Clean Prod 19 (8):867–876
61. ABS – Australian Bureau of Statistics (2000) Water account data 1993–94 to 1996–97. Electronic file, unpublished
62. Zhao X, Chen B, Yang ZF (2009) National water footprint in an input–output framework – a case study of China 2002. Ecol Model 220:245–253
63. Gumbo B, Forster L, Arntzen J (2005) Capacity building in water demand management as a key component for attaining millennium development goals. Phys Chem Earth 30:984–992
64. Esquerre KPO, Kiperstok A, Mattos MC, Cohim E, Kalid R, Sales EA, Pires VM (2011) Taking advantage of storm and waste water retention basins as part of water use minimization in industrial sites. Resour Conserv Recycl 55:316–324
65. Hammer MJ (2004) Water and wastewater technology, 5th edn. Prentice Hall, Upper Saddle River. ISBN 0-13-097325-4
66. Nathanson JA (2003) Basic environmental technology: water supply, water management, and pollution control, 4th edn. Prentice Hall, Upper Saddle River. ISBN 0-13-093085
67. El-Halwagi MM (1997) Pollution prevention through process integration: systematic design tools. Academic, San Diego
68. Wang YP, Smith R (1994) Wastewater minimization. Chem Eng Sci 49(7):981–1006
69. Chakraborty A (2009) A globally convergent mathematical model for synthesizing topologically constrained water recycle networks. Comput Chem Eng 33:1279–1288
70. Foo DCY (2008) Flowrate targeting for threshold problems and plant-wide integration for water network synthesis. J Environ Manag 88:253–274
71. Vidigueira F, Ferreira E (2011) Estratégias de Minimização de Efluentes na Síntesee Integração de Processos. http://repositorium.sdum.uminho.pt/bitstream/1822/3408/1/CHEMPOR-MSC25%5B1%5D.pdf. Accessed 10 Mar 2011
72. Valle EC (2005) Minimização de água e Efluentes com Considerações Econômicas e Operacionais via Programação Matemática, Dissertation. Federal University of Grande do Sul, Porto Alegre, 2005. http://www.lume.ufrgs.br/handle/10183/8170. Accessed 10 Mar 2011

73. El-Halwagi MM, Manousiouthakis V (1989) Synthesis of mass exchange networks. AIChE J 35(8):1233–1244
74. Bai J, Feng X, Deng C (2007) Optimization of single-contaminant regeneration reuse water systems. Trans IChemE, A, Chem Eng Res Des 85(A8):1178–1187
75. Mann JG, Liu YA (1999) Industrial water reuse and wastewater minimization. McGraw-Hill, New York
76. Castro P, Matos H, Fernandes MC, Nunes CP (1999) Improvements for mass-exchange networks design. Chem Eng Sci 54:1649–1665
77. Bandyopadhyay S, Cormos CC (2008) Water management in process industries incorporating regeneration and recycle through a single treatment unit. Ind Eng Chem Res 47 (4):1111–1119
78. Savelski M, Bagajewicz M (2000) Design of water utilization systems in process plants with a single contaminant. Water Manag 20(8):659–664
79. Anantha PR, Koppol MJ, Bagajewicz B, Dericks J, Savelski MJ (2004) On zero water discharge solutions in the process industry. Adv Environ Res 8(2):151–171
80. Bagajewicz M (2001) On the use of linear models for the design of water utilization systems in process plants with a single contaminant. Trans IChemE A 79:600–610
81. Feng X, Chu K (2004) Cost optimization of industrial wastewater reuse systems. Process Saf Environ Prot 82(B3):249–255
82. Gunaratnam M, Alva-Argaez A, Kokossis A, Kim J-K, Smith R (2005) Automated design of total water systems. Ind Eng Chem Res 44:588–599
83. Huang CH, Chang CT, Ling HC, Chang CC (1999) A mathematical programming model for water usage and treatment network design. Ind Eng Chem Res 38:2666–2679
84. Takama N, Kuriyama T, Shiroko K, Umeda T (1980) Optimal water allocation in a petroleum refinery. Comput Chem Eng 4:251–258
85. Xu DM, Hu YD, Hua B, Wang XL (2004) Optimum design of water-using network with regeneration reuse. Chinese J Chem Eng 18(2):202–207
86. Xu DM, Hu YD, Hua B, Wang XL (2003) Minimization of the flowrate of fresh water and corresponding regenerated water in water-using system with regeneration reuse. Chinese J Chem Eng 11(3):257–263
87. Feng X, Bai J, Zheng XS (2007) On the use of graphical method to determine the targets of single-contaminant regeneration recycling water systems. Chem Eng Sci 62:2127–2138
88. Rubio-Castro E, Ponce-Ortega JM, Serna-González M, Jiménez-Gutierrez A, El-Halwagi MM (2010) A global optimal formulation for the water integration in eco-industrial parks considering multiple pollutants. Comput Chem Eng. doi:10.1016/j.compchemeng.2011.03.010
89. Rosi OL, Casarci M, Mattioli D, De Florio L (2007) Best available technique for water reuse in textile SMEs (BATTLE LIFE Project). Desalination 206:614–619
90. Debik E, Kaykioglu G, Coban A, Koyuncu I (2010) Reuse of anaerobically and aerobically pre-treated textile wastewater by UF and NF membranes. Desalination 256:174–180
91. Casani S, Rouhany M, Knochel S (2005) A discussion paper on challenges and limitations to water reuse and hygiene in the food industry. Water Res 39:1134–1146
92. Notermans S, Zwietering MH, Mead GC (1994) The HACCP concept: identification of potentially hazardous microorganisms. Food Microbiol 11:203–214
93. Vourch M, Balannec B, Chaufer B, Dorange G (2005) Nanofiltration and reverse osmosis of model process waters from the dairy industry to produce water for reuse. Desalination 172:245–256
94. Fähnrich A, Mavrov V, Chmiel H (1998) Membrane processes for water reuse in the food industry. Desalination 119:213–216
95. Mavrov V, Fäihnrich A, Chmiel H (1997) Treatment of low-contaminated waste water from the food industry to produce water of drinking quality for reuse. Desalination 113:197–203
96. Álvarez PM, Pocostales P, Beltrán FJ (2011) Granular activated carbon promoted ozonation of a food-processing secondary effluent. J Hazard Mater 185:776–783

97. Matsumura EM, Mierzwa JC (2008) Water conservation and reuse in poultry processing plant – a case study. Conserv Recycl 52:835–842
98. Faria DC, de Souza AAU, de Souza SMAGU (2009) Optimization of water networks in industrial processes. J Clean Prod 17:857–862
99. Martins MAF, Amaro CAA, Souza LSS, Kalid RA, Kiperstok A (2010) New objective function for data reconciliation in water balance from industrial processes. J Clean Prod 18:1184–1189
100. Mirbagheri SA, Poshtegal MK, Parisai MS (2010) Removing of urea and ammonia from petrochemical industries with the objective of reuse, in a pilot scale: surveying of the methods of waste water treatment. Desalination 256:70–76
101. Gutterres M, Aquim PM, Passos JB, Trierweiler JO (2010) Water reuse in tannery beamhouse process. J Clean Prod 18:1545–1552
102. Gutterres M, Passos JB, Aquim PM, Severo LS, Trierweiler JO (2008) Reduction of water demand and treatment cost in tanneries through reuse technique. J Am Leather Chem Assoc 103(4):138–143
103. Aquim PM, Gutterres M, Trierweiler JO (2010) Assessment of water management in tanneries. J Soc Leather Technol Chem 94(6):253–258
104. Kanagaraj J, Chandra Babu NK, Mandal AB (2008) Recovery and reuse of chromium from chrome tanning waste water aiming towards zero discharge of pollution. J Clean Prod 16:1807–1813
105. Braeken L, Van der Bruggen B, Vandecasteele C (2004) Regeneration of brewery waste water using nanofiltration. Water Res 38:3075–3082
106. Tay JH, Jeyaseelan S (1995) Membrane filtration for reuse of wastewater from beverage industry. Resour Conserv Recycl I5:33–40
107. NCASI (2003) Memo report from Jay Unwin, April 23
108. Adewumi JR, Llemobade AA, Van Zyl JE (2010) Treated wastewater reuse in South Africa: overview, potential and challenges. Resour Conserv Recycl 55:221–231
109. CONSEMA Resolution 129/2006. Rio Grande do Sul State, Brazil. http://www.fepam.rs.gov.br. Accessed 10 Mar 2011
110. Council Directive 91/271/EEC. http://ec.europa.eu/environment/water/water-urbanwaste/index_en.htmljectivo. Accessed 10 Mar 2011
111. Hoeven VN (1998) Power analysis for the NOEC: what is the probability of detecting small toxic effects on three different species using the appropriate standardized test protocols? Ecotoxicology 7:355–361
112. Taking advantage of storm and waste water retention basins as part of water use minimization in industrial sites
113. USP. Espaço aberto. http://www.usp.br/espacoaberto/arquivo/2003/espaco32jun/vaipara.php?materia=0capa. Accessed Apr 2011
114. Van den Berg M, Birnbaum LS, Denison M, De Vito M, Farland W, Feeley M (2006) World Health Organization reevaluation of human and mammalian toxic equivalency factors for dioxins and dioxin-like compounds. Toxicol Sci 93:223–41
115. WHO (1989) Health guidelines for the use of wastewater in agriculture and aquaculture. WHO Technical Report Series 778, Geneva

Chapter 6
Hazardous Agents in Wastewater: Public Health Impacts and Treatment Options for Safe Disposal and Reuse

Keisuke Ikehata

List of Abbreviations

AOP	Advanced oxidation process
DBP	Disinfection byproduct
DDT	Dichlorodiphenyltrichloroethane
DPR	Direct potable reuse
E. coli	*Escherichia coli*
H_2O_2	Hydrogen peroxide
IPR	Indirect potable reuse
MBR	Membrane bioreactor
MF	Microfiltration
NDMA	*N*-Nitrosodimethylamine
NF	Nanofiltration
O_3	Ozone
PCB	Polychlorinated biphenyl
QMRA	Quantitative microbial risk assessment
QRRA	Quantitative relative risk assessment
RO	Reverse osmosis
UF	Ultrafiltration
U.S. EPA	United States Environmental Protection Agency
U.S. GAO	United States General Accounting Office
UV	Ultraviolet
WHO	World Health Organization

K. Ikehata (✉)
Pacific Advanced Civil Engineering, Inc., 17520 Newhope Street,
Suite 200, Fountain Valley, CA 92708, USA
e-mail: kikehata@pacewater.com

S.K. Sharma and R. Sanghi (eds.), *Wastewater Reuse and Management*,
DOI 10.1007/978-94-007-4942-9_6,

6.1 Introduction

In the water cycle, water continuously moves on, above, and below the surface of the planet Earth. As part of this global water cycle, humans withdraw raw water from the aquatic environment, use it in different activities, typically add some constituents to it, and release the used water back into the aquatic environment. The used water is generally called "wastewater" because it is considered a useless byproduct of human activities. Wastewater usually contains high levels of chemical and biological agents that are potentially harmful to public health if it is reused as is. Those chemical and biological agents are often considered as "contaminants" because their levels are higher than their original levels in raw water.

Because of the global water scarcity, more and more wastewater is being reused nowadays with or without treatment for different purposes from landscape and crop irrigation to industrial applications and other potable and non-potable domestic uses such as toilet flushing and dust control [1]. Wastewater is a readily available, drought-resistant water resource. Wastewater could also provide high nutrient values to crops and non-crop vegetations if it was used for irrigation.

In fact, wastewater reuse is not a new idea. Wastewater reuse has been practiced all over the world, intentionally and unintentionally, since the beginning of civilization. For example, wastewater irrigation for crop production has been a common practice in arid and semiarid areas [2]. Through discharge, wastewater has been and will be recycled in the water cycle. Consequently, some contaminants such as heavy metals may accumulate in soil and crops via wastewater irrigation, while other contaminants may be introduced into surface water and groundwater and transported to downstream (Fig. 6.1).

It has been a presumption of wastewater discharge that wastewater contaminants will dissipate and disappear by dilution, dispersion, settling, volatilization, and/or transformation in the natural environment by the time the next user touches the same water again. Unfortunately, natural cleansing processes could not be able to keep up with the speed of civilization and industrial development. Hazardous biological and chemical constituents in the wastewater effluent adversely affected and degraded the natural ecosystem, as well as the living environment of humans, which caused numerous outbreaks of deadly infectious diseases and acute and chronic toxicity events around the world. Wastewater treatment has become a critical component of safe and sustainable water resource management.

In addition to conventional wastewater constituents, a number of more persistent and/or potentially harmful chemical agents, such as pharmaceuticals, industrial chemicals, and personal care products, originated from wastewater have been found in receiving water, as well as in wastewater effluent and reclaimed water [3, 4]. Resistance of pathogenic microorganisms in wastewater toward conventional chlorine disinfection has also been realized [2]. It is evident that these persistent, potentially hazardous chemical and biological agents cannot be removed by conventional wastewater treatment processes.

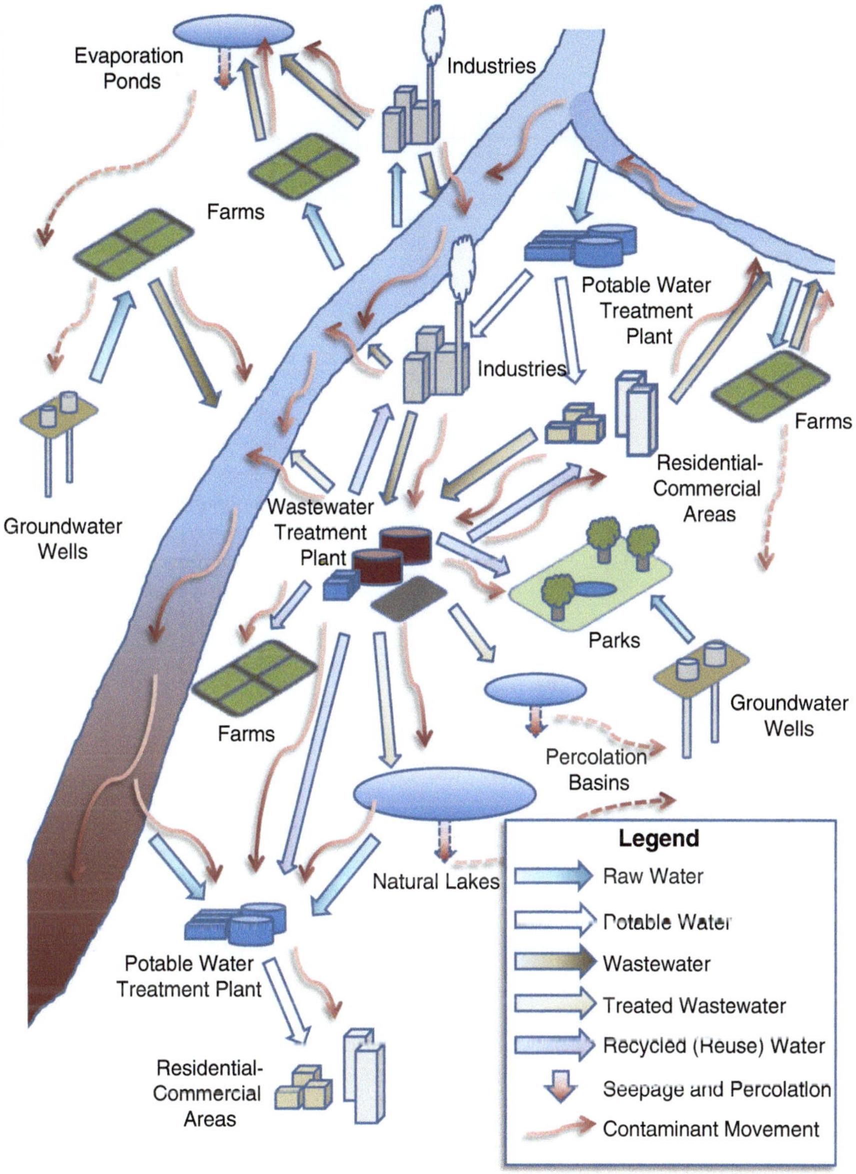

Fig. 6.1 Generation, treatment, disposal, and reuse of wastewater in the urban and rural aquatic environment.

Potential public health impacts of old and new wastewater constituents are especially important in modern water reuse scenarios, such as planned indirect potable reuse (IPR) and direct potable reuse (DPR) [5, 6], where the distances between reclaimed water producers and users will be much shorter than those in

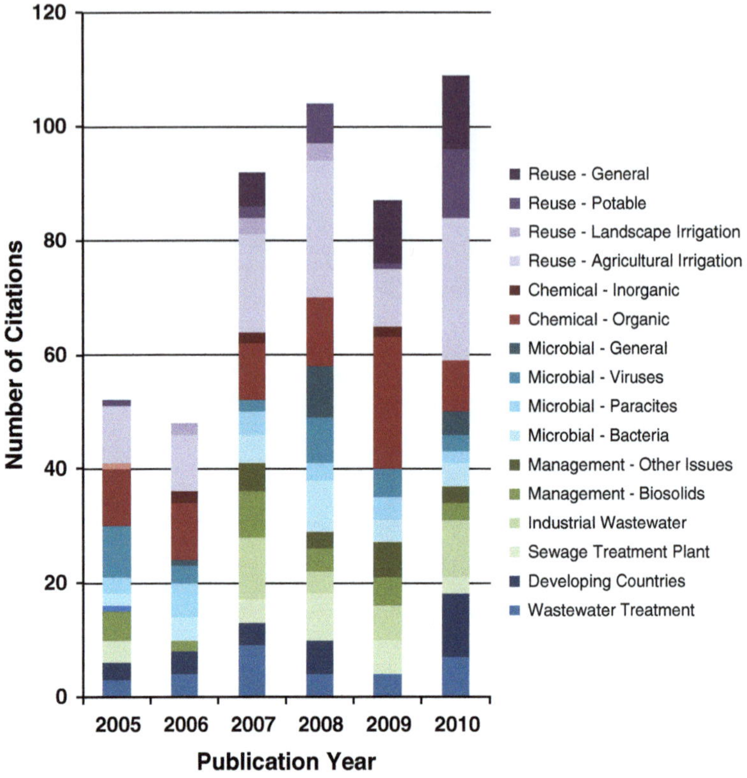

Fig. 6.2 Number of journal papers related to "health effects associated with wastewater treatment, reuse, and disposal" published between 2005 and 2010 (Sources: Refs. [7–12]).

indirect, unintentional water reuse within natural hydrologic cycles. In addition, in rural and peri-urban areas of many developing countries, the use of wastewater for irrigation is becoming more and more common. Concerns for human health, as well as for the environment, are the critical constraints to encourage and promote those wastewater reuse practices, as well as to ensure safe discharge to the environment, in both developed and developing countries.

Consequently, public health implications of wastewater management have become an important research subject. Based on the author's annual literature survey, the number of peer-reviewed publications dealing with the subject matter of wastewater and public health has been increasing in the past 6 years [7–12]. In particular, the number of research papers dealing with public health impact of wastewater reuse is increasing steadily (Fig. 6.2). The health effect of organic compounds, such as pharmaceuticals and personal care products (i.e., emerging contaminants), is another actively studied subject in the recent years.

The aim of this chapter is to provide an overview of public health implications of wastewater constituents with respect to various wastewater management scenarios, including treatment, disposal, and reuse. Wastewater treatment options to achieve various effluent/reclaimed water qualities for safe disposal and reuse are also presented. Regulatory aspects are not the main focus of this chapter, thus are not covered in detail.

6.2 Hazardous Agents in Wastewater and Their Health Effects

In addition to relatively benign or inert constituents such as biodegradable organics (e.g., fat, carbohydrate, and proteins) and inorganics (e.g., sodium, chloride, and silica), wastewaters contain a wide range of potentially harmful biological and chemical agents (Table 6.1). Examples of known adverse health risks associated with these wastewater constituents include gastroenteritis, meningitis, fevers, diarrhea, and other infectious diseases due to microbial infection and toxin production, and various acute and chronic toxicities of chemical agents, including cancers and potential endocrine disruption.

In order to assess the health risk associated with wastewater disposal and reuse, (1) chemical and microbial analyses and (2) either epidemiological data or development of quantitative risk assessment model for a particular constituent are required. Chemical and microbial analyses can quantify the concentrations of potentially hazardous agents, such as microbial pathogens and toxic chemicals, in raw and treated wastewater, including reclaimed water.

Epidemiological data, if available, can provide a historical evidence of actual hazard in a given exposed population during a health-related event. However, the availability of such information is limited, especially for chemical hazards present in wastewater. Thus, risk assessments are often conducted to estimate whether (qualitative) and/or how much (quantitative) damage or injury can be expected from an exposure to a given risk agent in a particular wastewater management scenario.

6.2.1 Microbial Hazards

A variety of human microbial pathogens are present in raw wastewater, including enteric bacteria, helminthes, protozoa, and viruses, as shown in Table 6.2. Epidemiological studies and quantitative microbial risk assessment (QMRA) have been used to measure and quantify the health risks associated with these microbial agents [1, 14].

A review of earlier epidemiological studies of wastewater reuse in agriculture is presented in WHO [14]. Based on the epidemiological data, it is clear that the use of untreated and partially treated wastewater to irrigate vegetables can pose a high risk of transmission of bacterial and helminth infections to the consumers and farm workers [15]. A significant risk of parasite infection (such as ascariasis

Table 6.1 Examples of wastewater constituents with known adverse health risks

Constituent	Example	Health risk	Exposure route
Biological			
Enteric bacteria	*E. coli*, *Vibrio cholerae*, *Salmonella* spp., *Shigella* spp.	Gastroenteritis, meningitis, toxin production, pneumonia	Contact, consumption, ingestion
Helminthes	*Ascaris* spp., *Ancylostoma* spp., *Necator* spp., *Hymenolepis* spp.	Intestinal obstruction, insomnia, stomach pains	Contact, consumption, ingestion, vectors
Protozoa	*Giardia* spp., *Cryptosporidium parvum*, *Entamoeba* spp.	Gastroenteritis, diarrhea, stomach pains, dehydration	Contact, consumption, ingestion
Viruses	Hepatitis A and E viruses, adenovirus, rotavirus, norovirus	Flu-like symptoms (fever, fatigue, abdominal pain, nausea)	Contact, consumption, ingestion, vectors
Chemical, inorganic			
Heavy metals	Mercury, lead, arsenic, cadmium	Bioaccumulation, acute toxicity	Consumption, ingestion
Nitrate	–	Methemoglobinemia (blue baby syndrome)	Ingestion
Perchlorate	–	Interfering iodine uptake	Ingestion
Chemical, organic			
Halogenated hydrocarbons	Chlorinated solvents, dioxins, PCBs	Bioaccumulation, acute and chronic toxicity including carcinogenicity	Consumption, ingestion, inhalation (volatile compounds only)
Pesticides	Organochlorine pesticides (DDT, aldrin, etc.), organophosphorus pesticides	Acute and chronic toxicity	Contact, consumption, ingestion
Antibiotics	Amoxicillin, tetracycline, sulfamethoxazole, erythromycin	Potential risk of side effects of antibiotics (gastrointestinal upset and diarrhea)	Consumption, ingestion
1,4-Dioxane	–	Probable human carcinogen	Ingestion
Phthalates and phenols	Alkylphenols, bisphenol-A, chlorophenols, diethyl phthalate, bis(2-ethylhexyl) phthalate	Potential acute toxicity, endocrine disruption	Inhalation, consumption, ingestion

Sources: Refs. [13, 14]

Abbreviations: *PCBs* polychlorinated biphenyls, *DDT* dichlorodiphenyltrichloroethane

Table 6.2 Hazardous chemical and biological agents in different types of wastewater

Wastewater origin	Chemical, inorganic	Chemical, organic	Biological
Domestic	Sulfides, ammonia, some metals	Trace organics (e.g., pesticides, pharmaceuticals)	Enteric bacteria, helminthes, protozoa, viruses
Industrial	Heavy metals, sulfides, ammonia, cyanides	Solvents, aromatic hydrocarbons	Vary
Agricultural	Sulfides, some metals	Pesticides, veterinary pharmaceuticals	Enteric bacteria, helminthes, protozoa, viruses

and giardiasis), cholera, typhoid, and shigellosis, as well as *Helicobacter pylori*, *Listeria* spp., *Salmonella* spp., enteropathogenic *Escherichia coli*, rotavirus, poliovirus, and other enterovirus infections, and nonspecific diarrhea associated with wastewater reuse and discharge has been reported.

The association between infectious disease outbreaks and inadequate sewage sanitation systems in both developed and developing countries has been reported. For example, Costa et al. [16] suggested a possible association of two large outbreaks of visceral leishmaniasis caused by a protozoan parasite, *Leishmania chagasi*, with inadequate sewage in Teresina, Brazil. Similar studies were reported on hepatitis A and D viruses in Bosnia and Herzegovina [17]; human norovirus in the Netherlands [18], Canada [19], and Japan [20]; *Helicobacter pylori* in India [21]; *E. coli* O157:H7 in Tanzania [22]; *Vibrio cholera* in Morocco [23]; and *Giardia* in Gasa Strip [24].

Quantitative risk assessment, such as QMRA, has been widely used to complement to epidemiological studies, which are more costly to perform. QMRA methodologies are described in detail in many publications, such as [25], [1], and [26]. In general, QMRA consists of four parts, namely,

- Hazard identification
- Exposure assessment
- Dose response
- Risk characterization

Critical data for constructing a QMRA model include pathogen dose (= concentration), exposure intensity, dose–response data, and infection duration [1]. Rotavirus, enteroviruses, *Giardia lamblia*, *Cryptosporidium*, *Campylobacter*, and *E. coli* are the common pathogenic microorganisms that have been evaluated for their potential public health risk in wastewater-related QMRA studies [27–33].

6.2.2 Chemical Hazards

There are three categories of chemical hazards in raw and treated wastewater, including (1) acute and chronic toxicity, (2) carcinogenicity, and (3) reproductive, developmental, and neurotoxicity. It is an assumption that dose thresholds do not exist for carcinogenic and neurotoxic effects. More than one toxic effect can be exerted by the same chemical substance. Examples of carcinogens that may be found in wastewater are arsenic, chloroform, 1,4-dioxane, dieldrin, heptachlor, and vinyl chloride [1]. In addition, potential health effects of trace emerging contaminants, such as pharmaceuticals and personal care products, in wastewater effluents are a rapidly evolving field [4, 34–36]. Many of these chemicals are biologically active at higher dosages; however, their chronic effects at the concentrations occurring in treated wastewater are still largely unknown.

While the health risks associated with microbial wastewater contaminants are relatively easy to measure and quantify using diarrhea and intestinal infections as general indicators, less evidence is available for the health risk associated chemicals in wastewater. This is because of the following reasons:

- Chemical exposures and their health outcomes are harder to measure.
- Many health effects of chemicals are chronic and take longer to develop.
- Observed health effect may be caused by not just one but a mixture of chemical agents in wastewater, as well as other environmental and genetic factors.

In general, there is little epidemiological data for chemicals available. Data from toxicological studies involving animal exposures, instead of those with human subjects, were usually used in chemical risk assessment. This requires extrapolation from animal data to human, which introduces a major uncertainty to the risk assessment model. Animal toxicological data are also limited to a small fraction of high-production-volume chemicals [37]. Extrapolation from high experimental dosages to extremely low environmental exposures is another source of uncertainty. A lack of appropriate indicators to monitor chemical exposures is also a major limitation in chemical risk assessment [38].

Similar to the case of microbial risks, quantitative risk assessment has been used to determine the relative risks of chemical wastewater constituents and to determine maximum tolerable concentrations of chemicals [39–43]. For example, Schwab et al. [44] conducted a human health risk assessment for 26 environmentally relevant pharmaceuticals and their metabolites in the United States. Weber et al. [45] performed a chemical risk assessment to determine the acceptable contaminant levels for three selected chemicals, including chloroform, 1,1,2-trichloroethane, and pyrene, in reclaimed wastewater for agricultural irrigation. More recently, a quantitative relative risk assessment (QRRA) methodology was used to evaluate the potential risk associated with reclaimed water for IPR in terms of emerging contaminants, such as pharmaceuticals and personal care products [46].

For wastewater reuse in agriculture, maximum tolerable soil concentrations of toxic chemicals may be used as numerical limits [14], whereas applicable drinking water standards and guidelines should be used for potable reuse applications. Because of uncertainties discussed above, risk assessment in many wastewater reuse applications will remain more qualitative than quantitative [1].

6.2.3 *Exposure Routes*

People are exposed to these wastewater constituents via a variety of routes, including direct skin contact, ingestion of contaminated water, inhalation of vapors and aerosols, consumption of contaminated crop and meat products, and person-to-person transmission. For example, those people who handle raw and treated wastewater, such as farmers, fishery workers, and wastewater workers are usually at the highest risk of helminth infections [14]. Certain population groups, such as pregnant women, infants, children under a certain age, elderly, immunocompromised patients, and lower socioeconomic classes are also more vulnerable and thus are at a considerably higher risk [47–49].

6.2.4 *Wastewater Sources*

Wastewater compositions vary significantly depending on the region, climate, time of day, and season, as well as the source and origin of wastewater, such as domestic, industrial, and agricultural (Table 6.2). Whereas the qualities of domestic wastewater are fairly similar from municipality to municipality, very wide variation in wastewater characteristics appears among different industries (e.g., mining, oil and gas, pulp and paper, chemical manufacturing, textile, and food processing) and agricultural activities (e.g., food and non-food crop production, fruits and vegetables production, livestock and dairy farming, and aquaculture). Thus, different hazardous agents are present in wastewaters from industries and agricultural activities.

In general, the major group of hazardous agents typically found in domestic wastewater is microbial pathogens, such as enteropathogenic *E. coli*, *Salmonella* spp., and *Giardia* spp., while industrial wastewater typically contains more toxic chemicals such as heavy metals, cyanides, and chlorinated solvents [50–52]. Agricultural wastewater may contain pesticides (e.g., insecticides, herbicides, and fungicides) and veterinary pharmaceuticals such as growth hormones and antibiotics, as well as microbial pathogens.

Hazardous agents may be removed or inactivated during wastewater storage and treatment, as well as after the discharge into the environment. As a result, the health risk changes significantly under different wastewater management scenarios. These issues will be discussed further in the subsequent sections.

6.3 Hazard and Wastewater Management Scenarios

The goal of wastewater management is to promote effective and responsible water use (including reuse), treatment, and disposal, as well as to protect the public and environmental health. Different types and levels of hazard exist in different wastewater management scenarios, such as:

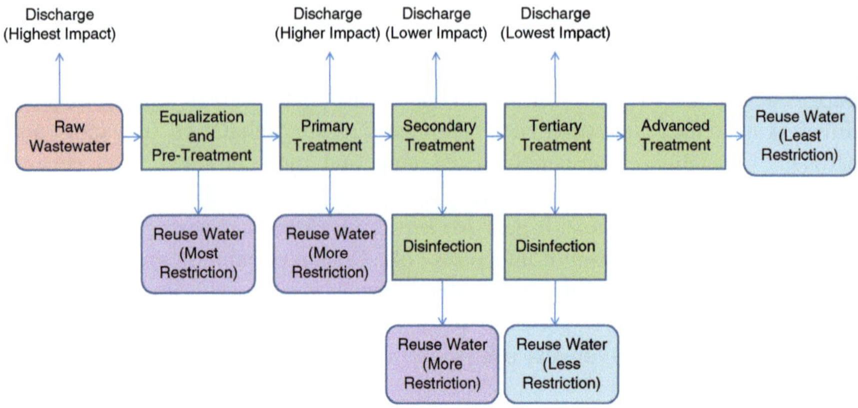

Fig. 6.3 Simplified wastewater treatment, discharge, and reuse schemes

1. Disposal only
   - No treatment
   - Pretreatment or primary treatment only
   - Secondary treatment and disinfection
   - Tertiary treatment and disinfection
2. Reuse (and disposal)
   - Raw wastewater reuse in agriculture and aquaculture
   - Primary or secondary treatment and reuse in agriculture and aquaculture
   - Tertiary treatment for urban non-potable reuse (e.g., park and golf course irrigation)
   - Advanced treatment for urban non-potable and potable reuse
   - Industrial reuse

Figure 6.3 illustrates these different wastewater treatment, discharge, and reuse schemes. Table 6.3 summarizes the relative risk severity among those affected populations under different wastewater management scenarios.

6.4 Disposal Only

In general, wastewater disposal will naturally affect the water body and downstream communities unless the wastewater is injected into a deep, confined aquifer or discharged into a very large water body, such as the oceans and large lakes, at a very low flow rate. Otherwise, raw and treated wastewater can contaminate rivers, lakes, groundwater aquifers, and oceans (Fig. 6.1). Although some of the wastewater constituents can be settled down, diluted, dispersed, and naturally degraded, some contaminants may remain unaltered and travel a long distance. As a result, those who use the affected water resources in downstream as drinking water, bathing water, and irrigation water may be at risk [53–56]

Table 6.3 Relative hazard severity among affected populations under different wastewater management scenarios

Scenario	Constituent of concern	Affected population	Relative hazard severity
Discharge			
Untreated wastewater or primary-treated effluent without disinfection	All pathogens and chemicals	Communities immediately downstream	Severe
		Communities further downstream	Intermediate
Secondary- or tertiary-treated effluent with disinfection	Persistent organic compounds, nitrate, perchlorate	Communities immediately downstream	Slight to intermediate
		Communities further downstream	Slight
Agricultural reuse			
Untreated wastewater or primary-treated effluent without disinfection	All pathogens and chemicals	Farmers and fishery workers	Severe
		Families of farmers and workers	Intermediate
		Communities adjacent to the facilities	Slight to intermediate
		Consumers	Intermediate
Secondary- or tertiary-treated effluent with disinfection	Persistent pathogens and organic compounds	Farmers and fishery workers	Intermediate
		Families of farmers and workers	Slight to intermediate (pathogens only)
		Communities adjacent to the facilities	Slight
		Consumers	Slight to intermediate
Other non-potable reuse	Persistent pathogens and organic compounds	Users of recreational facilities that use treated wastewater for irrigation, water features, and fishing lakes	Slight
Potable reuse			
Indirect potable reuse	(Similar to discharge scenarios)		
Direct potable reuse	Persistent organic compounds	Consumers	Slight
Industrial reuse			
Internal reuse	Vary	Workers at the industrial facility	Vary
External reuse	Vary	Users of reuse water	Vary

6.4.1 Agricultural and Non-potable Reuse

Reuse of treated and untreated wastewater in agriculture and aquaculture involves more complex exposure routes. Direct contact with wastewater may occur when farmers and fishery workers are handling the wastewater without protective equipment [14, 57, 58]. Transmission of pathogens such as helminthes within the families of those workers may also occur. Inhalation of aerosols and vapor can be a significant exposure route for farmers and fisher workers, as well as for people living in a close proximity to the facilities. The consumers may be exposed to microbial and chemical hazards by consuming contaminated crops and meat products. In addition, inappropriately managed runoff and effluent from the farms and fishery can contaminate the surface and groundwater sources.

Accumulation of toxic heavy metals, such as copper, zinc, mercury, cadmium, nickel, and chromium, as well as pesticides, such as chlorpyrifos and endosulfan, in wastewater-irrigated soils and vegetables has been noticed in many countries and regions, such as Zimbabwe [59], Ghana [60], China [61], Spain [62], Syria [63], and Pakistan [64]. Their levels sometimes exceeded the acceptable levels suggested by WHO [14].

Use of treated wastewater (i.e., reclaimed water) for non-potable reuse, such as landscape and golf course irrigation, water features, noncommercial fishing, and other recreational uses, can also pose potential risk to the people who come into contact with the reused water [1, 65–67]. In addition to direct skin contact, inhalation of aerosols and vapor and accidental ingestion may be possible exposure routes.

6.4.2 Potable Reuse

Although it has been largely overlooked, wastewater discharge to the surface water or groundwater that is used as a drinking water source is equivalent to IPR. Thus, the health risk associated with planned or unplanned IPR is similar to those of corresponding wastewater discharge scenarios. The reservoirs are considered natural buffer and *in situ* advanced treatment in indirect potable reuse scenarios [68], either intentionally or unintentionally. However, in DPR projects, which are still very rare in the present time, wastewater is treated to a very high level of purity to minimize potential health risk and then augmented with treated drinking water. Advanced treatment processes such as reverse osmosis (RO) and advanced oxidation processes (AOPs) were often employed in planned IPR and DPR projects [46, 68, 69].

6.4.3 Industrial Reuse

Industrial reuse scenarios may include onsite internal reuse (i.e., treatment and reuse of its own wastewater in various processes such as cooling, steam generation, oil extraction, and hydraulic fracturing) and external reuse (i.e., export of treated industrial wastewater to offsite for other reuse applications, such as agricultural and domestic irrigation, and reuse in other industries). In addition, treated domestic or agricultural wastewater may be imported to the industry for reuse. Internal reuse and imported water reuse only affect workers handling reuse water and have a minimal impact on public health directly, while domestic and agricultural reuse of industrial effluent may pose a significant health threat to farmers and other affected populations because of high heavy metal and other hazardous chemical contents [52, 61]. Since the compositions of constituents of concern vary among different industries and even within the same industry, it is hard to generalize relative health risks to the public in this case.

Obviously, wastewater treatment is a key to lower the health risks in all types of management scenarios. The impacts of different wastewater treatment processes are discussed in the following section.

6.5 Impact of Wastewater Treatment

Among the wastewater management efforts, wastewater treatment is the primary measure to lower health risk associated with wastewater disposal and reuse and to ensure sustainable water resource management. Wastewater treatment usually consists of a series of physical, chemical, and biological unit processes that can remove specific wastewater constituents in each step (Table 6.4). Different combinations of unit processes may be employed, depending on raw wastewater quality, objectives of treatment and reuse, technical and economic feasibility and constraints, and other site-specific restrictions and considerations. As shown in Fig. 6.3, higher levels of treatment (i.e., tertiary and advanced treatment) are required to produce better quality effluent that has less restriction for reuse and poses lower impact on the receiving environment and downstream communities. However, these higher-level treatment processes are more expensive and technologically demanding.

Alternatively, low-rate wastewater treatment systems, such as waste stabilization ponds, aerated lagoons, and constructed wetlands, may be used as low-cost wastewater treatment options. Because these low-rate treatment systems have a larger land requirement, they may not be suitable for the regions where the land cost is high. Despite their simpler construction and less technically involved operation and maintenance, these low-rate treatment systems can remove some of the hazardous wastewater constituents by sedimentation, biological degradation, and natural sunlight ultraviolet (UV) disinfection [14]. Therefore, these systems are often more appropriate, cost-effective wastewater treatment systems in developing countries [70–72].

6.5.1 *Pre- and Primary Treatment*

Pretreatment and primary treatment involve physical screening and gravity sedimentation of heavy solids or floatation of light solids, such as oil, grease, and fibers. Chemical coagulant and flocculant may be used prior to sedimentation/floatation to

Table 6.4 Typical unit processes and target wastewater constituents

Type	Unit process	Target constituent
Pretreatment	Flow equalization, maceration	–
	Screens, grit removal	Large solid objects (e.g., plant debris, stones, plastics)
Primary treatment	Sedimentation	Heavier solids
	Flotation	Lighter solids (e.g., oil and grease, algal cells, fibers)
Secondary treatment	Activated sludge, trickling filters, rotating biological contactors, membrane bioreactors	Biodegradable organics
Tertiary treatment	Filtration	Suspended solids, including pathogenic microorganisms
	Biological nutrient removal	Nitrogen and phosphorus, more recalcitrant organics
Advanced treatment	Chemical oxidation (ozonation, UV/H_2O_2 AOP)	Recalcitrant organics, pathogenic microorganisms
	Membrane filtration (NF, RO)	Dissolved and suspended solids, recalcitrant organics, pathogenic microorganisms
Disinfection	Chlorination	Pathogenic microorganisms
	Ultraviolet disinfection	Pathogenic microorganisms
Low-rate treatment	Waste stabilization ponds, aerated lagoons, constructed wetlands	Suspended solids, nutrients, biodegradable organics, some pathogenic microorganisms

Abbreviations: *NF* nanofiltration, *RO* reverse osmosis, *UV/H_2O_2 AOP* ultraviolet/hydrogen peroxide advanced oxidation process

enhance the solid–liquid separation. Although some pathogen removal can be achieved (up to 3-log reduction of helminth eggs by chemically enhanced sedimentation), primary treatment is generally insufficient for removing viruses and bacteria [14]. Moreover, only metals in a particulate form may be removed by gravity separation. More elaborate treatment (e.g., secondary or tertiary treatment followed by disinfection) and/or additional protection measures, including restricted crop irrigation, use of personal protective equipment, product washing and cooking, and regular monitoring for toxic chemical accumulation in irrigated soil and products, would be necessary when primary effluent was used for crop irrigation and aquacultures. Proper disposal of primary sludge is also required to prevent the spread of pathogens and other contaminants into soil and waterways.

6.5.2 *Secondary Treatment*

Secondary treatment consists of a combination of biological treatment and clarification. A wide variety of conventional and advanced biological treatment processes are commercially available, such as activated sludge, sequencing batch reactors, oxidation ditch, trickling filters, rotating biological contactors, and membrane bioreactors (MBRs) [1]. Biological digestion of biodegradable organics occurs in biological reactors (= bioreactors), and subsequently, biomass is clarified by secondary clarification. Pathogen reduction occurs in secondary clarification, while reduction of dissolved chemical constituents occurs through biological degradation. Up to 2-log reduction of viruses, bacteria, protozoan cysts and oocysts, and helminth eggs can be achieved by secondary bioreactors. Some toxic heavy metals and hydrophobic organics, such as organochlorine pesticides, may be adsorbed by the biomass [73, 74]. Because of superior particulate removal and pathogen reduction, MRBs are particularly suitable for urban water reuse projects [1].

When combined with proper disinfection, secondary-treated wastewater can be considered safe for discharge and for variety of non-potable applications, including crop irrigation, restricted recreational impoundments, and groundwater recharge via percolation basins [14, 75, 76]. However, certain restrictions are suggested when reclaimed water is used in a place where incidental direct contact or ingestion may occur, such as parks, golf courses, and recreational lakes, and for irrigation of food crop without processing.

Similar to primary sludge, secondary sludge also contains elevated levels of metals, organic constituents, and pathogens [77–79]. Proper treatment is required for the sludge disposal, land application, and reuse [80].

6.5.3 Tertiary Treatment

Tertiary treatment refers to the unit process(es) that polishes secondary-treated wastewater, such as chemical or biological nutrient removal and additional particulate and pathogen removal through granular media or membrane filtration, such as microfiltration (MF) and ultrafiltration (UF). Chemical coagulation followed by granular media filtration can achieve up to three-log reduction of viruses [14], which greatly reduces public health risk. Therefore, filtration is an essential component of current wastewater reclamation and reuse projects in developed countries as most of the regulatory agencies, such as the US EPA [76], suggest the use of filtration for reuse applications with fewer restrictions, such as urban irrigation and recreational impoundments.

Membrane filtration (MF and UF) is a rapidly growing tertiary treatment technology. Because of their very small pore sizes (<1 μm), both MF and UF membranes can remove virtually all pathogens, including viruses, to almost completely (>6-log reduction) [14]. In many urban water reuse projects, they are also used as a pretreatment of advanced treatment and in combination with secondary treatment [1].

6.5.4 Advanced Treatment

As noted earlier, a wide variety of organic wastewater constituents, such as pharmaceuticals, flame retardants, and personal care products, have been detected in secondary and tertiary effluent from wastewater treatment plants around the world [35, 81–84]. Their concentrations are usually very low (i.e., sub-microgram per liter levels), and little evidence of their adverse health effect is available. Because of their wide prevalence and unknown health impact, removal of these trace organic contaminants is desired in many urban wastewater discharge and reuse scenarios, especially those involved in planned IPR and DPR.

A number of advanced treatment technologies are available. For example:

- Ozonation (O_3), ozone/hydrogen peroxide (O_3/H_2O_2) AOP, UV/H_2O_2 AOP [85–87]
- Granular activated carbon adsorption [88, 89]
- Nanofiltration (NF) and RO [88, 90–92]

Advanced treatment of secondary- or tertiary-treated wastewater can produce reclaimed water with very high purity. However, these advanced treatment technologies are very expensive to construct, operate, and maintain. A number of other issues, such as RO and NF concentrate (i.e., brine) disposal, and removal of persistent contaminants and by-products [e.g., bromate and *N*-nitrosodimethylamine (NDMA), 1,4-dioxane, and tris (1-chloro-2-propyl) phosphate], must be addressed [38].

Another important aspect of highly purified reclaimed water, especially the water treated by RO, is its high purity and low ionic strength. Ironically, if placed in the environment, highly purified reclaimed water may be exposed to environmental contaminants and become contaminated [68]. In order to protect and maintain the quality of reclaimed water and public health, considering DPR scenarios may become more realistic in the future [6, 93].

6.5.5 Disinfection

Free chlorine, either in a form of chlorine gas or sodium or calcium hypochlorite, has been used as a primary disinfectant for wastewater for many years. The effectiveness of chlorine toward bacteria inactivation is very well known and understood [1]. Chlorine is also moderately effective toward viruses (up to 3-log reduction), but it is less effective toward protozoa cysts and oocysts (up to 1.5-log reduction) and helminth eggs (<1-log reduction) [14].

In addition to free chlorine, chlorine dioxide, ozone, and UV are also commonly used as primary disinfectants in wastewater treatment and reuse applications. These disinfectants are generally more effective toward inactivation of viruses and protozoa than free chlorine. UV disinfection is more commonly used in wastewater treatment because of economic considerations [1]. However, due to the recent awareness of trace organic contaminants in treated wastewater, the use of ozone and UV/H_2O_2 AOP is becoming an attractive option [69, 87]. Also, it has been reported that the use of chlorine and UV in combination can enhance the effectiveness of disinfection [94]. Chloramine (i.e., combined chlorine) is a weaker disinfectant compared with others and is usually used as a secondary disinfectant in a distribution system of reclaimed water.

As with the case of drinking water disinfection, formation of disinfection by-products (DBPs) is an important public health issue in reclaimed water disinfection [1]. Trihalomethanes and haloacetic acids are generated during chlorination, while nitrosamines (e.g., NDMA), chlorate, and bromate can be generated during chloramination, chlorine dioxide disinfection, and ozonation, respectively. The formation of these DBPs is of particular concern in potable reuse scenarios.

6.5.6 *Low-Rate Treatment*

Low-rate treatment systems such as waste stabilization ponds, lagoons, and constructed wetlands are low-cost alternatives to mechanical treatment systems discussed above. These treatment systems utilize sunlight for heat and UV source, gravity for sedimentation, and naturally occurring bacteria for biodegradation. If properly designed, waste stabilization ponds can achieve a 2- to 4-log removal of viruses, a 3- to 6-log removal of bacteria, a 1- to 2-log removal of protozoan cysts and oocysts, and an up to 3-log removal of helminth eggs [14]. Constructed wetlands are, on the other hand, usually not designed for the removal of pathogens [95]. Sunlight UV disinfection may be maximized by creating a wide, shallow channels at the end of wetlands.

Removal of organic and inorganic chemical constituents may be achieved in low-rate treatment systems via biological degradation, adsorption, UV photolysis, and phytoremediation [96, 97].

6.6 Treatment Options for Wastewater Safe Disposal and Reuse

There is no silver bullet in pollution control in general, and wastewater treatment is no exception. Thus, a wastewater treatment system is usually a combination of several processes that can provide multiple barriers toward different constituents being removed. In this section, several examples of wastewater treatment schemes are presented for different wastewater applications, including safe discharge, agricultural reuse, non-potable urban reuse, and potable reuse.

It should be noted that wastewater quality is highly site specific and that local discharge and reuse standards differ significantly from one place to another. Therefore, a treatment system will need to be examined individually and be tailored

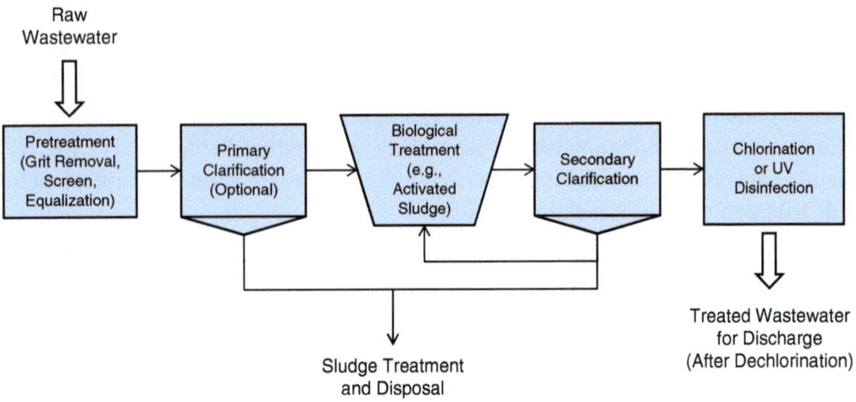

Fig. 6.4 Suggested treatment scheme for safe wastewater disposal (conventional secondary treatment system)

to meet specific treatment and reuse goals. In addition, economic consideration is often a major determinant of treatment process selection and overall system design. Thus, the following examples are conceptual and are presented for information purposes only.

6.6.1 Safe Disposal

For safe disposal of wastewater, the minimum requirements are the removal/inactivation of pathogenic microorganisms, heavy metals, and other toxic inorganic and organic contaminants. In addition, excess organic matter and nutrients, including nitrogen and phosphorus, need to be removed to lower environmental impacts. Figure 6.4 shows a flow scheme of a conventional wastewater treatment system for safe discharge.

Alternatively, where a large land is available at a relatively low cost, a combination of low-rate treatment processes, such as waste stabilization ponds and constructed wetlands, may be used (Fig. 6.5). Constructed wetlands will provide additional barrier to pathogens, toxic metals, and organic contaminants through phytoremediation, sunlight UV disinfection, and photolysis.

6.6.2 Agricultural Reuse

In addition to the secondary treatment and disinfection, tertiary filtration is recommended for unrestricted agricultural reuse (Fig. 6.6). While removal of nutrients may not be necessary, other water quality parameters such as salinity,

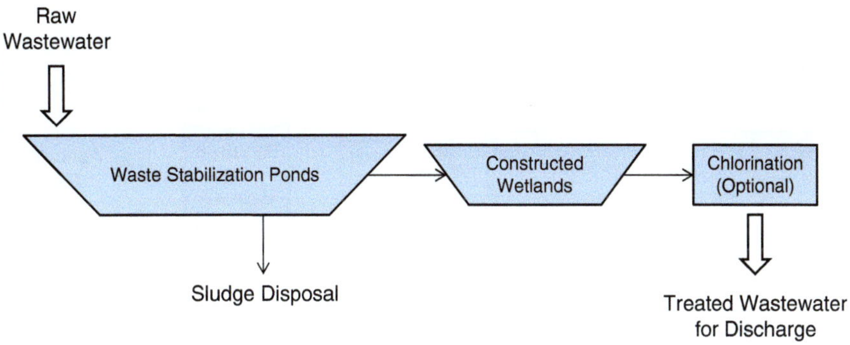

Fig. 6.5 Suggested treatment scheme for safe wastewater disposal (low-rate system)

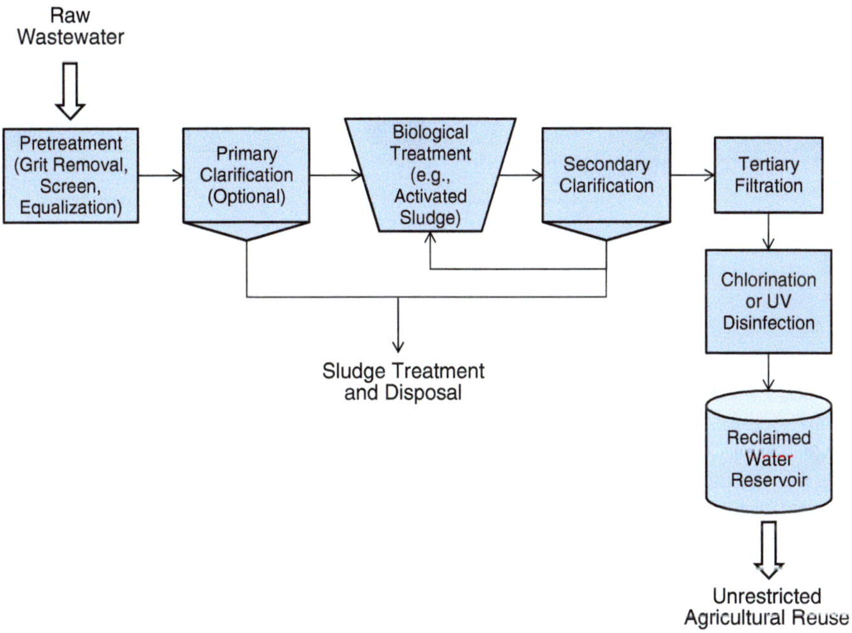

Fig. 6.6 Suggested treatment scheme for unrestricted agricultural reuse. Note: Combined chlorine may be used in the reclaimed water reservoir to prevent pathogen regrowth during storage

sodium adsorption ratio, and boron may require attention [14]. Where more controlled irrigation techniques, such as drip irrigation, are to be employed, tertiary filtration may be omitted. As suggested by WHO [14], waste stabilization ponds may be used for safe agricultural reuse in developing countries (Fig. 6.7).

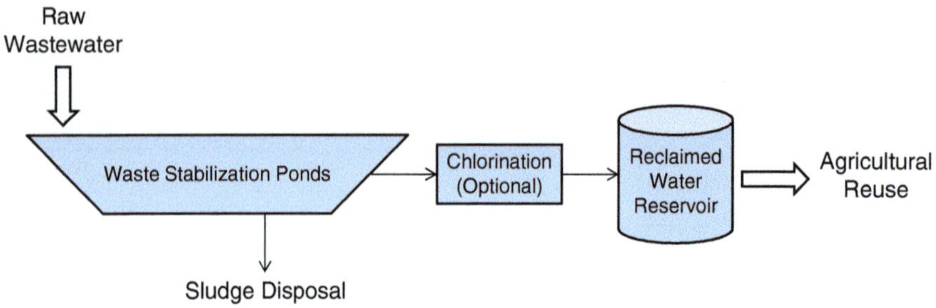

Fig. 6.7 Alternative treatment scheme for agricultural reuse in developing countries

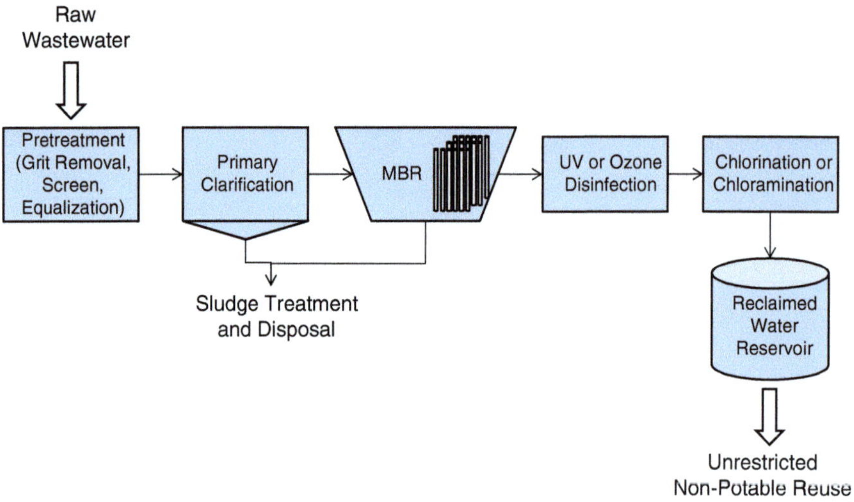

Fig. 6.8 Alternative treatment scheme for non-potable reuse using advanced technologies. Note: Nutrient removal may be required for reclaimed water impoundments. Dechlorination may be required before discharging into surface impoundment

6.6.3 *Non-potable Reuse*

For non-potable reuse, such as landscape irrigation, recreational impoundments, and toilet flushing, the treatment scheme for unrestricted agricultural reuse (Fig. 6.6) may be used. Alternatively, more advanced technologies, such as MBR and ozonation, may be used, as shown in Fig. 6.8. Nutrient removal may be necessary if the reclaimed water is discharged into surface impoundments for recreational use and/or for temporary storage.

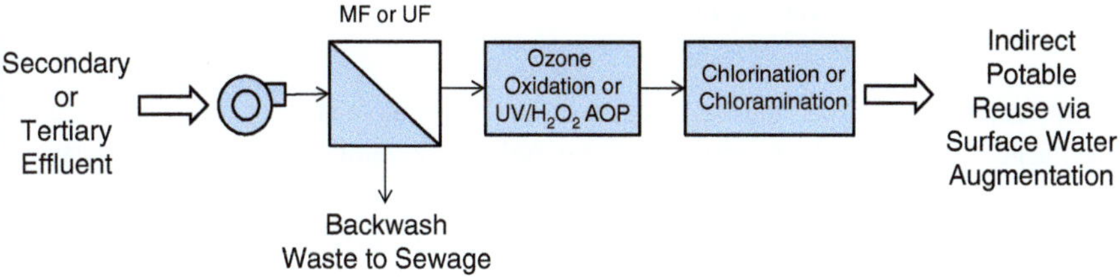

Fig. 6.9 Advanced treatment scheme for indirect potable reuse (IPR) via surface water augmentation. Note: Nutrient removal may be required. MBR may be used instead of the stand-alone MF/UF

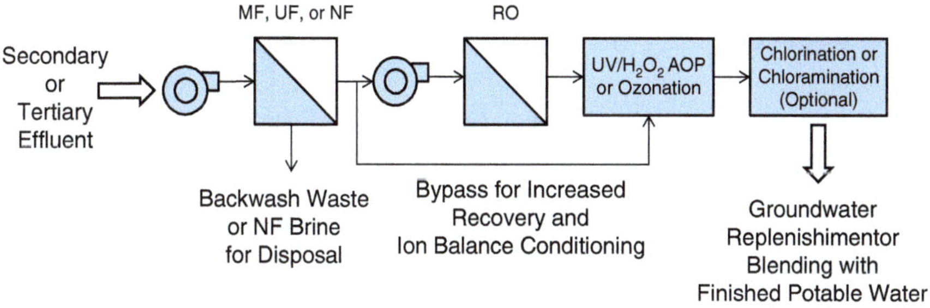

Fig. 6.10 Advanced treatment scheme for indirect potable reuse (IPR) via groundwater replenishment or direct potable reuse (DPR)

6.6.4 Potable Reuse

Depending on the desired potable reuse scenarios, required treatment can be different. For IPR via surface spreading and percolation into potable aquifers, the treatment scheme for safe discharge (Fig. 6.4) or unrestricted agricultural reuse (Fig. 6.6) may be sufficient. Advanced treatment such as ozonation and UV/H_2O_2 AOP, as well as nutrient removal, is desired for IPR via surface water augmentation where possible ecological impacts of trace organic contaminants and nuisance algal blooms may be of concern. The advanced treatment scheme shown in Fig. 6.8 may be used with some modifications such as shown in Fig. 6.9.

IPR via direct injection into potable aquifers (such as the Groundwater Replenishment System, Orange County Water District-Orange County Sanitation District, Fountain Valley, CA) and DPR will require high reclaimed water quality to ensure safety. Most of the current DPR and semi-direct potable reuse projects (e.g., NEWater, PUB Singapore) treat secondary- or tertiary-treated wastewater by a series of advanced treatment processes, including MF and RO followed by UV or UV/H_2O_2 AOP treatment (Fig. 6.10). UF or NF may be used instead of MF, and UV/H_2O_2 AOP may be replaced by ozonation or ozone-based AOPs such as O_3/H_2O_2 in this scheme.

6.7 Concluding Remarks

Wastewater is integral part of regional and global water cycle. Historically, discharge of untreated or improperly treated wastewater into natural waterways has been a major source of severe pollution in the environment. Hazardous biological and chemical constituents in wastewater effluent adversely affected and degraded the natural ecosystem, as well as the living environment of humans, which caused numerous outbreaks of deadly infectious diseases and acute and chronic toxicity events around the world.

At the same time, wastewater has been used as an important water resource for irrigating agricultural lands in both developed and developing countries. High nutrient value of wastewater has also been looked as a favorable side of wastewater irrigation. However, it has been demonstrated that raw wastewater is not safe for reuse because of high content of microbial pathogens, such as *Ascaris*, enteropathogenic *E. coli*, *Salmonella*, *Giardia*, and various viruses, and toxic chemicals, such as arsenic, lead, zinc, mercury, copper, organochlorine pesticides, and solvents, which represent a significant health risk to farmers and their families, as well as to consumers. Either for discharge or for reuse, wastewater treatment is extremely important for protecting public health and the environment and for ensuring the sustainable use of water resources.

In the past decade, a great deal of effort has been put into research to reveal the health effects associated with wastewater discharge and reuse. Due to global water scarcity, interest in wastewater reclamation and reuse has grown tremendously not only in the area of agricultural reuse but also in the area of other non-potable and potable reuse. A wide variety of wastewater treatment technologies are currently available to achieve different effluent/reclaimed water qualities. Advanced treatment processes, such as RO, NF, ozonation, and UV/H_2O_2 AOP, have become feasible for use in large-scale water reclamation projects. As a result, more aggressive reuse scenarios, such as DPR, are becoming more and more realistic.

However, some uncertainty still exists with respect to trace emerging contaminants (e.g., pharmaceuticals and personal care products, DBPs, flame retardants, etc.): risk assessment, monitoring, and control. Without clear understanding of their potential impacts on public health and control measures, a highly energy-intensive and costly, all-out approach like RO followed by UV/H_2O_2 AOP may still be needed. In addition, more work will be needed to develop robust, less energy-intensive alternative treatment technologies that can be used in developing countries.

Acknowledgments The author would like to thank his past and present colleagues, namely, Dr. Mohamed Gamal El-Din, Dr. Hongjing Fu, Dr. Ian D. Buchanan, Mr. Odell Pui, Dr. Yang Liu, Ms. Ruo Nan Sun (University of Alberta, Alberta, Canada), Ms. Lisa N. Gochenour (City of Omaha, Nebraska), Mr. Robert R. Murphy, Mr. Michael B. Nessl, Mr. Jacob D. Peterson (PACE, California), Ms. Yao Jin, and Ms. Ling Wang (University of California, Irvine-PACE), for their valuable contributions to the literature survey in the past 6 years. The editorial assistance of Ms. Kelly M. Huston and Mr. Carlos Gomez (PACE) is also gratefully acknowledged. The author would also like to thank Mr. Andrew T. Komor (PACE) for his support and encouragement.

References

1. Asano T, Burton FL, Leverenz H, Tsuchihashi R, Tchobanoglous G (2007) Water reuse: issues, technologies, and applications. McGraw-Hill, New York
2. WHO (2006) WHO guidelines for the safe use of wastewater, excreta and greywater, vol 1, Policy and regulatory aspects. World Health Organization, Geneva
3. Focazio MJ, Kolpin DW, Barnes KK, Furlong ET, Meyer MT, Zaugg SD, Barber LB, Thurman ME (2008) A national reconnaissance for pharmaceuticals and other organic wastewater contaminants in the United States - II) untreated drinking water sources. Sci Total Environ 402(2–3):201–216
4. Snyder SA (2008) Occurrence, treatment, and toxicological relevance of EDCs and pharmaceuticals in water. Ozone Sci Eng 30(1):65–69
5. Crook J (2010) Regulatory aspects of direct potable reuse in California – an NWRI white paper. National Water Research Institute, Fountain Valley
6. Tchobanoglous G, Leverenz H, Nellor MH, Crook J (2011) Direct potable reuse – a path forward., Alexandria
7. Fu H, Ikehata K, Buchanan ID, Gamal El-Din M (2007) Health effects associated with wastewater treatment, reuse, and disposal. Water Environ Res 79(10):2161–2169
8. Ikehata K, Gamal El-Din M, Gochenour LN (2006) Health effects associated with wastewater treatment, reuse, and disposal. Water Environ Res 78(10):2086–2096
9. Ikehata K, Liu Y, Sun RN (2009) Health effects associated with wastewater treatment, reuse, and disposal. Water Environ Res 81(10):2126–2146
10. Ikehata K, Murphy RR, Liu Y, Sun RN, Nessl MB (2010) Health effects associated with wastewater treatment, reuse, and disposal. Water Environ Res 82(10):2047–2066
11. Ikehata K, Pui O (2008) Health effects associated with wastewater treatment, reuse, and disposal. Water Environ Res 80(10):1980–1996
12. Peterson JD, Murphy RR, Jin Y, Wang L, Nessl MB, Ikehata K (2011) Health effects associated with wastewater treatment, reuse, and disposal. Water Environ Res 83 (10):1853–1875
13. US EPA (2009) Emerging contaminant – 1,4-dioxane fact sheet. US Environmental Protection Agency, Washington, DC
14. WHO (2006) WHO guidelines for the safe use of wastewater, excreta and greywater, vol 2, Wastewater Use in Agriculture. World Health Organization, Geneva
15. Raschid-Sally L, Carr R, Buechler S (2005) Managing wastewater agriculture to improve livelihoods and environmental quality in poor countries. Irrig Drain 54:S11–S22
16. Costa CH, Werneck GL, Rodrigues L Jr, Santos MV, Araújo IB, Moura LS, Moreira S, Gomes RB, Lima SS (2005) Household structure and urban services: neglected targets in the control of visceral leishmaniasis. Ann Trop Med Parasitol 99(3):229–236
17. Uzunovic-Kamberovic S, Durmisevic S, Tandir S (2005) Environmental risk factors for hepatitis A infection in the Zenica-Doboj Canton, Bosnia and Herzegovina. Clin Microbiol Infect 11(2):145–147
18. van den Berg H, Lodder W, van der Poel W, Vennema H, Husman AMD (2005) Genetic diversity of noroviruses in raw and treated sewage water. Res Microbiol 156(4):532–540
19. Greer AL, Drews SJ, Fisman DN (2009) Why "winter" vomiting disease? Seasonality, hydrology, and norovirus epidemiology in Toronto, Canada. Ecohealth 6(2):192–199
20. Ueki Y, Sano D, Watanabe T, Akiyama K, Omura T (2005) Norovirus pathway in water environment estimated by genetic analysis of strains from patients of gastroenteritis, sewage, treated wastewater, river water and oysters. Water Res 39(18):4271–4280
21. Ahmed KS, Khan AA, Ahmed I, Tiwari SK, Habeeb A, Ahi JD, Abid Z, Ahmed N, Habibullah CM (2007) Impact of household hygiene and water source on the prevalence and transmission of *Helicobacter pylori*: a South Indian perspective. Singapore Med J 48(6):543–549

22. Rajii MA, Minga UM, Machang'u RS (2008) Prevalence and characterization of verotocytoxin producing *Escherichia coli* O157 from diarrhoea patients in Morogoro, Tanzania. Tanzan J Health Res 10(3):151–158
23. Alaoui HL, Oufdou K, Mezrioui NE (2010) Determination of several potential virulence factors in non-o1 *Vibrio cholerae*, *Pseudomonas aeruginosa*, faecal coliforms and streptococci isolated from Marrakesh groundwater. Water Sci Technol 61(7):1895–1905
24. Yassin MM, Amr SS, Al-Najar HM (2006) Assessment of microbiological water quality and its relation to human health in Gaza Governorate, Gaza Strip. Public Health 120 (12):1177–1187
25. Fewtrell L, Bartram J (eds) (2001) Water quality: guidelines, standards, and health. IWA Publishing on behalf of the World Health Organization, London
26. WHO (2008) Guidelines for safe recreational-water environments, vol 1, Coastal and fresh waters. World Health Organization, Geneva
27. An YJ, Yoon CG, Jung KW, Ham JH (2007) Estimating the microbial risk of *E. coli* in reclaimed wastewater irrigation on paddy field. Environ Monit Assess 129(1–3):53–60
28. Donovan E, Unice K, Roberts JD, Harris M, Finley B (2008) Risk of gastrointestinal disease associated with exposure to pathogens in the water of the Lower Passaic River. Appl Environ Microbiol 74(4):994–1003
29. Mara DD, Sleigh PA, Blumenthal UJ, Carr RM (2007) Health risks in wastewater irrigation: comparing estimates from quantitative microbial risk analyses and epidemiological studies. J Water Health 5(1):39–50
30. Navarro I, Jimenez B, Lucario S, Cifuentes E (2009) Application of Helminth ova infection dose curve to estimate the risks associated with biosolid application on soil. J Water Health 7 (1):31–44
31. Ryu H, Alum A, Mena KD, Abbaszadegan M (2007) Assessment of the risk of infection by *Cryptosporidium* and *Giardia* in non-potable reclaimed water. Water Sci Technol 55 (1–2):283–290
32. Seidu R, Heistad A, Amoah P, Drechsel P, Jenssen PD, Stenstrom TA (2008) Quantification of the health risk associated with wastewater reuse in Accra, Ghana: a contribution toward local guidelines. J Water Health 6(4):461–471
33. ten Veldhuis JAE, Clemens F, Sterk G, Berends BR (2010) Microbial risks associated with exposure to pathogens in contaminated urban flood water. Water Res 44(9):2910–2918
34. Falconer IR (2006) Are endocrine disrupting compounds a health risk in drinking water? Int J Environ Res Public Health 3(2):180–184
35. Snyder SA, Westerhoff P, Yoon Y, Sedlak DL (2003) Pharmaceuticals, personal care products, and endocrine disruptors in water: implications for the water industry. Environ Eng Sci 20 (5):449–469
36. Wright-Walters M, Volz C (2009) Municipal wastewater concentrations of pharmaceutical and xeno-estrogens: wildlife and human health implications. In: Uzochukwu GA (ed) Proceedings of the 2007 national conference on environmental science and technology. Springer, New York, pp 103–113
37. US GAO (2001) Chemical risk assessment – selected federal agencies' procedures, assumptions, and policies. United States General Accounting Office, Washington, DC
38. Anderson P, Denslow N, Drewes JE, Olivieri A, Schlenk D, Snyder SA (2010) Monitoring strategies for chemicals of emerging concern (CECs) in recycled water. State Water Resources Control Board, Sacramento
39. Bruce GM, Pleus RC, Snyder SA (2010) Toxicological relevance of pharmaceuticals in drinking water. Environ Sci Technol 44(14):5619–5626
40. Cooper ER, Siewicki TC, Phillips K (2008) Preliminary risk assessment database and risk ranking of pharmaceuticals in the environment. Sci Total Environ 398(1–3):26–33
41. Cunningham VL, Perino C, D'Aco VJ, Hartmann A, Bechter R (2010) Human health risk assessment of carbamazepine in surface waters of North America and Europe. Regul Toxicol Pharmacol 56(3):343–351

42. Kumar A, Chang BA, Xagoraraki I (2010) Human health risk assessment of pharmaceuticals in water: issues and challenges ahead. Int J Environ Res Public Health 7(11):3929–3953
43. Levi Y (2009) Challenges in the assessment and management of health risks associated with emerging water micropollutants. Bull Acad Natl Med 193(6):1331–1341
44. Schwab BW, Hayes EP, Fiori JM, Mastrocco FJ, Roden NM, Cragin D, Meyerhoff RD, D'Aco VJ, Anderson PD (2005) Human pharmaceuticals in US surface waters: a human health risk assessment. Regul Toxicol Pharmacol 42(3):296–312
45. Weber S, Khan S, Hollender J (2006) Human risk assessment of organic contaminants in reclaimed wastewater used for irrigation. Desalination 187(1–3):53–64
46. Trussell RS (2012) Reclaimed water desalination technologies: a full-scale performance and cost comparison between electrodialysis reversal and microfiltration/reverse osmosis. WateReuse Association, Alexandria
47. Bellido JG, Barcellos C, Barbosa FD, Bastos FI (2010) Environmental sanitation and mortality associated with waterborne diseases in children under 5 years of age in Brazil. Revista Panamericana De Salud Publica-Pan Am J Public Health 28(2):114–120
48. Moubarrad FZL, Assobhei O (2007) Health risks of raw sewage with particular reference to *Ascaris* in the discharge zone of El Jadida (Morocco). Desalination 215(1–3):120–126
49. Qazi HA, Soomro JA, Hashmi A, Qadri MH, Rasheed F, Karim MT (2009) Public health problems in low and high socio economic areas of Karachi, Pakistan. Iran J Public Health 38 (3):105–108
50. Bhardwaj V, Singh DS, Singh AK (2010) Environmental repercussions of cane-sugar industries on the Chhoti Gandak river basin, Ganga Plain, India. Environ Monit Assess 171 (1–4):321–344
51. Gitipour S, Givehchi S, Aghaii A, Baghvand A (2007) Assessment of soil contamination due to leaching of industrial wastewater. Asian J Chem 19(3):1995–2001
52. Mapanda F, Nyamadzawo G, Nyamangara J, Wuta M (2007) Effects of discharging acid-mine drainage into evaporation ponds lined with clay on chemical quality of the surrounding soil and water. Phys Chem Earth 32(15–18):1366–1375
53. Al-Mikhlafi AS (2010) Groundwater quality of Yemen volcanic terrain and their geological and geochemical controls. Arab J Geosci 3(2):193–205
54. Silva IR, Pereira LCC, Guimaraes DD, Trindade WN, Asp N, Costa RM (2009) Environmental status of urban beaches in Sao Luis (Amazon coast, Brazil). J Coast Res SI 56:1301–1305
55. Sy I, Handschumacher P, Wyss K, Piermay JL, Tanner M, Cisse G (2009) Water and sanitation policies limits in Senegal cities: the case of Rufisque. Desalination 248(1–3):673–677
56. Wong M, Kumar L, Jenkins TM, Xagoraraki I, Phanikumar MS, Rose JB (2009) Evaluation of public health risks at recreational beaches in Lake Michigan via detection of enteric viruses and a human-specific bacteriological marker. Water Res 43(4):1137–1149
57. El Kettani S, Azzouzi E, Boukachabine K, El Yamani M, Maata A, Rajaoui M (2008) Intestinal parasitosis and use of untreated wastewater for agriculture in Settat, Morocco. East Mediterr Health J 14(6):1435–1444
58. Trang DT, Hien BTT, Molbak K, Cam PD, Dalsgaard A (2007) Epidemiology and aetiology of diarrhoeal diseases in adults engaged in wastewater-fed agriculture and aquaculture in Hanoi, Vietnam. Trop Med Int Health 12:23–33
59. Mapanda F, Mangwayana EN, Nyamangara J, Giller KE (2005) The effect of long-term irrigation using wastewater on heavy metal contents of soils under vegetables in Harare, Zimbabwe. Agric Ecosyst Environ 107(2–3):151–165
60. Amoah P, Drechsel P, Abaidoo RC, Ntow WJ (2006) Pesticide and pathogen contamination of vegetables in Ghana's urban markets. Arch Environ Contam Toxicol 50(1):1–6
61. Li P, Wang X, Allinson G, Li X, Xiong X (2009) Risk assessment of heavy metals in soil previously irrigated with industrial wastewater in Shenyang, China. J Haz Mat 161(1):516–521
62. Passuello A, Mari M, Nadal M, Schuhmacher M, Domingo JL (2010) POP accumulation in the food chain: integrated risk model for sewage sludge application in agricultural soils. Environ Int 36(6):577–583

63. Ramadan AA, Mandil H (2009) Wastewater irrigation and soil contamination effect on some leafy vegetables grown in Syrian Aleppo City. Asian J Chem 21(4):3243–3252
64. Murtaza G, Ghafoor A, Qadir M (2008) Accumulation and implications of cadmium, cobalt and manganese in soils and vegetables irrigated with city effluent. J Sci Food Agric 88 (1):100–107
65. Akoachere JF, Oben PM, Mbivnjo BS, Ndip LM, Nkwelang G, Ndip RN (2008) Bacterial indicators of pollution of the Douala lagoon, Cameroon: public health implications. Afr Health Sci 8(2):85–89
66. Brinks MV, Dwight RH, Osgood ND, Sharavanakumar G, Turbow DJ, El-Gohary M, Caplan JS, Semenza JC (2008) Health risk of bathing in southern California coastal waters. Arch Environ Occup Health 63(3):123–135
67. Crook J (2005) Irrigation of parks, playgrounds, and schoolyards with reclaimed water: extent and safety. WateReuse Association, Alexandria
68. Leverenz LH, Tchobanoglous G, Asano T (2011) Direct potable reuse: a future imperative. J Water Reuse Desalination 1(1):2–10. doi:10.2166/wrd.2011.000
69. Rosario-Ortiz FL, Wert EC, Snyder SA, Mezyk SP (2011) Optimization of advanced oxidation processes (AOP) for water reuse. WateReuse Association, Alexandria
70. Bastos RK, Bevilacqua PD, Silva CA, Silva CV (2008) Wastewater irrigation of salad crops: further evidence for the evaluation of the WHO guidelines. Water Sci Technol 57 (8):1213–1219
71. Drechsel P, Keraita B, Amoah P, Abaidoo RC, Raschid-Sally L, Bahri A (2008) Reducing health risks from wastewater use in urban and peri-urban sub-Saharan Africa: applying the 2006 WHO guidelines. Water Sci Technol 57(9):1461–1466
72. Keraita B, Jimenez B, Drechsel P (2008) Extent and implications of agricultural reuse of untreated, partly treated and diluted wastewater in developing countries. CAB Rev 3(058)
73. Bell JP, Tsezos M (1987) Removal of hazardous organic pollutants by biomass adsorption. J Water Pollut Control Fed 59(4):191–198
74. Rossin AC, Sterritt RM, Lester JN (1982) The influence of process parameters on the removal of heavy metals in activated sludge. Water Air Soil Pollut 17(2):185–198
75. Florida Department of Environmental Protection (1999) Reuse of reclaimed water and land application. Florida Department of Environmental Protection, Tallahassee
76. US EPA (2004) Guidelines for water reuse. U.S. Environmental Protection Agency and U.S. Agency for International Development, Washington, DC
77. Joss A, Keller E, Alder AC, Göbel A, McArdel CS, Ternes T, Siegrist H (2005) Removal of pharmaceuticals and fragrances in biological wastewater treatment. Water Res 39 (14):3139–3152
78. Lake DL, Kirk PWW, Lester JN (1983) Fractionation, characterization, and speciation of heavy metals in sewage sludge and sludge-amended soils: a review. J Environ Qual 13 (2):175–183
79. Straub TM, Pepper IM, Gerba CP (1993) Hazards from pathogenic microorganisms in land-disposed sewage sludge. Rev Environ Contam Toxicol 132:55–91
80. Ikehata K, Liu Y (2011) Land disposal of wastes. In: Nriagu JO (ed) Encyclopedia of environmental health. Elsevier, Amsterdam, pp 353–361
81. Belfroid AC, Van der Horst A, Vethaak AD, Schäfer AJ, Rijs GBJ, Wegener J, Cofino WP (1999) Analysis and occurrence of estrogenic hormones and their glucuronides in surface water and waste water in The Netherlands. Sci Total Environ 225(1–2):101–108
82. Drewes JE, Fox P, Jekel M (2001) Occurrence of iodinated X-ray contrast media in domestic effluents and their fate during indirect potable reuse. J Environ Sci Health A Toxic/Hazard Subst Environ Eng 36(9):1633–1645
83. Kolpin DW, Furlong ET, Meyer MT, Thurman EM, Zaugg SD, Barber LB, Buxton HT (2002) Pharmaceuticals, hormones, and other organic wastewater contaminants in US streams, 1999–2000: a national reconnaissance. Environ Sci Technol 36(6):1202–1211

84. Ternes TA (1998) Occurrence of drugs in German sewage treatment plants and rivers. Water Res 32(11):3245–3260
85. Ikehata K, Naghashkar NJ, Gamal El-Din M (2006) Degradation of aqueous pharmaceuticals by ozonation and advanced oxidation processes: a review. Ozone Sci Eng 28(6):353–414
86. Rosenfeldt EJ, Linden KG (2004) Degradation of endocrine disrupting chemicals bisphenol A, ethinyl estradiol, and estradiol during UV photolysis and advanced oxidation processes. Environ Sci Technol 38(20):5476–5483
87. Snyder SA, Wert EC, Rexing DJ, Zegers RE, Drury DD (2006) Ozone oxidation of endocrine disruptors and pharmaceuticals in surface water and wastewater. Ozone Sci Eng 28 (6):445–460
88. Snyder SA, Adham S, Redding AM, Cannon FS, Decarolis J, Oppenheimer J, Wert EC, Yoon Y (2006) Role of membranes and activated carbon in the removal of endocrine disruptors and pharmaceuticals. Desalination 202:156–181
89. Snyder SA, Wert EC, Lei H, Westerhoff P, Yoon Y (2007) Removal of EDCs and pharmaceuticals in drinking and reuse treatment processes. Awwa Research Foundation, Denver
90. Busetti F, Linge KL, Rodriguez C, Heitz A (2010) Occurrence of iodinated X-ray contrast media in indirect potable reuse systems. J Environ Sci Health Part A-Toxic/Hazard Subst Environ Eng 45(5):542–548
91. Rodriguez C, Cook A, Devine B, Van Buynder P, Lugg R, Linge K, Weinstein P (2008) Dioxins, furans and PCBs in recycled water for indirect potable reuse. Int J Environ Res Public Health 5(5):356–367
92. Yoon Y, Westerhoff P, Snyder SA, Wert EC (2006) Nanofiltration and ultrafiltration of endocrine disrupting compounds, pharmaceuticals and personal care products. J Membr Sci 270(1–2):88–100
93. Trussell RR, Anderson HA, Archuleta EG, Crook J, Drewes JE, Fort DD, Haas CN, Haddad BM, Huggett DB, Jiang S, Sedlak DL, Snyder SA, Whittaker MH, Whittington D (2012) Water reuse: potential for expanding the nation's water supply through reuse of municipal wastewater. The National Academies Press, Washington, DC
94. Munakata N, Tang C-C, Huitric S-J, Ackman P, Kuo J, Garcia A, Thompson S, Friess PL, Maguin SR (2009) Comparing free chlorine and chloramines in combination with UV for disinfection of wastewater effluents and reclaimed water. Proc Water Environ Fed Disinfect 2009(10):629–638
95. Graczyk TK, Lucy FE (2007) Quality of reclaimed waters: a public health need for source tracking of wastewater derived protozoan enteropathogens in engineered wetlands. Trans R Soc Trop Med Hyg 101(6):532–533
96. Maine MA, Suñe N, Hadad H, Sánchez G, Bonetto C (2006) Nutrient and metal removal in a constructed wetland for wastewater treatment from a metallurgic industry. Ecol Eng 26:341–347
97. Matamoros V, Bayona JM (2006) Elimination of pharmaceuticals and personal care products in subsurface flow constructed wetlands. Environ Sci Technol 40(18):5811–5816

Part II
Wastewater Treatment Options

Chapter 7
The Small-Scale Approach in Wastewater Treatment

Giovanni Libralato

7.1 Introduction

In the middle of the nineteenth century, engineers and natural scientists discovered the outbreak of some fatal diseases such as typhus, cholera and diarrhoea due to the direct contact of human beings with their own excreta containing pathogenic micro-organisms [1]. To protect the population and prevent infections, the sewer systems were substantially reinvented going further ancient civilisation experiences.

However, the transport of wastewaters out of human settlements got quickly insufficient because the quality of surface water, to whom the sewage was discharged, started to decrease (*e.g.* hypo/anoxia and pathogens), creating health risks for the population living downstream. So that, intensive and costly purification became necessary, pressing the birth of wastewater treatment facilities.

In general, the heavy industrialisation process and the continuous economic and population growth induced, time by time, the application of more strict water quality standards and the subsequent need to develop and implement wastewater treatment technologies able to satisfy the new requirements.

Normally, the main sources of wastewater originate not only from industrial activities (*e.g.* chemical syntheses, waste gas treatment systems, conditioning of utility water, bleed from boiler feed water systems, blow down from cooling cycles, back-washing of filters and ion exchangers, landfill leachates and rainwater from contaminated areas) but also from urban and commercial settlements, identifying potential

G. Libralato (✉)
Department of Environmental Sciences, Informatics and Statistics, University Cà Foscari Venice, Calle Larga Santa Marta, 2137, Dorsoduro, Venice 30123, Italy

ECSIN - European Center for the Sustainable Impact of Nanotechnology – Veneto Nanotech S.C. p.A., Viale Porta Adige, 45, I-45100 Rovigo, Italy
e-mail: giovanni.libralato@unive.it

S.K. Sharma and R. Sanghi (eds.), *Wastewater Reuse and Management*,
DOI 10.1007/978-94-007-4942-9_7,

impacts on hydraulic loads, content of pollutant substances, effect or hazardous potential on the receiving water body and effect on aquatic organisms as toxicity.

There are several wastewater treatment techniques to be used singly or in combination with others that should cover all potential needs for wastewater treatment. Separation techniques (*e.g.* grit separation, sedimentation, air flotation, filtration, micro- and ultrafiltration, oil-water separation) are mainly used in combination with other operations as a first or final treatment step. Physico-chemical treatment techniques (*e.g.* coagulation/precipitation/sedimentation/filtration, crystallisation, chemical and wet air oxidation, supercritical water oxidation, chemical reduction, hydrolysis, nanofiltration, reverse osmosis, adsorption, ion exchange, extraction, distillation/rectification, evaporation, stripping and incineration) are primarily used for non-biodegradable wastewaters, inorganic or hardly biodegradable organic contaminants, often as an upstream pretreatment; and biological treatment techniques (*e.g.* anaerobic/aerobic processes, nitrification/denitrification and central or decentralised biological treatments) are considered for biodegradable wastewaters [2]. Specifically, particular attention should be paid to membrane biological reactors and its various potential upgrading. Anyhow, the most challenging technologies are those able to easily adapt to specific situations, increasing the diversity of systems and stressing on their evolution. Indeed, experts are experiencing that there is no system able to cope all the situations such as it occurs in the case of the matter of scale [3].

At the moment, sustainable development is at the forefront of today's policy agendas, and technology developers are focusing on sustainable and best available technologies (BATs) on wastewater treatments and nutrient cycling, not only on a centralised basis but mostly on a small- and medium-sized configuration. Indeed, centralised and decentralised (small- and medium-sized) plants coexisted over the past years, but industrialised countries started to be fond of the small- and medium-scale alternatives [1], due to the poor efficiency of the on-site traditional applications (*e.g.* septic tanks, ponds and wetlands), even though costs could be of main concern. Their legitimation is a crucial task and may strongly influence decision makers and thus subsequent pollution control standards, efforts to preserve resources, maintain and strengthen hygienic standards, increase waste material recovery, recycle and reuse, changes in water consumption and innovative technological applications [4].

The Newman's vademecum [5] provided some wastewater-oriented goals for sustainability such as the ocean and river outfalls made redundant, the recycling of water, nutrients and organics for various uses, the reduction for large pipes requirement and the increase of soft surfaces to reduce urban sprawl for storm water retention. Besides, other options are related to a more far-reaching eco-friendliness as well as energy and nutrient recovery [4]. Currently, the main emphasis is towards the grey water recycling [6] considering various levels of technology, especially best available technologies (BATs), ranging from minimal surface filtration devices with short residence times [7], purely physical membrane processes and to advanced and integrated biological systems [8].

7.2 Does Decentralisation Make Sense?

7.2.1 A Preliminary Brainstorm

Today, in the urban context, centralisation is still the norm in the developed world, while in the developing world, the opposite is generally the case. In the latter, wastewater from houses, businesses and industry remain untreated or are frequently treated on-site and discharged (whether treated or not) to the ground or nearby drains and watercourses [3]. The question facing the communities in the developing world is, however, the same, that is, whether they should install a centralised or decentralised system if they want to deal with their wastewater. Indeed, at now, decentralisation processes seem to be able to satisfy all traditional centralised treatment requirements, presenting at the same time some added values mainly related to the ability of minimising potential residual effluent contamination as well as ecosystem disruption by removing emerging micropollutants such as metals and pharmaceuticals and personal care products [9]. Jefferson et al. [6] highlighted the fact that small wastewater treatment plants (WWTPs) have been starting to play an important role at global level in the management of water quality (*i.e.* rivers, lakes, estuaries and aquifers). Indeed, they are frequently characterised by a greater numerical growth compared to centralised systems [2]. Certainly, in some countries, the total amount of small plants is able to treat a greater volume of wastewater than the existing centralised ones [10].

The international debate upon decentralisation is including several international governmental and non-governmental organisations considering policymakers, local authorities and stakeholders. Actually, this kind of approach involves various economic, social, technological and environmental constraints in the centralisation/decentralisation dichotomy resulting in the fact that there is no possibility to accept or refuse one of them *a priori*, being the necessity to proceed on a case-by-case basis [3]. Moreover, apart for the fact that the major costs in centralisation are absorbed by the collection system [11] and in decentralisation by the suggested treatment technology [12], it might be highlighted that the budgetary aspects cannot be generalised [3].

7.2.2 Small- and Medium-Sized WWTPs: A Step Beyond Centralisation

Small- and medium-sized wastewater treatment plants (SMTPs) are playing a major role, globally, in the local management of water quality for rivers, lakes, estuaries and groundwater, constituting a potential growth market for the next millennium [6] for decentralised treatments as recognised by IPPC [2] in opposition to centralised ones. They are not only very numerous, but in some countries, a much greater proportion of wastewater is already treated in small systems than in large plants

[10] because they seem to have the capacity to make waste treatment more sustainable; certainly, large-scale systems need management, administrative and large monetary requirements to develop the plant, but their effluents are, at the moment, still more cost-effective than SMTP discharges. However, a decentralised system is considered competitive only if it can provide an advanced wastewater treatment, that is highly effective, robust, easy to operate and low in costs, and an adequately trained personnel for its management is available [1].

The matter of scale is strictly related to local regulations and traditional assumptions of the involved categories such as engineers and environmental scientists. As reported in Libralato et al. [3], the Directive 91/271/EEC defined the treatment ability classification at European level basing it on the treated organic load expressed as person equivalent. So, small treatment plants present no unique definition. For the English Institute of Water Pollution Control, a WWTP is small when less than 1,000 p.e. can be treated, whereas the threshold is set at 10,000 p.e. for the US Environmental Protection Agency [13]. De Fraja Frangipane and Pastorelli [13] and Avezzù et al. [14] considered 2,500 p.e. as maximum level to consider a WWTP as small, but evidencing that until 5,000 p.e., the same WWTP configurations could be maintained. Moreover, Libralato et al. [3] evidenced that decentralisation processes are not categorised as well. Indeed, Ho and Anda [15] indicated that a decentralised WWTP would be able to supply <5,000 p.e. that is one order of magnitude greater than the definition for small systems arbitrarily set by the IWA Specialist Group for Small Water and Wastewater Systems as systems treating less than 100,000 L day^{-1}, but still two orders of magnitude smaller than for a centralised system. Besides, the previous definition for small systems given by the IWA Specialist Group on Design and Operation of Small Wastewater Treatment Plants was confuted, meaning that a treatment ability <2,000 p.e. or having an average daily flow <200 m^3 [16] could not be assumed as exclusively referenced to a decentralised WWTP. In decentralised systems, raw wastewater is frequently treated next to the source [1], meaning that wastewater still requires being collected, but the use of large and long pipes and expensive excavations works are avoided [3]. Additionally, decentralised processes could also be classified in three main categories according to Orth [4]: (a) simple sanitation systems (*i.e.* toilets) (*e.g.* pit latrines, pour-flush toilets, composting toilets and aquaprivies), (b) small-scale mechanical-biological treatment plants and (c) recycling systems. The simple-based technology and relative inexpensive sanitation systems (*i.e.* toilets) are asked to minimise sanitary problems by retaining faecal matter and discharging the liquid phase, with control of water pollution being of minor significance. Small-scale mechanical-biological treatment plants present at least a mechanical and a biological treatment step or, otherwise, might offer a natural-like treatment such as ponds and wetlands (*e.g.* phytoremediation pond). Thus, water pollution is contained, and other facilities enhancing nutrient removal, disinfection or solids removal might be upgraded to the system, also by means of membranes. About recycling systems, the top priority regards environmental protection. Particularly, within this approach, it might be stressed on supporting the diversion of wastewater flows while complying with modern hygienic standards, the production of high-quality fertilisers and,

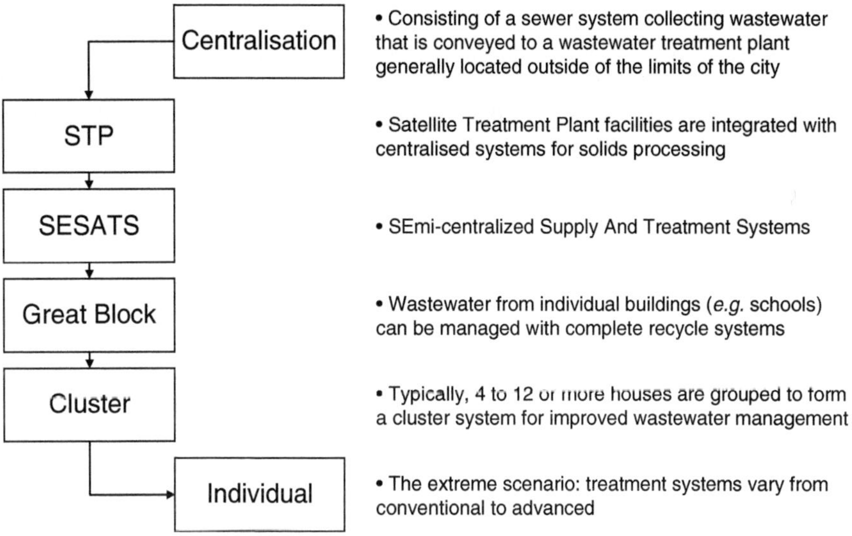

Fig. 7.1 Transition from centralisation to decentralisation [3]

eventually, biogas (*i.e.* anaerobic wastewater treatment processes), as well as the possibility of treated wastewater reuse for non-potable purposes.

Certainly, there still remains a great potential for wastewater flow separation besides urine (yellow wastewater) and faecal matter (brown wastewater) diversion, which are both components of black wastewater. Indeed, it is possible to organise and support the separation, treatment and reuse of white (rain/storm water) and grey wastewater (*e.g.* kitchen, bathtub, washing machine) [17–21].

Furthermore, it must be detailed that frequently the term "decentralised" does not comply with "small". Indeed, according to Libralato et al. [3], decentralisation can consist of from one to several scaling levels: from individual on-site systems to a series of larger clusters or semi-centralised WWTPs (*e.g.* cluster systems grouping some houses, great block WWTPs, semi-centralised supply and treatment systems, and satellite treatment plants) before approaching the traditional concept of centralised wastewater treatment system as shown in Fig. 7.1.

Several circumstances are strongly influencing the next future of SMTPs such as the population increase in rural areas and developing countries, the increasing impairment of surface water quality, the development of heavily polluted high-rise buildings in metropolitan areas, the development of planned but somewhat isolated communities and the growing shortages of water resources [22]. For example, a possible solution for small communities could be to take into consideration extensive systems with intensive processes. Anyhow, some challenging questions with no obvious or unique answer must be asked before taking the final decision: local or regional solutions for wastewater treatment? If a local solution is chosen, intensive or extensive processes? If an extensive one is preferred, which technological facility is to be selected to meet the goal of regulatory standards? [23].

About 25% of the population in USA (over 60 million) are serviced by on-site and small-scale WWTPs [24], especially those living in rural areas or not served by sewerage and centralised wastewater collection and treatment systems, but also at what time large WWTPs failed or there is a lack of action or capacity by the central governing body. Actually, after several years from the Clean Water Act, it was recognised that complete USA sewerage may never be possible due to geographic and economic constraints [25].

Similarly in smaller countries such as Ireland, the domestic wastewater of over one third of the population is treated by on-site systems, and specific regulations have been implemented to strengthen the environment safeguard (*i.e.* especially of groundwater quality), thanks to the attention of local policymakers on this topic [26, 27].

There has been a general acceptance by stakeholders and professionals of small- and medium-sized wastewater treatment (SMTs) applications because they help the improvement of public and environmental health by minimising the impact of wastewaters, being, potentially, sustainable considering equally, economic, environmental and social topics. Of course, a series of factors needs to be taken into consideration to select the SMTs, such as the population density, the location, the technology and its efficiency, the investment, the operation of maintenance, the protection of environmental quality, the conservation of resources (including the energy use), the water reuse, the nutrient recycling, the protection of public health, the convenience security, the government policy and the human settlement planning [28].

The design methodology of SMT systems is based on the combining effective treatment processes with the commercial demand of the market and the achievement of a high-quality effluent suitable for direct discharge to surface waters.

At the moment, the majority of SMTs are designed to remove carbonaceous compounds and suspended solids, but there is an increasing demand for nitrification and phosphorus removal [29]. Some of the applications of small-scale high-quality wastewater treatment could be for localised storm water treatment and recycling and for water harvesting for local water supply purposes and water appliances; definitely, fittings and technologies could facilitate the achievement of quality goals [30].

The main general characteristics of SMTs are related to economic costs, sustainable resource uses, ecological and human health impacts, system reliability and resilience, and social and institutional implications [31]. They are relative inexpensive, including capital, installation and running costs [32] depending on the technology used, making the use of recovered wastes more practical and cost-effective, especially for small nutrients and water cycles. They can maximise the opportunities for the reuse of sewage components (*i.e.* reuse of treated wastewaters for toilet flushing and watering lawns) [33], decrease the final discharge volume reducing the cumulative impacts to water environments due to the increased potential for reuse [31] and lower the level of pathogen risk to the total community than equivalent centralised reuse [34]. They can make the source stream separation easily applied allowing treatment and reuse of different waste streams (grey and black water and urine) and greater efficiencies [15] saving energy [18]. At the same time, they are able to minimise the potential for nutrients containing residuals to be contaminated with metals and other toxins. They are suitable for remote locations where access to the main drainage is difficult; they are compact, having a small footprint and minimal depth [35], reducing the aesthetic impact (*i.e.* visual, noise and odour) [1] and without any potential for catastrophic impacts on the contrary of large-scale centralised infrastructures [31]. There is no need of extensive piping and pumping systems, but easy operation and maintenance are led on [12]. They have greater flexibility and adaptability to changes and to changing service requirements [36]. Even though the community and ownership are widely involved, there are some negative aspects such as the uncertain regulatory framework in many jurisdictions, the organisational challenges for utilities managing multiple dispersed assets and the perceived actual fluctuating reliability of decentralised systems [31].

Package treatment systems are, generally, preferred for biological processes. There are many process configurations and technologies that could be applied to small- and medium- scale WWTPs such as rotating biological contactors (RBCs); extended aeration activated sludge (EAAS); activated sludge sequencing biological reactors (AS-SBRs); membrane biological reactors (MBRs) and submerged aerated media systems [29, 37]; wet composting and vermicomposting tanks; anaerobic treatments; sand, soil and peat filters; constructed wetlands; and grey, black and urine water separation systems [38]. In particular, modern membrane reactors integrating a suspended growth bioreactor with a membrane filtration device (microfiltration/ultrafiltration, MF/UF) could provide efficient barriers achieving high removal rates: indeed, the membranes retain suspended solids, colloidal and macromolecular materials, including bacteria.

Currently, small-scale applications in the wastewater treatment sector are gaining a great attention both in the developed and developing countries. Brown et al. [39] reported that the city of Melbourne (Australia) with an approximate population of 3.9 million is studying a portfolio of decentralised and on-site design concepts of WWTPs. This strategy was intended to cope with the potential uncertainty in future sewage production and its reuse and the need to prepare for integrated water cycle planning. The existing Melbourne sewerage system is largely centralised, thus about 90% of sewage discharges are conveyed to two major centralised plants. However, several small satellite treatment plants service local urban areas generally more distant from the centralised system. The use of decentralised WWTPs in Melbourne is still rare, but the aim of the future integrated water planning is to combine centralisation with various levels of decentralisation as well as on-site operations.

Orth [4] presented the case study of the Ruhr district (Germany) as an example of how the sewage disposal may evolve according to a sort of ecologically sound adaptive ability. Initially, centralisation represented the first step to solve intolerable hygienic conditions. Afterwards, the need to control pollution and support the regional development besides the existing WWTPs led to a series of about 4,200 small-scale treatment systems creating a composite system made of centralised and decentralised elements.

Libralato et al. [3] highlighted that in Italy, 6% of the population is served by WWTPs with <2,000 p.e., which represent 73% of existing Italian WWTPs mainly due to the country morphology. Likewise, Venice (Italy) may be considered as an interesting case study about decentralisation. Indeed, this well-known ancient city that was built on a series of 119 islands located in the middle of a 540-km^2 lagoon with an average depth of 0.5 m does not have a real sewage collection and treatment system due to its peculiar urban characteristics, but has a huge number (4,493) of on-site decentralised WWTPs [40] that are mostly remote controlled. Their installation, supported by policy and lawmakers, has been reducing the total load of inorganic and organic contamination, enhancing the general health and environmental status of the lagoon.

Peter-Fröhlich et al. [21] presented a pilot project at European level to separate yellow (*i.e.* urine), grey and black wastewater supporting both environmental and economic-related topics. Indeed, this new approach could allow wastewater treatment in remote areas, having no centralised sewage collection systems, or in new developing or refurbishing urban areas where the existing systems could not be cost-effectively upgraded.

Actually, the main added value is related to the increase of environmental sustainability via treated wastewater and nutrient recovery and reuse such as in the case of public and private garden watering or agricultural uses reducing potable water consumption in the perspective of near future water scarcity. The safeguard of water resources is a really challenging topic that has been frequently stressed on a decentralised basis. Nolde [41] reported the results from a 15-year project developed in Germany about grey wastewater recovery and reuse within great block of buildings (*e.g.* commercial centres) and mainly devoted to toilet flushing. It was evidenced that the technological investment mortgage is of about 5–7 years in the case of the adoption of membrane biological reactors.

Furthermore, it was observed that the separation of flows at source (*i.e.* white, yellow, brown, grey and black wastewaters) might sustain the management of emerging micropollutants such as in the case of pharmaceuticals, which are mainly contained in urine rather than in faecal matter [42]. Indeed, nevertheless, urine contains 15% of the total COD of black wastewater [43]; it was discovered that about 64% of pharms and pharmaceutical residues could be found in urine after a quantitative study about 212 drugs [44]. An extreme application of source separation is represented by the NoMix approach that operates a direct urine separation at source [18, 28, 45, 46]. It is already successfully applied in China in the dry version with more than 700,000 users [45]. Palmér Rivera et al. [47] proposed other interesting options promoting source-separating sanitation systems supporting the design of a blackwater separation system with vacuum toilets in a local folklore centre. The sanitation system consisted of three vacuum toilets and one water-free urinal connected to a collection tank, and greywater treatment in a sludge sedimentation tank before vertical sand filter.

7.2.3 To Decentralise or Not to Decentralise: That Is the Question

Today, the debate about the centralisation/decentralisation dichotomy in wastewater treatment science showed that they might be reciprocally unsuitable on a case-by-case basis: one approach cannot exclude the other and vice versa [3]. The possibilities for wastewater treatment are quite huge, and a sort of transition exists in decentralisation processes moving from individual on-site treatment, to cluster or community type (*e.g*. great blocks), and to satellite treatment and semi-centralised WWTPs. Each type is substantially related to the characteristics and volumes of wastewater to be treated as well as to the possibility of flow separation at source.

When talking about decentralisation, it is not possible to refer to just one technical approach such as the NoMix toilet or any other extreme individual treatment system, but to a range of wastewater treatment facilities presenting a strong scaling transition. In most cases, the adoption of a decentralised approach in highly dense populated areas with an already existing sewage collection system could not be a viable alternative to the centralised treatment. A general clue is not yet available on how to approach decentralisation due to the high number of conditioning variables and the creativity of engineers puzzling their brains to make alternatives to traditional wastewater treatment modes [3]. The most probable suggestion is to support the coexistence of various levels of centralisation and decentralisation in WWTPs considering the potentiality of the full series of decentralised approaches that are currently showing a highly realistic appeal, mainly in the case of great blocks (*i.e*. hospitals, shopping centres, airports, schools) and refurbished urban areas, especially in relation to the new trend of treated wastewater recovery and reuse mainly due to climate change-related phenomena (*i.e*. water drought). Moreover, it is of extreme interest to speculate upon the forthcoming necessity of current centralised WWTPs substitution as well as refurbishment and upgrading of their collection system due to their ageing. Major interventions are estimated to be required every 50–60 years, but no unique solutions have been found whether to keep and/or substitute the pristine centralised facilities, generating potential traffic and other public utility disruption or to introduce some kind of decentralisation in the system. Today, this second option appears to be sometimes more suitable, mainly due to the continual growth of urban areas as well as the increasing demand for water resources. It seems that the use of decentralised and satellite systems will allow treated wastewater recovery and reuse, making them a stable and sustainable water source, especially for those areas that historically suffer or have recently been suffering from water scarcity [3].

Since now, pros and cons of centralisation and decentralisation were elucidated in a scattered way evidencing the fact that economic, environmental, social and technological issues may strongly influence decision makers within their potential choices about the implementation of one approach rather than another. Particularly, Brown et al. [39] highlighted that some releases are of main importance such as the life span of system elements, the estimated capital, the maintenance and operating costs, and the (re)use of energy, residuals and water as well as nutrient budgets [48]. On the basis of Libralato et al. [3], the main advantages and disadvantages on centralisation and decentralisation have been assessed also according to Brown et al. [39] issues and listed as follows. On centralisation, the subsequent statements may be provided from a series of authors:

- The wastewater treatment cost per unit volume is still competitive compared to decentralisation where the wastewater collection system already exists [11, 15, 28, 49, 50].

- Around 80–90% of the capital costs are related to the collection system with potential economies of scale associated to densely populated areas [11, 50, 51].
- It is predicted that the whole collection system or of part of it has to be renewed every 50–60 years, besides the required periodic maintenance, potentially generating disruptions to traffic and other public utilities [50].
- Wastewater treatment generally means "to sanitise", but nutrients and other micropollutants might not be removed [52–54].
- Potential eutrophication phenomena may occur in the receiving water body due to the large volumes of treated wastewater discharged [1, 15, 28, 52, 55];
- Rainwater is frequently drained from residential areas by infiltration into the collection system, potentially causing the lowering of the aquifer [15, 28, 49].
- Diluted wastewater requires more expensive treatment approaches [15, 28, 50, 55].
- Heavy rainfall events or contamination by industrial wastewater may generate overflow phenomena [15, 55].
- Natural disasters such as earthquakes and terroristic attacks may cause disruptions to the system generating strong pollution phenomena in the receiving water body [1, 15].
- Diseconomies of scale are possible where long distances have to be covered or as a consequence of rainwater infiltration [11, 50].
- There is a strong dependency on electrical energy supply that might not be adequate due to an economic or political crisis [1, 11, 50].
- Huge volumes of potable water are required to keep the sewage system clean [1, 11, 50].

Furthermore, on decentralisation, the following reports may be delivered from a series of authors:

- Decentralisation may respond to suburban areas and rural centres, industrial, commercial, residential and (re)developing areas in addition to population growth in rural areas and developing countries [1, 15, 25, 28, 33, 39, 42, 56].
- It tends to stop the decrease of surface water quality [25, 33, 39].
- It may be of some help in the case of great block construction in metropolitan areas, pretreating/treating and reusing wastewater, even if in part, thus limiting the volume of discharged wastewater into the existing sewage collection system and obviating its upgrading to support greater volume loads [15, 28, 39, 56].
- It may contribute in the planning of isolated community development [9, 12, 39, 42] and support treated wastewater recovery and reuse [9, 12, 15, 25, 28, 33, 39, 56–58].
- It reduces or excludes the inconveniences related to the collection of discharges, with much smaller and shorter pipes compared to centralisation [15, 25, 28, 33, 39, 57], being applicable to various levels from individual to community [9, 11, 25, 33].
- Small WWTPs are considered as viable if a medium-high technological level is implemented that is efficient, robust and easy to manage and maintain [25, 33, 58], although some unexpected bad performance was experienced mainly due to their managing [59].

- Small WWTPs are eligible to be easily remote controlled facilitating their management [40].
- Much of the cost could be related to possible economies of scale that could be achieved organising wastewater treatment on a cluster basis such as in Australia [31].
- The cost of technologies in decentralisation is becoming comparable to that of centralisation per unit of treated organic load [31].
- Small WWTPs may assure a greater level of environmental sustainability by supporting the potential reuse of treated wastewater as well as nutrient recovery [9, 25, 31, 33, 39, 48, 55, 57, 58].
- The potential contamination of nutrients to be reused by metals and xenobiotic substances in general could be greatly limited [55, 58].
- Eutrophication events may be reduced [1, 9, 15, 28, 31, 39, 52, 55, 57, 58].
- It may support an easier urine source separation, reducing/removing micropollutants such as metals and other emerging compounds (*e.g.* pharmaceuticals and personal care products) [9, 25, 33, 57, 58].
- Decentralised small WWTPs allow the separation of domestic wastewater and rainwater, avoiding dilution phenomena [15, 28].
- It is possible to operate a separation of contaminants at source, easing their treatment and potential reuse and at the same time increasing treatment efficiency and saving energy [9, 25, 33, 39, 57].
- It is possible to exclude the possibility of domestic wastewater contamination by industrial wastewater as well as the relative sludge produced [9, 11].
- It is possible to maximise the *in situ* reuse of treated wastewater, as a consequence of diminishing the final discharge volume and the potential cumulative impacts on the receiving water bodies [25, 33, 39].
- It is possible to considerably reduce the health risk for the population, also by preventing catastrophic events [11, 25, 33, 39, 55].
- Small WWTPs are suitable for isolated or scattered settlements or in the case where only a small amount of space is available for the installation [11, 39].
- Small WWTPs are generally compact, with highly flexible operating conditions and reduced aesthetic impact [11, 39, 40, 55].

Moreover, other aspects that are directly related to decentralisation approaches involve national security concerns. Centralised WWTPs can be seen as an easy-to-attack target that could seriously affect life in some urban areas, for example, due to the physico-chemical and microbiological contamination of surface water preventing its use as a drinking water source. A series of decentralised WWTPs may considerably reduce the risk and potential impact to the receiving water body without compromising the system functions. Moreover, decentralisation processes might reduce the impact of natural disasters such as a flood, tornado, hurricane, volcanic eruption, earthquake, or landslide that could in this way affect only a limited part of the territory keeping the rest safe [3].

7.2.4 *Having a Decentralised Perspective in Land Planning Activities*

Decentralisation might be supported during decision-making processes if cost, flexibility of land use, maintenance and environmental protection are taken into account with special reference to small communities within which reuse scenarios are generally more feasible [60]. Resource efficiency in wastewater management means not only to recover and reuse treated water but also a matter of optimising the management of resources spent on treatment and transport, the natural resource to protect as well as the capital [61]. Particularly, Chung et al. [62] proposed a model to assess the suitability of decentralisation, showing that it fits best to territories with mixed morphology and scattered urban centres. Comparing a centralised WWTP and a series of satellite WWTPs for a community of 1.2 million inhabitants, economies of scale showed to favour centralisation, except for areas with significantly different heights above sea level, making decentralisation a more suitable option. Innovators must tackle the costs of initial investments in infrastructures, and concerted actions are required with the involved stakeholders. Since now, the centralised model has received huge public capital investments making it sustainable due to the economies of scale that are created. Besides, reducing the technological investment compared to initial expectations could save the major costs that could be required to terminate the collection system. At the moment, in Europe, the water sector management is already oriented towards considering the full costs of the resource and the potential effects of water policy changes, which will be more evident in the near future. Nevertheless, the risk of overestimation of such resources is not risky because the benefits obtained from a good quality aquatic environment are still considered as intangible, which is the main reason for the lack of support for innovation in this sector [3]. Indeed, when water resources are in some way related to public health, tourism, education and research, greater funds might be obtained. These uses that require a high water quality would justify the support for innovation and innovators [63].

7.3 Best Available Technologies

The expression "best available technologies", which is defined in Article 2(11) of the 96/61/EC Directive [64], is very similar to that of the US Clean Water Act (1972) and its following amendments being "*the most effective and advanced stage in the development of activities and their methods of operation which indicate the practical suitability of particular techniques for providing in principle the basis for emission limit values designed to prevent and, where that is not practicable, generally to reduce emissions and the impact on the environment as a whole, where:*

- *'Techniques' shall include both the technology used and the way in which the installation is designed, built, maintained, operated and decommissioned;*

- *'Available' techniques shall mean those developed on a scale which allows implementation in the relevant industrial sector, under economically and technically viable conditions, taking into consideration the costs and advantages, whether or not the techniques are used or produced inside the Member State in question, as long as they are reasonably accessible to the operator;*
- *'Best' shall mean most effective in achieving a high general level of protection of the environment as a whole".*

In more general terms, BATs or best available technologies not entailing excessive costs (BATNEECs) can be defined as the best viable mix of technologies, processes, services and ways of management in order to reduce the pollution and increase the efficiency [65]. According to this definition, they should pervade all economic activities and sectors, cutting costs and improving competitiveness by saving energy and reducing resource consumption and so creating fewer emissions and less waste. This definition is essentially in line with that given in Chapter 34 of Agenda 21 [66] for environmentally sound technologies, stating that "*Environmentally sound technologies protect the environment, are less polluting, use all resources in a more sustainable manner, recycle more of their wastes and products, and handle residual wastes in a more acceptable manner than the technologies for which they were substitutes. Environmentally sound technologies in the context of pollution are process and product technologies that generate low or no waste, for the prevention of pollution. They also cover end of the pipe technologies for treatment of pollution after it has been generated. Environmentally sound technologies are not just individual technologies, but total systems which include know-how, procedures, goods and services, and equipment as well as organisational and managerial procedures*" [67]. As a consequence, the selection and the application of the best available technologies should be the right union between growth and the environment. Indeed, [...] *growth and the environment are not opposites, they complement each other. Without adequate protection of the environment, growth is undermined; but without growth it is not possible to support environmental protection* [68].

On the basis of Article 9(4) of the 96/61/EC Directive [64], emission limit values, equivalent parameters and technical measures must be based on BATs, without prescribing the use of any technique but considering the technical characteristics of the installation, its geographical location and the local environmental conditions. Under this point of view, there is the need to strengthen the efforts to improve testing (*e.g.* toxicity bioassays), verify performances and standardise environmental technologies, in order to establish a mechanism to validate objectively the performance of these products, increasing the purchasers' confidence in new environmental technologies (*e.g.* nanotechnologies applied to wastewater treatment). Furthermore, standardisation, preferably at the international level, can stimulate innovation and environmentally friendly practises. A meaningful example is that related to "*the lack of European standards for wastewater reuse that is one of the main barriers to the market uptake of membrane bio-reactors for municipal waste*

Table 7.1 BAT-associated emission levels for final treated discharge without dilution. Only the strictest values are reported ([2] mod.)

Parameter	Performance rates (%)	Emission levels (mg L^{-1} day^{-1})
SS		10[a]
COD	96	30
Total inorganic N		5
Total P		0.5

[a]Monthly average

water treatment. Membrane bioreactors have several environmental advantages over conventional activated sludge plants: they remove recalcitrant micropollutants more effectively and they reduce the amount and toxicity of the resulting sewage sludge. They are currently more expensive but provide an effluent that is ready for reuse. However, as this reuse is not encouraged by any kind of regulation or standard, the market is not as developed as it could be" [65].

Besides, the IPPC [2] provided the basic knowledge about the measures for the determination of the BATs for wastewater and waste gas treatment supporting process-integrated methods for preventing and reducing the contamination. In most cases, they are production- or process-specific measures whose applicability requires a special assessment. The use of BATs in wastewater treatment is especially focused to the end-of-pipe approach, optimising treatment procedures, preventing or minimising mixing of contaminated and uncontaminated wastewaters and considering the fact that it is the application (*e.g.* serial combination of some technologies) and the way of management that could make the difference.

At what time wastewater is discharged into surface water (*i.e.* river, lake or sea, and all other kind of surface water bodies), BATs must present a suitable combination of [2]:

- Avoiding a discharge situation such as excessive hydraulic load or toxic wastewater (*i.e.* potential damages to the river bed, the embankment or the biosphere).
- Choosing the discharge point in order to optimise the dispersion.
- Balancing the wastewater not coming from central wastewater treatment plants (WWTPs) to reduce the impact on the receiving water body and to meet discharge requirement before discharging it.
- Implementing a monitoring system to check the water discharge with adequate monitoring frequency.
- Performing toxicity assessments as a complementary tool in order to obtain more information on the effectiveness of the control measures and on the hazard assessment for the receiving body. Generally, a case-by-case basis application is required.

Some emission level requirements for final wastewaters discharged without any dilution into surface water after BAT treatment are reported in Table 7.1.

7.4 Advanced Technologies Potentially Supporting On-Site Water Reuse

The contaminants in wastewater can be removed by physical, chemical and biological means according to a number of different treatment, disposal and reuse alternatives and their relative optimum combination.

Physical unit operations are characterised by the prevalence of physical forces during the treatment operations and were firstly used for wastewater treatment. The main physical processes are screening, mixing, flocculation, sedimentation, flotation and filtration.

In chemical unit processes, the wastewater treatment is obtained via the addition of chemicals or by other chemical reactions. Coagulation, precipitation, gas transfer, disinfection and adsorption are the main chemical unit processes. Both physical and chemical unit operations are, mainly, used to treat non-biodegradable wastewaters and inorganic or hardly biodegradable organic contaminants. On the other hand, biological treatment is primarily used to remove the biodegradable organic substances, basically, converting them into gases released to the atmosphere and into cells removed by settling. Specially, the activated-sludge process is a treatment process that is running since 1914 when Arden and Lockett firstly approached it. Many versions of the original process are in use today, but fundamentally, they are all similar [69].

Generally, unit operation and processes are grouped together to provide what is known as primary, secondary and tertiary (or advanced) treatments. Essentially, the term primary refers to physical unit operations, secondary refers to chemical and biological unit processes and tertiary refers to the potential combination of all three [69].

In this section, the attention is focused on the description of the basic knowledge of some secondary (chemical coagulation/precipitation, activated-sludge sequencing batch reactor, ultrafiltration membrane biological reactor) and, for the most part, tertiary processes (nanofiltration, reverse osmosis and activated carbon) that might support treated wastewater recovery and reuse. These wastewater treatment technologies can also be ranked according to their biodegradability-dependency. In fact, biological treatments are highly dependent from the biodegradable fraction of wastewater, while little influence is exerted on physical (*i.e.* filtration, adsorption, Reverse Osmosis) and chemical processes (*i.e.* precipitation).

7.4.1 Chemical Coagulation/Precipitation

Chemical coagulation and precipitation are well-known methods of wastewater treatment for soluble non-biodegradable contamination since the end of nineteenth century. In most cases, lime is used as a precipitant, generally, in combination with calcium chloride, magnesium chloride, alum, ferric chloride, ferrous sulphate or charcoal. Nowadays, the chemical coagulation and precipitation are mainly used as a means of improving the performance of primary settling facilities, a basic step in the independent physical-chemical treatment of wastewater (*i.e.* heavy metals removal) and for the removal of phosphorus [69]. The chemical coagulation and precipitation are treatment techniques recognised as a BAT in relation to some specific applications [2].

7.4.2 Activated Sludge Sequencing Batch Reactor (AS-SBR)

In recent years, AS-SBR technology received increasing attention worldwide, and a great number of plants have been built [70]. The AS-SBR has been accepted as an alternative to more conventional activated-sludge systems for a wide range of industrial and non-toxic biodegradable wastewater treatments [71, 72], even though it can present some low removal rates for some recalcitrant organic compounds [73]. Anyway, it is seen as one of the most promising processes from SMTPs [70, 74–77]. Goronszy [78] showed that AS-SBR is especially suited for wastewater treatment application characterised by intermittent flow and loading conditions, or in decentralised drainage locations for small population equivalent wastewaters [29, 37].

The AS-SBR is an activated-sludge process; this means that the removal of carbonaceous BOD, the coagulation of non-settleable colloidal solids and the stabilisation of organic matter are accomplished biologically using a variety of microorganisms, principally bacteria. The microorganisms are used to convert the colloidal and dissolved carbonaceous organic matter into various gases and into cell tissue. Therefore, the cell tissue, having specific cell gravity slightly greater than that of water, can be removed from the treated liquid by gravity settling. Other applications are related to nitrification/denitrification and stabilisation processes. Operationally, the organic wastewater is introduced into a reactor where an aerobic/anaerobic bacterial culture is maintained in suspension, constituting the mixed liquor. In the reactor, the bacterial culture carries out the degradation of the wastewater organic fraction, and, after a specified period of time, the settling phase takes place to allow the separation of the cells (as sludge) and of the treated wastewater. Generally, the level at which the biological mass is kept in the reactor depends on the desired treatment efficiency and growth kinetics and is monitored taking into consideration the mixed liquor suspended solids (MLSS) and the mixed liquor volatile suspended solids (MLVSS, *i.e.* a rough approximation of the amount of organic matter present in the solid fraction of wastewater).

The AS-SBR can accomplish the tasks of primary clarification, bio-oxidation, and secondary clarification within the confines of a single reactor in the time-oriented configuration [35], whereas the space-oriented is characterised by at least two reactors where all the treatment processes are serially administered. In particular, the desired effluent quality can be delivered operating on fill/reaction ratios, aeration periods and mixing cycles [79]. The operating cycle of AS-SBR is determined both by degradation and settling performances [80].

Generally, a full AS-SBR sequence is composed of five serial steps: feed, mixing, aerobic reaction, settling and drawing. During the feed and the mixing period, denitrification and phosphorus release take place, while phosphorus uptake, carbonaceous BOD removal and nitrification occur in the following oxidation stage. In addition, endogenous denitrification should take place during the settling phase [70]. Generally, the SS concentration is around 10 g L^{-1} with a sludge retention time from 5 days to more than 30 days [73].

The AS-SBR is able to mitigate the effects due to physico-chemical variations in the wastewater feed flow because the reactor acts also as a surrogate of an equalisation tank; the clarification phase is quicker because the sedimentation is allowed in the same reaction basin (*i.e.* in the AS-SBR time-oriented configuration), and there is an increase in the O_2 transfer during the feeding phase due to the entrance of untreated wastewater rich in biodegradable organics [81]. Manning and Irvine [82] stated that microorganism activity in AS-SBR is greater than through flow processes, and maintenance and management are reduced to a minimum [81] due to the absence of sedimentation basins, the possibility of controlling bulking phenomena modifying the feeding phase [83] and the potential for saving energy, especially, for the treatment of domestic wastewaters produced by small- and medium-sized communities [84].

In particular, the AS-SBR technology at small- and medium-sized municipal treatment plants has increased since 1970s [71], also for retrofitting small works [85] and for some applications to industrial wastewaters (dairies and wineries) [72] or for dealing with piggery effluents [86]. The AS-SBR exists both in the anaerobic/aerobic configurations, providing effluents very low in organic compounds and nutrients, potentially, meeting strict effluent standards. Moreover, this technology could be suitable for rural areas, where experts in WWTP management are rather limited, via the introduction of automatic remote controls [70].

7.4.3 Membrane Biological Reactor (MBR)

Research studies about the application of membranes to biological processes for wastewater treatment started more than 40 years ago, and since about 20 years, they are commercially available. The membrane bioreactor is one of the most promising large-scale newer technologies for providing high-quality effluents (MBR) that has already been classified as BAT by IPPC [2] for its physico-chemical performances and for its potential for retrofitting existing WWTPs.

MBRs coupled to various filtration devices (*e.g.* MF and UF) are a further development of the conventional ASP, where the secondary clarifier is replaced by a membrane filtration [87], for domestic, urban and industrial wastewater treatment. Effluents from MBR plants could cover a range of reuse applications such as irrigation (*e.g.* agriculture and landscape), recreation and environmental use (*e.g.* lakes and ponds), groundwater recharge, if toxicity free, and industrial use [31].

Nevertheless, the MBR has been around since 1960s; it has seen significant advances since the early 1990s. It can be defined as a combined activated-sludge biological (previously described for the AS-SBR) and physical process technology for solid/liquid filtration, integrating suspended growth reactors with membrane filtration, which can be considered as the bottleneck of the process. In a conventional secondary clarifier, only the fraction of the activated sludge that settles as flocks can be retained, but in a MBR, all components of the biomass that are larger than the membrane cutoff are retained. However, the fundamental differences in the biology of an MBR compared to an activated-sludge process are not yet clear since a limited amount of information is available on the way in which descriptive variables such as the flock structure, respiration rate, species and off gas production are affected by the changes in operation [69, 87].

The pressure gradient across the membrane is the driving force accomplishing the separation. As a result, the separation of biomass from the treated wastewater is independent of biomass sedimentation qualities [88].

The MBR has recognised advantages over conventional biological treatments such as higher quality effluent (particle free), higher mixed liquor suspended solids and less excess biosolids [89], absolute control of solids and hydraulic retention times [88], compactness with smaller footprint, a modular nature which suits small scale and system expansion and upgrade [89, 90].

A small-scale MBR WWTP flow sheet could be characterised by a pretreatment step (screening phase), a hybrid membrane process (*e.g.* MF/UF-MBR) followed by an oxidative post-treatment to yield water for non-potable reuse or a second high retention membrane process plus post-treatment for indirect potable use (*e.g.* toilet flushing). The main costs, benefits and issues for membrane-based technologies for decentralised WWTPs can be summarised in the following list [6, 91]:

- Costs of membrane dropped over the past decade, but membrane technology can still be considered relatively expensive – further cost reductions are expected from machine fabrication and an increasing market for small MBRs.
- Negative economy of scale, although balanced by the modular applicability of that technology allowing plant size from single dwellings to clusters and to suburb plant size.
- Weaknesses related to energy demand and subsequent greenhouse gases emission; energy demand is high than a conventional plant (+20%) because fouling needs to be controlled by air scouring or high shear velocity [31].
- Energy consumption could be reduced via the use of an anaerobic MBR which could be a net energy producer due to biogas generation [92], even though at small scale, it could be probably too complex, but suitable for clusters or medium-sized plants.
- Persistent organic pollutants (*e.g.* endocrine disruptors, pharmaceuticals and hormones) are only partially removed, so post-treatment processes (*e.g.* AOP) could be required to enhance the removal of these species.
- Management problems related to unqualified personnel (*e.g.* the owner itself without any in-depth knowledge).

Anyway, further investments to apply membranes successfully to SMTs are required [6, 69, 87, 91, 93] in order to lower system costs and increase affordability, minimise energy demand, maximise nutrient removal for beneficial use, develop disposal strategies from remaining residuals, develop integrity monitors that are low cost, effective and reliable, establish remote management systems, develop regulatory framework for decentralised non-potable effluent reuse in urban areas and provide wastewater system managers with planning tools that account for the advantages of decentralised systems based on membrane technologies.

There are two main potential configurations of MBRs: the immersed (submerged or outside-in filtration) and the side stream one (inside-out filtration). A submerged system operates at a lower transmembrane pressure (TMP) than side stream membranes and therefore at a lower flow.

In most membrane processes, it can be said that there are three main streams: a feed, a retentate (unpermeated product) and a permeate. If there is no retentate, the filtration is termed dead-end, vice versa it is termed crossflow.

The membrane productivity is expressed as the permeate flow through the membrane that is reduced along the operating time due to fouling [94]. The fouling control in the submerged membranes is achieved by an air scour at the membrane surface by coarse bubble crossflow aeration. The movement of the bubbles close to the membrane surface causes the necessary shear velocity. At side stream filtration membranes, a high water velocity achieves the fouling control across the filtration channel. In particular, the side stream crossflow configuration employs a tangential flow across the membrane surface which provides a continuous scouring action and hence reduces the membrane fouling layer due to feed stream debris and macromolecules, whereas frequent backwashing is required for the dead-end configuration. Crossflow membranes can provide high system utilisation with minimal plant cleaning downtime. In both cases, the membrane productivity can also be re-established after mechanical or chemical cleaning [91]. However, periodically, chemical cleaning is generally required, and, depending on the type of membrane and treated wastewater, some chemicals such as NaOCl, H_2O_2 and citric acid are used [87, 95]. Its frequency and intensity depend on the loading rate of the wastewater, the bacterial yield, the microbial production of extracellular polymer substances and the retention of non-biological solids in the biofilm matrix [96]. For example, Brindle et al. [97] signalled that membranes do not require any cleaning for over 172 days during the treatment of an ammonia-rich synthetic solid-free wastewater.

Submerged membranes are applied in municipal and industrial WWTPs and can be placed either inside the aeration tank or in an external filtration tank, while the side stream one is mostly used for industrial wastewater treatment. In Venice (Italy), this technology showed to be particularly suitable also for domestic purposes because of its really small footprint, even for real small-scale plants, and its maintenance and management costs [3, 98]. Indeed, there exist more than 140 small-decentralised WWTPs as well as a huge number of septic tanks [40, 99]. Besides wastewater treatment processes, the crossflow side stream membrane filtration can be considered as a mature technology that is also regularly employed as a standard

Table 7.2 Main differences between an activated sludge and an UF-MBR domestic wastewater configuration [87]

Parameter		Activated sludge	UF-MBR
Sludge age	Day	20	30
COD removal	%	94.5	99.0
DOC removal	%	92.7	96.9
SS removal	%	60.9	99.9
Ammonia-N removal	%	98.9	99.2
Total P removal	%	88.5	96.6
Sludge production	KgVSS/COD day	0.22	0.27
Mean flock size	mm	20	3.5

technique for liquid processing to effect clarification, product isolation, concentration and separation duties in a large number of manufacturing industries [100].

The main performance differences between an activated sludge and an UF-MBR domestic wastewater configuration are displayed in Table 7.2. The main advantage of MBR over a traditional suspended growth reactor is the high concentration of MLSS that can be treated leading to small- and high-scale treatment system with reduced sludge production from longer SRTs [91]. The main benefits of membrane technology applied to MBR are related to the selective and consistent separation abilities, the increased product yield, the fact that is an almost well-established technology, and no additives, flocculating agents or pre-coat chemicals are required; besides, large variations in feed quality have little influence on permeate quality; the retrofitting is easy and low maintenance is needed [87].

Membranes with pore size of 0.001–1 μm are typically used [100] with a permeate flow ranging from 15 to 80 $L\ m^{-2}\ h^{-1}$ depending of the type of the membrane. Anyway, the pore size of membranes is too large to separate single molecules or ions. The MLSS is independent of the sedimentation behaviour of the sludge, so that it can be increased significantly. Typical values of MLSS are in the range 12–15 $g\ L^{-1}$ at submerged MBRs and up to higher values for side stream systems [87, 101, 102], keeping in mind the problems related to the oxygen gas-liquid mass transfer. Anyhow, operating MBRs at these conditions can, frequently, create serious problems because high aeration rates are required to provide adequate oxygen supply and effective membrane scouring is difficult to achieve due to the increased mixed liquor viscosity [103]. Besides, by the loss of the secondary clarifier and increase of MLSS in the aeration tank, the footprint of the treatment plant is significantly smaller.

The hydraulic load and the achievable flow are the key parameters for the design of the membrane surface in order to allow the membranes to permeate the maximal flow. The food to microorganism ratio is the key parameter, and as high MLSS can be achieved, the resulting tank volumes are smaller. One advantage of this process is the complete removal of suspended solids and bacteria, including greater viruses [87]. Actually, MF and UF membrane provide from log 2 to log 5 virus removal, respectively, [104] and >log 5 removal protozoa [105]. Indeed, virus removal in wastewater treatment is receiving increasing attention because of the epidemiological significance of viruses as waterborne pathogens [106–108]. Generally, secondary

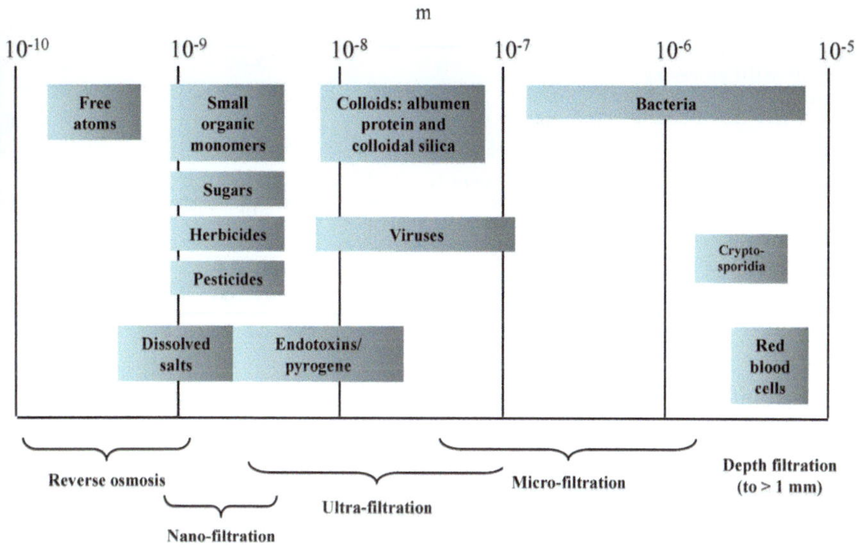

Fig. 7.2 Rejection capability of membrane separation process ([87] mod.)

treatments such as physico-chemical coagulation/precipitation (chemical coagulation/precipitation and sedimentation) and biological treatment are able to remove 99.9% of faecal coliforms, whereas the removal of viral indicators (*e.g.* bacteriophages) is much lower [109, 110].

About energy consumption, the MBR performances are dependent on many factors such as the plant design, the operational philosophy and the plant size. MBR power requirements come mostly from pumping feed water, recycling retentate, the occasional suction of permeate and the aeration [111]. Generally, the overall energy for submerged systems tend to be lower than for side stream operations [87]. For example, Lesjean et al. [112] stated that a hollow-fibre submerged MBR requires an amount of energy of 0.3–0.6 kWh m^{-3} with a design flow of 20–30 L m^{-2} h^{-1} and a crossflow side stream of 2 –10 kWh m^{-3} with a design flow of 70–100 L m^{-2} h^{-1}.

7.4.3.1 Micro-, Ultra-, Nano- and Reverse Osmosis Filtration

Membrane filtration processes can be divided into groups, based on separation of the respective particle diameters. Fig. 7.2. presented an overview of the diameter application range for the different filtration techniques from Stephenson et al. [87] modified. The following classification is also used in order to categorise crossflow membrane filtration. There is no fixed demarcation between each group. For example, a tight UF membrane made by one manufacturer may be regarded by a second one as a loose NF and so on. All MBRs are generally classified in accordance with the filtration unit separation capabilities.

Table 7.3 Treatment efficiency of MF- and UF-MBR processes [87, 95]

Parameter	MF (%)	UF (%)
BOD	75–90	≈81
COD	46–70	70–85
SS	95–98	97–99.5
Total N		≈12
Ammonium-N	5–15	
Total P	≈14	≈26
Turbidity	92–99	>99
Total coliforms	90–100	100
Faecal coliforms	95–100	100

Microfiltration is a membrane filtration process designed to retain particles in the range 0.10–5 μm with typical operating pressures from 0.5 to 3 bar and is mainly considered as a clarification technique [113] for separation of suspended solids from water and wastewater. Common applications include degreasing processes, metal particle recovery, metal plating wastewater treatment and sludge separation after activated-sludge process [87].

Ultrafiltration membranes have a pore size in the range 0.005–0.1 μm and typical operating pressures range from 0.5 to 5 bar. They are used for separation of both large, dissolved solute molecules and suspended colloidal particles. In particular, its applications include removal of non-toxic degradable and toxic nondegradable pollutants, segregation of oil-in-water emulsions, separation of heavy metals after complexation/precipitation, separation of components not readily degradable in STWs and pretreatment step prior to reverse osmosis or ion exchange [87]. Both MF and UF present high separation efficiency with a high flexibility in usage due to the modular system composition. Sometimes, clogging phenomena can occur, like compaction events due to the presence of softening agents. The MF- and UF-MBR treatment efficiencies for some parameters are reported in Table 7.3. [2].

All membranes are available in several materials, which are directly related to the specific nature of the wastewater, since the resistances are material-specific, and to the required pore size. MF and UF membrane materials are mainly composed of cellulose acetate, glass fibre, polyamide, polycarbonate and polyvinylidenfluoride (PVDF). PVDF membranes have the advantage to be easily cleaned with strong acids, caustic soda and bleaches and, after that, to be immediately ready for reuse [2, 87].

Nanofiltration is an end-of-pipe filtration technique formerly called "leaky reverse osmosis", mainly, for the removal of soluble non-biodegradable and inhibitory contaminants, and already recognised as BAT by IPPC [2] and GURI [73] for some specific applications.

The separation is achieved through a combination of charge rejection, solubility-diffusion and sieving (0.01–0.001 μm). NF has high separation efficiency, modular systems (*i.e.* flexible in usage), low operating temperatures and possibility of fully automatic operation enabling the immediate recycling of permeate. Conversely, clogging, plugging and fouling processes can occur, like compaction in the presence of softening agents, and high pressures are required producing low permeate flows [2]. NF membranes are composed of cellulose acetate and polyamide [2, 87].

It is used for the removal of dissolved materials in the molecular range of 100–500 D molecular weight. Monovalent species are transmitted through the membrane preferentially. Besides wastewater applications, it is used for partial desalination, for final removal of degradable and toxic components and heavy metals, for removal of sucrose and egg albumin, for segregation of pollutants with the aim of concentrating or further processing them and for blood osmosis and blood filtration [87]. Active nanofiltration might also refer to the application of engineered nanomaterials for wastewater treatment. Indeed, active nano-based wastewater treatments showed to be able to administer not only a physical but also a chemical treatment such as in the case of nano-TiO_2 or nano-zerovalent iron upgraded membranes [114].

Conventionally, osmosis is defined as the net movement of water across a selectively permeable membrane by a difference in osmotic pressure across the membrane itself. Conversely, RO uses hydraulic pressure to oppose and exceed the osmotic pressure of an aqueous feed solution to produce purified water [115]. RO, classified as BAT by IPPC [2] and GURI [73] for some specific applications, is one of the several demineralisation techniques applicable to the production of water suitable for reuse, presenting the added benefit of removing dissolved organics but showing, generally, greater costs and a lack of operating experience on the treatment of domestic wastewater than more traditional techniques [69].

The basic components of a RO unit are the membrane (0.01–0.0001 μm), a membrane support structure, a containing vessel and a high-pressure pump. RO units can be arranged either in series or in parallel and provided of various types of membrane configuration supports (spiral-wound, tubular, plate and frame and hollow-fibre).

RO processes are generally operated crossflow on an end-of-pipe basis, that is, the permeate is directly perpendicular to the feed flow. So that, the impurities remain in the feed which, reducing in volume, leaves the membrane system as a concentrated waste stream [2]. A very high-quality feed wastewater is required to prevent clogging phenomena and improve the efficiency of the membrane, so that pretreatment of a secondary effluent with filtration and AC are necessary, like the removal of iron and manganese in order to decrease the scaling potential. Also, the pH should be adjusted in the range 4.0–7.5 for the same reasons.

RO presents several applications such as the treatment of outflows containing colourings with their possible recovery, oily emulsions, latex and electrophoretic paints and wastewater from the metal-finishing industry with recovery of concentrated solutions of metal salts and reuse of the water in cleaning. Moreover, some industrial sectors, such as precision microelectronics, use the RO process together with treatment using resin exchangers to obtain very pure water [87].

Advantages and disadvantages are more or less the same of NF, with some more restrictions like the treatment of salt solutions with low solubility tending to precipitate and thus causing fouling or contaminants tending to polymerise for the same reason. Moreover, solutions with too high osmotic pressures frequently exceed the operating pressure, and their treatment becomes not economically viable [2].

An example of a multi-application of membrane technologies is that reported by Fane and Fane [31] and summarised in Fig. 7.3. The flowchart shows a pretreatment

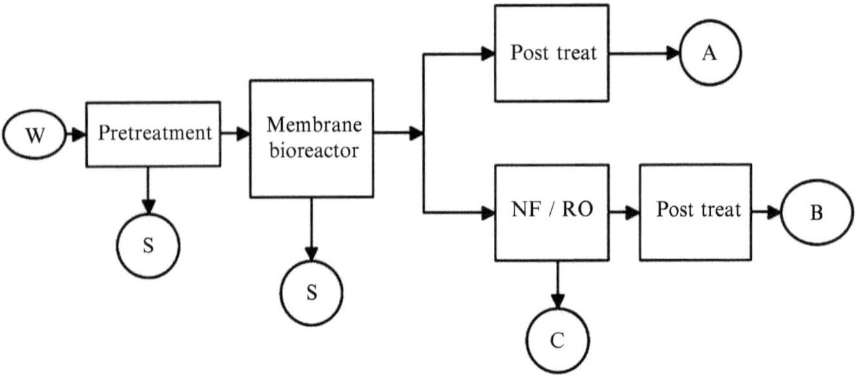

Fig. 7.3 Generic view of wastewater treatment for reuse with membrane processes; usage *A* (high quality); usage *B* (very high quality); *C* concentrate; *S* solids; *W* wastewater [31]

step, tailed by a membrane process (*i.e.* MF/UF-MBR) that is followed by either oxidative post-treatment to support water for reuse (A) (*i.e.* irrigation and non-potable uses) or by a second membrane process (NF/RO) plus post-treatment for further reuse potentiality (B) (*i.e.* non-potable or indirect potable).

7.4.4 Adsorption via Activated Carbon (AC)

Activated carbon is an adsorption medium used to allow the transfer of substances in the water phase to a fixed surface and is classified as one of the end-of-pipe techniques. It is commonly used to remove non-polar organic contaminants, humic substances [95] and refractory compounds to very low concentrations [69]. The adsorption of substances onto AC can be predicted on the basis of their K_{ow} coefficient; substances with $K_{ow} < 0$ are not retained by AC [95]. In Europe, little long-term continuing research has been performed looking at the application of AC filtration for the removal of organic contaminants, pesticides, hormone disrupters and pharmaceuticals, but not clear conclusions are available [95].

In the AC filtration, the effluent is led over a bed of AC in the granular form (GAC) or mixed within the reactor in the powdered one (PAC). In the case of GAC, several filters are serially used. Depending on the bonds between adsorbate and adsorbent, the process could be reversible or irreversible; however, the adsorbent has a finite capacity for each compound to be removed. When the capacity is exhausted, the adsorbent is spent and has to be replaced by fresh material. The spent material can be regenerated or incinerated and finally disposed. Pretreatment for the removal of SS is always required to prevent clogging of the column [95].

The adsorption via AC is a treatment technique that has been recognised as a BAT with respect to some specific applications [2, 73].

7.4.5 Advanced Oxidation Processes (AOPs)

Advanced oxidation processes are taken into consideration to oxidise complex and recalcitrant organic constituents into simpler end products. Generally, it has not required a full oxidation process, but frequently, partial treatments are carried on in order to render specific compounds more amenable to the following activated-sludge biological treatments, so that they may be also considered as a sort of pretreatment stage. The administration of AOPs might lead to (a) primary degradation (*i.e.* structural change in the parent compound), (b) acceptable degradation (*i.e.* structural change in the parent compound with a decrease in the general toxicity level), (c) ultimate degradation (*i.e.* conversion of organic carbon into CO_2) and (d) unacceptable degradation (*i.e.* structural change in the parent compound with an increase in the general toxicity level) [69].

Typically, AOPs involve the generation and use of hydroxyl free radical (OH^-) that is a strong oxidant allowing the oxidation of compounds with which conventional oxidants (*e.g.* oxygen, ozone, and chlorine) failed. Generally, an excess of hydroxyl free radical starts to react in a nonselective way at normal temperature and pressure with the dissolved constituents until they are completely mineralized. Commonly used technological approaches involve ozone/UV, ozone/hydrogen peroxide and hydrogen peroxide/UV that are able to oxidise refractory organic compounds exerting a disinfecting action at the same time [69].

Acknowledgements The author strongly acknowledges Prof. Francesco Avezzù for his great support and for his contributions and suggestions in the revision of this chapter.

References

1. Wilderer PA, Schreff D (2000) Decentralised and centralised wastewater management: a challenge for developers. Water Sci Technol 41:1–8
2. IPPC (Integrated Pollution Prevention and Control) (2003) Reference document on best available techniques in common wastewater and waste gas treatment/management systems in the chemical sector. European Commission, Brussels
3. Libralato G, Volpi Ghirardini A, Avezzù F (2012) To centralise or to decentralise: an overview of the most recent trends in wastewater treatment management. J Environ Manage 94:61–68. doi.10.1016/j.jenvman.2011.07.010
4. Orth H (2007) Centralised versus decentralised wastewater systems? Water Sci Technol 56:259–266
5. Newmann PWG (1993) Prioritising urban water research – an urban planning viewpoint. In: Proceedings of urban water research forum AWWA, Gold Coast
6. Jefferson B, Laine AL, Judd SJ, Stephenson T (2000) Membrane bioreactors and their role in wastewater reuse. Water Sci Technol 41:197–204
7. Sayers D (1998) A study of domestic grey water recycling. In: Interim report, National Water Demand Management Centre, Environment Agency, Worthing, UK
8. Holden B, Ward M (1998) An overview of domestic and commercial reuse of water. In: Presented at water recycling and effluent reuse, London, UK, 16 December

9. Borsuk ME, Maurer M, Lienert J, Larsen TA (2008) Charting a path for innovative toilet technology using multicriteria decision analysis. Environ Sci Technol 42:1855–1862. doi:10.1021/es702184p
10. Deininger A, Wilderer PA (1999) Small wastewater treatment plants IV. Water Science and Technology 41(1):vii
11. Bakir HA (2001) Sustainable wastewater management for small communities in the Middle East and North Africa. J Environ Manag 61:319–328. doi:10.1006/jema.2000.0414
12. Hong SW, Choi Y-S, Kim SJ, Kwon G (2005) Pilot-testing an alternative on-site wastewater treatment system for small communities and its automatic control. Water Sci Technol 51:101–108
13. De Fraja Frangipane E, Pastorelli G (1997) Piccoli Impianti di Depurazione e Manuale di Progettazione. CIPA, Milan
14. Avezzù F, Faletti L, Riganti V (2010) La gestione dei piccoli impianti di depurazione. CIPA, Milan
15. Ho G, Anda M (2004) Centralised versus decentralised wastewater systems in an urban context: the sustainability dimension. In: Second IWA Leading Edge on Sustainability in Water-Limited Environments, pp 81–89. IWA Publishing, London, UK
16. Ødegaard H (1997) Preface: small wastewater treatment plants III. Water Sci Technol 51:vii
17. Nolde E (2007) Possibilities of rainwater utilisation in densely populated areas including precipitation runoffs from traffic surfaces. Desalination 215:1–11. doi:10.1016/j.desal.2006.10.033
18. Otterpohl R, Braun U, Oldenburg M (2003) Innovative technologies for decentralised wastewater management in urban and peri-urban areas. Water Sci Technol 48:133–138
19. Gruening H, Hoppe H, Messmann S, Giga A (2011) Cost effectiveness of centralised and decentralised storm water treatment. Water Sci Technol 63:2598–2604
20. Li F, Wichmann K, Otterpohl R (2010) Review of the technological approaches for grey water treatment and reuses. Sci Total Environ 407:3439–3449. doi:10.1016/j.scitotenv.2009.02.004
21. Peter-Fröhlich A, Pawlowski L, Bonhomme A, Oldenburg M (2007) EU demonstration project for separate discharge and treatment of urine, faeces and greywater – part I: results. Water Sci Technol 56:239–249
22. Randall CW (2003) Changing needs for appropriate excreta disposal and small wastewater treatment methodologies or The future technology of small wastewater treatment systems. Water Sci Technol 48:1–6
23. Ronen T (2011) Extensive systems with intensive process for decentralized communities. In: Proceedings IWA conference on small sustainable solutions for water, Venice, Italy, 18–22 April 2011
24. UNEP (2002) Environmentally sound technologies for wastewater and stormwater management – an international Source book, IWA Publishing/UNEP – International Environmental Technology Centre, London/Osaka. http://www.unep.or.jp
25. Tchobanoglous G, Ruppe L, Leverenz H, Darby J (2004) Decentralized wastewater management: challenges and opportunities for the twenty-first century. Water Sci Technol Water Supply 4:95–102
26. CSO (2007) Census 2006, principal demographic results. Central Statistics Office, Government of Ireland. Stationary Office, Ireland
27. Gill LW (2011) Recent developments in on-site wastewater treatment legislation for Ireland. In: Proceedings of IWA conference on small sustainable solutions for water, Venice, Italy, 18–22 April 2011
28. Ho G (2005) Technology for sustainability: the role of on site, small and community scale technology. Water Sci Technol 51:15–20
29. Dee T, Sivil D (2001) Selecting package wastewater treatment works. In: CIRIA project report 72, Construction Industry Research and Information Association, London, UK
30. Hunt J, Anda M, Mathew K, Ho G, Priest G (2005) Emerging approaches to integrated urban water management: cluster scale application. Water Sci Technol 51:21–27
31. Fane AG, Fane SA (2005) The role of membrane technology in sustainable centralised wastewater systems. Water Sci Technol 51:317–325

32. Griffin P (2001) Small works strategy – the role of package plants. In: Proceedings from IWEX 2001 – technology seminar on package treatment plants: solutions for AMP3 and beyond, pp 26–32. School of Water Sciences, Cranfield University, UK
33. Tchobanoglous G (2003) The strategic importance of decentralised wastewater management in the twenty-first century. In: Proceedings IDA conference on water reuse and desalination, Singapore, 25–26 February
34. Fane SA, Ashbolt N, White SB (2002) Decentralised urban water reuse: the implications of system scale for cost and pathogens risk. Water Sci Technol 46:281–288
35. Del Solar J, Hudson S, Stephenson T (2005) On-site treatment of a motorway service area wastewater using a package sequencing batch reactor (SBR). Water Sci Technol 51:311–316
36. Jeffrey P, Seaton R, Parsons S, Stephenson T (1997) Evaluation methods for the design of adaptive water supply systems in urban environments. Water Sci Technol 35:45–51
37. Daude D, Stephenson T (2004) Cost-effective treatment solutions for rural areas: design and operation of a new package treatment plant for single households. Water Sci Technol 48:107–114
38. Green W, Ho G (2005) Small scale sanitation technologies. Water Sci Technol 51:29–38
39. Brown V, Jackson DW, Khalifé M (2010) 2009 Melbourne metropolitan sewerage strategy: a portfolio of decentralised and on-site concept designs. Water Sci Technol 62:510–517. doi:10.2166/wst.2010.296
40. MAV (Magistrato alle Acque di Venezia) (2007) In: Ferrari G, Tromellini E (eds) Venezia, una scelta obbligata. I trattamenti individuali di depurazione. Marsilio, Venice
41. Nolde E (2005) Greywater recycling systems in Germany – results, experiences and guidelines. Water Sci Technol 51:203–210
42. Lamichhane KM (2007) On-site sanitation: a viable alternative to modern wastewater treatment plants. Water Sci Technol 55:433–440
43. Larsen TA, Maurer M, Udert KM, Lienert J (2007) Nutrient cycles and resource management: implications for the choice of wastewater treatment technology. Water Sci Technol 56:229–237
44. Lienert J, Burki T, Escher BI (2007) Reducing micropollutants with source control: substance flow analysis of 212 pharmaceuticals in faeces and urine. Water Sci Technol 56:87–96. doi:10.2166/wst.2007.560
45. Larsen TA, Alder AC, Eggen RIL, Maurer M, Lienert J (2009) Source separation: will we see a paradigm shift in wastewater handling? Environ Sci Technol 43:6121–6125. doi:10.1021/es803001r
46. McCann B (2010) Exploiting wastewater potential through nutrient recovery research. Water 21:39–41
47. Palmér Rivera M, Ridderstolpe P, Reimer K-A (2011) Blackwater separation and reuse with vacuum toilets in a local activity centre in Sweden. In: Proceedings IWA conference on small sustainable solutions for water, Venice, Italy, 18–22 April 2011
48. Beler Baykal B, Kocaturk NP (2011) Recovery of plant nutrients from dilute solutions of human urine. In: Proceedings IWA conference on small sustainable solutions for water, Venice, Italy, 18–22 April 2011
49. Crites RW, Tchobanoglous G (1998) Small and decentralized wastewater management systems. MacGraw-Hill, New York
50. Maurer M, Rothenberger D, Larsen TA (2006) Decentralised wastewater treatment technologies from a national perspective: at what cost are they competitive? Water Sci Technol 5:145–154
51. Otis R (1996) Small diameter gravity sewers: experience in the Unites States. In: Mara D (ed) Low-cost sewerage. Wiley, Chichester, pp 123–133
52. Crites RW, Middlebrooks EJ, Reed SD (2006) Natural wastewater treatment systems. Taylor and Francis, Boca Raton
53. Tidaker P, Sjoberg C, Jonsson H (2007) Local recycling of plant nutrients from small-scale wastewater system to farmland – a Swedish scenario study. Resour Conserv Recycl 49:388–405

54. Van Lier JB, Lettinga G (1999) Appropriate technologies for effective management of industrial and domestic wastewaters: the decentralised approach. Water Sci Technol 40:171–183
55. Libralato G, Anselmi M, Avezzù F (2008) Criteri di scelta tra centralizzazione e i piccoli impianti decentralizzati. In: Proceedings of the 36th Giornata di Studio di Ingegneria Sanitaria-Ambientale: La Gestione dei Piccoli Impianti di Depurazione, Piacenza, Italy, 11 April
56. Weber B, Cornel P, Wagner M (2007) Semi-centralised supply and treatment systems for (fast growing) urban areas. Water Sci Technol 55:349–356. doi:10.2166/wst.2010.189
57. Rauch W, Brockmann D, Peters I, Larsen TA, Gujer W (2003) Combining urine separation with waste design: an analysis using a stochastic model for urine production. Water Res 37:681–689. doi:10.1016/S0043-1354(02)00364-0
58. Ronteltap M, Maurer M, Gujer W (2007) The behaviour of pharmaceuticals and heavy metals during struvite precipitation in urine. Water Res 41:1859–1868. doi:10.1016/j.watres.2007.01.026
59. Liang X, van Dijk MP (2010) Financial and economic feasibility of decentralized wastewater reuse systems in Beijing. Water Sci Technol 61:1965–1973. doi:10.2166/wst.2010.105
60. Engin GO, Demir I (2006) Cost analysis of alternative methods for wastewater handling in small communities. J Environ Manage 79:357–363. doi:10.1016/j.jenvman.2005.07.011
61. Larsen TA (2011) Redesigning wastewater infrastructure to improve resource efficiency. Water Sci Technol 63:2535–2541
62. Chung G, Lansey K, Blowers P, Brooks P, Ela W, Stewart S, Wilson P (2008) A general water supply planning model: evaluation of decentralized treatment. Environ Modell Softw 23:893–905. doi:10.1016/j.envsoft.2007.10.002
63. Krozer Y, Hophmayer-Tokich S, van Meerendonk H, Tijsma S, Vos E (2010) Innovations in the water chain – experiences in the Netherlands. J Clean Prod 18:439–446
64. EC, European Commission (1996) Council directive 96/61/EC of 24 September 1996 concerning integrated pollution prevention and control. Off J L 257:26–40
65. COM (Commission of the European Communities) (2004) Stimulating technologies for sustainable development: an environmental technologies action plan for the European union, 38 final, Brussels, 28 January 2004
66. Rio de Janeiro Conference (1992) United Nations Conference on Environment and Development (UNCED) 3–14 June. Rio de Janeiro, Brasil
67. UNCED, United Nations Department of Economic and Social Affairs (1993) Agenda 21: earth summit – the United Nations programme of action from Rio – transfer of environmentally sound technology, cooperation and capacity-building. United Nations, United States, 23 April 1993
68. Lomborg B (2001) The skeptical environmentalist. Cambridge University Press, Cambridge, MA
69. Metcalf, Eddy Inc (2003) Wastewater engineering: treatment and reuse, 4th edn. McGraw-Hill, New York
70. Kazmi AA, Furumai H (2000) Field investigation on reactive settling in an intermittent aeration sequencing batch reactor activated sludge process. Water Sci Technol 41:127–135
71. Irvine R, Ketchum L (1989) Sequencing batch reactors for biological wastewater treatment. Critic Rev Environ Control 18:255–294
72. Wilderer PA, Irvine R, Goronszy M (2001) Sequencing batch reactor technology. In: Scientific and technical report no. 10. IWA Publishing, London
73. GURI (Gazzetta Ufficiale della Repubblica Italiana) (2007) Emanazione delle line guida per l'individuazione e l'utilizzazione delle migliori tecniche disponibili in materiali di gestione dei rifiuti, per le attività elencate nell'allegato I del decreto legislativo 18 febbraio 2005, n. 59, GURI 133, 7 giugno 2007
74. Furumai H, Nagasaka M, Sato Y (1998) Modelling of nitrogen removal in sequencing batch reactors treating domestic sewage. Indian J Eng Mater Sci 5:153–181
75. Hayakawa N, Tsuji J, Hamamoto Y (1986) Simultaneous nitrogen and phosphorus removal by intermittent cyclic process. Water Sci Technol 18:319–326
76. Rusten B, Eliassen H (1993) Sequencing batch reactors for nutrient removal at small wastewater treatment plants. Water Sci Technol 28:233–242

77. Yuyama Y, Fujino K, Miyamoto Y, Oonishi R (1993) Treatment system of wastewater from rural settlements with batch activated sludge process. Water Sci Technol 28:223–232
78. Goronszy MC (1979) Intermittent operation of the extended aeration process for small systems. J Water Pollut Control Fed 51:274–287
79. Irvine R, Ketchum L, Arora M, Barth E (1985) An organic loading study of full-scale sequencing batch reactors. J Water Pollut Control Fed 57:847–853
80. Wett B, Rauch W (1996) Simulation of discontinuous treatment strategies. In: Preprint of first IAWQ specialised conference on sequencing batch reactor technology, Munich, Germany, 18–20 March
81. Arora ML, Barth EF, Umphres MB (1985) Technology evaluation of sequencing batch reactors. Water Pollut Control Fed J 57:874
82. Manning JF, Irvine RL (1985) The biological removal of phosphorus in a sequencing batch reactor. J Water Pollut Control Fed 57:87–94
83. Chiesa SC, Irvine RL (1985) Growth and control of filamentous microbes in activated sludge: an integrated hypothesis. Water Res 19:471–479. doi:10.1016/0043-1354(85)90039 9
84. De Fraja Frangipane E, Pastorelli G (1993) Impianti di depurazione di piccole dimensioni. CIPA, Milan
85. Melcer H, Bedford WK, Topnik BH, Schmidtke NK (1987) Conversion of small municipal wastewater treatment plants to sequencing batch reactors. J Water Pollut Control Fed 59:79–85
86. Edgerton BD, McNevin D, Wong CH, Menoud P, Barford JP, Mithchell CA (2000) Strategies for dealing with piggery effluent in Australia: the sequencing batch reactor as a solution. Water Sci Technol 41:123–126
87. Stephenson T, Judd S, Jefferson B, Brindle K (2000) Membrane bioreactors for wastewater treatment. IWA Publishing, Cornwall
88. Ng HY, Hermanowicz SW (2005) Membrane bioreactor operation at short solids retention times: performance and biomass characteristics. Water Res 39:981–992. doi:10.1016/j.watres.2004.12.014
89. Fane AG, Chang S (2002) Membrane bioreactors: design and operational options. Filtr Sep 39:26–29. doi:10.1016/S0015-1882(02)80168-9
90. Caetano A (1995) Membrane technology: applications to industrial wastewater treatment. Kluwer Academic Publishers, Boston
91. Gander MA, Jefferson B, Judd SJ (2000) Membrane bioreactors for use in small wastewater treatment plants: membrane materials and effluent quality. Water Sci Technol 41:205–211
92. Parameshwaran K, Fane AG, Cho BD, Moosbruger R, Kim KJ (2002) Low energy treatment system for brewery water reuse. In: Proceedings of 2nd IWA world water congress, Melbourne, Australia
93. Eckenfelder WW Jr (2000) Industrial water pollution control, 3rd edn. McGraw-Hill, New York
94. Culp RL (1978) Handbook of advanced wastewater treatment. Van Nostrand Reinhold, New York
95. STOWA, Institute of Applied Water Research – NL (2005) Explanatory study for wastewater treatment techniques and the European Water Framework Directive. Report 34, STOWA, Utrecht, Netherland
96. Wanner O, Debus O, Reichert P (1994) Modelling the spatial distribution and dynamics of a xylene-degrading microbial population in a membrane-bound biofilm. Water Sci Technol 29:243–251
97. Brindle K, Stephenson T, Semmens MJ (1998) Nitrification and oxygen utilisation in a membrane aeration bioreactor. Water Environ Res 71:1347–1354. doi:10.1016/S0376-7388(98)00047-7
98. Libralato G, Volpi Ghirardini A, Avezzù F (2010) Toxicity removal efficiency of decentralised sequencing batch reactor and ultra-filtration membrane bioreactors. Water Res 44:4437–4450. doi:10.1016/j.watres.2010.06.006

99. McCann B (2008) Centralised control for Venice's decentralised wastewater treatment. Water 21(10):51–52
100. Cornel P, Krause S (2006) Membrane bioreactors in industrial wastewater treatment – European experiences, examples and trends. Water Sci Technol 53:37–44. doi:10.2166/wst.2006.074
101. Houten R, Eikelboom D (1997) High performance membrane bioreactors: a physiological approach. In: MBR1 – proceedings of the first international meeting on membrane bioreactors for wastewater treatment, p 7. Cranfield University, Cranfield, UK
102. Muller EB, Stouthamer AH, van Verseveld HW, Eikelboom D (1995) Aerobic domestic wastewater treatment in a pilot plant with complete sludge retention by cross-flow filtration. Water Res 29:1179–1189. doi:10.1016/0043-1354(94)00267-B
103. Shimizu Y, Uryu K, Okuno Y, Watanabe A (1996) Crossflow microfiltration of activated sludge using submerged membrane with air bubbling. J Ferment Bioeng 81:55–60
104. Fane AG (1996) Membranes for water production and wastewater reuse. Desalination 106:1–9. doi:10.1016/S0011-9164(96)00085-9
105. Trussell R, Adham S, Tchobanoglous G (2003) Testing protocols for membrane selection: equivalency and fate, organism/pathogen removal. In: Proceedings of IDA conference on water reuse and desalination, Singapore, 25–26 February
106. AUEPA, Environment Protection Authority (1995) Managing sewage discharges to inland waters, publication 473. State Government of Victoria, Melbourne
107. Ottoson J, Hansen A, Björlenius B, Norder H, Stenström TA (2006) Removal of viruses, parasitic protozoa and microbial indicators in conventional and membrane processes in a wastewater pilot plant. Water Res 40:1449–1457. doi:10.1016/j.watres.2006.01.039
108. Rose JB, Gerba CP (1991) Assessing potential health risks from viruses and parasites in reclaimed water in Arizona and Florida, USA. Water Sci Technol 23:2091–2098
109. Fleischer J, Schlafmann K, Otchwemach R, Botzenhart K (2000) Elimination of enteroviruses, other enteric viruses, F-specific coliphages, somatic coliphages and *E. coli* in four sewage treatment plants of southern Germany. J Water Supply Res Technol 49:127–138
110. Jacangelo JG, Loughran O, Petrik B, Simpson D, McIlroy C (2003) Removal of enteric viruses and selected microbial indicators by UV irradiation of secondary effluent. Water Sci Technol 47:193–198
111. Owen G, Bandi M, Howell JA, Churchoues SJ (1995) Economic assessment of membrane processes for water and wastewater treatment. J Membr Sci 102:77–91. doi:10.1016/0376-7388(94)00261-V
112. Lesjean B, Rosenberger S, Schrotter JC (2004) Membrane aided biological wastewater treatment – overview of some applied systems and their fields of application. Membr Technol 8:5–10. doi:10.1016/S0958-2118(04)00200-9
113. Cheryan M (1998) Ultrafiltration and microfiltration handbook. Technomic Publishing AG, Basel
114. Liu Z, He Y, Li F, Liu Y (2006) Photocatalytic treatment of RDX wastewater with nano-sized titanium dioxide. Environ Sci Pollut Res Int 13:328–32. doi:10.1065/espr2006.08.328
115. Sourirajan S (1970) Reverse osmosis. Academic Press Inc, New York

Chapter 8
Hospital Wastewaters: Quali-Quantitative Characterization and for Strategies for Their Treatment and Disposal

P. Verlicchi, A. Galletti, and M. Al Aukidy

8.1 Introduction

During recent years, the issue of hazardous substances in wastewater (WW) has become a major concern with respect to both human health and the environment. This has lead to the launch of several studies into the monitoring of urban wastewaters (UWws) (among them: Yu et al. [1]; Lishman et al. [2]; Santos et al. [3]; Terzic et al. [4]). However, few studies have yet been devoted to the monitoring hospital effluents in terms of either conventional pollutant or micropollutant load [5–7].

With few exceptions, hospital wastewaters (HWws) are considered to possess the same pollutant load as UWws and are therefore discharged into the same sewage network and conveyed for co-treatment at the nearest municipal wastewater treatment plant (WWTP).

Nevertheless, HWws contain a great variety of toxic or persistent substances, such as pharmaceutical compounds (PhCs), radionuclides, heavy metals, solvents and disinfectants, not generally present in large quantities in UWws. In HWws, these substances, products of laboratory and research activities as well as general hospital activity and excretion of medicines, may be present in a wide range of concentrations.

Most PhCs belong to the group of so-called *emerging contaminants*, that is, pollutants that are quite often unregulated as yet but may be candidates for future regulation, depending on the results of research monitoring their occurrence and assessing their potential effects on health [8]. Their main characteristic is that they do not need to persist in the environment to cause negative effects, since their high transformation/removal rates are counterbalanced by their continual introduction into the environment [9].

P. Verlicchi (✉) • A. Galletti • M. Al Aukidy
Department of Engineering, University of Ferrara, Via G. Saragat 1,
Ferrara I-44122, Italy
e-mail: paola.verlicchi@unife.it

S.K. Sharma and R. Sanghi (eds.), *Wastewater Reuse and Management*,
DOI 10.1007/978-94-007-4942-9_8,

Once PhCs have been administered, their active components are metabolized in the body, but only to a certain extent. The unmetabolized active substances are excreted, largely in the urine (55–80% of the total with few exceptions, as reported by Jjemba [10] and Ternes and Joss [11]) and partially in the faeces, either as the unchanged substance, as a mixture of their metabolites or conjugated with an inactivating compound attached to the molecule [12], thereby entering the water cycle. Although many of these PhCs are specifically consumed in hospitals, a fraction of the administered dose could be excreted at home by outpatients and thereby enter the municipal sewage system. These compounds include anti-neoplastics, such as ifosfamide and cyclophosphamide, or contrast media like iopamidol or iopromide.

Special attention has been paid to PhCs for various reasons: they are continuously emitted into the environment via effluents from WWTPs, and their high transformation/removal rates are counterbalanced by their continuous introduction into the environment. Furthermore, as they are developed with the intention of provoking a biological response, they need to be persistent in order to avoid inactivation of the compound before it can exert its therapeutic effect. Moreover, they need to be lipophilic in order to be able to pass through cell membranes.

Recent environmental risk assessments have shown that the concentrations of pharmaceutical and personal care products (PPCPs) in aquatic environments may exceed their predicted no-effect concentrations (PNEC) [3, 13]. Furthermore, although the detection of multiple contaminations is common, little is known about the potential interactive effects, such as synergistic or antagonistic toxicity, that may result from complex mixtures or metabolites [14].

Thus, the wide variety of PhC chemical structures, combined with their excellent water solubility, their surface-active nature and the persistence of some of their known metabolites, make them a group of environmental pollutants whose behaviour needs to be promptly addressed [15].

8.2 Comparison of Annual and Daily Hospital and Urban Water

The quantity of water required by hospitals on a daily basis is invariably large and dependent upon the various activities performed within, as well as the range of services they provide. Likewise, the quantity of WWs produced by a hospital also depends on several factors: bed numbers; hospital age; access to water; general services present inside the structure (kitchen, laundry and air conditioning); number and type of wards, departments and units; institutional management policies and attention to safeguarding the environment; climate; and cultural and geographic factors. Thus, there is no clear correlation between specific hospital consumption (expressed as L bed^{-1} day^{-1}) and hospital size (i.e. the bed number), as shown by the data reported in Fig. 8.1, which refer to hospitals in different countries around the world [5–7, 16–26]. In this graph, data are spread between 200 and 1,200 L bed^{-1} day^{-1}, with the highest values pertaining to industrialized countries and the lowest to developing nations, where consumption is generally around 0.2–0.4 m^3 bed^{-1} day^{-1}.

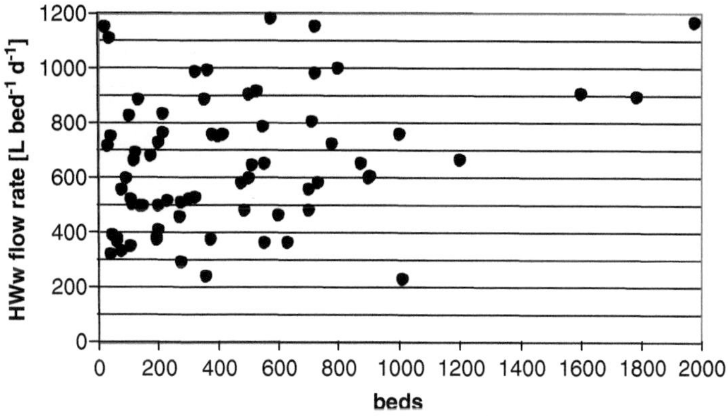

Fig. 8.1 Water consumption per bed and per day with respect to hospital size

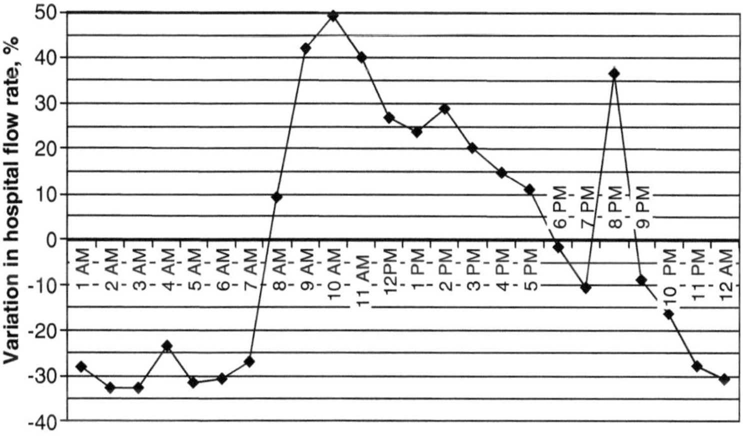

Fig. 8.2 Percentage variation in hourly hospital consumption (flow rate) with respect to its corresponding average value

Commonly adopted values for water consumption in urban settlements, on the other hand, are in the range of 150–300 L p.e.$^{-1}$ day^{-1} for industrialized areas and 50–100 L p.e.$^{-1}$ day^{-1} for developing countries. However, further literature data for specific water consumption (L pro capita^{-1} day^{-1}) by commercial, institutional and recreational facilities show that the rate for hospitals is far greater than for other specific consumption types and that it has a much wider range [27].

As can be expected, the hourly hospital consumption (flow rate) increases progressively throughout the morning from 5 to 6 a.m., reaching a maximum at about 10 a.m. It then decreases steadily to a minimum at about midnight and subsequently remains stable until about 7–8 a.m. The graph in Fig. 8.2 illustrates the findings of an investigation carried out by Boillot et al. [5] on a French hospital with about 700 beds and an average flow rate of 27.3 m^3 h^{-1}. The highest variations, that is, the peaks in water consumption, were observed mid-morning

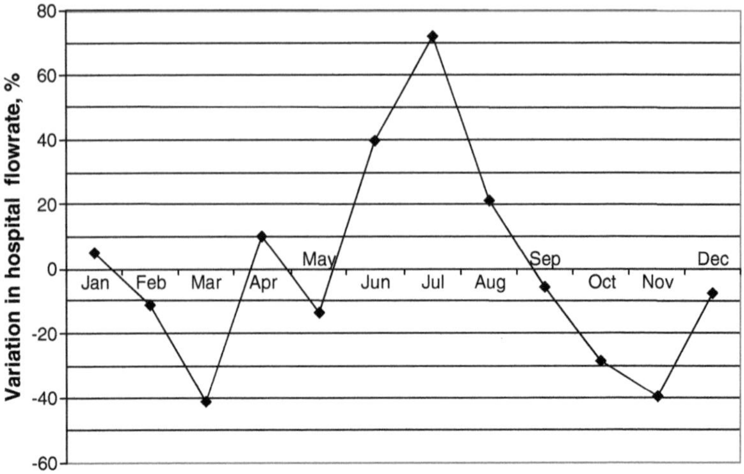

Fig. 8.3 Percentage variation in monthly hospital flow rate with respect to its corresponding average value. Data refer to the consumption measured in an Italian hospital of medium size (450 beds)

and mid-afternoon and generally coincided with the hours when clinic patients were treated and laboratory analyses performed.

With reference to variations in flow rate over the course of the year, the highest values are expected during the hot season, due to a greater consumption of water required for health care. The graph in Fig. 8.3 shows the percentage variation in flow rate over the year with respect to its average monthly value at a medium-sized Italian hospital (450 beds) with an average monthly flow rate of 9,400 m^3 $month^{-1}$. Data were collected in 2008 and are reported in [28]. The minimum flow rate occurred in March (−41%) and the maximum in July (+72%). These trends in water consumption are similar to those reported by other studies: Joss et al. [29], Mohee [19], Boillot et al. [5] and Verlicchi et al. [30].

Variations in urban flow rate over the course of the year are also quite marked and are influenced by the size of the resident population, the climate and the different activities (industry, tourism, recreation) carried out in the area served by the WWTP [31].

In urban settlements, the WW flow rate also varies over the course of the day according to the specific activities carried out in the area and fluctuations in the density of the resident population. As regards diurnal variations, two patterns, representative of small and large urban settlements, showing the typical percentage variation in UWw production flow rate over the course of a typical (dry) day are shown in the graph in Fig. 8.4. Data were derived, respectively from [31] (small urban centre <10,000 inhabitants) and from [32] (large urban settlement of about 250,000 inhabitants). It should be noted that in the case of combined sewage, during rain events, the urban flow rates consistently increase.

As expected, higher discrepancies occur in a small centre as compared to a large town, where activities are not restricted to the daytime, thereby resulting in a

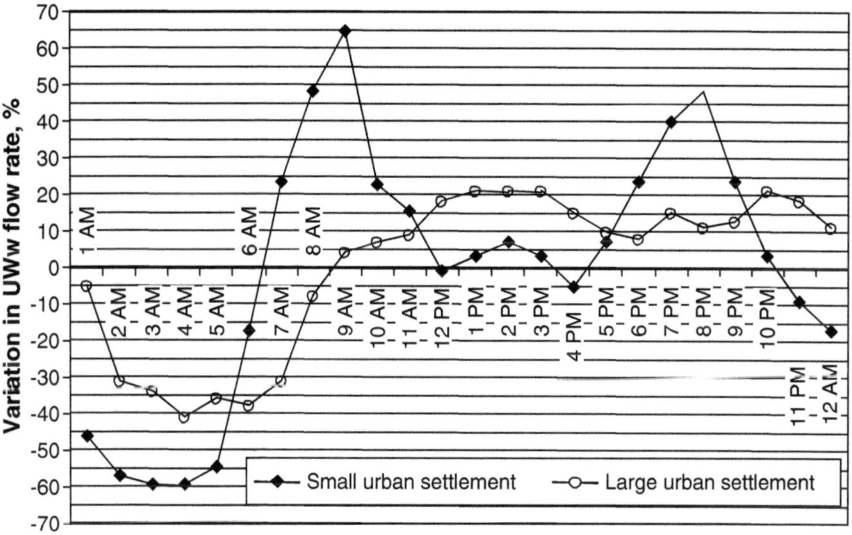

Fig. 8.4 Percentage variation in hourly UWw flow rate with respect to its corresponding average value for a small and a large urban settlement

Table 8.1 Peak coefficient for hospital effluents and urban influent to a small WWTP

Peak coefficient	Hospital effluent	Small urban WWTP influent
Monthly	1.5–1.8	1.2–2
Daily	2–2.8	2–5
Hourly	3.5–4	3–4

smaller variation with respect to the average value throughout the night. Interestingly, the trend in water consumption for a small town (Fig. 8.4) appears to be quite similar to that of a hospital of medium size (Fig. 8.2). In fact, peak coefficients for hospital flow rates are fairly analogous to those generally assumed for the influent to a small WWTP (<10,000 inhabitants or population equivalent, p.e.), as reported in Table 8.1 [31, 33].

8.3 Comparison of Physical, Chemical and Microbiological Characteristics (Macropollutants) of Hospital and Urban Effluents

Analysis of the physical and chemical characteristics of hospital and urban effluents was carried out via a comprehensive review of the literature. The ranges of variability of the main measured parameters in both hospital and urban raw WWs [5, 7, 34–39] are compiled in Table 8.2.

Table 8.2 Range of variability in concentrations of the main physical, chemical and microbiological parameters in HWws and UWws

Parameters	HWws	UWws
pH	6.9–9.18	7.5–8.5
Conductivity, μS cm^{-1}	750–1,000	420–1,340
Redox potential, mV	850–950	100
SS, mg L^{-1}	120–400	120–350
VSS/SS	0.45–0.75	0.65–0.8
COD, mg L^{-1}	450–2,300	500–600
BOD_5, mg L^{-1}	150–603	100–400
TKN, mg L^{-1}	30–100	20–70
NH_4, mg L^{-1}	10–55	12–45
P_{tot}, mg L^{-1}	3–8	4–10
Chlorides, mg L^{-1}	80–400	30–100
Fats and oils, mg L^{-1}	13–60	50–150
Total detergents, mg L^{-1}	3–7.2	4–8
COD/BOD_5	1.4–6.6	1.7–2.4
Total coliforms, MPN/100 mL	10^6–10^9	10^7–10^8
Faecal coliforms, MPN/100 mL	10^3–10^7	10^6–10^7
E. coli, MPN/100 mL	10^3–10^6	10^6–10^7
Streptococci, MPN/100 mL	10^3–10^5	10^3–10^5

As regards BOD_5, COD and SS, taking into consideration their usual concentrations in the influent to municipal WWTPs, it can be observed that the parameters are consistently two to three times higher in HWws than in UWws. In some cases, the ratio COD/BOD_5 is very high in the former, especially in hospitals including a haemodialysis ward [40]. The resulting specific contributions for each patient are:

- 160–180 g BOD_5 patient^{-1} day^{-1}
- 260–300 g COD patient^{-1} day^{-1}
- 120–150 g SS patient^{-1} day^{-1}

corresponding to about two to three times those typically assumed for the population equivalent, respectively:

- 60 g BOD_5 p.e.$^{-1}$ day^{-1}
- 100–120 g COD p.e.$^{-1}$ day^{-1},
- 70–90 g SS p.e.$^{-1}$ day^{-1} [27]

Moreover, great differences also occur for TKN, redox potential and chlorides.

In hospital effluent, pH is mainly subject to fluctuations during the morning, presumably linked to floor- and surface-cleaning activities using large volumes of detergent/disinfectant products [5].

Roughly 30 years ago, Peou et al. [41] did not find significant differences between hospital and urban effluent in terms of bacterial concentrations and in terms of percentage of resistance to antibiotics. In the last years, other studies ([42, 43, 67]) have shown that the variability ranges are not consistently different, as reported in Table 8.2. But they found an important difference: several bacterial strains carrying different resistance genes are present in hospital effluents.

Furthermore, Chitnis et al. [44] found that the effluents from some Indian hospitals contain various pathogens, including salmonella, shigella and multi-drug resistance (MDR) coliforms. The patterns of resistance noted included simultaneous resistance to ampicillin, cephalosporins, aminoglycosides, quinolones, co-trimoxazole, tetracycline and chloramphenicol, that is, to the majority of available antibiotics. Furthermore, they found that residential sewage samples had extremely small MDR populations with respect to investigated hospital effluents. In fact, samples from the municipal sewage stream just upstream of the hospital effluent influx showed that it did not harbour MDR bacteria. But after the mixing with the urban wastewater the resistant population persisted to an extent of 0.5% up to a length of 100 m and downstream to a level of 0.06% at a distance of 2 km.

Hartemann et al. [37], Leprat et al. [45] and Boillot et al. [5] also found far fewer microorganisms (bacteria and viruses) in French hospital effluent samples than in their ones. Nevertheless, they did raise the issue of the increasing quantity of bacteria resistant to antibiotics, and sometimes to disinfectants, being released into the environment due to inadequate treatment of hospital effluent. Moreover, Prado et al. [46] reported that HWws may contain approximately 2 log higher levels of enteric viruses, mainly rotavirus A (RV-A) and human adenovirus (HAdV), than mean values quantified in urban raw WWs.

8.4 Comparison of Chemical Characteristics (Micropollutants) of Hospital and Urban Effluents

HWws contain a large number of compounds that are rarely found, or found at far lower concentrations, in UWws. Examples of these chemicals include AOX (organic halogens adsorbable onto active carbon), free chlorine, disinfectants, detergents, freon 113, alcohols, acetone, formaldehyde and phenols. Heavy metals have also been discovered in HWw, in particular: mercury, copper, lead and arsenic [5, 30].

PhCs are used both within and outside of the hospital; while numerous PhCs are mainly (only) administered to inpatients (some antibiotics, analgesics, anaesthetics, anti-neoplastics, etc.), some, like lipid regulators, anti-hypertensives or beta blockers, need to be administered to outpatients over long periods of time or, in some cases, for instance, anti-diabetics, over the entire lifetime. Still, other PhCs, for example, those used in day-hospital therapy (such as anti-neoplastics) are administered inside the health-care structure but excreted outside.

Once administered to human body, PhCs are only partially metabolized, and the remainder (varying between 10 and 95%) is excreted mainly through the renal system (urine), the biliary system (faeces) or a combination of both, depending on the nature of the compound and the individual in question. As a consequence, PhCs residues join WWs and thereby enter the water cycle, as unchanged substances (also called parent compounds), as a mixture of their metabolites or conjugated with an inactivating compound [10, 12].

A brief description of the main groups of micropollutants typically present in HWws is reported below.

AOX (Adsorbable Organohalogens): This class includes a great number of adsorbable, organically bound halogens, which may be present in PhCs and their metabolites. Other sources of AOX are chlorine-forming disinfectants, used in cleaning activities, and halogen-containing solvents, used in laboratories, as well as other chemical substances like ethidium bromide [34]. Frequently these chemicals are nonbiodegradable, allowing them to spread widely into the aquatic environment and thereby enter the food chain. The presence of AOX is correlated to free chlorine: their concentration has been found to range from 150 to 7,760 μg L^{-1} in HWws, in stark contrast to the range 0.04–0.2 μg L^{-1} measured in UWws.
Total and free chlorine have been found in HWws at an average daily concentration of 0.59 and 0.55 mg L^{-1}, respectively, resulting in a combined chlorine concentration (including chloramines and different organochloride compounds) of 0.040 mg L^{-1} [5]. According to Liu et al. [47], many hospitals in China, and presumably other countries, still use excessive disinfectants to ensure thorough sterilization, which can lead to extremely high residual chlorine levels in the effluent: up to 128 mg L^{-1}.

Detergents: Nonionic detergents are present in relatively low concentrations between 5 p.m. and 5 a.m. (0.5–0.65 mg L^{-1}), and at higher concentrations, 2.8–5.8 mg L^{-1}, at other times of the day [5].

Disinfectants: The most used compounds are sodium hypochloride, peracetic acid, ammonium quaternary salts, some aldehydes, alcohols and phenol compounds. Their average daily concentration generally varies between 2 and 200 mg L^{-1}, according to the size of the hospital and its consumption of disinfectants [37]. Concentrations of *glutaraldehyde*, a disinfectant which was widely used in the past but has now, in many cases, been replaced by other compounds with a smaller impact on WWTP biological processes, range between 0.5 and 3.72 mg L^{-1} [48]. Due to their pronounced bacterial toxicity and their potential to foster resistance, all these disinfectants require special attention [49].

There are also many *organic compounds* including acetaldehyde, ketones, alcohols and acetates whose quantities vary from hospital to hospital. In general, however, they exhibit remarkably homogeneous daily fluctuations. Indeed, it is possible to group the periods 9 a.m. to 1 p.m. and 1 p.m. to 5 p.m. as being the most polluting, and the periods 5 to 11 p.m. and 11 p.m. to 5 a.m. as being the least polluting as regards this type of substance. The period 5–9 a.m. tends to be a transient period, albeit more similar to the less polluting time spans [5, 30]. The most prevalent compounds in this group are ketones (up to 7,000 μg L^{-1}), ethanol (1,700 μg L^{-1}) and methanol (760 μg L^{-1}).

VHOC (volatile halogenated organic compounds) have average daily concentrations in HWw ranging between <0.5 and 2.4 μg L^{-1} [5].

The main *heavy metals* found in HWws are *platinum*, resulting from excretion by oncological patients treated with cis-platinum and carboplatinum or other cytostatic agents; *mercury*, usually found in diagnostic agents, active ingredients of disinfectants and diuretic agents and *gadolinium*, which is present in iodinated contrast media (ICM) and, due to its high magnetic moment, used in magnetic resonance imaging (MRI). Following administration, the organic complexes of these metals are very quickly excreted as parent compounds. They are not degradable and are highly toxic in some oxidative states. According to Hartemann et al. [37], hospitals are the main source of Hg in WW; indeed, the concentration of Hg in the raw effluent has been reported to range between 0.04 and 5.03 $\mu g\ L^{-1}$. Concentration ranges in HWw cited for other heavy metals are Gd between 1 and 100 $\mu g\ L^{-1}$, Pt between 0.01 and 41.3 $\mu g\ L^{-1}$, Ag from 150 to 437 $10^3\ \mu g\ L^{-1}$, As from 0.8 to 11 $\mu g\ L^{-1}$, Cu from 50 to 230 $\mu g\ L^{-1}$, Ni from 7 to 71 $\mu g\ L^{-1}$, Pb from 3 to 19 $\mu g\ L^{-1}$, and Zn from 70 to 670 $\mu g\ L^{-1}$ [5, 50]. In contrast, Hg in UWws is generally <0.5 $\mu g\ L^{-1}$, Cu is in the range 27–33 $\mu g\ L^{-1}$, Pb in the range 30–144 $\mu g\ L^{-1}$ and Zn ranges from 135 to 226 $\mu g\ L^{-1}$.

Pharmaceutical Compounds: A great number of PhCs are used in hospitals for treatment, diagnosis and research purposes. Table 8.3 reports the most widely investigated compounds, grouped according to their specific therapeutic use. In this table, the third and the fourth columns report the range of concentrations of each of the compounds cited, measured in HWws and, for comparison, in UWws [5, 39, 52].

In 2003, Kummerer and Henninger highlighted great differences between the PhCs administered in hospitals and households, in particular, as regards antibiotics. The in-depth comparative analysis performed by Verlicchi et al. [39] also revealed variations in the ratio between the average concentrations of the different therapeutic classes in HWws and UWws of between 8 and 15 for analgesics and anti-inflammatories, 5–10 for antibiotics, 4–10 for anti-neoplastics, 70–150 for contrast media and 7–15 for AOX. The ratios for hormones and beta blockers ranged between 1 and 4.

Among the groups of PhCs present in HWws, scientists and researchers have, in recent years, paid particular attention to antibiotics and anti-neoplastics. This focus has been spurred on by two main considerations. The first of these is the desire to reduce the risk of discharging great quantities of antibacterials into the environment, as this can significantly contribute to the development of antibiotic-resistant strains of bacteria [37, 53, 54]; the second is that different studies have shown that, owing to the presence of cytostatics and certain antibiotics like fluoroquinolones, hospital effluents can possess mutagenic properties and be toxic to bacteria [55–57].

Few studies have explicitly quantified the PhC load (PhC mass referred to a time period, for instance g $year^{-1}$) in HWws as compared to the total PhC load in WWTP influents to the hospital catchment area [25, 58–61]. Furthermore, these investigations have been limited to a small number of compounds and, instead of measuring the WW flow on-site to determine acute loads, have based their calculations on assumed rates of water flow.

Table 8.3 Average range of concentrations measured in HWws and UWws for the most commonly investigated PhCs, grouped according to their therapeutic class

Therapeutic class	Investigated compounds	Range in HWw (μg L^{-1})	Range in UWw (μg L^{-1})
Analgesics/anti-inflammatories	Codeine	0.02–50	0.1–35
	Diclofenac	0.24–15	0.01–28
	Ibuprofen	0.07–43	0.04–84
	Naproxen	9.8–11	0.04–18.78
	Paracetamol	5.4–1,368	0.04–483
	Salicylic acid	23.4–70.1	34–160
Antibiotics	Ciprofloxacin	0.03–125	0.01–5.88
	Clarithromycin	0.2–3	0.1–4.8
	Coprofloxacin	0.85–2	
	Doxycycline	0.1–6.7	
	Erythromycin	27–83	0.04–2.7
	Lincomycin	0.3–2	
	Metronidazole	0.1–90	
	Norfloxacin	0.029–44	0.01–0.96
	Ofloxacin	0.353–35.5	0.01–31.70
	Oxytetracycline	0.01–3.75	
	Penicillin G	0.85–5.2	0.03
	Sulfamethoxazole	0.04–83	0.01–6.0
	Tetracycline	0.01–4.2	0.01–1.30
	Trimethoprim	0.01–15	0.02–7.90
Psychiatric drugs	Carbamazepine	0.54–2	0.02–3.11
Anti-hypertensives	Diltiazem	0.71–1.6	0.01–1.6
Anti-neoplastics	5-Fluorouracil	5–124	
	Methotrexate	1	
	Cyclophosphamide	0.019–4.5	
	Ifosfamide	0.01–1.9	0.01–7.0
Beta blockers	Metoprolol	0.419–25.1	0.03–6.50
Hormones	17β-Estradiol, E2	0.028–0.043	0.0–3.0
	Estriol, E3	0.353–0.502	0.0–0.32
	Estrone, E1	0.017–0.034	0.0–0.8
	Ethinylestradiol, EE2	0.015–0.018	0.001–0.007
Contrast media	Iopromide	0.2–2,500	0.2–22
	Iomeprol	0.01–1,392	
Anti diabetics	Glibenclamide	0.048–0.113	0.05–0.159

PhC consumption varies from country to country, especially given the restrictions on the use of some pharmaceuticals (for instance, vancomycin is widely used as a first-line antibiotic in the United States, whereas its use in European countries is highly restricted), and from year to year, due to the progress in the development of new active ingredients. Furthermore, PhC concentrations in effluents may change throughout the year and from 1 year to the next. Indeed, analysis of the distribution of PhC consumption throughout the year evidences that there are *critical* months, featuring greater administered quantities, especially of antibiotics [30]. These aspects are discussed in the following paragraphs.

8.5 Pollutant Concentration Over Time (Day and Year)

8.5.1 Hourly Variation in COD Concentration in HWws and UWws

Due to the variation in hospital water consumption (as reported in Fig. 8.2) and the different kinds of activities that characterize the hospital (care, diagnosis, research activities, laboratories, kitchen, laundry, etc.) at different times of the day, the macropollutant load may differ. Figure 8.5 reports the hourly evolution of the COD on the basis of the investigations into an Italian 300-bed hospital with an average daily water consumption of 600 L bed^{-1} day^{-1} [28]. Similar trends have been found by Emmanuel et al. [62] in a French infectious and tropical disease department, and by Boillot et al. [5] in a French hospital of about 700 beds.

The variation in COD concentration in the influent of a municipal WWTP over the course of the day is reported in Fig. 8.6 as the percentage variation with respect to the average hourly value. Data are from [27] and refer to a hypothetical urban settlement. As expected, the maximum concentrations of COD occur around 10–11 a.m. and 7–8 p.m. A similar trend was found by Joss et al. [29] when evaluating the diurnal variation of the organic load in the influent to a medium/ large Swiss WWTP.

8.5.2 Diurnal Variations in PhC Concentration in HWws and UWws

For most PhCs, measured concentrations were consistently fairly low during the night, and peaks were noted in the morning as well as in the afternoon, presumably caused by different consumption and excretion patterns. Figure 8.7 shows the concentrations of several PhCs, AOX and a rare element in instantaneous HWw samples [5, 6, 29, 34, 50, 63].

Joss et al. [29] and Gobel et al. [64] evaluated the diurnal load to a municipal WWTP in Swiss for some PhCs and found that the highest concentrations of these substances occur between 8 a.m. and 4 p.m., with a few exceptions, for example, roxythromycin, which has higher concentrations between 4 p.m. and 12 a.m. due to its longer excretion time. Details are reported in Fig. 8.8.

8.5.3 Annual PhC and Concentration in HWws and UWws

It is quite difficult to ascertain the distribution patterns of PhC consumption inside the hospital and in households over the course of the year, especially as data on the monthly distribution of water consumption in hospitals and households are not

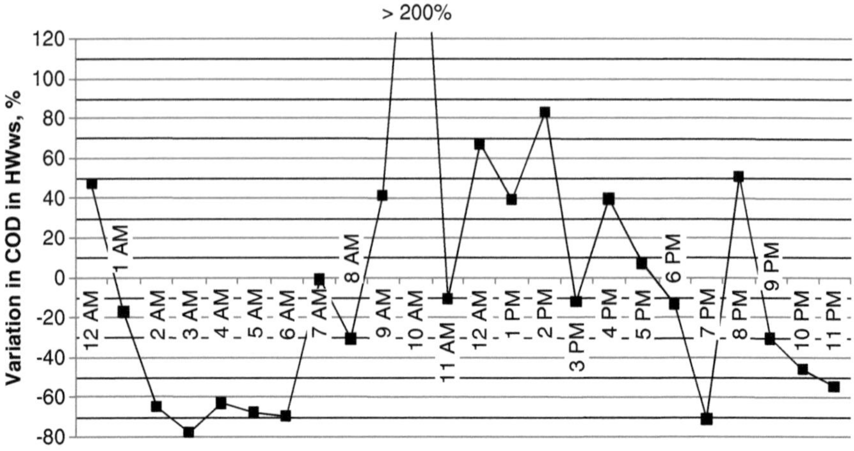

Fig. 8.5 Variation in COD concentration in HWw throughout the day with respect to the average hourly concentration

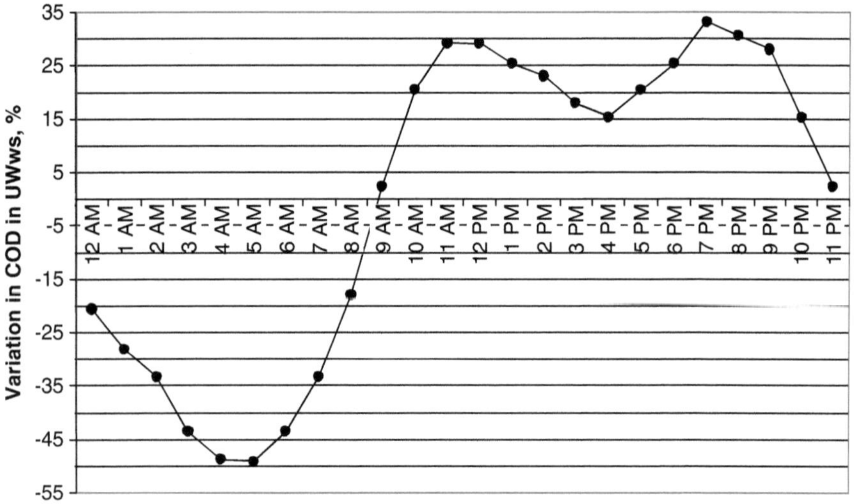

Fig. 8.6 Diurnal variation in the concentration of COD in the influent to an urban WWTP

readily available. Moreover, the consumption, use and application of PhCs may vary considerably over time and from country to country. Further variations in annual trends are correlated to the introduction of new active pharmaceutical ingredients (API), while others fall into disuse following medical advances. Although UWws have been thoroughly investigated and year-long experimental campaigns have been carried out, few data are available regarding the distribution of PhC consumption within hospitals.

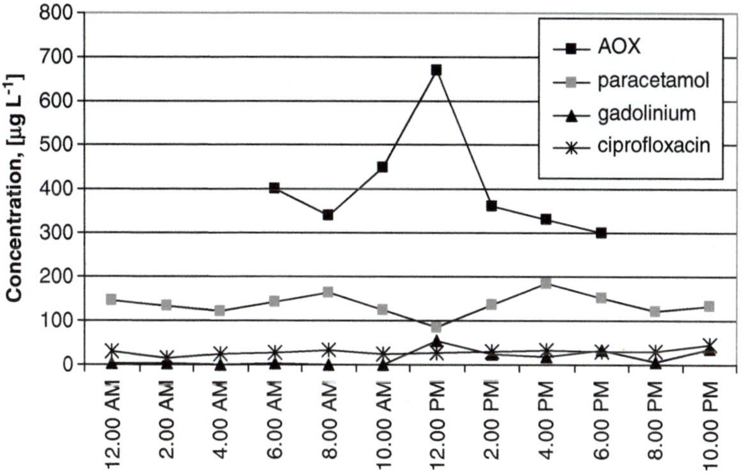

Fig. 8.7 Variation in PhC concentrations in HWws during daytime

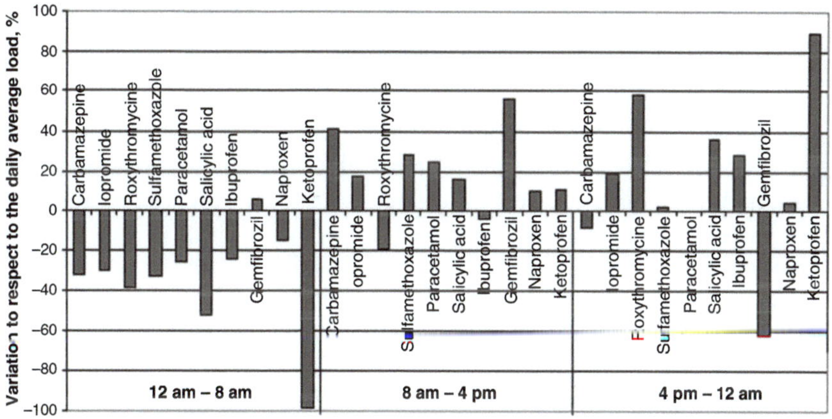

Fig. 8.8 Percentage variation in several PhCs in UWws with respect to the average daily load

However, Fig. 8.9 illustrates the consumption of all antibiotics in both an Italian 300-bed hospital and its WW catchment area (5,000 inhabitants) in 2008 [28]. It shows the percentage variation with respect to the average monthly value, and consistent differences in consumption between the two different users are evident: in the hospital, the most critical months were January and February, whereas the catchment area featured critical periods corresponding to February, March and July.

Skadsen et al. [65] and Alexy et al. [66] confirm this trend for household consumption of PhCs, evidencing higher values in some months, resulting in a higher concentration in the (municipal or hospital) wastewater.

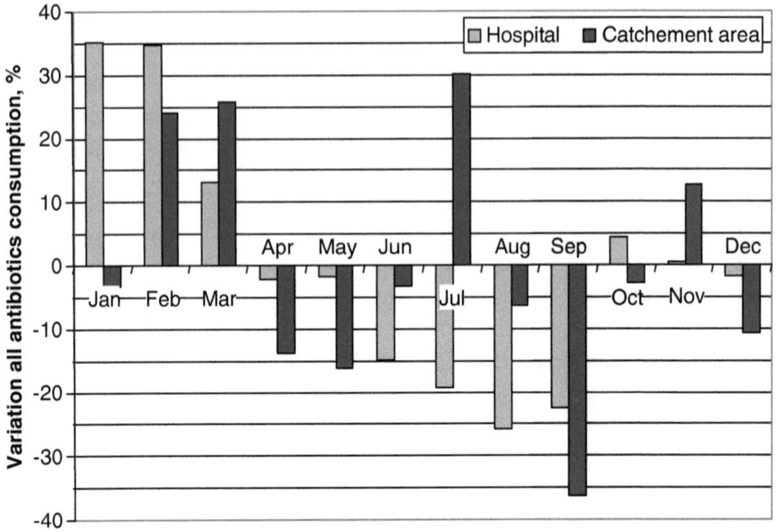

Fig. 8.9 Percentage variations in the consumption of all antibiotics in an Italian 300-bed hospital and its catchment area (5,000 inhabitants) with respect to the average monthly value. Data refer to 2008

It is possible to make a rough estimation of the predicted environmental concentration (PEC) of PhCs in hospital effluents, as well as in UWws, on the basis of the quantity of PhCs administered, their excretion rate and the flow rate of the WW in question. This can permit comparison with the corresponding PNEC. Indeed, Kummerer and Henninger [67] evaluated the predicted environmental concentration (PEC) of a great number of antibiotics for German hospitals of different sizes and found that the amount of antibiotics emitted into hospital effluent may reach, and indeed exceed, the minimum inhibitory concentrations (MIC_{50}) of susceptible pathogenic bacteria. The PEC/PNEC ratio is highest for HWws and frequently >1 for UWws. These authors concluded that steps should be taken to reduce risks by ensuring proper handling of antibiotics and their residues in hospitals and in private homes.

8.6 Regulation

As already emphasized, the frequent discharge of antibiotics into waterways can lead to the development of resistant bacteria against antibiotics and consequently render treatment of human bacterial infections ineffective. Therefore, HWws should be classified as hazardous materials [67].

However, hospital effluents are quite often considered to possess the same pollutant load as UWws. As a result, the former are generally discharged into public sewage and conveyed to municipal WWTPs, where they are subjected to co-treatment with UWws. In some cases, local regulations dictate the use of a pre-disinfection step (quite often pre-chlorination) of the raw HWws before their emission into the municipal sewage [30, 47], but in general, no specific requirement is set either for the micropollutant concentration in the pre-discharge hospital effluent or the concentration of PhCs in the final WWTP effluent before its discharge into a surface water body.

In contrast, great attention is paid to the *source* of PhCs. Indeed, stringent guidelines in the USA [68] and Europe (2001/83/EC for human PhCs and 2001/82/EC for animal PhCs) impose environmental assessment of *new* PhCs before they can be put on the market. Furthermore, a registration, evaluation, authorization and restriction of chemicals system (REACH) came into force in Europe on 1st June 2007, with the aim of improving protection of human health and the environment from the risks posed by dangerous chemicals, such as persistent, very persistent, bioaccumulative and toxic compounds with hazardous properties. This system makes industries responsible for assessing and managing the risks posed by chemicals and for providing their users with appropriate safety information.

Nevertheless, despite the fact that the scientific community has repeatedly highlighted the risk posed by PhCs in the aqueous environment, the major route for water surface contamination with PhCs being WWTP discharge, these chemicals have not yet been included in any priority list, either in Europe or in the USA. Furthermore, limits and regulation regarding PhCs and new compounds have not thus far been specifically set as regards water and wastewater treatment criteria.

However, the Stockholm Convention on persistent organic pollutants (which came into force in May 2004) established that governments must develop a programme setting out their plans to reduce or eliminate the release of persistent pollutants into the environment by the use of best available technologies (BATs) and the application of best environmental practices (BEPs). Referring to European countries, the Water Framework Directive 2000/60/EU [69] requires all its member states to establish an acceptable chemical and biological water status by 2015. In this case, chemical status refers to specific pollutants (priority substances or priority hazardous substances) for which environmental quality standards have been proposed. Nonetheless, many PhCs were not taken into consideration in either of these sets of guidelines and no legal limits regarding their emission into public sewage or concerning the final effluent of the WWTP treating it have been established.

Despite the absence of legislative guidance, some attempts to reduce the emission of PhCs into the environment have been taken. In Germany, for example, in the Rhine catchment area, the value of 100 ng L^{-1} has been chosen as the target value for each PhC in the effluent from a dedicated treatment plant for HWws if it is discharged directly into the waterways [70]. This value was set by the International Association of Waterworks in the Rhine catchment area [71] and corresponds to the recommendation by the German Ministry for the Environment [72] as the target value for water systems that feed drinking water supplies.

Moreover, the Global Water Research Coalition [8] carried out a study aimed at developing a list of representative priority PhCs that are most likely to be found in water supplies and those that may have significant impacts on human and

Table 8.4 The three classes set by GWRC [8] showing the high-priority PhCs (Class 1), priority PhCs (Class 2) and lower priority PhCs (Class 3)

Class	Pharmaceutical compounds
1	Carbamazepine, sulfamethoxazole, diclofenac, ibuprofen, naproxen, bezafibrate, atenolol, ciprofloxacin, erythromycin, gemfibrozil
2	Paracetamol, acetyl salicylic acid, clofibric acid, furosemide, iopromide, amidotrizoic acid, diazepam, lincomycin, amoxicillin, (hydro)chlorothiazide, metoprolol, ranitidine, trimethoprim, sotalol, codeine, clarithromycin
3	Iomeprol, iopamidol, metformin, dilantin, doxycycline, enalapril, fluoxetine, norfluoxetine, oxazepam, salbutamol, simvastatin β-hydroxy acid, cefalexin, cimetidine, clotrimazole, diltiazem, valproic acid

environmental health. Based on 25 publications, it took into consideration 153 compounds, which were categorized according to the seven following criteria: (1) regulation; (2) consumption/sales; (3) physico-chemical properties; (4) human toxicity/ecotoxicity; (5) occurrence in surface water, groundwater, drinking water and wastewater; (6) degradability/persistence; and (7) resistance to treatment. Three main classes of these compounds, itemized in Table 8.4, were thereby defined:

- Class 1 – High-priority pharmaceuticals, including PhCs mentioned in five or more of the base documents and that fulfil more than four of the seven criteria
- Class 2 – Priority pharmaceuticals, including those mentioned in more than two of the base documents cited and that fulfil more than two criteria
- Class 3 – Lower priority pharmaceuticals, including those mentioned in two of the documents cited and fulfil two or more criteria

The compounds of Class 1 represent the "*minimum*" group of compounds that should be considered in any study on PhCs in water management; those of Class 2 and 3 represent secondary targets.

The lists compiled in Table 8.4 derive from the compounds included in the base documents and could therefore be time-restricted. Indeed, an update of the classification could be necessary as more data becomes available.

8.7 Considerations on the Respective Contributions of a Hospital and Its Catchment Area to the Flow Rate and the Pollutant Load

Proper management of hospital effluent should take into consideration the characteristics of the catchment area in which the hospital is situated, that is, (1) size of area, number of inhabitants (p.e.) and average and maximum urban flow rate; (2) industrial activities (type, flow rate, pretreatments, final disposal, co-treatment) present in the area; (3) characteristics of existing WWTPs (nominal capacity, residual capacity, treatment sequence); (4) characteristics of the receiving water body (hydraulic regime, auto-depurative capacity, irrigation, recreational and industrial uses); and (5) legal and regulatory constraints.

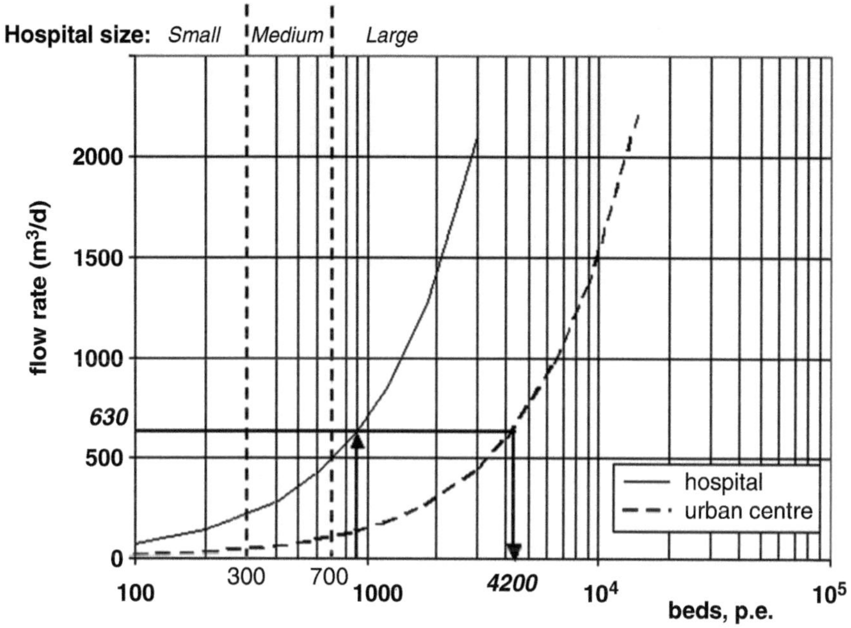

Fig. 8.10 Flow rates for hospitals and urban centres of different sizes

In the case of co-treatment, it is important to evaluate the percentage of hospital flow rate with respect to the total WWTP influent flow rate. This value depends on hospital size (small size with <300 beds, medium size with 300–700 beds and large size with >700 beds) and on the size of the resident population in the urban centre.

A representative situation is depicted in Fig. 8.10 based on the study by Verlicchi et al. [61], where hospital and urban flow rates are plotted, respectively, against the number of beds or population equivalent. These flow rates are based on a specific water consumption of 700 L bed^{-1} day^{-1} for hospitals and 150 L $p.e.^{-1}$ d for resident populations. A large hospital of 900 beds without internal laundry, producing a daily flow rate equal to 630 m^3 day^{-1}, has the same hydraulic load as an urban centre of 4,200 p.e. If this hospital is situated in a large town (for instance 100,000 p.e., corresponding to 15,000 m^3 day^{-1}) and its effluent is co treated at the same WWTP, the hospital flow rate corresponds to $630/(630 + 15{,}000) \times 100 = 4\%$ of the total WWTP influent. In contrast, if the same quantity of hospital effluent is collected in a small WWTP receiving UWws from a 2,000 p.e. urban centre (300 m^3 day^{-1}), its percentage contribution to the total WWTP influent flow rate increases to $630/(630 + 300) \times 100 = 68\%$.

Hartemann et al. [37] postulated that, if a hospital of 1,000 beds has an internal laundry (hypothesis not always verified), it is as polluting as a town with a population of 10,000 people, meaning that its contribution to the influent of a co-treatment WWTP could be even higher than the figures cited.

However, it should be noted that the characteristics of the catchment area will be quite different from case to case. Thus, the parameter "number of beds per 1,000 inhabitants", also called the *bed density*, is used to give an idea of the resident population in areas surrounding the health-care structure and to provide useful information about the pollutant load into the environment. For instance, if a large, 1,000-bed hospital were situated in a town of 130,000 inhabitants, the bed density would be 7.7, while if it were placed near an urban settlement of 6,000 inhabitants, the same parameter would become 167. In the first case, the hospital would not be the greatest source of PhCs, while in the second it would be the main contributor. The value of the bed density may vary between 0.4 and 170 beds/1,000 inhabitants [25, 58–61, 73, 74].

Concerning the pollutant load, a mass balance of the most critical PhCs (for instance, some antibiotics, the most administered analgesics/anti-inflammatories and the psychiatric drug carbamazepine, considered an anthropogenic marker in wastewaters) on a local level provides useful information about the contribution of the hospital effluent with respect to that of the catchment area. Hospital effluent contributes to the emission of the different PhCs to a different extent with respect to the UWws generated by the surrounding catchment area, depending on the specific compound. In fact, it can reach as high as 94% for some antibiotics (ciprofloxacin), as found by Beier et al. [70]. According to Kummerer [75], on the other hand, up to 25% of the antibiotics administered in Germany are used in hospitals.

Langford and Thomas [76] evaluated the contribution of two medium-/large-size hospitals (600–1,200 beds) in a large catchment area (610,000 p.e.) in Norway, for different PhCs. Only ciprofloxacin, present in HWws in extremely high concentrations, contributed more than 38% to the influent at the WWTP. The contribution of the other PhCs varied in the range 0.5–10%.

A similar study was carried out by Ort et al. in 2010 [25]. These authors made an in-depth analysis of the fraction of 59 pharmaceutical residues in wastewater originating from a hospital (200 beds) and from its catchment area (45,000 inhabitants) discharged in a region of Australia. They found that five compounds (the lipid regulator atorvastatin, the antidepressant citalopram, the anti-inflammatory diclofenac, the analgesic oxycodone and the biocide triclosan) were only detected in HWws, while 24 were detected neither in the HWw nor in the total WWTP influent. Only two substances, trimethoprim (18%) and roxythromycin (56%), displayed a maximum observed hospital contribution of >15%. On average, the contribution of the hospital for the compounds investigated in both hospital and sewage treatment plant influents was small and fairly constant.

8.8 Current Treatments of Hospital Effluents

In most European countries, hospital effluents are discharged into the public sewage system and conveyed to municipal WWTPs, which were originally built, and have more recently been upgraded, with the aim of removing carbon, nitrogen and phosphorus compounds, in addition to microbiological organisms and pollutants that commonly arrive at the plant in concentrations to the order of several mg L^{-1}, at least 10^5 CFU/100 mL. As a result, the efficiency in removing micropollutants, typically present in quantities of μg L^{-1}, is in general very low.

HWw flow rates quite often only amount to a small percentage of the total influent flow rate for co-treatment at a municipal WWTP. Consequently, dilution of HWws with UWws usually results in a decrease in the PhC content in the final effluent (from μg L^{-1} to ng L^{-1}), although not in the total load, that is to say, the quantity released daily into the receiving water body.

Pretreatment of HWws by coagulation/flocculation and flotation seems to be a suitable option for removing lipophilic compounds, such as fragrances and diclofenac, although it is unable to eliminate many other common hydrophilic PhCs, including carbamazepine, iopromide, diazepam and antibiotics in general, from the liquid phase [7].

Some recent literature reports dedicated WWTPs adopted for HWws. For example, at the Waldbrol hospital (350 beds) near Cologne in the Federal State of North Rhine-Westphalia in Germany, a full-scale membrane biological reactor (MBR) has been in operation since April 2006. It is equipped with *microfiltration* (MF) membranes and was designed for a flow rate of 130 m^3 day^{-1}. Beier et al. [24] investigated the removal rates for 11 PhCs: bezafibrate, bisoprolol, carbamazepine, clarithromycin, ciprofloxacin, diclofenac, ibuprofen, metronidazole, moxifloxacin, telmisartan and tramadol. The removal rates on average were very scarce (<20%) for bezafibrate, carbamazepine and telmisartan, roughly 40% for clarithromycin, between 60 and 80% for diclofenac and ciprofloxacin, and >80% for bisoprolol, metronidazole, moxifloxacin, tramadol and ibuprofen. Based on mass balance in an appropriate large-scale case study [70], it was found that the proportion of antibiotics detected in the municipal wastewaters originating from the hospital was, on average, 27–84% .

Another example of dedicated HWw treatment was reported by Verlicchi et al. [61] and describes the case study of a large hospital situated in a small catchment area in the first outskirts of Ferrara, Italy. In this case, the treatment consisted of a full-scale MBR equipped with *ultrafiltration* membranes, followed by an advanced oxidation process (O_3/UV). This multi-barrier treatment was chosen because the treated effluent discharges into an effluent-dominant river used for irrigation in the summer season.

In many developing countries, direct chlorination, or primary treatment followed by chlorination, is the most widely used basic method of treating HWws and has the aim of preventing the spread of pathogenic microorganisms, the causal agents of nosocomial infectious diseases [47, 61]. The widespread medical use of chlorine as a disinfectant is due to its (potential) capacity to remove bacteria, virus and fungi and to its ease of application. However, high quantities of this type of disinfectant are required and only a modest elimination rate can be achieved due to the organic load of the wastewater under treatment. Moreover, residual toxicity in the treated effluent is unavoidable due to the formation of chloramines and other undesirable disinfection by-products (DBPs).

Thus, in the last decade, China, due to its more stringent regulations and to the application of wastewater reuse strategies, has seen the application of MBR plants, mostly exploiting submerged technology and with capacities ranging from 20 to 2,000 m^3 day^{-1}, for the treatment of HWws [47].

8.9 Best Strategies in HWw Treatment

As previously mentioned, HWws are generally co-treated with domestic effluents in conventional WWTPs and are then released into the environment. However, many pharmaceuticals are resistant to conventional treatments, and therefore the efficiency of several municipal WWTPs in eliminating these compounds has been investigated by many authors [3, 4, 77–82]. Considering all the pharmaceuticals investigated, the overall average removal rates reported range between 10 and 90%.

As there is no specific treatment able to remove, to a high percentage, the many kinds of micropollutants typically found in HWws, due to their differing behaviour during treatments, and as many PhCs are resistant to conventional treatments, innovative solutions to this problem are required. Different operational configurations should be developed and calibrated, in order to provide information for potential practitioners about the financial aspects and overall risks associated with putative treatments of HWws [83].

A dedicated treatment for HWws is always desirable, especially for large hospitals in rural areas, where its treated effluent may be indirectly reused for irrigation after its discharge into a surface water body. In fact, although co-treatment with UWws at a municipal WWTP is common practice, it has several fundamental drawbacks. In the first place, dilution of HWws with UWws is not the correct procedure, as some substances in the hospital effluents may cause inhibition of the treatment plant biomass and thereby reduce the removal efficiency.

Regarding the best available technologies for removing PhCs from UWws, recent studies, generally investigating urban or synthetic WWs, have show that, as yet, there is no unique treatment sequence able to remove all the various kinds of micropollutants, due to their complex molecules and differing behaviours. However, their main removal mechanisms are biodegradation, favoured by a long sludge age (more than 30 days) and adsorption onto sludge flocks [84–89]. Chemical-physical treatments like coagulation/flocculation/precipitation, on the other hand, are only efficacious for lipophilic compounds [11], while secondary biological treatments are considered to be an effective barrier for most PhCs due to the metabolic and co-metabolic processes [90–93] that can take place in these systems. Moreover, Zhang and Farahbakhsh [32] showed that the MBR system (nominal pore size 0.04 μm) is capable of a high rate of coliphage removal irrespective of variations in coliphage feed concentrations. In fact, the removal rates measured were higher than those achieved by a conventional activated sludge system followed by an advanced tertiary treatment.

Furthermore, as many micropollutants tend to adsorb/absorb to the biomass flocks, efficient solid/liquid separation can greatly improve their removal from wastewater and, at the same time, guarantee a consistently good effluent quality. MBRs have been suggested for this purpose by many authors [83, 91, 93], some of whom found that ultrafiltration (UF) membranes are more efficient than MF membranes [24, 39]. MBR processes have also been suggested as better alternatives for the removal of pathogenic microorganisms, including some viruses [32, 94].

Ozonation and advanced oxidation processes (AOPs) are also promising candidates for efficient degradation of pharmaceuticals in water and wastewaters [40, 95–100]. In fact, treatment with ozone can reduce the concentration of many pharmaceuticals: 15 mg L^{-1} of ozone at 18 min contact time could be an adequate dose [101]. However, as AOPs are not affordable at many municipal WWTPs, Kim et al. [102] proposed that prolonging the SRT in biological WWTP may be the best practicable solution to reducing levels of pharmaceuticals in treated wastewaters.

An alternative to end-of-pipe upgrading of treatment plants, and an effective precautionary measure, could be source control. As reported above, administered pharmaceutical compounds are excreted from the human body via faeces and urine at a percentage that depends on the compounds in question. Although it will never be the perfect solution, separate collection of urine can contribute to keeping these substances away from wastewaters. Furthermore, source separation of urine (NoMix technology) can be conveniently adopted for other reasons, for instance, limitation of nutrient pollution of water. In this case, facilitated removal of pharmaceuticals could be a very welcome side effect [12]. In fact, Larsen et al. [103], Winker et al. [104] found that source separation of urine, which contains many of the pharmaceuticals and their transformation products from human metabolism, may offer the most effective solution to the problem of pharmaceuticals contaminating the environment. Due to the higher concentrations of micropollutants, biological as well as physical processes are expected to be more efficient for urine than for diluted wastewater. However, economic and practical feasibility must be carefully evaluated.

In particular, correct management of hospital effluents requires that discharge from toilets used by patients undergoing nuclear medicine therapy must be collected into separate tanks and treated as appropriate, thereby avoiding emission of radioactive compounds into the hospital sewage and from there into the public wastewater [22, 105].

8.10 Conclusion

Hospital effluents are generally considered to possess the same pollutant load as UWws, but many studies have demonstrated that this is not the case. Indeed, HWws tend to have higher concentrations of both macropollutants and micropollutants (PhCs and their metabolites, detergents, disinfectants, AOX, heavy metals, etc.). In fact, the average concentration of the main therapeutic classes (analgesics and anti-inflammatories, antibiotics, beta blockers, contrast media and hormones) reported for HWws are 4–150 times the corresponding average concentrations measured for UWws. For most PhCs, the peak concentration periods in the effluent are in the morning and in the afternoon, according to their different consumption and excretion patterns.

Hospitals require a significant quantity of water for the different activities and services taking place within the structure. In general, a hospital of 1,000 beds with an internal laundry pollutes to the same extent as a town with a population of 10,000 p.e., although hospital effluents contribute to different extents to the emission of the different PhCs with respect to the UWws generated by the surrounding area. Specifically, the proportion depends on the specific compound. Antibiotics have the highest values: 18% for trimethoprim, 38% for ciprofloxacin and 56% for roxythromycin, and the value for many other PhCs is <15%.

HWws are generally co-treated with UWws in conventional WWTPs designed and then upgraded for removing carbon, nitrogen and phosphorus compounds in addition to microbiological organisms, the pollutants regularly arriving at the plant in concentrations of the order of mg L^{-1} and at least 10^5 CFU/100 mL. As a result, the efficiency in removing micropollutants, typically present at μg L^{-1}, is in general very low, and persistent compounds readily enter the water cycle.

In recent years, great attention has been added to the monitoring of antibiotics in wastewaters and in the environment, as these drugs significantly contribute to the development of antibiotic-resistant bacteria strains, as well as anti-neoplastics, which have mutagenic and bacteria-toxic characteristics. Many scientists report that the careless discharge of HWws into the environment can expose the public to health risks from resistant infections.

A dedicated treatment for HWws is always a good solution, especially in the case of a large hospital in a scarcely populated area. The treatments that seem to be the most appropriate are a combination of biological, physical and chemical mechanisms such as (ultrafiltration) MBR, followed by advanced oxidation processes by means of O_3/UV.

References

1. Yu JT, Bouwer EJ, Coelhan M (2006) Occurrence and biodegradability studies of selected pharmaceuticals and personal care products in sewage effluent. Agric Water Manag 86:72–80
2. Lishman L, Smyth SA, Sarafin K, Kleywegt S, Toito J, Peart T, Lee B, Servos M, Beland M, Seto P (2006) Occurrence and reductions of pharmaceuticals and personal care products and estrogens by municipal wastewater treatment plants in Ontario, Canada. Sci Tot Environ 367:544–558
3. Santos JL, Aparicio I, Alonso E (2007) Occurrence and risk assessment of pharmaceutically active compounds in wastewater treatment plants. A case study: Seville city (Spain). Environ Int 33:596–601
4. Terzic S, Senta I, Ahel M, Gros M, Petrovic M, Barceló D, Muller J, Knepper T, Marti I, Ventura F, Jovancic P, Jabucar D (2008) Occurrence and fate of emerging wastewater contaminants in Western Balkan Region. Sci Total Environ 399:66–77

5. Boillot C, Bazin C, Tissot-Guerraz F, Droguet J, Perraud M, Cetre JC, Trepo D, Perrodin Y (2008) Daily physicochemical, microbiological and ecotoxicological fluctuations of a hospital effluent according to technical and care activities. Sci Tot Environ 403:113–129
6. Duong H, Pham N, Nguyen H, Hoang T, Pham H, Ca Pham V, Berg M, Giger W, Alder A (2008) Occurrence, fate and antibiotic resistance of fluoroquinolone antibacterials in hospital wastewaters in Hanoi, Vietnam. Chemosphere 72:968–973
7. Suarez S, Lema JM, Omil F (2009) Pre-treatment of hospital wastewater by coagulation-flocculation and flotation. Bioresour Technol 100:2138–2146
8. GWRC (2008) Development of an international priority list of pharmaceuticals relevant for the water cycle. Available at the site http://edepot.wur.nl/138086. Last access on 1 Mar 2011
9. Barceló D (2003) Emerging pollutants in water analysis. TrAC Trends Anal Chem 22(10): xiv–xvi
10. Jjemba PK (2006) Excretion and ecotoxicity of pharmaceuticals and personal care products in the environment. Ecotoxicol Environ Saf 63(1):113–130
11. Ternes TA, Joss A (2006) Human pharmaceuticals, hormones and fragrances. The challenge of micropollutants in urban water management. IWA Publishing, London
12. Lienert J, Burki T, Escher BI (2007) Reducing micropollutants with source control: substance flow analysis of 212 pharmaceuticals in faeces and urine. Water Sci Technol 55(5):87–96
13. Stuer-Lauridsen F, Birkved M, Hansen LP, Holten Lutzhùft HC, Halling-Sorensen B (2000) Environmental risk assessment of human pharmaceuticals in Denmark after normal therapeutic use. Chemosphere 40:783–793
14. Kolpin DW, Furlong ET, Meyer MT, Thurman EM, Zaugg SD, Barber LB, Buxton HT (2002) Pharmaceuticals, hormones, and other organic wastewater contaminants in U.S. streams, 1999–2000: a national reconnaissance. Environ Sci Technol 36(6):1202–1211
15. Furhacker M (2008) The water framework directive – can we reach the target? Water Sci Technol 57(1):9–17
16. Wangsaatmaja S (1997) Environmental action plan for a hospital. In: MS thesis in engineering, Asian Institute of Technology, Bangkok, Thailand
17. Laber J, Haberl R, Shrestha R (1999) Two stage constructed wetland for treating hospital wastewater in Nepal. Water Sci Technol 40(3):317–324
18. Altin A, Altin S, Degirmenci M (2003) Characteristics and treat-ability of hospital (medical) wastewaters. Fresenius Environ Bull 12:1098–1108
19. Mohee R (2005) Medical wastes characterisation in healthcare institutions in Mauritius. Waste Manag 25:575–581
20. Rezaee A, Ansari M, Khavanin A, Sabzali A, Aryan MM (2005) Hospital wastewater treatment using an integrated anaerobic aerobic fixed film bioreactor. Am J Environ Sci 1(4):259–263
21. Sarafraz S, Khani M, Yaghmaeian K (2007) Quality and quantity survey of hospital wastewater in Hormozgan province. Iran J Health Sci Eng 4(1):43–50
22. Regione Emilia Romagna, Guidelines of Regione Emilia Romagna on the Managing of Sanitary Institutions wastes and wastewaters, DGR 1155 27 Luglio 2009 (Linee guida per la gestione dei rifiuti e degli scarichi idrici nelle aziende sanitarie dell'Emilia-Romagna, In Italian). Available at the site: https://worksanita.regione.emilia-romagna.it/sites/seas/gestamb/Linee%20guida%202009/AdapterHTTP_atto_completo.pdf. Last access on 24 July 2012
23. Mesdaghinia AR, Naddafi K, Saeedi R, Zamanzadeh M (2009) Wastewater characteristics and appropriate method for wastewater management in the hospitals. Iran J Public Health 38(1):34–40
24. Beier S, Koster S, Veltmann K, Schroder H, Fr Pinnekamp J (2010) Treatment of hospital wastewater effluent by nanofiltration and reverse osmosis. Water Sci Technol 61(7): 1691–1698
25. Ort C, Lawrence M, Reungoat J, Eagleham G, Carter S, Keller J (2010) Determination of the fraction of pharmaceutical residues in wastewater originating from a hospital. Water Res 44:605–615
26. CTC - Clean Technology Consultant (1994) Design criteria of wastewater treatment plant. Ratchawithi Hospital, Bangkok, Thailand

27. Metcalf, Eddy (1991) Wastewater engineering. Treatment, disposal, re-use, III edn. McGraw-Hill, Singapore
28. Galletti A (2011) Pharmaceutical compounds in waters. Investigations on hospital effluents as a source of environmental contamination and on their treatability. In: PhD thesis in science of engineering, University of Ferrara, Italy
29. Joss A, Keller E, Alder AC, Gobel A, McArdell CS, Ternes T, Siegrist H (2005) Removal of pharmaceuticals and fragrances in biological wastewater treatment. Water Res 39:3139–3152
30. Verlicchi P, Galletti A, Masotti L (2008) Caratterizzazione e trattabilità di reflui ospedalieri: indagine sperimentale (con sistemi MBR) presso un ospedale dell'area ferrarese. In: Proceedings of Sidisa Conference Florence, Italy
31. Masotti L, Verlicchi P (2005) Depurazione delle acque di piccole comunità. Hoepli, Milano. ISBN 88-203-2963-8
32. Zhang K, Farahbakhsh K (2007) Removal of native coliphages and coliform bacteria from municipal wastewater by various wastewater treatment processes: implications to water reuse. Water Res 41:2816–2824
33. Mersi A, Marraccini E, Rapaci G, Rubini P (1993) Requisiti normativi e tecnici dei reflui ospedalieri. L'Igiene moderna 100:1443–1450
34. Kummerer K, Erbe T, Gartiser S, Brinker L (1998) AOX-emissions from hospital into municipal wastewater. Chemosphere 36(11):2473–2445
35. Wen X, Ding H, Huang X, Liu R (2004) Treatment of hospital wastewater using a submerged membrane bioreactor. Proc Biochem 39:1427–1431
36. Emmanuel E, Perrodin Y, Keck G, Blanchard JM, Vermande P (2005) Ecotoxicological risk assessment of hospital wastewater: a proposed framework for raw effluents discharging into urban sewer network. J Hazard Mater A117:1–11
37. Hartemann P, Hautemaniere A, Joyeux M (2005) La problématique des effluents hospitaliers. Hygiène 13(5):369–374
38. Tsakoma M, Anagnostopoulou E, Gidarakos E (2007) Hospital waste management and toxicity evaluation: a case study. Waste Manag 27:912–920
39. Verlicchi P, Galletti A, Petrovic M, Barcelò D (2010) Hospital effluents as a source of emerging pollutants: an overview of micropollutants and sustainable treatment options. J Hydrol 389:416–428
40. Chiang C, Tsai C, Lin S, Huo C, Lo K (2003) Disinfection of hospital wastewater by continuous ozonization. J Environ Sci Health A Tox Hazard Subst Environ Eng A38 (12):2895–2908
41. Peou S, Blech MF, Hartemann P (1981) Etude de la fréquence des bacttéries résistantes aux antibiotiques dans les eaux résiduaires hospitalièrers et urbaines. Environ Toxicol Lett 2:347–356
42. Schwartz T, Kohnen W, Jansen B, Obst U (2003) Detection of antibiotic-resistant bacteria and their resistance genes in wastewater, surface water, and water biofilms. FEMS Microbiol Ecol 43:325–335
43. Adelowo O, Fagade OE, Oke AJ (2008) Prevalence of co-resistance to disinfectants and clinically relevant antibiotics in bacterial isolates from three hospital laboratory wastewaters in South-western Nigeria. World J Microbiol Technol 24:1993–1997
44. Chitnis V, Chitnis S, Vaidya K, Ravikant S, Patil S, Chitnis DS (2004) Bacterial population changes in hospital effluent treatment plant in central India. Water Res 38:441–447
45. Leprat P, Chedevergne E, Camus A, Pacheco A, Mounier M (1996) Diagnostic physico-chimique et microbiologique des rejets hospitaliers. État des lieux à l'hôpital Dupuytren CHU de Limoges. Tech Hosp 612:35–38
46. Prado T, Silva DM, Guilayn WC, Rose TL, Gaspar AMC, Miagostovich MP (2011) Quantification and molecular characterization of enteric viruses detected in effluents from two hospital wastewater treatment plants. Water Res 45:1287–1297
47. Liu Q, Zhou Y, Chen L, Zheng X (2010) Application of MBR for hospital wastewater treatment in China. Desalination 250:605–608

48. Leprat P (1999) Caractéristiques et impacts des rejets liquides hospitaliers. Tech Hosp 634:56–57
49. Lopez N, Deblonde T, Hartemann Ph (2010) Les effluents liquid hospitaliers. Hygiène 18(6): 405–410
50. Kummerer K, Helmers E (2000) Hospital effluents as a source of gadolinium in the aquatic environment. Environ Sci Technol 34:573–577
51. Kosma CI, Lambropoulu DA, Albanis TA (2010) Occurrence and removal of PPCPs in municipal and hospital wastewaters in Greece. J Hazard Mater 179:804–817
52. Mullot JU,Karolak S, Fontova A, Levi Y (2010) Modeling of hospital wastewater pollution by pharmaceuticals: first results of Mediflux study carried out in three French hospitals. Wat Sci Technol 62(12):2912–2919
53. Lenz K, Koellensperger G, Hann S, Weissenbacher N, Mahnik SN, Fuerhacker M (2007) Fate of cancerostatic compounds in biological wastewater treatment of hospital effluents. Chemosphere 69:1765–1774
54. Lenz K, Mahnik SN, Weissenbacher N, Mader RM, Krenn P, Hann S, Koellensperger G, Uhl M, Knasmuller S, Ferk F, Bursch W, Fuerhacker M (2007) Monitoring, removal and risk assessment of cytostatic drugs in hospital wastewater. Water Sci Technol 56(12):41–149
55. Hartemann A, Golet E, Gartiser S, Alder AC, Koller T, Widmer RM (1998) Identification of fluoroquinolone antibiotics as the main source of humic? Genotoxicity in native hospital wastewater. Environ Toxicol Chem 17(3):377–382
56. Jolibois B, Guerbet M (2005) Evaluation of industrial, hospital and domestic wastewater genotoxicity with the Salmonella fluctuation test and the SOS chromotest. Mutat Res 565(2): 151–162
57. Ferk F, Misik M, Grummt T, Majer B, Fuerhacker M, Buchmann C, Vital M, Uhl M, Lenz K, Grillitsch B, Parzefall W, Nersesyan A, Knasmuller S (2009) Genotoxic effects of wastewater from an oncological ward. Mutat Res 672(2):69–75
58. Heberer T, Feldmann D (2005) Contribution of effluents from hospitals and private households to the total loads of diclofenac and carbamazepine in municipal sewage effluents – modeling versus measurements. J Hazard Mater 122(3):211–218
59. Thomas KV, Dye C, Schlabach M, Langford KH (2007) Source to sink tracking of selected human pharmaceuticals from two Oslo city hospitals and a wastewater treatment works. J Environ Monitoring 9(12):1410–1418
60. Feldmann DF, Zuehlke S, Heberer T (2008) Occurrence, fate and assessment of polar metamizole (dipyrone) residues in hospital and municipal wastewater. Chemosphere 71(9): 1754–1764
61. Verlicchi P, Galletti A, Masotti L (2010) Management of hospital wastewaters: the case of the effluent of a large hospital situated in a small town. Water Sci Technol 61(10):2507–2519
62. Emmanuel E, Keck G, Blanchard JM, Vermande M, Perrodin Y (2004) Toxicological effects of disinfections using sodium hypochlorite on aquatic organisms and its contribution to AOX formation in hospital wastewater. Environ Int 30:891–900
63. Kummerer K, Helmers E, Hubner P, Mascart G, Milandri M, Reinthaler F, Zwakenberg M (1999) European hospitals as a source for platinum in the environment in comparison with other sources. Sci Tot Environ 225:155–165
64. Gobel A, Thomsen A, Mcardell C, Joss A, Giger W (2005) Occurrence and sorption behavior of sulfonamides, macrolides, and trimethoprim in activated sludge treatment. Environ Sci Technol 39:3981–3989
65. Skadsen JM, Rice BL, Meyering DJ (2004) The occurrence and fate of pharmaceuticals, personal care products and endocrine disrupting compounds in a municipal water use cycle: a case study in the city of Ann Arbor: City of Ann Arbor. Available at: http://www.a2gov.org/government/publicservices/water_treatment/Documents/EndocrineDisruptors.pdf. Last access on 1 Mar 2011
66. Alexy R, Sommer A, Lange FT, Kummerer K (2006) Local use of antibiotics and their input and fate in small sewage treatment plant – significance of balancing and analysis on a local scale vs. nationwide scale. Acta Hydrochem Hydrobiol 34:587–592

67. Kummerer K, Henninger A (2003) Promoting resistance by the emission of antibiotics from hospital and households into effluent. Clin Microbiol Infect 9:1203–1214
68. US FDA (1998) Guidance for industry: environmental assessment of human drug and biologies applications. CMC 6 Revison 1. Food and Drug Administration, Rockville, MD. http://www.fda.gov/downloads/Drugs/GuidanceComplianceRegulatoryInformation/Guidances/ucm070561.pdf. Last access on 1 Mar 2011
69. Directive 2000/60/EC of the European Parliament and of the Council establishing a framework for the community action in the field of water policy. 23 Oct 2000
70. Beier S, Cramer C, Koster S, Mauer C, Palmowski L, Schroeder H, Fr Pinnekamp J (2011) Full scale membrane bioreactor treatment of hospital wastewater as forerunner for hot-spot wastewater treatment solutions in high density urban areas. Water Sci Technol 63(1):66–71
71. International Association of Waterworks in the Rhine Catchment Area, IAWR (2008) Danube, Meuse and Rhine-Memorandum, 1–20
72. UBA 2008 The Federal Environment Agency (ed) (2008) Public drinking water supply – evaluation of organic micropollutants. Letter to the ministry of the environment and conservation, agriculture and consumer protection of the state of North Rhine-Westphalia from March 14th 2008
73. Schuster A, Hadrich C, Kummerer K (2008) Flows of active pharmaceutical ingredients originating from health care practices on a local, regional, and nationwide level in Germany – is hospital effluent treatment an effective approach for risk reduction? Water Air Soil Pollut Focus 8:457–471
74. Verlicchi P, Al Aukidy M, Galletti A, Petrovic M, Barceló D (2012) Hospital effluent: investigation of the concentrations and distribution of pharmaceuticals and environmental risk assessment. Sci Tot Environ 430:109–118
75. Kummerer K (2009) Antibiotics in the aquatic environment – a review – part II. Chemosphere 75:435–441
76. Langford KH, Thomas KV (2009) Determination of pharmaceutical compounds in hospital effluents and their contribution to wastewater treatment works. Environ Int 35:766–770
77. Carballa M, Omil F, Lema J, Llompart M, Garcia-Jares C, Rodriguez I, Gomez M, Ternes T (2004) Behavior of pharmaceuticals, cosmetics and hormones in a sewage treatment plant. Water Res 38:2918–2926
78. Nakada N, Tonishima T, Shinohara H, Kiri K, Takada H (2006) Pharmaceutical chemicals and endocrine disrupters in municipal wastewater in Tokyo and their removal during activated sludge treatment. Water Res 40:3297–3303
79. Batt A, Kim S, Aga D (2007) Comparison of the occurrence of antibiotics in four full-scale wastewater treatment plants with varying designs and operations. Chemosphere 68:428–435
80. Oppenheimer J, Stephenson R, Burbano A, Liu L (2007) Characterizing the passage of personal care products through wastewater treatment processes. Water Environ Res 79(13): 2564–2577
81. Vieno N, Tuhkanen T, Kronberg L (2007) Elimination of pharmaceuticals in sewage treatment plants in Finland. Water Res 41:1001–1012
82. Teske S, Arnold R (2008) Removal of natural and xeno-estrogens during conventional wastewater treatment. Rev Environ Sci Biotechnol 7:107–124
83. Pauwels B, Verstraete W (2006) The treatment of hospital wastewater: an appraisal. J Wat Health 4(4):405–416
84. Jones OAH, Voulvoulis N, Lester NJ (2005) Human pharmaceuticals in wastewater treatment processes. Crit Rev Environ Sci Technol 35:401–425
85. Castiglioni S, Bagnati R, Fanelli R, Pomati F, Calamari D, Zuccato E (2006) Removal of pharmaceuticals in sewage treatment plants in Italy. Environ Sci Technol 40:357–363
86. Lindberg R, Olofsson U, Rendahl P, Johansson M, Tysklind M, Andersson B (2006) Behaviour of fluoroquinolones and trimethoprim during mechanical, chemical, and active sludge treatment of sewage water and digestion of sludge. Environ Sci Technol 40(3): 1042–1048

87. Pauwels B, Verstraete W (2006) The treatment of hospital wastewater: an appraisal. J Water Health 4(4):405–416
88. Matamoros V, Garcia J, Bajona JM (2008) Organic micropollutant removal in a full-scale surface flow constructed wetland fed with secondary effluent. Water Res 42:653–660
89. Miège C, Choubert JM, Ribeiro L, Eusèbe M, Coquery M (2008) Removal efficiency of pharmaceuticals and personal care products with varying wastewater treatment processes and operating conditions – conception of a database and first results. Water Sci Technol 57(1): 49–56
90. Kreuzinger N, Clara M, Droiss H (2004) Relevance of the sludge retention time (SRT) as design criteria for wastewater treatment plants for the removal of endocrine disruptors and pharmaceuticals from wastewater. Water Sci Technol 50(5):149–156
91. Daigger GT, Rittmann TE, Adams S, Andreottola G (2005) Are membrane bioreactors ready for widespread applications? Environ Sci Technol 39(19):399A–406A
92. Kimura K, Hara H, Watanabe Y (2007) Elimination of selected acidic pharmaceuticals from municipal wastewaters by activated sludge systems and membrane bioreactors. Environ Sci Technol 41:3708–3714
93. Radjenovic J, Petrovic M, Barceló D (2009) Fate and distribution of pharmaceuticals in wastewater and sewage sludge of the conventional activated sludge (CAS) and advanced membrane bioreactor (MBR) treatment. Water Res 43:831–841
94. Ottoson J, Hansen A, Bjorlenius B, Norder H, Stenstrom TA (2006) Removal of viruses, parasitic protozoa and microbial indicators in conventional and membrane processes in a wastewater pilot plant. Water Res 40:1449–1457
95. Huber MM, Canonica S, Park YG, von Gunten U (2003) Oxidation of pharmaceuticals during ozonation and advanced oxidation processes. Environ Sci Technol 37:1016–1024
96. Balcioglu IA, Otker M (2003) Treatment of pharmaceuticals wastewater containing antibiotics by O3 and O3/H2O2 processes. Chemosphere 50:85–95
97. Ternes TA, Stuber J, Herrmann N, McDowell D, Ried A, Kampmann M, Teiser B (2003) Ozonation: a tool for removal of pharmaceuticals, contrast media and musk fragrances from wastewaters? Water Res 37:1976–1982
98. Machado EL, Schmidt LT, Hoeltz JM, Dalbert D, Alcayaga ELA (2007) Secondary hospital wastewater detoxification and disinfection by advanced oxidation process. Environ Technol 28:1128–1143
99. Zimmermann SG, Salhi E, Koepke S, Hollender J, Hammes F, Gansner E, Koch M, Traber J, Ort C, Siegrist HR, von Gunten U (2008) Assessment of a full scale ozonation to reduce micropollutants concentrations in municipal wastewater. In: Proceedings of LET conference, Zurich, CH
100. Zwiener C, Frimmel FH (2003) Short-term tests with a pilot sewage plant and biofilm reactors for the biological degradation of the pharmaceutical compounds clofibric acid, ibuprofen and diclofenac. Sci Tot Environ 309:201–211
101. Gagnon C, Lajeunesse A, Cejka P, Gagn F, Hausler R (2008) Degradation of selected acidic and neutral pharmaceutical products in a primary-treated wastewater by disinfection processes. Ozone Sci Eng 30(5):387–392
102. Kim S, Weber AS, Batt A, Aga DS (2008) Removal of pharmaceuticals in biological wastewater treatment plants. In: Aga DS (ed) Fate of pharmaceuticals in the environment and in water treatment systems. CRC Press, Boca Raton, pp 349–361
103. Larsen TA, Lienert J, Joss A, Siegrist H (2004) How to avoid pharmaceuticals in the aquatic environment. J Biotechnol 113:295–304
104. Winker M, Faika D, Gulyas H, Otterpohl R (2008) A comparison of human pharmaceutical concentration in raw municipal wastewater and yellow water. Sci Tot Environ 399:96–104
105. Heinzmann B, Schwarz RJ, Schuster P, Pineau C (2008) Decentralized collection of iodinated x-ray contrast media in hospitals-results of the feasibility study and the practice test phase. Water Sci Techno 57:209–215

Chapter 9
Advances in Mechanical Dewatering of Wastewater Sludge Treatment

Akrama Mahmoud, Jérémy Olivier, Jean Vaxelaire, and Andrew F.A. Hoadley

9.1 Introduction

Over the last decades, the increase in municipal and industrial wastewater purification activities has been confronted with a dramatically increasing flow of sewage sludge. A common characteristic of different type of sludge is the very high water content, the colloidal and compressible nature of the sludge. Waste-activated sludge is an important class of these waste products and has to be treated and disposed of. After gravitational thickening, sewage sludge still contains only as little as about 1–5% (wt%) on a wet basis of dry solids, with the remaining fraction being water (95–99%) [1]. Because the bulk of the sludge is water, the cost of transporting the sludge is prohibitive. Moreover, sewage sludge often requires supplementary bulking agent to assist the composting process. The high water content also means that the energy content is low. Therefore, it is both economically and technologically feasible to decrease the water content. This facilitates the possible use of sludge as a fuel, or as organic matter for composting and reduces transport and disposal costs.

It is well documented that sludge dewatering is one of the most challenging technical tasks in the field of wastewater engineering [2–4]. In fact, when compared with thermal (evaporative processes) for water reduction, mechanical dewatering is often selected due to its low energy requirement [5]. This operation not only reduces the total waste volume but also increases the caloric value of the product [6, 7]. The amount of water that can be removed depends on the mechanical dewatering process and the status of the water in the materials. Figure 9.1a shows

A. Mahmoud (✉) • J. Olivier • J. Vaxelaire
Laboratoire de Thermique Energétique et Procédés (EAD 1932), ENSGTI, rue Jules Ferry, BP 7511, 64075 Pau, France
e-mail: akrama.mahmoud@gmail.com; amahmoud@gmx.fr

A.F.A. Hoadley
Department of Chemical Engineering, Monash University, Building 35, Clayton Campus, Victoria 3800, Australia

S.K. Sharma and R. Sanghi (eds.), *Wastewater Reuse and Management*,
DOI 10.1007/978-94-007-4942-9_9,

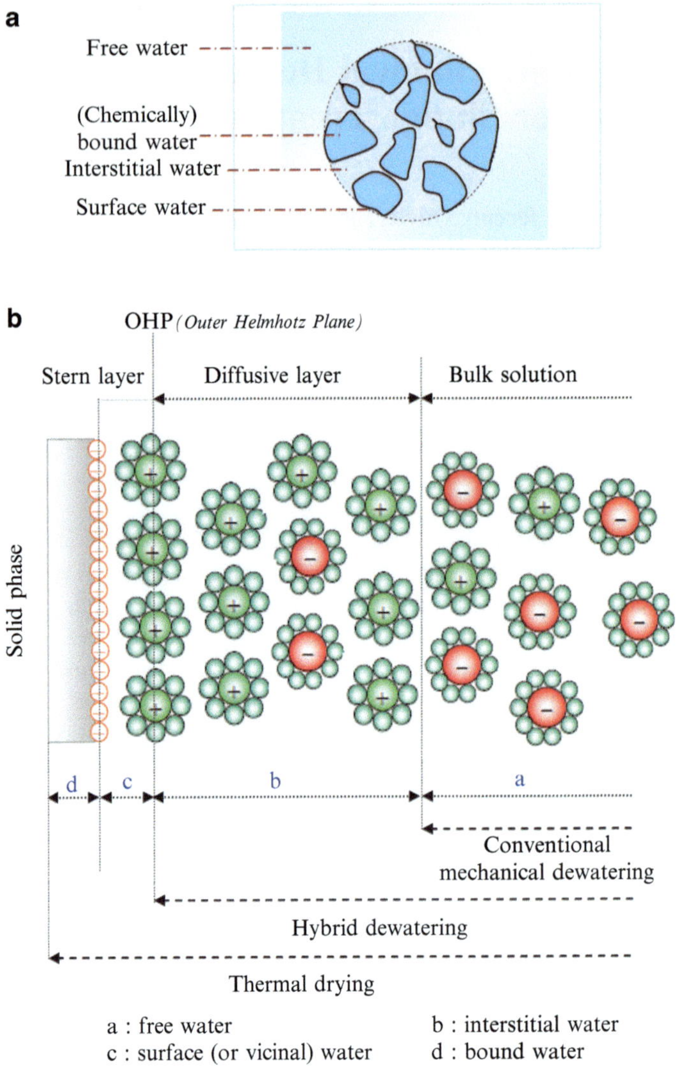

Fig. 9.1 (**a**) Water distribution in wastewater sludge [8–10]. (**b**) Dewatering methods in relation to water distribution in the materials [11]

a general conceptual visualisation of water distribution in wastewater sludge [8–11]. Various interpretations of the types of water found in fine-particle suspensions or flocculated systems are offered [8, 12–19]. The categorisation proposed by Vesilind and his co-workers [8, 18–20] and more recently by Vaxelaire and Cézac [21] is often used as reference for wastewater sludge, with the water existing as (a) free water that is not attached to the particles (including void water not affected by capillary force), (b) interstitial water that is trapped within the flocs, (c) surface (or vicinal) water that is held on the surface of the solid particles by adsorption and adhesion, and (d) intercellularly and chemically bound water. Although this classification is a simplification of the true behaviour of wastewater sludge [21], it is generally sufficient for dewatering considerations. Indeed, it is generally admitted that the water that can be removed relatively easily using mechanical dewatering included parts of free and interstitial water. The spatial location of water in the material intrinsically determines the magnitude of energy input for water removal.

Dewatering is mainly performed by mechanical techniques based on gravitational settling centrifugation, or filtration/compression, e.g. by belt or filter presses. However, the colloidal and compressible nature of the sludge will hamper its dewatering without pretreatment. Moreover, the presence of organic components, mainly bacterial cells and EPS (extracellular polymeric substances), in the sludge makes it very difficult to dewater even at high pressure. Therefore, prior to dewatering, most sludge is conditioned using flocculants in order to improve the dewaterability with regard to dewatering kinetics and attainable dry solids content. Yet, wastewater sludge remains hard to dewater, and for many applications, it cannot achieve sufficiently low water content. Therefore, an improvement in the traditional mechanical dewatering (MDW) equipment is desirable. As a consequence, current research tends to propose potential alternatives to enhance the dewatering ability of conventional processes [11, 22].

The intensification of the dewatering process can take several forms: (1) the combination of mechanical and thermal effects (pressure and temperature) [23–37]; (2) the combination of mechanical force and an electric field [11, 38–40], possibly pulsed [41–44]; (3) the superimposition of ultrasounds and/or magnetic fields [45–49]; (4) the combined fields (e.g. electric and ultrasonic) applied simultaneously [22, 50]; (5) both shear and compressive forces [51–56]; (6) and microwave-assisted dewatering [57]. Proper choice of combined fields for a given system is essential, but a good choice can lead to higher product recovery (solids or liquid), less moisture in the resulting cakes, and lower energy and environmental costs.

Most mechanical dewatering processes involve two stages; the first is the filter cake formation stage, and the second is the compression stage where further water is squeezed from the cake by the application of a mechanical force. The application of additional fields can be applied to either or both dewatering stages, or as a pre- or post-treatment of the dewatering process. Results have been reported for many of these options from laboratory-scale and pilot-scale experiments. The operating conditions in the intensification of the dewatering process are sufficient to remove a significant proportion of the water that cannot be removed using mechanical dewatering technologies alone. Thus, intensification of the dewatering process has the potential to be viable for a range of slurries, which either could not be sufficiently dewatered or would otherwise require extreme conditions using conventional dewatering devices. The established dewatering techniques and their applicability are shown in Fig. 9.1b. Looking more closely at the mechanism of the hybrid dewatering process, a theory for the conventional dewatering steps and different possible assisted fields is presented and discussed herein.

9.2 Theoretical Background

The dewatering of sludge is a mechanical unit operation used to reduce the moisture content of sludge for one or more of the following reasons [58]: (1) to reduce the transportation costs to drying and disposal sites, (2) to increase the calorific value of the sludge and thus makes it more suited for incineration, (3) to reduce leachate production at the landfill site upon landfilling, and (4) to improve the handling properties of the sludge by removing excess water.

The following considerations will prevail when evaluating the various dewatering processes: what is the rate at which water can be removed and how much water can be separated? [59].

In practical applications, these two considerations will play a major role in the equipment selection process and optimisation actions. A number of techniques are used in dewatering devices for removing moisture. In mechanical dewatering devices, mechanically assisted physical means, such as centrifugal forces, shear, and pressure, are used to dewater the sludge at a fast rate and up to relatively high solids content.

9.2.1 Conventional Dewatering Technologies

There are established techniques that can concentrate or thicken dilute suspensions of fine particles. These include gravitational settling, mechanical dewatering, and thermal drying. Due to the low solids content of the thickening and the high-energy cost of the thermal treatment, the mechanical dewatering devices are widely used as an intermediate process prior to some form of thermal drying. Mechanical dewatering techniques are usually based on a single non-thermal driving force, such as vacuum, pressure, or centrifugal force.

On the other hand, it is difficult to compare the dewatering results from different mechanical dewatering devices when these are obtained for sludge of different origins. Very few pilot or full-scale comparative studies are reported in the scientific literature, which compare the performance of different dewatering techniques on the same sludge. An interesting exception is the work by Andreasen and Nielsen [60] who tested four types of mechanical dewatering devices on three types of sludge. They reported that the diaphragm filter presses are by far the most powerful dewatering devices, followed by centrifuges, as shown in Table 9.1. Belt filter presses deliver more moderate dewatering results. The selection of a given dewatering technique will also depend on other factors, such as polymer consumption, maintenance, labour, and waste disposal cost.

The efficiency of these techniques depends on the particle size of the product being dewatered. Conventional techniques are inefficient for dewatering of colloidal and gelatinous sludge as they rely on hydraulic flow [61], a bulk property that falls off drastically with decreasing particle and pore size [62]. As mentioned

Table 9.1 Dry solids content of dewatered sludge (wt%) for different mechanical dewatering devices [60]

Device	Dilute waste activated sludge	Thickened activated sludge	Mixed thickened waste activated sludge with anaerobically digested biosorption sludge
Solid bowl centrifuge			
M and J Decanter 142	18–20	17–19	23–24
Belt filter press			
Bellmer VP08F	21–22	22	22–26
Press centrifuge			
Alfa-Laval DSNX4230	21–23	20–22	30–31
KHD Centripress 2-1	23–25	19–21	27–29
Diaphragm filter press			
Lenser	25–27	15–23	30–36

A activated sludge, *WAS* waste activated sludge, *AnD* anaerobically digested sludge

earlier, wastewater sludge is not simply a system of particles surrounded by water [59], but usually contains water within their structure. These structures are known as flocs, as shown in Fig. 9.2, containing a large variety of microorganisms, dead cells, organic and inorganic matters, and the large amounts of entrapped water cause them to have a low density and a high susceptibility to compression, which results respectively in low settleability and difficult mechanical dewaterability. Moreover, the presence of dissociated molecules on the surface of the individual constituents of a sludge floc, the so-called extracellular polymeric substances (EPS), creates a high osmotic swelling pressure inside the sludge floc, which counteracts the imposed mechanical driving force during dewatering [63–66].

Thermal drying is one commonly used method for water removal from wastewater sludge. The advantage of drying is that it can produce a completely dry solid product. However, it also consumes a large amount of energy, which is proportional to the amount of water which must be evaporated. Therefore, new techniques have been explored in the last decade, which provide an improved mechanical dewatering prior to the sludge being fed to a thermal dryer. These techniques include conditioning the sludge or preparing the sludge by thermal treatment, mechanical disintegration using ultrasounds, mills, homogenizers and others, magnetic and/or electromagnetic, microwave irradiation, electrical, as shown in Fig. 9.3.

Conditioning methods may be divided into two types: chemical conditioning, in which one or more chemical additives are used to alter sludge properties, and physical conditioning, in which temperature or other physical properties are used to effect changes in sludge properties. Chemical conditioning is the more common process. In either case, the conditioning process must fundamentally alter the microscopic structure of sludge so that the solid and liquid portions are more easily separable.

One of the steps generally involved in thickening and dewatering is the flocculation of sludge with inorganic (ferric or aluminium chloride) or organic flocculant (synthetic polyelectrolyte), called conditioning. In this way, an inorganic or organic flocculant may be added to flocculate the particle networks, resulting in an

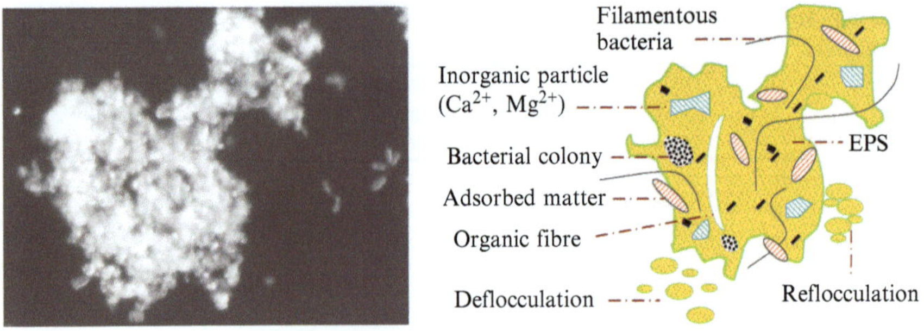

Fig. 9.2 Image of an activated sludge floc (*left*) and its composition (*right*) [66]

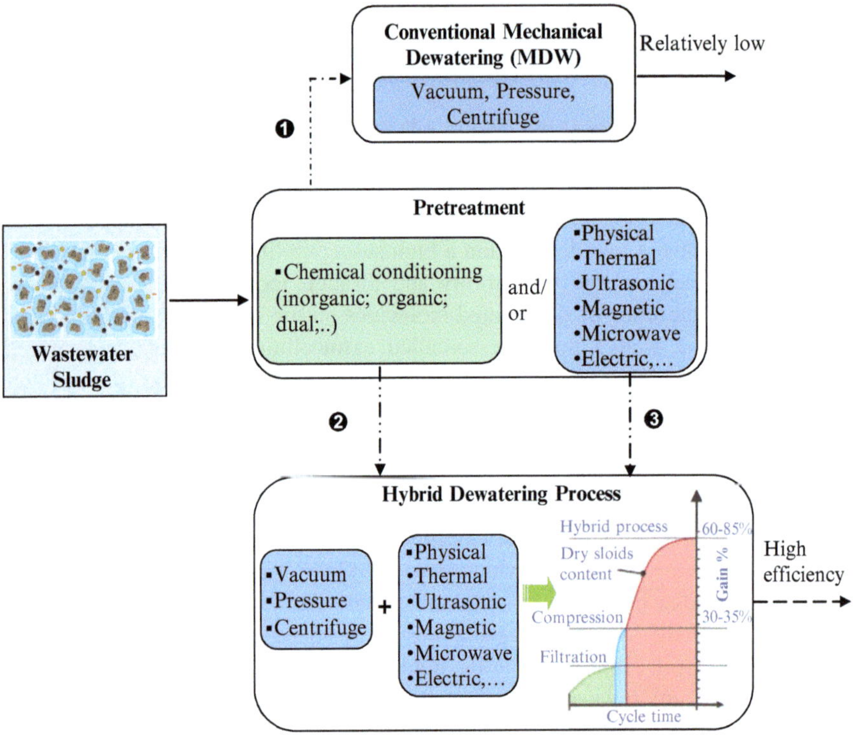

Fig. 9.3 Design concept for improving the dewatering efficiency

improved structure with reduced water retention. Charged organic polymers, also called polyelectrolytes, have gained a large market share over the last decades, since they can be dosed in much lower quantities than inorganic flocculants such as lime and iron chloride [67–69]. The selection of the appropriate chemical and dosage is quite difficult because it depends on both the sludge composition and the dewatering device and remains widely empirical [6, 70].

Recent research in chemical conditioning prior to mechanical dewatering methods includes, among others, pH reduction and the use of surfactants [71], acidification or alkalisation and consecutive oxidation by H_2O_2 [72, 73], coagulation with metal salts followed by flocculation with an amphoteric polymer [74], coagulation with cationic polyelectrolytes followed by flocculation with nonionic polyelectrolytes [75], conjunction of lignite with polyelectrolytes [76], by adding fly ash material generated from power plants [77], etc.

Conditioning of sludge without the use of chemicals offers the advantage of avoiding ongoing purchases of expensive flocculating agents. Different types of physical conditioning or pretreatment prior to dewatering processes may be considered. Thermal pretreatment in the range (40–180 °C) causes hydration of proteins, polysaccharides, lipids, and other intracellular macromolecules secreted from disrupted sludge cells. Consequently, the vicinal water content of sludge increased, while the overall free water content increased by release of some interstitial water originally trapped inside sludge cells [78, 79]. New approaches involving quick heating (up to about 220 °C and quenching after 10–30 s) or combined thermal and acid or alkaline treatment seem to improve the overall efficiency of the process [80–82].

During freezing and thawing, the floc structure will be irreversibly changed into more compact form; thus, the bound water content will be reduced, and therefore, the sludge dewatering characteristics can be significantly improved. This may be related to the fact that organisms are subjected to mechanical stress caused by the expanding ice crystals, as well as to osmotic pressures caused by the salts dissolved in the diminishing liquid water pools outside the cells, leading to irreversible structural damage [79, 82, 83]. The freeze/thaw conditioning does not only increase the sludge dewaterability but also reduces pathogenic microorganisms in sludge [84]. It is also considered as a low-cost sludge treatment technique suitable for moderate to cold climates [85].

Methods involving high-energy applications such as mechanical, electrical, or sonic impulses or microwave irradiation have been shown to be effective in some cases. In general, the use of additional forces to dewatering has gained an increasing recognition in recent years. Magnetically assisted filters are now widely accepted as technical alternatives in the choice of the liquid/solids separation equipment. Electrically assisted filters are commercially available, but have not been widely adopted, whereas ultrasonically assisted filters have till now only been demonstrated in laboratory studies. Although these pretreatment methods have given significant dewatering improvements, they are not discussed in this chapter. The interested readers may refer to the [79, 86–111] for a comprehensive coverage of the numerous new alternative methods of sludge conditioning and/or pretreatment prior to dewatering.

The intensification of dewatering process by application of an additional force field, such as heat, shear, acoustic or ultrasonic, magnetic, microwave, or electricity, is shown in Fig. 9.3. These coupled processes are our concern in this chapter, with the emphasis on their applications to wastewater sludge dewatering. They are discussed in greater detail in the following sections.

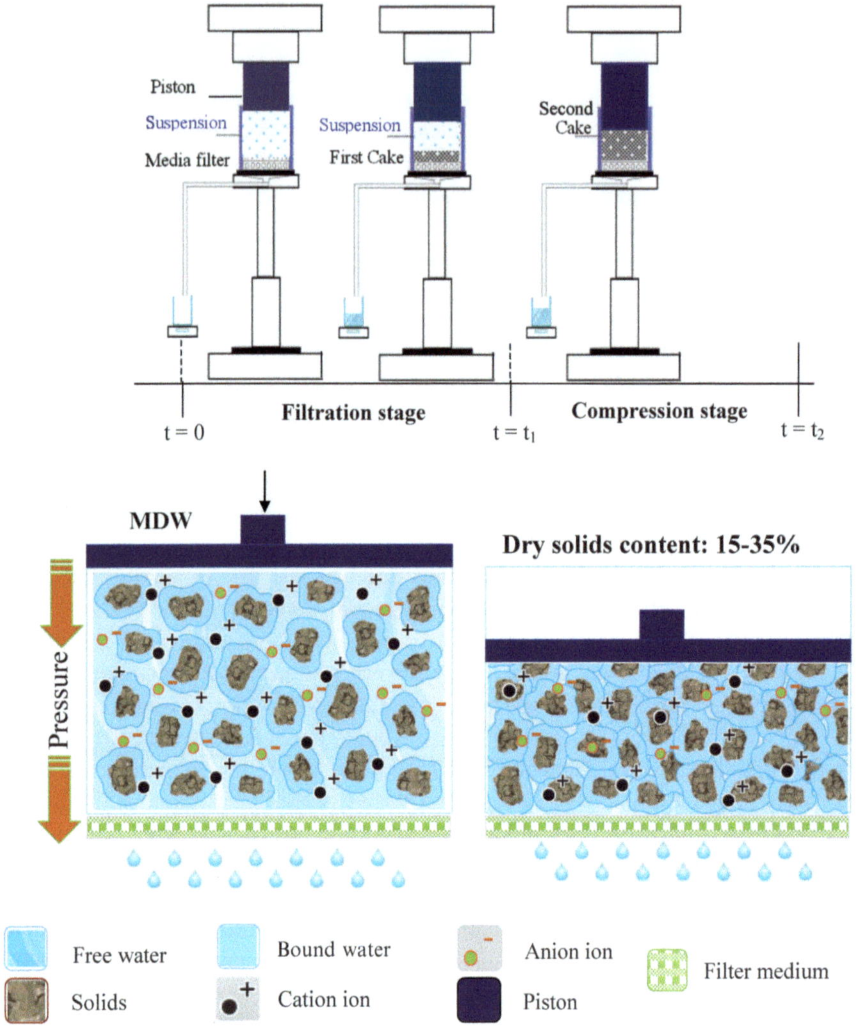

Fig. 9.4 Schematic representation of the different stages in a piston type dewatering device with medium filter at the *bottom* side of the dewatering device [36, 265]

9.2.2 Mechanical Dewatering Steps (Filtration–Compression)

Figure 9.4 shows a typical laboratory arrangement for the pressure filtration of a suspension in a cylindrical cell. Pressure is exerted on the suspension by pressing down on it using a piston, while the liquid is expelled from the bottom through a permeable membrane or filter medium. This medium which can be a cloth or polymer membrane has the property of retaining the solids while allowing the liquid to pass through. As mentioned earlier, during the mechanical dewatering processes, two steps can be distinguished: the so-called filtration stage and the compression stage. In the filtration stage, the compressive piston pushes the slurry in the direction of the filtering media, and a filter cake builds up. The filtrate flows throughout the porous cake and the filter medium. The ratio of the pressure drop across the filter medium to the pressure drop across the cake decreases with time. As the cake builds up, the hydraulic pressure near the filter medium approaches zero (when the filter resistance is negligible), and the hydraulic pressure at the top of the cake and in the suspension remains constant [36, 112, 113].

The cake compression stage begins when the compressive piston comes in contact with the filtration cake. During this stage, the cake is squeezed into a more compact form which leads to lower final moisture content. The filterability of the cake and its ability to express the liquid are essential parameters for the separation process. They largely depend on the cake structure, which is fixed by both the operating conditions and the suspension properties [114].

9.3 Assisted Mechanical Dewatering Devices

Several approaches have been suggested to improve dewatering in conjunction with mechanical dewatering processes. These techniques are presented and discussed in greater detail in the following sections.

9.3.1 Thermal Mechanical Dewatering (TMD)

The mechanical dewatering is, in principle, limited by the resistance of the solid structure against deformation. An essential acceleration of dewatering can be achieved by increasing the temperature [27–29, 115–117]. Mechanical and thermal effects, resulting in the reduction of water content, are a combination of different mechanisms [30, 31, 33–37, 118, 119]: (1) mechanical dewatering reduces the volume of the pores and squeezes the water out of the product. The yield stress of the solid matrix reduces with increasing temperature, and therefore, the solid matrix is more easily compressed [120]. (2) The application of elevated temperatures reduces the liquid density, viscosity, and the surface tension, and this leads to a reduction in the amount of water that can be held in the pores, so that water drains from the pores by gravity (thermal dewatering). (3) At more elevated temperatures, thermal decomposition of cellular matter leads to hydrophobisation through the elimination of polar functional groups. This in turn leads to the electrostatic forces of attraction between water and particles being more easily overcome [117].

Coupling between pressure and temperature can be achieved in different ways. It can occur during the filtration regime only, the compression regime only, during both regimes, or as a pre- or post-treatment of the dewatering stage. Various processes have been tested at the lab and pilot scale, but most of the time, the operating conditions lead to a vaporisation of the solvent.

Figure 9.5 summarises a few selected examples of the research and development of the combination of mechanical and thermal effects. The purpose of this section is not to provide an exhaustive listing of all the processes developed but to provide a brief overview of the core technologies, so as to give the reader a flavour of the variety of these techniques.

For a few years now, Arlabosse and her co-workers from Centre RAPSODEE, Ecole des Mines Albi, France [30, 31, 36, 37], specifically investigate a thermally assisted mechanical dewatering (TAMD) process, which couples mechanical dewatering at low pressure (usually $P_{\text{applied}} < 3000$ kPa) with a moderate heating ($T < 80\ °\text{C}$) of the walls of the apparatus in contact with the product. The proposed combination of pressure and temperature leads to the reduction of the water expressed as a liquid and, for some products, results in a significant energy saving on the overall separation process. Mahmoud et al. [36] showed that, after processing bentonite sludge clay suspension used as a sealing agent in the construction industry under moderate operating conditions ($T_{\text{wall}} = 90\ °\text{C}$ and $P_{\text{applied}} = 340$ kPa, followed by a final pressure step to 1650 kPa), the dry solids content reaches 81.43%. Under the same operating pressures but at ambient temperature, the dry solid content of the bentonite sludge would reach only 30%, insufficient for landfilling according to the regulation. For this specific application, the energy consumption of the TAMD process was estimated at 69.8 kWh m^{-3} [37, 121]. The effectiveness of the TAMD process for a specific biomass (alfalfa) dewatering was also investigated successfully [37]. They illustrated that the TAMD can be used to remove up to 83% of the inherent water from alfalfa under moderate processing conditions, the enhancement being less than 43% in a conventional mechanical dewatering process. The reduction of the liquid content in the alfalfa press cake, with an energy consumption of less than 150 kWh m^{-3}, leads to an energy saving of at least 35% on the overall separation chain. Unfortunately, this was not the case for wastewater sludge. From preliminary tests, it was found that although the benefits of this combination, the TAMD process appeared limited for wastewater sludge dewatering.

The TAMD technology developed by Arlabosse and her co-workers is slightly different from the mechanical thermal expression (MTE) process, first investigated in the mid-1990s by Strauss and his co-workers [24] and recently extended by Hoadley and his co-workers in Monash University [33, 34]. In its current design, the MTE comprises a preheating step prior to filtration and consolidation. The processing temperature is set above the normal boiling point of the water. To prevent evaporation, the process is held under a sufficient back pressure. Consequently, upon exposure to atmospheric pressure, a flash evaporation occurs and contributes to further moisture reduction. The MTE process has been successfully applied to low rank coals and brown coals as well as paper waste [116–118, 122, 123]. For an MTE plant with heat recovery operating at 200 °C, the moisture in the lignite could be reduced from 60 to 28 wt% with an estimated energy requirement of approximately 20% of that required in an evaporative process without heat

Process/Group

TREFLE (Bordeaux, France)

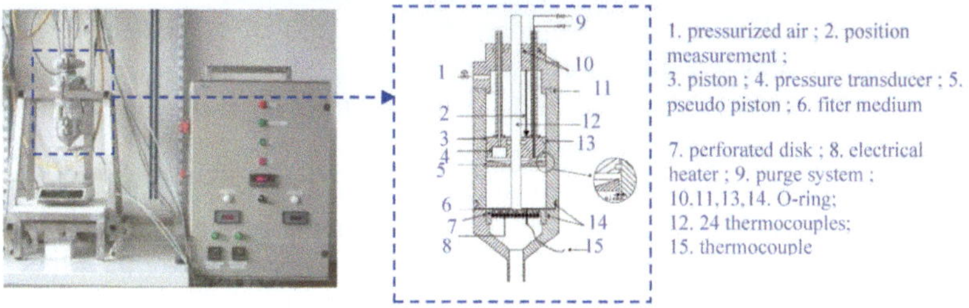

(Piston cell; Filtration + Thermal (20-200 °C; 155 kPa); and/or Compression + Thermal (20-200 °C; 550 kPa); (Talc, Kaolin, wastewater sludge). [26, 35, 129]

Process/Group

CRC for Clean Power from Lignite (Monash, Australia)

Mechanical Thermal Expression (MTE)

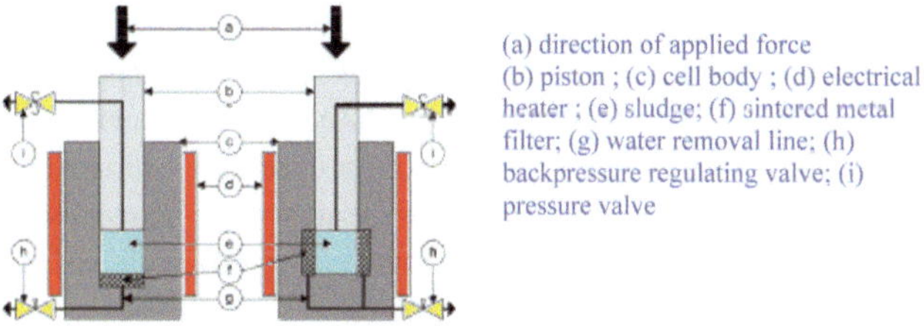

(Piston cell; (1) Filtration + Thermal (1500-24000 kPa; 20-200 °C); (2) Compression + Thermal then (3) Flash Evaporation. Back-pressure is maintained during filtration and compression stages to prevent evaporation. (Lignite, Biosolides, Bagasse). [33, 264]

Process/Group

Centre RAPSODEE (Albi, France)

Thermally Assisted Mechanical Dewatering (TAMD)

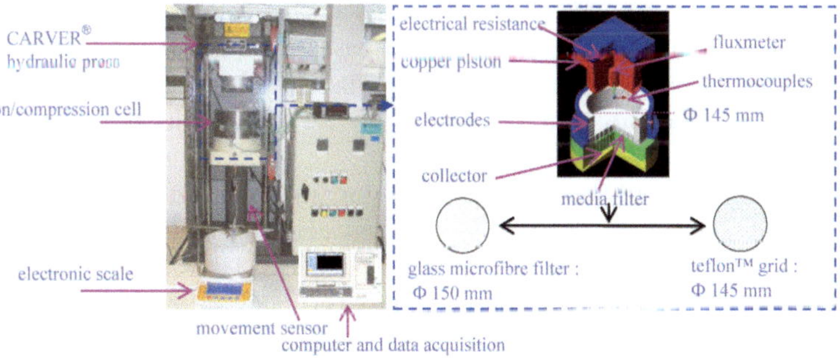

(Piston cell; Filtration/Compression; Filtration/Compression + Thermal;100-3000 kPa; 20-90 °C); (Bentonite, Biomass (alfalfa),..). [30, 32, 36, 37]

Fig. 9.5 Some selected examples of the research and development of the combination of mechanical and thermal effects recovery, i.e. 160 kWh m^{-3}. The effectiveness of MTE dewatering of sewage biosolids and bagasse was also investigated successfully [33].

For bagasse, MTE can remove approximately 55% of the inherent water at moderate processing conditions of 150 °C and 12 MPa. The main part of the moisture reduction was attributed to the collapse of the internal material porosity, resulting from the temperature increase.

Another enhancement is the so-called hot filter press or vapour pressure dewatering technique developed at the Technical University of Karlsruhe, Germany [124–128]. In this hybrid process, steam, which is produced within the filter cake, is used to displace the pore liquid from a heating plate to the filter medium. More recently, another enhancement mechanism has been developed at the Laboratoire TREFLE, Bordeaux, France [26]. The main idea of this study is to supply a heating flux from the filter medium during compression phase. It is characterised by an effect of "cake collapse" which reduces the average porosity of the final cake and consequently leads to a dryer sludge. This phenomenon is due to a vaporisation of pore liquid in the lower part of the cake. The mechanisms involved are various, and their coupling is complex. In the first approach, it seems that heat supply leads to a vaporisation of pore liquid at the interface of the filter medium. Due to density difference, the steam progresses within the cake and reaches colder layers where it condenses. This creates weak spaces which lead to micro-collapses. These micro-collapses then induce a macro-scale reduction of the cake volume [26].

The magnitude of the collapse depends on the material compressibility and on the cake average porosity and thickness. The effectiveness of this process for talc and kaolin dewatering was investigated successfully [26, 35, 129, 130]. They showed that the "cake collapse" provides a significant decrease of cake liquid content. The more compressible the material is, the more significant the collapse is. Unfortunately, this was not the case for flocculated biological sludge testing. The vaporisation of water leads to an increase of liquid pressure in the cake corresponding to a small increase thickness. The behaviour may be attributed to a layer of low permeability at the bottom of the cake which limits vapour transfer [129, 130].

A modification to the hot filter press or vapour pressure dewatering technique is the so-called steam pressure filtration process and recently steam pressure and centrifugation process developed at the Technical University of Karlsruhe, Germany [25, 131–134]. In this hybrid process, pure steam, coming from an external source, is used for dewatering of the filter cake. After steam breakthrough, the heated cake becomes dewatered by compressed air flowing through the cake, which accelerates drying and reduces steam consumption.

In more recent work, Lee [32] investigated the performance of the thermal dewatering (TDW) process for unconditioned wastewater sludge. In his experiments, heating plates were placed between membranes plates, which were heated by hot water supplied at the heat generator. When the hot water temperature of heating plate reaches 95 °C, the sludge is pumped into the filter chamber under a pressure of 490 kPa. When the cake formation is accomplished inside the chamber, the sludge supply is cut and then a hydraulic pressure is applied to the membrane plates at a pressure of about 1471 kPa. He showed that final water content of cake produced by MDW is 82.3% (wt%) and by the TDW is 66.2% (wt%). The energy consumed to treat the dry solids of 1 kg using TDW is 0.156 kWh/kg_{DS}. Therefore, energy consumption of TDW would be economically valid, and the TDW could increase the dry solids content of cake, compared with MDW, by 16.1%.

Other enhancements have been developed around the world, but these are not discussed in this chapter. The interested readers may refer to [134].

9.3.2 *Shear Force-Assisted Mechanical Dewatering*

The effect of shear force superimposed on the compressive force in a piston press has essentially been studied at laboratory scale [51–53]. A conventional filtration/compression cell was modified to allow the piston to rotate at a controlled speed. It should be noted that in belt filter presses these combined forces occur due to the differential speed between belts around the pressing rollers.

The shear force contributes to dewatering in the following way. The slowly rotating piston not only pushes the liquid through the filter medium to form a cake but also reduces the porosity of the cake by changing the mode of deposition of the particles in suspension (during cake formation) and altering the packing structure of the cake (during expression). This consequently decreases the specific cake resistance making the dewatering process more effective.

Wakeman and his co-workers [51, 52] investigated the shear force-assisted mechanical dewatering for materials such as mica, China clay, and anatase suspensions for the different processing conditions (20 °C, 300 kPa, 0–16 rpm). They showed that shearing increases significantly the dryness of the filter cake. These experimental results are in line with findings by [53] who investigated the additional effect of a shear force applied during compression phase on the dewatering course of mineral suspensions such as kaolin and talc for the different processing conditions (20 °C, 500 kPa, 10 rpm). However, Olivier [53] reported that the application of shear force-assisted mechanical dewatering for biological sludge did not show any beneficial effect to the dewatering results for the processing conditions (20 °C, 200–1200 kPa, 2–30 rpm). This behaviour has been attributed mainly to the fact that biological sludge dewatering is markedly influenced by a low permeability skin at the cake/medium interface, which is not perturbed by shear forces, because of cake thickness. Moreover, the weak rigidity of solids in this particular case also limits the reorientation of the particles [53].

Rather than using a rotating piston, another variant proposed by Gundogdu et al. oscillates the piston vertically to enhance dewatering of polyvinyl chloride particles [54–56]. They reported that the fluctuating vertical motion applied to the slurry produces a shear force and results in an increase in the solids volume fraction in the vicinity of the medium. This shear force is expansive, and therefore, it counteracts the compressive stress in the cake, which is normally present. Gradually increasing the vibration amplitude thus leads to a point at which the compressive intergranular force is entirely balanced by the oscillatory induced stress. At this point, the particle–liquid mixture in the vicinity of the medium becomes fluidised. Consequently, the specific cake resistance decreases leading to an increase in permeability and better dewatering.

9.3.3 *Acoustic Mechanical Dewatering*

Acoustic principles are very general term. These principles are discussed briefly, in the following section, to provide a basis for incorporating acoustic techniques into dewatering systems. Ultrasound is divided into three regions: power ultrasound (20–100 kHz), high-frequency ultrasound (100 kHz–1 MHz), and diagnostic ultrasound (1–500 MHz). Audible acoustic waves are in the range 20 Hz and 20 kHz. Ultrasound is a form of mechanical vibratory energy that propagates in material media as elastic or stress waves. The wave velocities, dispersion, absorption, reflection, scattering, and other interactions depend on frequency, mode, elastic, inertial, and molecular properties of the transmitting media. Acoustic waves can cause very high inertial and elastic forces at a solids–liquid interface due to large differences between the acoustic impedances of the two phases. These forces have the capability of reducing the effective viscosity and surface tension, which in turn eases diffusion and migration of the liquid from the particles and improves overall dewatering [86]. In fact, ultrasonic stresses produce a kind of sponge effect and facilitate the migration of moisture through natural channels or other channels created by wave propagation.

Application of acoustic energy to dewatering has been investigated by a number of researchers [45, 46, 50, 135–140]. They reported that the acoustic energy, in the form of ultrasonic vibrations, has been found to promote solids/liquid separation through various bulk and surface mechanisms.

Some of the ways in which ultrasonics can contribute to dewatering are listed below:

1. Attenuation, or absorption, of the acoustic energy leads to heating, which will cause some vaporisation, and lower the viscosity of the fluid.
2. Cavitation can generate free ions, erode or break, and generate local high temperatures and pressures.
3. The shear wave can cause a change in viscosity of non-Newtonian fluids.
4. Stresses can be induced in the filter cake which can cause the cake to fracture, opening new channels for fluid to escape.
5. The vibration motion will cause a net displacement or rectified diffusion of the liquid through the cake in the direction of any gradient, such as gravity, pressure, or electric field.
6. Filter surface will be kept clean, by increasing the particle size through agglomeration which helps to prevent clogging.

The primary mechanisms for ultrasonic agglomeration are (1) particle collision due to orthokinetic, radiation, oseen, and stokes forces, and (2) attraction between particles due to Bernoulli's effect [141]. Collision of particles helps form agglomerates, and Bernoulli's effect further enhances this process. It has been reported by Battelle Memorial Institute researchers [141] that ultrasonic agglomeration takes place at intensities typically below 1 W/cm^2; at higher intensities, ultrasonic dispersion takes place. Hence, it is important to carefully control the ultrasonic conditions to achieve the desired (agglomeration or dispersion) effect.

In addition to the agglomeration effects, ultrasonics can also create micro-bubbles within micropores, and capillaries can grow from these due to diffusion or the displacement of the trapped liquid. This water would normally be very difficult to remove, but ultrasonics makes it available as bulk water. Lastly, ultrasonics can reduce the apparent viscosity of a shear-thinning suspension such as wastewater sludge, and this can enhance the dewatering rate.

Optimisation of ultrasonics for a given sludge is critical as shown by two studies which provided negative dewatering results. Wang et al. [142] demonstrated that the sludge dewaterability diminished following the ultrasonic disintegration of waste-activated sludge. Both the specific resistance of filtration (SFR) and capillary suction time (CST) increased. Dewil et al. [101] observed that increasing the applied ultrasonic energy dosages decreased sludge dewaterability. Similarly, both CST and filtration time increased as the duration of the ultrasonic treatment was prolonged. The poorer dewatering behaviour was blamed on a considerably smaller floc size, offering an extended water-binding surface area.

The conclusions from these previous studies remain controversial, and thus further studies are necessary to assess whether ultrasonication ultimately improves sludge dewaterability and to establish the optimal dewatering conditions.

Very little work has been done specifically in acoustic mechanical dewatering of wastewater sludge. However, the work that has been reported in literature, related to the application of acoustic energy to assist liquid/solids filtration processes, illustrates to an extent the variety of ways in which dewatering may be enhanced acoustically. Acoustic energy is used in combination with filtration processes in which the primary function of acoustic energy is to cause the solids particles to move constantly, thus allowing free flow of fluid through the filter under vacuum.

Because the frequency, power intensity, feed properties, pressure, etc., are very important factors, several research teams have devoted considerable effort to studying the effect of these various parameters in acoustic-assisted filtration processes. For example, Tuori [143] studied the influence of the frequency on the filtration efficiency in gravity filtration and in vacuum filtration of bio-fibre sludge with 1% of dry solids content originating from the wastewater plant of a paper mill. It was found that the lowest frequency was the best in both filtration systems and the best filtration efficiencies were almost 20 times higher than the reference.

Vacuum filtration tests were also investigated by Riera-Franco de Sarabia et al. [45]. They filtered a TiO_2 suspension with 30% initial mass concentration and obtained a 6% increase in solids concentration of the cake by using electric power of 100 W applied to the transducer. The time of treatment was 10 s.

Gallego-Juárez et al. [144] applied power ultrasonic during cake deliquoring stage to improve conventional rotary vacuum filtration. Their experimental system consisted of one (or various) power transducers with an extensive area radiator which is placed parallel and very close to the filter surface in such a way that the radiating surface touches the cake during operation. The mechanical contact between the ultrasonic plate vibrator and the cake creates an acoustic coupling which favours the effective penetration of the ultrasonic energy.

The ultrasonic technique was tested with different acoustic powers to determine its performance. Based on the comparative results from TiO_2 slurries, the application of ultrasonic energy is shown to improve the solid concentration of the cake by around 6% with a relatively low power applied to the transducer (60 W) and a short treatment time (2.5 s). The initial mass concentration of particles and the mean size of particles were of 30% and 0.3 μm, respectively. By increasing the treatment time, the dewatering effect can be as much as doubled. Such result represents a substantial additional improvement on the conventional dewatering effect.

Swamy et al. [145] studied the effects of sound waves on the dewatering of granular materials with different particle sizes including calcium carbonate, magnesite, sand, and sawdust prior to drying. Acoustic energy from a stem-jet whistle was applied in combination with a basket centrifuge. Comparative results for centrifuge dewatering showed that a 25–95% reduction in the final moisture can be obtained depending upon the conditions of the superimposing sonic waves. They concluded that a similar process can be used in the case of filtration to produce a low moisture-containing cake.

9.3.4 Magnetic Mechanical Dewatering

The combination of two classical separation methods, cake filtration and magnetic field-driven separation, results in positive synergetic effects. Either a homogenous or inhomogeneous magnetic field can provide a torque that can align or anti-align magnetic particles. Thus, the systematic use of homogeneous or inhomogeneous magnetic fields results in magnetic volume forces and interparticle interactions that causes structural modification of the filter cake. The magnetic force causes a decoupling of solids phase motion from liquid phase flow and leads to an additional degree of freedom regarding the solid phase motion.

In general, there are two major effects [48, 49, 146, 147]: (1) in an inhomogeneous external magnetic field, the counterwise orientation of differential pressure and magnetic force leads to a slower cake formation; (2) interparticle magnetic forces lead to formation of chain-like agglomerates in the direction of the external magnetic field [148], as shown schematically in Fig. 9.6. The magnetic force can be oriented either in the same direction as the mechanical force (Fig. 9.6b) or in the opposite direction to the mechanical force (Fig. 9.6c). The permeability of a cake with this channelled structure is obviously higher than a cake consisting of smaller pores, broadly distributed in size. These effects together result in a strong

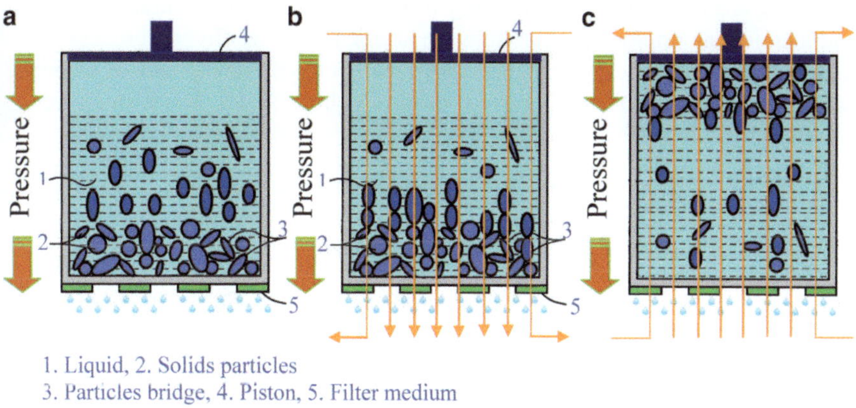

Fig. 9.6 Magnetic field enhanced sludge dewatering [48, 152]. (**a**) Classical mechanical dewatering. (**b**) Magnetic force in the same direction as the mechanical force. (**c**) Magnetic force in the opposite direction to the mechanical force

improvement of filtration kinetics due to a reduction of the filtration resistance. Moreover, Chaplin [149] noted that the magnetic radiation causes water to reorient thus breaking hydrogen bonds in water clusters causing the clusters to become smaller. The water subsequently becomes more reactive with other species, and its viscosity is reduced.

Some of the typical magnetic separation techniques are low-intensity magnetic separation (LIMS), high-intensity magnetic separation (HIMS), low-gradient magnetic separation (LGMS), high-gradient magnetic separation (HGMS), magnetic gravimetric separation (MGS), and magnetohydrostatic separation (MHS). Typical particle size and feed concentration range <40–4,000 μm and 5–20% (wt%) (LIMS or HIMS), and <400 μm and <10% (wt%) (HGMS) [150].

Not much work had been done in the past on the application of this combination to wastewater sludge. However, an interesting device is the "multi-force dewaterer" developed by Watson and Gardner [151]. It involves chemical flocculation in a magnetic field and the application of a vacuum to remove the water from waste pulps and steel plant sludge containing magnetic particles. A cross section of rare earth magnet has an arc of 120° designed and built by Eriez Magnetics of Erie, PA, USA. It was fitted within the vacuum drum filter to produce a radial magnetic field extending into the filter tank. In this manner, magnetic particles would be attracted to the drum surface by both the vacuum force and by the magnetic force.

Finally, from literature, it was learned that a new magnetic field-enhanced cake filtration has been developed at the University of Karlsruhe [48, 49, 152, 153]. In their hybrid process "magnetic field-enhanced press filtration" of ferromagnetic iron oxide particles (Fe_3O_4) in a lab-scale filter press, the kinetics of dewatering is significantly influenced by magnetophoresis and magnetic self-assembly in the filter cake. Magnetophoresis hereby decreases the rate of cake formation, and magnetic self-assembly increases the permeability of the cake by structural changes. The result is a big increase of the overall filtrate mass flow and therefore an improvement of filtration kinetics. This new hybrid process offers high potential in the field of magnetic bio-separation with functionalised magnetic particles. However, it may also be of future interest for wastewater sludge dewatering.

9.3.5 Microwave Mechanical Dewatering

According to our knowledge, no work had been done in the past on the combination of microwave and mechanical dewatering, and this hybrid process may offer opportunities for enhanced dewatering as well. Therefore, for a comprehensive coverage, it seems appropriate to present the basic principles and potential advantages of using imposed microwave force as a physical pretreatment method prior to dewatering in this section.

Microwaves are electromagnetic wave whose frequency ranges lie between that of radiowaves and infrared radiation. Domestic and industrial ovens generally operate at a frequency of 2.4 GHz corresponding to a wavelength of 12.2 cm and energy of 1.02 10^{-5} eV [154]. However, not all materials can be heated rapidly by microwaves. Only material capable of absorbing microwave energy, termed dielectrics, are heated by microwaves. Thus, microwave heating is also referred to as dielectric heating.

Microwave heating is a volumetric heating that, unlike the conventional heating method, does not depend on heat transfer during the heating process. Theoretically, the microwave heating has much higher heat efficiency than conventional heating, because the former does not have any losses associated with heat transfer. Moreover, microwave heating is more rapid as heat transfer occurs throughout the sludge rather than at the surface. Indeed, it is generally admitted that the microwave heating is a clean source of heat energy, because there are no secondary waste streams generated. It is documented that the microwave irradiation is considered to be an efficient tool for improving sludge dewaterability because of its "thermal" and "athermal" effects [57, 100, 108, 109, 155]. The ways in which microwave can contribute to dewatering are described below. Though the quantum energy of microwave radiation (1.02 10^{-5} eV) is too low to break the chemical bonds, still some structures can be altered by the energies carried by microwaves. These include for instance hydrogen-bonded structures in which very low energy intensities may cause displacement of protons [156].

Due to high water content, sewage sludge absorbs microwave radiation. The sludge is a multiphase medium, containing water, mineral and organic substances, proteins, and cells of microorganisms. Microwave irradiation causes kinetic energy of water dipoles to increase and reach the boiling point value [157–159]. Then the cells of microorganisms burst, and the bound water is released. The ability of microwave radiation to interact with cells of microorganisms has been already proven. Microwave radiation has been used for inactivation of pathogens in food industry [160]. The investigations of NiederwÖhrmeier et al. [161] proved that microwave radiation (2.4 GHz) substantially reduced the number of *Escherichia coli*, *Salmonella senftenberg*, *Erysipelothrix rhusiopathiae*, and *Yersinia enterocolitica* in water and liquid manure. Woo et al. [162] reported inactivation of *Escherichia coli* and *Bacillus subtilis* colonies exposed to microwave radiation. Recently, Hong et al. [163] reported that microwave irradiation of biosolids sludge appeared to be a viable and economical method of destructing pathogens. They observed that biosolids sludge irradiated with microwaves showed less faecal coliforms and *Salmonella spp*. densities.

Generally, the use of electricity produced from fossil fuel requires about three times as much energy as fossil fuel which is used to heat drying air directly. Thus, to be competitive, microwave drying using electricity needs to provide about three times as efficient, or the increased cost needs to be justified on the basis of increased value of the product, safety of the operation or protection of the environment.

Very few studies are reported in the scientific literature, which compare the performance and the power consumption of microwave dewatering of wastewater sludge. An interesting exception is the work by Collins et al. [57] and Fu et al. [164] who evaluated the feasibility, efficiency and cost-effectiveness of using the heat generated with microwave irradiation for drying waste-activated sludge and municipal biological sludge, respectively. Collins et al. [57] concluded that the microwave sludge dewatering, in the traditional dewatering range from about 5 to 30% solids content, involves a large electricity cost, which greatly exceeds the total cost of other conventional processes. They showed that the water removal by microwave heating in the traditional sludge drying range, from the current limits of physical-–chemical processes (30–35%) to the optima for disposal, may be feasible and might also offer other advantages. On the other hand, in their experimental results, at lab-scale, Fu et al. [164] showed that the microwave sludge drying method has its advantages. They demonstrated that, comparing to hot air sludge drying, the microwave heated drying to reduce the sludge water content from 75 to below 40% may be cost-effective, as shown in Fig. 9.7. The treatment cost can further be reduced if the steam produced during the drying process is reused, and the dried sludge is used as an agricultural resource.

9.3.6 Electrical Mechanical Dewatering

As the fine-particle suspensions possess a surface charge, usually negative, they are surrounded by a layer with a higher density of positive charges, the electric double layer. When an electric field is applied, the usually negative charged particles move towards the electrode of the opposite charge. The water, commonly with cations, is driven towards the negative electrode. Electro-dewatering thus involves the well-known phenomena of electrophoresis, electro-osmosis, and electromigration, as shown in Fig. 9.8. Figure 9.8 represents the process in which suspended solids move towards the filter media to form a cake when mechanical forces are applied.

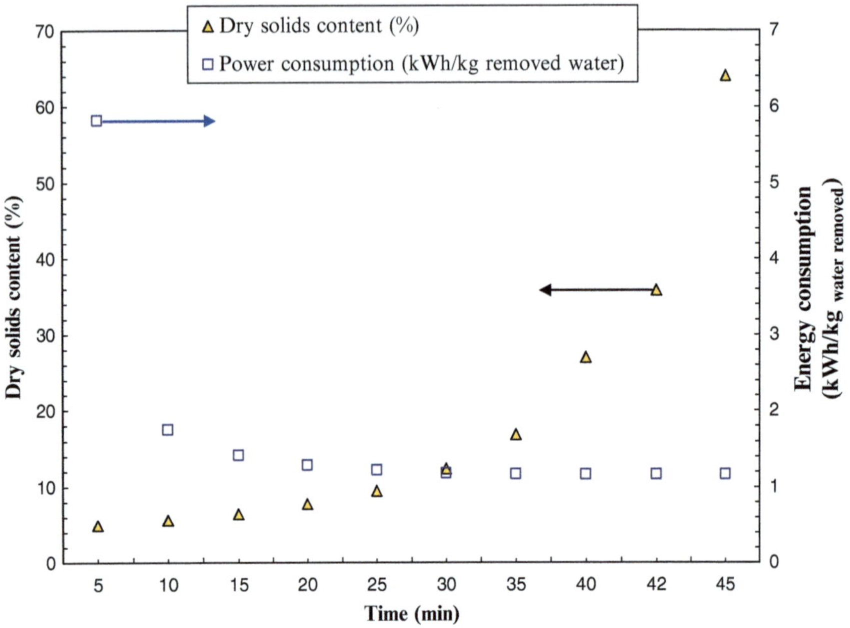

Fig. 9.7 Dry solids content and energy consumption for drying using microwave heating [164]. Experimental conditions were: 800 g of sludge with 4.9% initial solids content; the input and output power were 1,200 and 750 W, respectively

It also depicts the movements of suspended solids in electro-dewatering where an electric field is applied.

Although electromechanical dewatering is still not fully understood, the mechanical and electrical effects, resulting in the reduction of water content, are usually thought to comprise the following steps [165]:

1. The mechanical dewatering reduces the volume of the pores and squeezes the water out of the product.
2. The charged particles are still free to move in the fluid suspension. They tend to migrate towards the electrode carrying the opposite charge.
3. When the cake has formed, the particles are locked in position and hence unable to move. Water-containing ions are transported by viscous and/or molecular interactions and/or as a hydration sheath.
4. Electrochemical reactions at the electrodes are essential to restore charge equilibrium that enables electro-dewatering to proceed beyond the transient stage.
5. Finally, water ceases to be the continuous phase in the cake, and the electrical resistance rises leading to ohmic heating. The resulting reduction in viscosity facilitates the removal of some of the remaining water.

Electro-dewatering has been extensively studied in recent years for its proven advantages in terms of high efficiency, low energy consumption and being able to avoid the blockage of the filter media, which is a common problem in

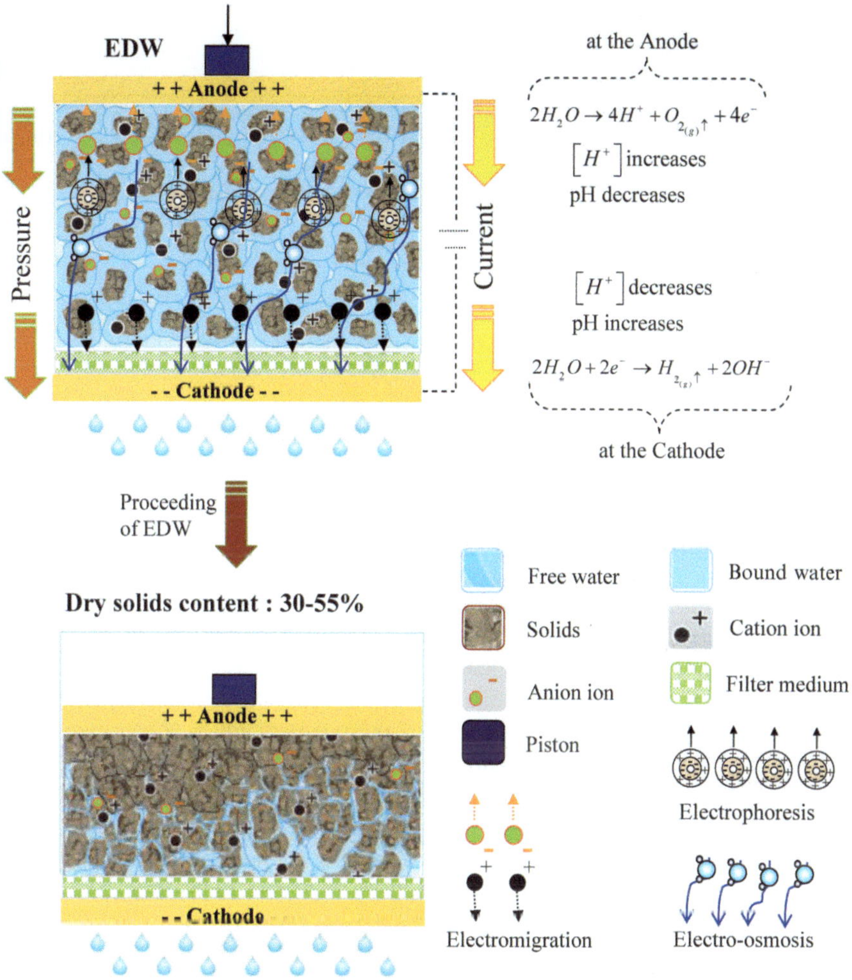

Fig. 9.8 Schematic representation of electro-dewatering with the different mechanisms occurring when negatively charged particles are placed in an electrical field [11, 265]

filtration [166]. Most electro-dewatering techniques developed to date use a process known as pressurised electro-dewatering. This process involves a vertical electric field, which is usually applied in conjunction with vacuum filtration [167–173] or an imposed mechanical pressure acting on the upper anode [166, 169, 174–183]. The combination of the electric field and mechanical compression is complementary and leads to efficient dewatering [165, 169, 178, 184–193]. In typical electric field-assisted dewatering operations, the applied voltages are in the range of a few volts to several tens of volts DC or slowly alternating AC. The electro-dewatering system is essentially an electrochemical cell. The application of an electric field produces electrochemical reactions which affect the performance of the electro-dewatering process. The possible cathode and anode reactions are:

At the cathode,

$$2H_2O + 2e^- \rightarrow H_{2(g)}\uparrow + 2OH^- \qquad E_0 = -0.83\,V \tag{9.1}$$

$$2H_3O^+ + 2e^- \rightarrow 2H_2O + H_{2(g)} \qquad E_0 = 0\,V \tag{9.2}$$

$$M^{n+} + ne^- \rightarrow M \tag{9.3}$$

where M^{n+} represents any kind of cationic species that can be reduced. E_0 is the standard electrode potential at a temperature of 298 K.

At the anode,

$$M \rightarrow M^{n+} + ne^- \tag{9.4}$$

$$2H_2O \rightarrow O_{2(g)}\uparrow + 4H^+ + 4e^- \qquad E_0 = 1.23\,V \tag{9.5}$$

It is clear that these electrode reactions would be affected not only by the materials of the electrode but also the ions in the electrolyte. As a result, the pH near the cathode increases, while the pH near the anode decreases (Fig. 9.8). Electrochemical reactions play a significant role in affecting the electro-dewatering process [11, 194–196]. The gas (oxygen and hydrogen) evolution at both electrodes leads to the appearance of void spaces within the bed and increases the electrical resistance of the system. This may produce an electrically insulating layer. On the other hand, it may lead to an explosive mixture, causing a potential safety risk. Fortunately, this risk can easily be reduced by simple ventilation measures [197]. It is sometimes believed that the generated gas may cause an additional driving force for liquid expulsion in piston-type dewatering devices causing the cake to be only partially saturated [198].

Utilisation of an electrical field has received great attention from researchers around the world. The application of an electrical field during mechanical dewatering was significantly studied in the case of wastewater sludge treatment. It is discussed in greater detail in the following sections. The effects of the electrical field applications and the electrode arrangements and configurations on dewatering process are shown, and the usage of the EDW is focused to wastewater sludge.

9.3.6.1 Design and Operational Considerations

Electro-dewatering techniques have been used in conjunction with different common dewatering devices, such as belt filter presses [182, 199–201], and filter presses or diaphragm filter presses [175, 192, 202], both on laboratory and pilot or industrial. However, technical problems have hampered its widespread application to date. These problems include the requirement for corrosion-resistant electrode materials and high-power consumption [182]. Process parameters such as the timing of the electric field application, total energy input and energy input distribution in time will affect the dewatering results and overall energy efficiency. Different modes of operation are possible, such as keeping the voltage, current or electric field constant or allowing it to vary according to a certain pattern, such as a sine or block wave. Short to intermediate interruptions to the electric field have also been reported to enhance the process efficiency. Also, salts, flocculants, suspension types, conductivity, various electrodes materials, electrode surface area, etc., have been studied [166, 172, 180, 191–193, 203–209]. Most of these studies are empirical and carried out at lab-scale devices. The equipment used for most electro-dewatering studies in lab-scale comprised a number of test cells, columns and tanks of circular and rectangular shape and a variety of sizes. A number of these electro-dewatering cells, with different horizontal, vertical, coaxial, or other electrode arrangements, have been engineered by researchers in the field and have reached varying degree of development.

The purpose of this section is not to provide an exhaustive listing of all the devices developed but to provide a brief description of a cross section of these devices, so as to give the reader a flavour of the variety in cell design. Electro-dewatering devices can be classified into four broad categories:

1. Vertical device for electrically enhanced vacuum or pressure dewatering
2. Vertical device for electrically enhanced compression dewatering stage
3. Vertical multistage electro-dewatering device
4. Horizontal electro-dewatering device

Direct electric field, D.C., has been usually used for electro-dewatering. However, the electrical resistance of the cake formation at the anode is increased considerably with the process of dewatering. Such a circumstance brings to an end effective electro-dewatering. A number of technical solutions have been proposed to overcome this limitation, and the following survey provides a partial summary of some of these methods.

The application of an A.C. electric field with periodic reversals of the electrode polarity has been tested by Yoshida et al. [207], as one means of overcoming the electrical resistance of the dewatered filter cake at the electrodes. Their experiments for white clay electro-dewatering were carried out in the region of very low frequency (such as 0.01 and 0.001 Hz) for the electrode polarity reversals of A.C. with constant voltage (20 and 40 V), and both rectangular and sine waves were used. They concluded that the efficiency of electric power consumption for the same amount of water removed under A.C. was lower than that under D.C., except near the end of dewatering. In addition, the excessive increase of the electrical contact resistance between the electrode and sludge bed was reduced because the water in the bed under A.C. moved periodically in both directions. Moreover, the changes of pH and water content distributions within the bed were reduced, thus the dewatering process could proceed more effectively.

Rabie et al. [206] demonstrated that electro-dewatering of clay (e.g. bentonite) using interrupted power can remove more water for the same energy consumption than electro-dewatering using continuous power. Their results showed that the interrupted dewatering process provided a 20% improvement in water removal compared with the electro-dewatering using continuous direct current. Moreover, the metal ions contained in sludge can be precipitated as insoluble metal hydroxide on the alkaline side (cathode side), and this precipitates on the cathode surface, and filter media is detrimental to electro-dewatering. This effect can be minimised by reversing the polarity of the electrodes for a few minutes which is long enough for the acid filtrate to dissolve and remove the precipitate [175, 176]. However, depending on the type of sludge, the reversal of polarity may not dissolve the precipitate on the filter media. In particular, the sludge containing a large amount of calcium can blind the filter media in a short time [210].

Lightfoot and Raghavan [189, 211], from their studies on electrically enhanced compression of seaweed (*Nereocystis luetkeana*), observed that the energy costs for producing dried kelp meal were significantly lower if dewatering precedes thermal drying. Electro-dewatering was found to result in significant energy savings over compression-only dewatering if dewatering is followed by drying to produce the finished product. In their laboratory electro-dewatering equipment, they tested currents of 72.4, 146.9 and 219 A/m^2 and pressures of 7.46, 34.6 and 104 kPa.

Friehmelt and Gidarakos [212] reported that the essential features of the screw press are the concentric arrangement of electrodes and the continuous mode of operation with a spiral conveyor. In their experiments, the filter cell was composed of two concentric, cylindrical strainer baskets. Both strainer baskets were made of stainless steel from triangular rods, with a slot width of 0.10 mm to enable the filtrate to flow off. The outer basket was fixed, with the spiral conveyor mounted on the rotating inside basket. The sludge was transported between the inside and outside basket through a 20 mm gap. By adjusting the speed of the drive motor connected to the inside basket, the throughput could be varied between 2 and 30 kg/h. The mechanically pre-dewatered sludge was pumped into the apparatus using a spiral pump which pressurises to 50–150 kPa. Both baskets were electrically insulated. The outer basket was directly connected to the DC power supply. Sliding contacts applied voltage to the rotating inside basket. Filtrate is collected from both baskets separately. At a current of 20 A, the current density at the inside basket is 11.7 mA/cm^2 and at the outside basket 8.1 mA/cm^2. In many tests for wastewater sludge electro-dewatering, they showed that a solids content starting in a range of 20–30% could be increased by up to 37%.

Recently, Watanabe et al. [213] proposed a new method of electro-dewatering in which a third electrode called the "gate" electrode is placed between the anode and the cathode in order to simulate a field effect transistor (FET). In their experiments, the so-called "gate" electrode (connected to the positive of the power supply, with negative pole grounded) is in fact a sort of second anode. By keeping the voltage constant between the main anode and cathode (10 V D.C.), they applied an additional voltage at gate electrode ranging from -15 to $+$ 15 V in order to control the current. They concluded that when the gate voltage is positive, the dewatering is enhanced. On the other hand, when the gate electrode is negative, it works in the opposite direction and leads to a reduction compared with the initial dewatering voltage of 10 V between the main anode and the cathode.

Vijh [214] investigated also the concept of the gate electrode and pointed out that this system is not much different from the multistage dewatering method proposed by Yoshida [166] which involved three perforated upper electrodes vertically arranged at regular intervals within the sludge bed. The "gate" electrode method may be a promising new avenue for electro-dewatering under certain experimental conditions. Other aspects that have been examined by Weber and Stahl [215] are the locations and the new electrode arrangement in a pilot filter press. Their experiments for quartz sand SF800 were carried out in two different electrode arrangements: (1) outer arrangement, in which the electrodes are situated outside the filtration chamber behind the filter clothes, and (2) middle electrode, in which a third electrode (anode) is inserted between the filter plates and the cathodes are placed behind the filter clothes. They observed that the overall filtration kinetics was accelerated remarkably by the use of the outer electrodes compared to mere pressure dewatering. On the other hand, the use of the middle electrodes results in a further acceleration. As a result of this operation mode, the anode was embedded in the cake, and the cake needed to be scraped off after the dewatering run.

Focusing on the gradual discontinuity of the liquid state in the dewatering material, due to the formation of the unsaturated layer at the anode, which is caused by the electrical blockage rendering electro-dewatering no longer effective, Ho and Chen [216] proposed a mixing effect by having the anode rotating. They designed an experimental apparatus with an anode able to rotate at rates up to 300 rpm. Bentonite slurry, with 9.1 wt% initial solids content, was used as the test model substance. The experimental results showed that, by using a flat plate as the rotating anode, the water removed from the slurry increased significantly at the beginning with the rotational speed and then reached a plateau at 240 rpm. This was believed to be due to the solid cake which collects at the anode being removed by the rotation. The solid content of the cake could be 23.0 wt%, 60% better than the performance of normal electro-dewatering. Besides, the excessive suspension added on the top of the anode was found to have a beneficial effect for both water removal and the suppression of dry (unsaturated) solid at the anode. It was also found that a constant thickness of the cake formed (about 5 mm), and this was more efficient than in electro-osmotic dewatering. The combined effects of anode rotation and dry cake thickness control could remove 56% of the initial moisture within the suspension, and this was 37% better than normal electro-dewatering.

Zhou et al. [217] investigated a new electro-dewatering apparatus characterised by the application of a horizontal electric field in order to facilitate the dissipation of gases produced at the electrodes and to keep the anode soaked in water during the dewatering process. They found that it was particularly applicable to the removal of water from activated sludge. Compared to the conventional electro-dewatering techniques operated in a vertical electric field, the horizontal electro-dewatering techniques have advantages in terms of simplicity in construction, high efficiency and ease of operation. These make the electro-dewatering in a horizontal electric field promising for large-scale applications. These experimental observations are in line with findings by Larue et al. [180] who investigated, at lab-scale filter press, the electro-dewatering of a highly conductive silica suspension under constant pressure and constant electric field.

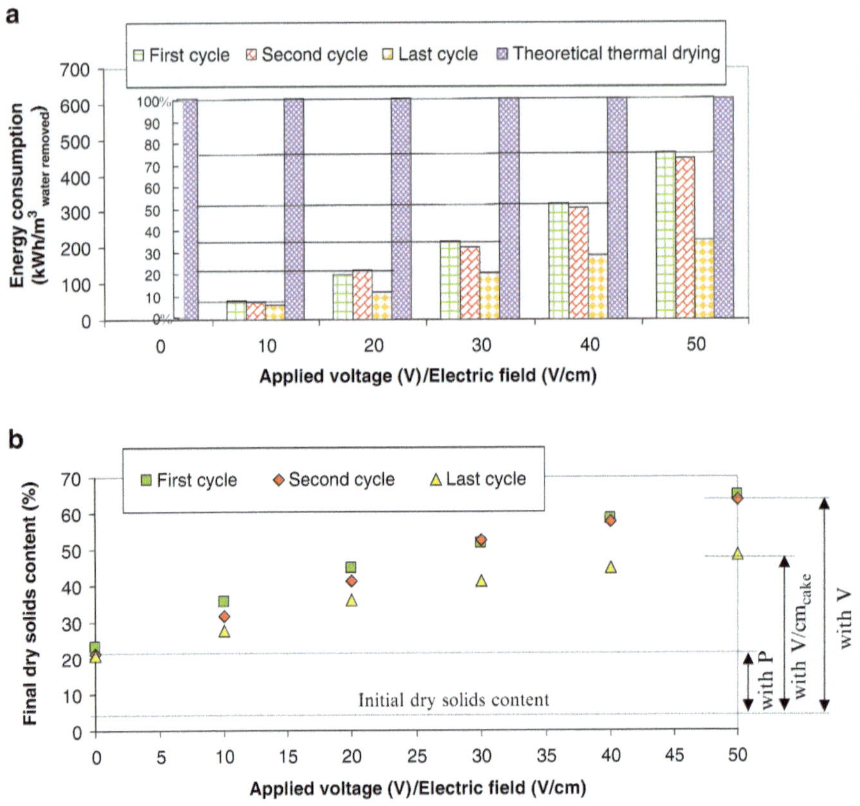

Fig. 9.9 Energy consumption (**a**) and final dry solids content (**b**) of waste activated sludge as functions of voltage or electric field for three electric cycles for electro-dewatering [192]

Saveyn et al. [192] undertook a bench-scale evaluation study of sludge electro-dewatering in which the electric field is applied after the onset of the compression phase. In their experiments, three electric regimes (cycles) were tested to investigate the best conditions to electro-dewater activated sludge. The first cycle consisted of applying a constant voltage over the electrodes once the 700 kPa pressure was set. The second cycle consisted of applying a constant voltage on electrodes once the full dewatering cycle at 1600 kPa was executed. The last cycle consisted of applying a constant electric field, and it was selected once the 700 kPa pressure was achieved. Figure 9.9, adapted from [192], gives an overview of the different specific electric energy consumptions based on water removal, calculated as the difference between the amount of water removed by pressure-only dewatering and the amount of water removed by the electro-dewatering process. As shown in Fig. 9.9, it was found that high voltage/or electric field resulted in the best dewatered cakes. It was also found that the electro-dewatering technique requires much less energy than thermal drying techniques. For example, this technique is much more energy efficient requiring less than 10 and 25% of the theoretical thermal drying energy for the low and moderate voltages/or electric fields cases, respectively.

Yoshida and Okada [218] replaced the upper electrode with several rod-type electrodes which are inserted into the sludge bed. Using rod-type electrodes, the total area of the upper electrodes which are in contact with the sludge becomes smaller than with the plate-type electrode used ordinarily. In their experiments on $CaCO_3$ sludge, the influence of such electric field application as decreasing one-sided area of electrodes on the dewatering process was investigated in terms of total amount of water removed and electric power consumption. Half of the cross-sectional area of the sludge bed was employed for the upper electrode area, and the electric field application was operated by the constant electric current condition. Then the experimental results were compared with the upper electrode with the same area as the cross-sectional area of the bed. They demonstrated that the increase in electrical contact resistance, in the half area electrode compared with the same area electrode, can be reduced to some extent, and as a consequence, an increase in both the total amount of water removed and the electric power efficiency can be expected.

Glendinning et al. [3] used polymer-based materials containing conducting elements. These have been used to minimise the problem of electrode corrosion and create a sludge treatment system that can produce dry solids contents in excess of 30%. By encasing the metallic filaments in a relatively inert material, electrode corrosion is effectively managed or eliminated because the oxidation reactions at the anode are very much reduced and designed so that the anode remains effective over its intended design life. In their experiments for humic sludge electro-dewatering, electrodes were cut from woven polyester material and electrified with carbon fibre strips at spacing of 5, 10, and 20 mm. In this way, these electrodes had very similar electrical and filtration properties to those that would be used in an operating belt press. They used two separate voltages of 15 and 30 V applied to a 15-mm-thick sample with a back pressure of 70 kPa for humic sludge. The tests lasted for 20 min. They found that although the best improvements in solids content were obtained when the conducting elements were spaced at 5 mm and a potential of 30 V, dewatering performance was more sensitive to applied voltage than to element spacing.

De-Gang and Shu-Ting [219] proposed to use porous material to adsorb the moisture near the cathode instead of gravity separation. In their studies, the adsorptive material was soft and bibulous and easy to regenerate by pressure. It was found that, due to adsorptive material applied in EDW process, much more water from the sludge can be easily separated and removed; meanwhile, the current through sludge is depressed lowering the overall power consumption. Their experiments showed that when voltages of 10, 20, and 30 V were applied to decrease the water content of a mixture of sludge (primary and secondary) from 77 to 70% w/w in EDW using adsorptive material, the power consumption was, respectively, 0.025, 0.041, and 0.053 $kWh/kg_{removed\ water}$, and the processing times were 4.5, 1.5, and 0.8 min, respectively.

Yu et al. [220] investigated the influence of filter cloths on the cathode on the electro-osmotic dewatering performance. They reported that thicker filter cloth on the cathode electrode leads to lower water removal in the sludge cake. In their studies, the influence of filter cloth on the cathode on the dewatering efficiency was explained by analysing the intrinsic resistance of filter cloth/stainless steel net and the ratio of electric resistance of filter cloth to the total resistance. They observed that the application of cathode electrode net with small pore size without filter cloth on the cathode electrode remarkably improved the dewatering efficiency and reduced the electric energy consumption in the electro-osmotic dewatering process. The average electric energy consumption increased with the increase of electric potential gradient applied. Without filter cloth on the cathode at the potential gradient of 16, 20, and 24 V/cm, the average energy consumption was 0.043, 0.058, and 0.075 kWh/kg removed water, respectively. They concluded that it was more favourable to remove water at lower electric potential gradient and without a filter cloth on the cathode electrode.

Finally, the application of an electric field seems to be an interesting technique to enhance wastewater sludge. However, further work is needed to determine the optimal operating conditions and to select the technology best suited to wastewater sludge. At the present time, it is difficult to compare the electro-dewatering results from different electro-dewatering devices when these results a specific to different sludge systems. Very few pilot or full-scale comparative studies are reported in the scientific literature, which compare the performance of different electro-dewatering techniques on the same material. The next section is devoted to electro-dewatering studies of wastewater sludge.

9.3.6.2 Electro-Dewatering of Wastewater Sludge

Dewatering of wastewater sludge is a difficult process. The difficulty has been attributed mainly to the fact that the particles are very fine, colloidal in nature, and possess a gel-like structure due to polymeric flocculation. In order to tackle the limitations in wastewater sludge dewatering, new technologies have been developed in recent years. Some technologies such as wastewater sludge digestion, wastewater sludge mineralisation, or peroxidation are allowed to reduce the amount of wastewater sludge to be dewatered, or the dewaterability of the sludge, by changing the biochemical composition [69]. Nevertheless, wastewater sludge remains hard to dewater, and therefore, an improvement in the traditional pressurised dewatering equipment is desirable. Different options have been investigated to enhance wastewater sludge dewatering, and one of the most researched used is the pressure dewatering assisted by an electrical field (D.C. or A.C.).

For more than 20 years, laboratory and pilot studies are reported on electric field-assisted dewatering of wastewater sludge using non-pressurised electro-dewatering, pressurised (vacuum or pressure) electro-dewatering, and compression electro-dewatering. The non-pressurised electro-dewatering was applied on its own (1) to aerobic or anaerobic wastewater sludge [183, 217, 221–223] and (2) to a composted wastewater sludge [224]. Other applications are pre-dewatered sludge, in which some water was first removed by mechanical dewatering [225–228]. Results showed that the final dry solids content of the sludge can reach more than 40%, but a very long application time of the electric field (4 h to 4 days) is required. In comparison, the dry solids content reached after 1 min, using a gravity belt thickener [229], is the same as the solids content after 1 h of the electric field application obtained by Zhou et al. [217].

However, the literature summarised in Table 9.2 shows that kinetics of both pressurised and compression electro-dewatering are much better than those of non-pressurised electro-dewatering. Besides, the final dry solid content obtained from both pressurised and compression electro-dewatering is significantly higher than those obtained by conventional mechanical dewatering.

It was also highlighted that increasing the pressure was beneficial to the electro-dewatering performances [4, 175, 178, 181, 202, 230]. Not only is more liquid removed by the higher pressure, but it seems that also the electrical contact is improved and thus the energy efficiency. In addition, an increase of the voltage (or the current density) leads to a dryer sludge cake [4, 165, 166, 175, 178, 181, 192, 199–202, 230, 231] and better dewatering kinetics. The performances of electro-dewatering are really interesting if the additionally power consumption remains significantly low compared to thermal processes. The minimal drying energy requirement refers to water vaporisation enthalpy that is about 0.617 kWh/kg [232]. In an industrial device, it can reach as high as 1.2 kWh/kg [11, 178, 233].

However, concerning electro-dewatering, the data on power consumption reported in the literature is difficult to compare and analyse (Table 9.2). Several authors employed the additionally electrical energy consumed per unit mass of the dry solids (kWh/kg_{DS}), and therefore, it is difficult to compare with the data reported for a drying process. Moreover, the electro-dewatering energy expressed as kWh/kg_{DS} strongly depends on the initial cake dryness which is different in each study.

On the other hand, authors analysed the electro-dewatering efficiency in terms of the dewatering time and power consumption at the target (desired) cake dryness. One of the best ways to compare the power consumption is to calculate the energy per additional mass of water removed (in comparison to dewatering without an electrical field). Using this datum [178, 226] and [192] clearly showed that electro-dewatering energy consumption is lower than drying energy consumption for municipal wastewater sludge.

In recent years, several researchers have investigated the relationship between the energy consumption and the dry solids content, or the amount of filtrate additionally removed during the electro-dewatering run. For example, a linear relationship between filtrate yield and energy consumption was obtained by Laursen and Jensen [225]. They reported that the energy consumption was proportional to the cake thickness and the current density. Miller et al. [202] reported an exponential relationship between the dry solids content and the energy consumption expressed in kWh/kg_{DS}. Indeed, Miller et al. [202] and Gingerich et al. [230] demonstrated that the energy consumption is not uniform during the

Table 9.2 Overview of electro-dewatering studies related to wastewater sludge

Reference	Exp. device	Sludge	Pressure (kPa)	Operation mode	[a]Gain (%)	Time	Power consumption
[175]	Filter-press pilot	P	25–400	0–50 V	20 V 4–7 50 V 26-38	N.D.	N.D.
		A, AnD	–	–	–	–	0.72–0.83[b] (S > 40%)
[176]	Filter-press pilot	A	390	40 V	20.7	10 min (0V) then 40 min	0.40[b] (S = 30%) 0.52[b] (S = 35%)
[166]	Lab piston cell	A	122–1906	0–50 V	20 V 4–6 50 V 25–27	–	–
[178]	Lab piston cell	AnD	100–600	0–20 V/cm	20 V/cm 3.9–7	2 h (0V) then 30 min	–
		N.D.	400	15 V/cm	4.4–13	2 h (0V) then 30 min	0.06–0.2[c]
[199]	Belt filter press	A	–	0–20 V	20 V 5.2–5.9	4 min of gravity drainage then 10 min	0.31–0.6[d]
		AnD	–	0–20 V	20 V 8.1–11.3		0.33–1.22[d]
[234]	Lab piston cell	N.D.	25–100	0–200 A/m^2	13–27	0–20 min	–
[212]	Continuous spiral conveyor	–	50–150	30–150 A/m^2	6	35–40 min	0.1–0.4[c]
[231]	Lab piston cell	A	1000	100 A/m^2	25.4	20–30 min	1.80[b] (S ≃ 40%)
[202]	Lab piston cell	A	300–1000	100 A/m^2	30–39	60 min	2.85–2.9[b] (S = 40%)
		AnD			11–26		1.57–3.02[b] (S = 40%)
[165]	Lab piston cell	A	300	100 A/m^2	22	30 min	1.47[b] (S = 40%)
[230]	Lab piston cell	A	52	0–60 V	20 V 16.7 60 V 36	5 min of gravity drainage then 20 min	– 1.66[b] (S = 40%)
		AnD	52–214	0–60 V	20 V 6.9–16.9 60 V 26.7–40.8		– 4[b] (S = 40%)
[200]	Belt filter press	A	–	0–25 V	20 V 3.5	N.D.	–
		AnD	–	20 V (progressive)	4–5	N.D.	–
[181]	Lab piston cell	N.D.	98–392	0–120 V/cm	80 V/cm 15–16	15 min	1.47[b] (S = 38%)
[182]	Lab piston cell	AnD	250	30 V	29	17 min	–
[201]	Belt filter press	A	200–300	0–45 A/m^2	40 A/m^2 18–19	–	–
		AnD			43 A/m^2 14-20	8–9 min	–

[192]	Lab piston cell	A	200 (28 min) then 700 (40 min) then 1600	0–50 V when P≥700	20 V 21.5	Until the end of process	0.13[c]
					50 V 41.5		0.46[c]
				0-50 V at the end of compression	20 V 19.9		0.14[c]
					50 V 42.2		0.45[c]
				0–50 V/cm when P≥700	20 V/cm 15.1		0.08[c]
					50 V/cm 27.4		0.22[c]
	Diaphragm filter press pilot	A	700 (45 min) then 1600 (30 min)	N.D.	18.6–20.8	45 min (0V) then 30 min	0.22–0.38[c] (S > 40%)
			700 (45 min) then 1500 (50 min)	N.D.	19.7–33.4	45 min (0V) then 30 min	0.31–0.36[c] (S > 40%)
[219]	Lab piston cell	P+A	6.4	10–30	10 V 5.4	7 min	0.03[d]
					20 V 9.9	3 min	0.07[d]
					30 V 14	2 min	0.13[d]
[220]	Lab piston cell	A	7	10–24 V/cm	10 V/cm 14.8	5 min	0.04[d]
					20 V/cm 16		0.06[d]
					30 V/cm 19		0.08[d]
[4]	Lab piston cell	Humic sludge	25	10–15 V	2.3–3.8	60 min	0.13–0.20[c]
			25	1–2 A	3.3–3.7		0.18–0.36[c]
			50		3.8–12.4		0.10–0.17[c]
			75		5–13.7		0.08–0.16[c]
[193]	Lab piston cell	A	200–1200	10–50 V	400 kPa 10 V 6.1	120 min (0V) thin 120 min with (V)	0.03–0.06[c]
					20 V 22.0		0.12–0.16[c]
					30 V 29.6		0.16–0.29[c]
					40 V 42.7		0.26–0.40[c]
					50 V 49.6		0.33–0.45[c]

P primary sludge, *A* activated sludge, *AnD* anaerobically digested sludge, *S* dry solids content, S_f final dry solids content, *N.D.* no data given

[a]the gain corresponds to the difference in the final dry solids content between EDW and MDW

[b]kWh/kg_{DS}

[c]$kWh/kg_{\text{additionally water removed}}$

[d]$kWh/kg_{\text{removed water}}$

electro-dewatering process, but increases with the augmentation of cake dryness. This is caused by the fact that the last water fractions become increasingly difficult to dewater because of the increasing binding strength. At the same time, the desiccation of the cake may also deteriorate the electrical contact between the cake and the electrodes, and the increasing contact resistance causes a drop in the electric field, and thus a drop in the driving force, inside the cake. More recently, a similar linear relationship between the energy consumption and the amount of filtrate additionally removed during the electro-dewatering run was also reported by Saveyn et al. [191].

The electro-dewatering performance (rate, cake dryness, power consumption) largely depends on both the duration of the electric field application and the point at which it is applied. The results with belt filter presses showed that short contact times were not able to induce a satisfactory dewatering effect [199, 200]. Yet most studies [165, 230] showed that an extended application of the electric field was not efficient either. It is therefore necessary to find the optimum time as a function of the wastewater sludge properties that gives the best electro-dewatering performance.

On other hand, the energy required to raise the solids content increases with current density (or voltage) and must be optimised with regard to the desired throughput. The required energy can be clearly reduced when an extensive pre-dewatering is achieved with a mechanical process.

Friehmelt et al. [234] stated that in order to be an economically feasible process for sludge, pressurised electro-dewatering has to be preceded by pressure dewatering. This finding was confirmed by Miller et al. [202], Barton et al. [165], and Lee et al. [181] who noticed that application of the electric field at the start of the whole dewatering run in the filtration stage did not show any beneficial effect to the dewatering result and that no electrophoretic effects could be observed. This may be related to the fact that conditioned sludge particles networks are too large to be transported by electrophoresis. Moreover, well-conditioned sludge shows a smooth and fast filtration phase, and the filtrate flow due to the pressure is much higher than the electro-osmotic flow [70, 191].

According to Miller et al. [202], the required power consumption to achieve 12% of dry solids content is halved compare to the power consumption to reach 50% of dry solids content. They reported that to make energy savings, it is useful to delay (or pause) the application of the electric field by 5–15 min during the dewatering process. Nevertheless, when the voltage (or current) application is delayed, the results showed that a 20-min electric field application is not enough to reach the same cake dryness compared to experiments without a delay [202, 230]. Moreover, Miller et al. [202] also showed that the electro-dewatering process is accelerated after 20–30 min of current application. On the other hand, they demonstrated that the delay did provide power savings. For example, there was a large reduction in power consumption to reach 40% solids content with a 30-min delay. These observations seem to be confirmed by the results of Saveyn et al. [192]. Therefore, if the electric field application time is long, there would appear to be significant benefits in using a delay in the application of the electric current during electro-dewatering.

Gingerich et al. [230] and Yuan and Weng [228] made an economic analysis of the electro-dewatering process taking account of the additional energy costs and possible savings by transport and disposal and came to the conclusion that electro-dewatering to intermediate solids contents is most cost-efficient. Therefore, it seems preferable to implement the electric field technique in dewatering devices such as filter presses that have an intermediate contact time and that can operate at high pressure [202].

As mentioned earlier, conditioning prior to the dewatering is generally required to reach an acceptable efficiency. The chemical conditioning method, either with inorganic or organic chemicals or even dual conditioning, is the more common process. The selection of the appropriated chemical and dosage is quite difficult because it depends on both the sludge composition and the dewatering device and remains widely empirical [6].

However, the effect of the flocculent dose on the electro-osmotic dewatering performance has not yet been clarified. Most authors agree on the fact that a minimum polyelectrolyte dose is necessary in order to guarantee fast filtration or gravity drainage. Miller et al. [202], Gingerich et al. [230] and Citeau et al. [235] noticed that if this condition was fulfilled, polyelectrolyte dosing had no distinct effect on the electro-osmotic dewatering performance. Smollen and Kafaar [199] claimed that addition of polyelectrolyte provides an opposite excess charge in the liquid phase, to balance the surface charge of the sludge particle, which improves the efficiency of the applied voltage gradient responsible for the electro-osmotic flow. Snyman et al. [200] reported that the performance of the electro-osmotic belt filter press was not as sensitive to polyelectrolyte dosages as traditional belt presses. Finally, Laursen and Jensen [225, 236], who investigated the electro-osmotic water transport through preformed sludge cakes, reported that this transport was independent of the conditioning history. They explained their experimental results by assuming a picture of the filter cake as an exocellular polymeric substances matrix with sludge particles embedded and residual pore solution in isolated pores. This idea is in line with the findings by Mikkelsen [237] that the majority of the surface charges inside the sludge remain unaffected by conditioning, even at pronounced polyelectrolyte overdosing. Saveyn et al. [191] investigated the additional effect of an electric field applied during compression phase on the dewatering course of polyelectrolyte-conditioned sludge compared to mere pressure dewatering. They showed that the application of an electric field markedly improved the dewatering kinetics, regardless of the conditioning treatment. They also found that the polyelectrolyte characteristics and dose have a major effect during pressure-only dewatering of sludge but do not have a significant effect on the electro-osmotic water transport efficiency during the compression phase.

Nevertheless, the addition of the optimum dose of flocculent can reduce the electro-dewatering energy requirement for the same dewatering performance, but the choice of flocculant, as in pressure dewatering, must be chosen to match the specific sludge. For example, Kondoh and Hiraoka [175, 176] observed that by using the polyaluminum chloride PAC injection method, the clogging or the blinding of the filter media can be avoided and the treatability of the electro-dewatering is improved about 1.5–2 times and the power consumption is decreased by around 50% compared with electro-dewatering without PAC injection. On the other hand, Miller et al. [202] showed a possible effect of different compounds such as paper fibre, hydrochloric acid, magnesium sulphate, sulphuric acid which allow the electro-dewatering energy to be reduced while maintaining the same dewatering performance.

Tuan et al. [183] showed that an electric field application induced the migration of negatively charged organic matter such as fatty acids and humic substances from the cathode to the anode, and it was removed from sludge by the water flow in pressure-driven experiments. Therefore, the chemical oxygen demand (COD) and total organic carbon (TOC) concentrations in the filtrate were higher at the anode side and lower at the cathode side.

Using a similar approach, electro-dewatering may potentially be used for decontamination purposes. Hwang and Min [201] observed a large reduction of the heavy metal concentration such as Zn, Mn, Pb, Cd, and Ni in the sludge cake after electro-dewatering. As these metals often limit the use of sludge cake in agriculture, any reduction in heavy metals could be extremely useful. Acar and Alshawabkeh [238] discussed the need to use different enhancement techniques to overcome precipitation of the cationic species close to the cathode electrode. They reported that the use of a weak organic acid such as acetic acid to neutralise and/or depolarise the hydroxyl ions generated as result of cathode reaction successfully alleviates the precipitation problem for most heavy metals and transports the species to the cathode compartment [239, 240]. Also, to enhance the removal of heavy metals, other acid solutions including citric acid, hydrochloric acid, sulphuric acid or complexing substances such as ethylenediaminetetraacetic acid (EDTA) were added in the electrokinetic systems [241–245].

Recently, Tuan and Sillanpää [7, 246] studied migration of ions and organic matter during electro-dewatering of anaerobic sludge. They showed that concentration of Na^+ and K^+ was reduced by 51 and 78% in the sludge cake in comparison to blank experiments. Fe ions, Ca^{2+} and Mg^{2+} concentrations were found to be lower in sludge at the anode and higher at the cathode.

Other studies investigated the effect of electrokinetic processes on the removal of water and oil and grease from the oily sludge/wastes [247]. Oil, a group of hydrophobic hydrocarbons, is not only a contaminant in the natural environment but also a reusable resource. Their experiments were carried out with two different configurations of electro-dewatering cells, i.e. vertical and horizontal positioned electrodes. They reported that the electro-dewatering process offered potential for enhanced oil recovery as well as the reduction of the water content of the sludge by more than 40%.

Esmaeily et al. [223] noticed that it is possible to inactivate coliform in the final cake with non-pressurised electro-dewatering combined with the addition of a conditioning liquid containing nitrates and ammonia. These observations seem to be confirmed by the results of [248]. In their experiments, ten bench-scale electrokinetic cells were set up for the inactivation of *Salmonella spp.* bacteria and faecal coliforms in different types of sludge (primary, combined primary and secondary

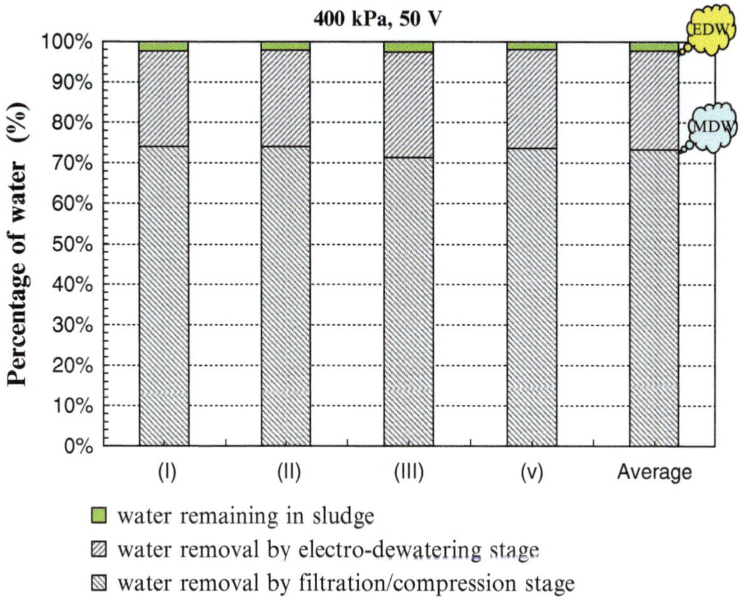

Fig. 9.10 Column chart illustrating the percentage of water removal in each EDW-process stage and the water remaining after processing conditions. The EDW tests for the processing conditions (400 kPa, 50 V) were repeated four times (*I, II, III, V*)

and anaerobically digested sludge). They showed that the electro-dewatering process achieved the dry solids content three to four times higher than conventional techniques. They also demonstrated that the pathogen indicator inactivation always exceeded three-log reduction. On the other hand, they reported that the anode area achieved a higher log reduction than the cathode area. It was assumed that the acidic front yielded the highest log reduction of *Salmonella spp*. bacteria.

Since many authors [1, 165, 181, 202, 234] noticed that the application of the electric field from the start of the dewatering test (in the filtration stage) did not show any beneficial effect to the dewatering result, Mahmoud et al. [193] investigated the application of the electric field during the compression stage. In their experiments on activated sludge, an analysis of the extent of water removal associated with each stage of the dewatering process was investigated for the different processing conditions (200–1200 kPa, 10–50 V). They showed that the dry solids content was increased from 13 to 56% (wt%), noting that the original solid content of the conditioned activated sludge was just 2.8% (wt%). Their analysis of the percentage of water removal in each dewatering stage, as well as the amount of water remaining after processing is shown in Fig. 9.10. In the course of the MDW with a processing pressure of 400 kPa, the percentage of water removal was more than 70%. The simultaneous combination of an applied voltage (50 V) with an applied pressure (400 kPa) removed 24% of additional water, which could not be accessed by the pressure forces alone.

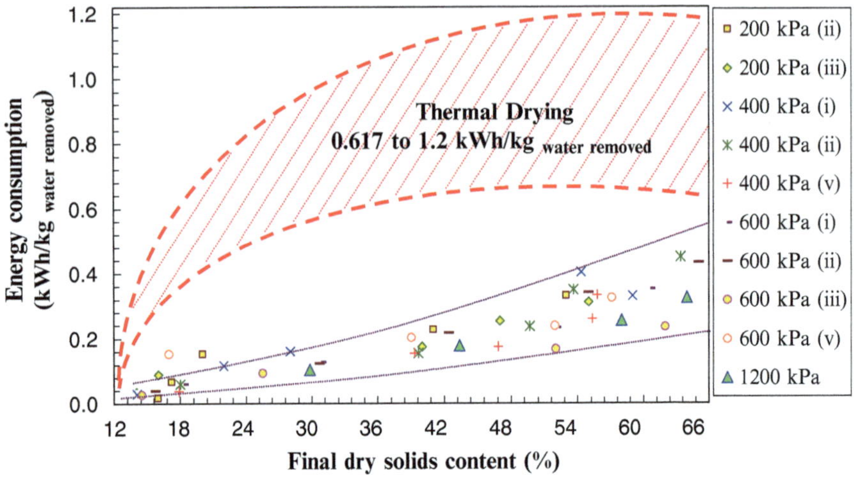

Fig. 9.11 Energy consumption during EDW as a function of the final dry solids obtained for the processing conditions 200–1200 kPa, 10–50 V compared to thermal drying techniques. The EDW tests (5 experiments per series) for the chosen operating conditions, pressure 200–1200 kPa and voltage 10–50 V, were repeated 3 or 4 times (*I, II, III, V*) except for the series of 1200 kP, 10–50 V

In their studies, it was found that EDW can remove approximately from 10 to more than 24% of additional water, which cannot be accessed by the pressure force alone (MDW). These experimental observations are in line with findings by Miller et al. and Saveyn et al. [192, 202, 249, 250]. Moreover, they reported that at low levels of voltage, an increased processing pressure has a relatively pronounced effect on the water removal. Not only is more liquid removed by the higher pressure, but it seems that also the electrical contact is improved and thus the energy efficiency.

On the other hand, it was highlighted that a clear non-linear relationship was found between the energy consumption and the final dry solids content obtained after the EDW run, as shown in Fig. 9.11. This relationship is a complex sum of several interdependent parameters which determine the behaviour of the sludge during an EDW test and thus the overall shape of the curve. However, it is possible to discern some trends from Fig. 9.11. When regarding the theoretical thermal drying energy requirement, which is about 0.617 kWh/ kg evaporated water [232], it is clear that the electro-dewatering technique is much more energy efficient, where for some low-voltage cases, less than 10% of thermal energy is required.

9.3.6.3 Industrial Development of Wastewater Sludge Electro-Dewatering Apparatus

Some examples of wastewater sludge electro-dewatering equipments, quite recently developed and slated for practical use around world, are described briefly in the following section.

A number of electro-dewatering processes have been reported in the literature by Bureau of Mines (USA), Commonwealth Scientific and Industrial Research Organisation (CSIRO, Australia), Central Electricity Generating Board (CEGB, UK), Battelle Memorial Institute (Switzerland), Fuji Electric Co., Ltd. (Japan) and Monsanto Enviro-Chem Systems, Inc. (USA). Fuji Electric Co. 1987 is the producer of most commercialised machines. Their electro-osmotic dewatering system uses a continuous press [251, 252] consisting of a rotary drum and a circulating caterpillar belt. The surface of the rotary drum is used as one electrode, and the caterpillar belt is another electrode. The wastewater sludge is fed into the space between the drum and the caterpillar belt. The sludge is dewatered electro-osmotically under the combined electrical field and expression. The experimental results showed that the water content in the discharge cake of wastewater sludge is remarkably reduced and the smallest value of the water content reached about 50%. Typically, the operating pressure is less than about 400 kPa, and the applied voltage is ranging from 0 to 120 V. The energy consumption of this type of machine is about 0.33–0.44 $kWh/kg_{removed\ water}$.

The Shinko Pantec Co. 1987 has also developed an electro-osmotic dehydrator (Super Filtron) of the semi-continuous filter press type installed with electrodes [175, 253]. After the filter chamber is filled with sludge, a fluid pressure is applied to the space between the membrane and the filter plate in order to inflate the membrane. A voltage is also applied between chamber's electrodes at the same time. Thus, the electro-osmotic dewatering can be combined with compression, and high-performance dewatering is obtained under a low pressure of about 200–400 kPa. The results showed that the Super Filtron can easily decrease the water content by about 50–60% using only half the electricity required for a conventional mechanical dewatering process.

Another industrial EDW machine is from ELODE [254] (ACE Korea Incorporation Co., Ltd.). For this machine, the dewatered sludge cake is fed between the anode drum and cathode caterpillar of the ELODE machine. A three-phase alternating current (A.C.), converted to direct current (D.C.) is applied to the anode drum and cathode caterpillar creating an electrical potential difference. This leads to a strong migration of sludge particles to the anode drum and water to the cathode caterpillar. Thus, highly efficient dehydration with a solids content above 40 wt% is achieved.

Although electro-dewatering techniques have been examined for decades and their advantages compared with conventional mechanical dewatering are well documented, commercial applications are still few. Typically, EDW is not as cost-effective as a stand-alone operation. It is necessary to use it in combination with pressure or vacuum dewatering. Notably, the equipment capital cost is higher than conventional dewatering equipment. Future research and development will hopefully allow cost-effective development of EDW-enhanced equipment for sludge dewatering.

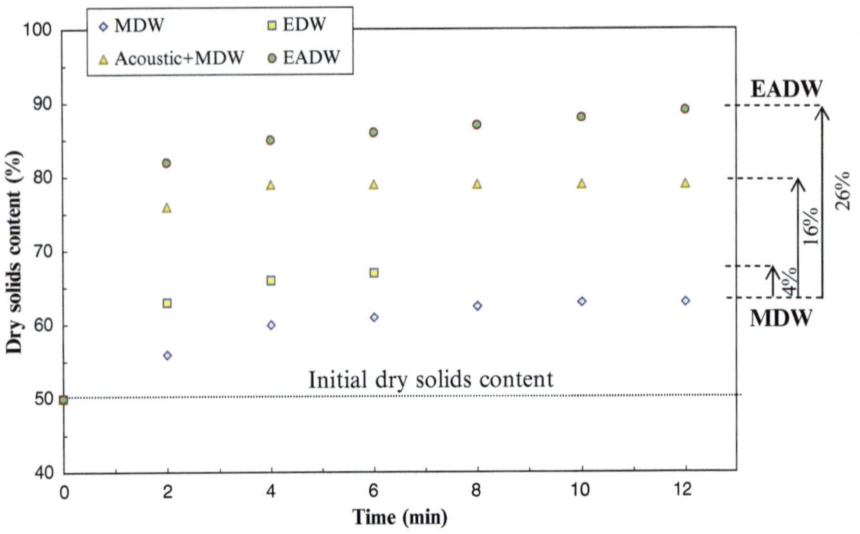

Fig. 9.12 Comparison of conventional mechanical dewatering and other dewatering techniques for a sample of 200 mesh coal slurry [257]

9.3.7 Electro-Acoustic Mechanical Dewatering (EADW)

The basis of the EADW process is a synergistic effect of combining electric and ultrasound fields in concert with conventional mechanical dewatering (vacuum, pressure). The relative effects of these 3 fields are dependent on the hydrodynamic, rheological, surface chemical and electrokinetic properties of suspensions.

In fact, at combined field conditions, ultrasonic energy can assist the electro-osmotic phenomena in the following ways: (1) by providing improve contact between water and electrodes, (2) by helping moisture pass through the filter, and (3) by assisting in compacting the cake and providing electrical continuity through a continuous moisture path to the point of optimum moisture removal. The electrical field allows for increased electro-osmosis and electrophoresis, while the acoustical forces help to maintain electrical continuity through the material by helping to agglomerate and compact the cake as well as keep the filter material from blinding [46, 50, 255–262].

Battelle Memorial Institute researchers have studied EADW for many years and have demonstrated successful scale up of its electro-acoustic dewatering units [255–257]. They tested the applicability of over 50 different types of suspensions and established wide applications of this process. Their experiments for coal slurry with 50 wt% initial dry solids content were carried out at a lab-scale. The dewatering efficiency for this material is compared to vacuum dewatering alone and other dewatering enhancements, as shown in Fig. 9.12. They observed that when vacuum and ultrasonic were combined with electric field, the dry solids content was increased from 63 to about 89%, corresponding to a 26% increase of

Table 9.3 Typical results for Battelle's electro-acoustic dewatering of a number of suspensions and wastes (at lab scale) [256, 257]

Material	Initial dry solids content (%)	Final dry solids content (%)		Gain[a] = $(S_f^{EADW} - S_f^{MDW})$ (%)	Power consumption (kWh/kg $_{removed\ water}$)
		MDW	EADW		
Peat	10	20	52	32	N.D.
Corn gluten slurry	13.1	35	46.3	11.3	0.033
Brewer's yeast suspension	5	22	41	19	0.004
Kaolin clay	27	60	70	10	N.D.
Ceramic slurry	27	50	60	10	0.066
Phosphate sludge	5	5	40	35	0.022
Mineral sludge	20	20	48	28	0.044
Coal slurry	50	63	89	26	0.055
Sewage sludge					
1. P + secondary	2	15–22	43	21–28	0.066
2. AnD	2	15–22	45	23–30	N.D.
3. A, and centrifuged	5	15–22	38	16–23	N.D.

P primary sludge, *A* activated sludge, *AnD* anaerobically digested sludge, *N.D.* no data given
[a]the gain corresponds to the difference in the final dry solids content between EADW and MDW

the dry solids content originally present in the coal slurry (50% wt%). The specific energy requirement to produce this solids level was only about 0.055 kWh/kg $_{removed\ water}$. Moreover, a continuously operating EAD device has now also been designed. This EADW press is utilised after conventional dewatering devices and can remove up to 50% of the water from dewatered sludge cake at a fraction of the cost incurred by existing thermal dry devices. As the material moves around the anode and as the water content in the cake is reduced, the electro-osmosis declines due to drop in liquid continuum from anode to cathode. The application of ultrasonic field at this point aids electro-osmosis by consolidation of the filter cake and thus allowing the release of water that would otherwise not be liberated. In addition, Battelle scientists presented some interesting research in dewatering sludge and other materials, particularly using acoustic energy in combination with electro-osmosis and electrophoresis.

Table 9.3 lists a number of suspensions and wastes studied by Battelle using a variety of equipment types, such as vacuum, belt presses, filter presses, drum filters and centrifuges. These data clearly indicate that EADW is more effective than the state-of-the-art mechanical dewatering processes in terms of achieving a high final dry solids content with low energy consumption compared to thermal drying or the EDW process. It is known that electro-osmotic flow quickly declines with the loss of the liquid continuum from the anode to the cathode as dewatering proceeds. This result is usually reflected in a rapid decline of current. However, the ultrasonic action retards the loss of the water continuum, thus increasing the current for a given potential gradient. This basically increases the electro-osmotic flow, without increasing the voltage gradient. This is also the reason why the energy requirement of EADW is less than the energy use for EDW only.

Ultrasonic was also reported to help keep the cathode clean, which reduces belt blinding metal hydroxides [263]. Moreover, Smythe and Wakeman [46] observed that the application of an acoustic field increases the measured conductivity of the electrolytic and rutile suspensions used as the test material. Thus, ultrasound enhances the effect of an applied electric field close to the isoelectric point of the suspension by reducing the effective particle size and increasing electrophoretic velocities.

Finally, the high efficiency and the significant energy saving of the EADW are encouraging and may show great promise in wastewater sludge dewatering.

9.4 Conclusions

The results presented in this chapter show the potential benefits in combining multiple strength field to enhance the dewatering ability of conventional processes, to increase the final dry solids content and/or to accelerate the dewatering process with relatively low energy consumption compared to thermal drying. Assisted separation processes which combine thermal, shear or vibration force, acoustic, magnetic, microwave, electrical, and the combined fields (e.g. electric and acoustic) forces with conventional mechanical force can be used to enhance wastewater sludge dewatering.

Except for the combination of microwave and mechanical dewatering, for which no work had been done in the past, the results presented in the scientific literature show that these hybrid processes have been successfully applied to several different mineral suspensions. Significant improvements in the dewatering rate and final solids content were obtained. However, the data on power consumption reported per unit mass of dry solids (kWh/kg_{DS}) is difficult to compare and analyse with data reported for drying processes.

Moreover, very few studies are reported in the scientific literature specifically in acoustic mechanical dewatering and magnetic mechanical dewatering of wastewater sludge. It is the authors' opinion that these hybrid processes appear to be very promising techniques. In addition, the interesting results obtained by Battelle Memorial Institute researchers point to the potential for the electro-acoustic dewatering (EADW) in the application of wastewater sludge dewatering.

On the other hand, the application of shear force-assisted mechanical dewatering for wastewater sludge did not show any beneficial effect to the dewatering results.

When regarding the thermal mechanical dewatering processes, it was found, from the few data reported, that although the benefits of this combination, the TAMD process and TREFLE process [26] processes appeared limited for wastewater sludge dewatering, while the data reported for both the MTE process and the TDW process were encouraging.

Finally, the research and development on electro-dewatering of wastewater sludge carried out to date and described in this chapter shows that the electro-dewatering was assessed to have potential for improving the efficiency of dewatering operations by achieving levels of water removal and final solids contents, which cannot be achieved by conventional dewatering equipment. Moreover, the results revealed that the electro-dewatering accelerates the dewatering rate with low energy consumption compared to thermal drying. The results are very encouraging and attractive enough to justify further development.

References

1. Saveyn H, Meersseman S, Thas O, Van Der Meeren P (2005) Influence of polyelectrolyte characteristics on pressure-driven activated sludge dewatering. Colloids Surf A 262:40–51
2. Tchobanoglou G, Burton FL, Stensel HD (2003) Wastewater engineering: treatment and reuse. Metcalf and Eddy Inc./McGraw-Hill, New York
3. Glendinning S, Lamont-Black J, Jones CJFP (2007) Treatment of sewage sludge using electrokinetic geosynthetics. J Hazard Mater 139:491–499
4. Glendinning S, Mok CK, Kalumba D, Rogers CDF, Hunt DVL (2010) Design framework for electrokinetically enhanced dewatering of sludge. J Environ Eng 136:417–426
5. Vaxelaire J, Bongiovanni JM, Puiggali JR (1999) Mechanical dewatering and thermal drying of residual sludge. Environ Technol 20:29–36
6. Vaxelaire J, Olivier J (2006) Conditioning for municipal sludge dewatering. From filtration compression cell tests to belt press. Dry Technol 24:1225–1233
7. Tuan PA, Sillanpää M (2010) Migration of ions and organic matter during electro-dewatering of anaerobic sludge. J Hazard Mater 173:54–61
8. Tsang KR, Vesilind PA (1990) Moisture distribution in sludge. Water Sci Technol 22:135–142
9. Lee DJ, Lee SF (1995) Measurement of bound water content in sludge: the use of differential scanning calorimetry (DSC). J Chem Technol Biotechnol 65:359–365
10. Chen G, Yue PL, Mujumdar AS (2002) Sludge dewatering and drying. Dry Technol 20:883–916
11. Mahmoud A, Olivier J, Vaxelaire J, Hoadley AFA (2010) Electrical field: a historical review of its application and contributions in wastewater sludge dewatering. Water Res 44:2381–2407
12. Bouyoucos GJ (1921) A new classification of soil moisture. Soil Sci 11:33–48
13. Vesilind PA (1979) Treatment and disposal of wastewater sludge. Ann Arbor Scicnce, Michigan
14. Smollen M (1988) Moisture retention characteristics and volume reduction of municipal sludges. Water SA 14:25–28
15. Smollen M (1990) Evaluation of municipal sludge drying and dewatering with respect to sludge volume reduction. Water Sci Technol 22:153–161
16. Vesilind PA, Martel CJ (1990) Freezing of water in wastewater sludge. J Environ Eng 116:854–862
17. Robinson J, Knocke WR (1992) Use of dilatometric and drying techniques for assessing sludge dewatering characteristics. Water Environ Res 64:60–68
18. Smith JK, Vesilind PA (1995) Dilatometric measurement of bound water in waste-water sludge. Water Res 29:2621–2626
19. Vesilind PA, Hsu CC (1997) Limits of sludge dewaterability. Water Sci Technol 36:87–91
20. Vesilind PA (1994) The role of water in sludge dewatering. Water Environ Res 66:4–11

21. Vaxelaire J, Cézac P (2004) Moisture distribution in activated sludges: a review. Water Res 38:2215–2230
22. Tarleton ES (1992) The role of field-assisted techniques in solid/liquid separation. Filtr Sep 29:246–252
23. Mahmood T, Zawadski M, Banerjee S (1998) Pilot study of impulse drying industrial sludge. Environ Sci Technol 32:1813–1816
24. Strauss K (1998) Method and device for reducing the water content of water containing brown coal. Eur Pat EP784660
25. Peuker UA, Stahl W (2001) Steam pressure filtration mechanical thermal dewatering process. Dry Technol 19:807–847
26. Couturier S (2002) Etude de la déshydration mécanique assistée thermiquement. PhD dissertation thesis, Université de Bordeaux I, France
27. Hulston J, De Kretser RG, Scales PJ (2004) Effect of temperature on the dewaterability of hematite suspensions. Int J Miner Process 73:269–279
28. Hulston J, Favas G, Chaffee AL (2005) Physico-chemical properties of Loy Yang lignite dewatered by mechanical thermal expression. Fuel 84:1940–1948
29. Hulston J, Chaffee A, Bergins C, Strauß K (2005) Comparison of physico-chemical properties of various lignites treated by mechanical thermal expression. Coal Prep 25:269–293
30. Fernandez A, Arlabosse P, Descoins N (2005) Thermally assisted mechanical dewatering: state of the art and new developments. Chem Eng Trans 7:737–742
31. Fernandez A, Auduc B, Arlabosse P, Mahmoud A (2009) Method for mechanical dehydration with thermal assistance. Eur Pat EP2212637
32. Lee EL (2006) Thermal dewatering (TDW) to reduce the water content. Dry Technol 24:225–232
33. Clayton SA, Scholes ON, Hoadley AFA, Wheeler RA, Mcintosh MJ, Huynh DQ (2006) Dewatering of biomaterials by mechanical thermal expression. Dry Technol 24:819–834
34. Clayton SA, Wheeler RA, Hoadley AFA (2007) Pore destruction resulting from mechanical thermal expression. Dry Technol 25:533–546
35. Couturier S, Valat M, Vaxelaire J, Puiggali JR (2007) Enhanced expression of filter cakes using a local thermal supply. Sep Purif Technol 57:321–328
36. Mahmoud A, Fernandez A, Arlabosse P (2008) Thermally-assisted mechanical dewatering (TAMD) of suspensions of fine particles: analysis of the influence of the operating conditions using the response surface methodology. Chemosphere 72:1765–1773
37. Mahmoud A, Arlabosse P, Fernandez MA (2011) Application of thermally assisted mechanical dewatering process to biomass. Biomass Bioenerg 35:288–297
38. Lee JE, Lee JK, Choi HK (2007) Filter Press for electro-dewatering of waterworks sludge. Dry Technol 25:1649–1657
39. Curvers D, Maes KC, Saveyn H, De Baets B (2007) Modelling the electro-osmotically enhanced pressure dewatering of activated sludge. Chem Eng Sci 62:2267–2276
40. Mujumdar AS, Yoshida H (2008) Electro-osmotic dewatering (EOD) of bio-materials. In: Vorobiev E, Lebovka N (eds) Electrotechnologies for extraction from food plants and biomaterials, 1st edn. Springer, New York
41. Lebovka NI, Praporscic I, Vorobiev E (2003) Enhanced expression of juice from soft vegetable tissues by pulsed electric fields: consolidation stages analysis. J Food Eng 59:309–317
42. Praporscic I, Muravetchi V, Vorobiev E (2004) Constant rate expressing of juice from biological tissue enhanced by pulsed electric field. Dry Technol 22:2395–2408
43. Gachovska T, Ngadi MO, Raghavan GSV (2006) Pulsed electric field assisted juice extraction from alfalfa. Can Biosyst Eng 48:33–37
44. Grimi N, Praporscic I, Lebovka N, Vorobiev E (2007) Selective extraction from carrot slices by pressing and washing enhanced by pulsed electric fields. Sep Purif Technol 58:267–273

45. Riera-Franco de Sarabia E, Gallego-Juárez JA (2000) Application of high-power ultrasound to enhance fluid/solid particle separation processes. Ultrasonics 38:642–646
46. Smythe MC, Wakeman RJ (2000) The use of acoustic fields as a filtration and dewatering aid. Ultrasonics 38:657–661
47. Tao D, Parekh BK (2000) Enhanced fine coal beneficiation using ultrasonic energy. Miner Metall Process 17:252–258
48. Stolarski M, Fuchs B, Bogal Kassa S, Eichholz C, Nirschl H (2006) Magnetic field enhanced press-filtration. Chem Eng Sci 61:6395–6403
49. Eichholz C, Stolarski M, Gortz V, Nirschl H (2008) Magnetic field enhanced cake filtration of superparamagnetic PVAc-particle. Chem Eng Sci 63:3193–3200
50. Wakeman RJ, Smythe MC (2000) Clarifying filtration of fine particle suspensions aided by electrical and acoustic fields. Chem Eng Res Des 78:125–135
51. Wakeman RJ, Tarleton ES (1999) Filtration. Equipment selection modelling and process simulation. Elsevier advanced technology, Oxford
52. Wakeman RJ, Zhang GM, Koenders MA (2000b) Effect of rotational flow induced in a rotary piston press on the formation and properties of filter cakes. In: Proceedings of 8th world filtration congress, pp 1121–1124
53. Olivier J (2003) Etudes des filtres à bandes pour la déshydratation mécanique des boues résiduaires urbaines. PhD dissertation thesis, Université de Pau et des Pays de l'Adour, Pau, France
54. Gundogdu O, Koenders MA, Wakeman RJ, Wu P (2003) Permeation through a bed on a vibrating medium: theory and experimental results. Chem Eng Sci 58:1703–1713
55. Gundogdu O, Koenders MA, Wakeman RJ, Wu P (2003) Vibration assisted dead-end filtration: experiments and theoretical concepts. Chem Eng Res Des 81:1–9
56. Gundogdu O, Koenders MA, Wakeman RJ, Wu P (2003) Permeation with vibrated media: experiments and modelling. Filtr Soc 3:106–113
57. Collins AG, Mitra S, Pavlostathis SG (1991) Microwave heating for sludge dewatering and drying. Res J Water Pollut Control Fed 63:921–924
58. Tchobanoglous G, Burton FL (1991) Wastewater engineering treatment, disposal, and reuse. McGraw-Hill, New York
59. Novak JT (2001) Dewatering. In: Spinosa L, Vesilind PA (eds) Sludge into biosolids: processing, disposal and utilization. IWA Pub, London
60. Andreasen I, Nielsen B (1993) A comparative-study of full-scale sludge dewatering equipment. Water Sci Technol 28:37–45
61. Lockhart NC (1986) Electro-dewatering of fine suspensions. In: Muralidhara HS (ed) Advanced in solid-liquid separation. Battelle, Columbus
62. Lawler DF, Chung YJ, Hwang SJ, Hull BA (1986) Anaerobic digestion: effects on particle size and dewaterability. J Water Pollut Control Fed 58:1107–1117
63. Li DH, Ganczarczyk JJ (1990) Structure of activated-sludge flocs. Biotechnol Bioeng 35:57–65
64. Snidaro D, Zartarian F, Jorand F, Bottero JY, Block JC, Manem J (1997) Characterization of activated sludge flocs structure. Water Sci Technol 36:313–320
65. Saveyn H (2005) Modelling and optimization of sludge conditioning and electric field assisted dewatering. PhD dissertation thesis, Ghent University, Belgium
66. Govoreanu R (2004) Activated sludge flocculation dynamics: on-line measurement methodology and modelling. PhD dissertation thesis, Ghent University, Belgium
67. Novak JT, Agerbaek ML, Sorensen BL, Hansen JA (1999) Conditioning, filtering, and expression waste activated sludge. J Environ Eng 125:816–824
68. Dentel SK (2001) Conditioning. In: Spinosa L, Vesilind PA (eds) Sludge into biosolids: processing, disposal and utilization. IWA Pub, London
69. Saveyn H, Hendrickx PMS, Dentel SK, Martins JC, Meeren PV (2008) Quantification of hydrolytic charge loss of DMAEA-Q-based polyelectrolytes by proton NMR spectroscopy and implementations for colloid titration. Water Res 42:2718–2728

70. Saveyn H, Curvers D, Thas O, Meeren PV (2008) Optimisation of sewage sludge conditioning and pressure dewatering by statistical modelling. Water Res 42:1061–1074
71. Chen Y, Yang H, Gu G (2001) Effect of acid and surfactant treatment on activated sludge dewatering and settling. Water Res 35:2615–2620
72. Vosteen B, Weissenberg HG (2000) New conditioning process for the dewatering of wastewater treatment sludge. Chem Eng Technol 23:677–681
73. Neyens E, Baeyens J (2003) A review of classic Fenton's peroxidation as an advanced oxidation technique. J Hazard Mater 98:33–50
74. Watanabe Y, Tanaka K (1999) Innovative sludge handling through palletization/thickening. Water Res 33:3245–3252
75. Lee CH, Liu JC (2000) Enhanced sludge dewatering by dual polyelectrolytes conditioning. Water Res 34:4430–4436
76. Thapa KB, Qi Y, Clayton SA, Hoadley AFA (2009) Lignite aided dewatering of digested sewage sludge. Water Res 43:623–634
77. Lee JE, Lee JK, Kim DS (2010) A study of the improvement in dewatering behavior of wastewater sludge through the addition of fly ash. Korean J Chem Eng 27:862–867
78. Erdincler A, Vesilind PA (2000) Effect of sludge cell disruption on compactibility of biological sludges. Water Sci Technol 42:119–126
79. Neyens E, Baeyens J (2003) A review of thermal sludge pre-treatment processes to improve dewaterability. J Hazard Mater 98:51–67
80. Zheng JH, Graff RA, Fillos J, Rinard I (1998) Incorporation of rapid thermal conditioning into a wastewater treatment plant. Fuel Process Technol 56:183–200
81. Neyens E, Baeyens J, Creemers C (2003) Alkaline thermal sludge hydrolysis. J Hazard Mater 97:295–314
82. Neyens E, Baeyens J, Dewil R, De Heyder B (2004) Advanced sludge treatment affects extracellular polymeric substances to improve activated sludge dewatering. J Hazard Mater 106:83–92
83. Örmeci B, Aarne Vesilind P (2001) Effect of dissolved organic material and cations on freeze-thaw conditioning of activated and alum sludges. Water Res 35:4299–4306
84. Sanin FD, Vesilind PA, Martel CJ (1994) Pathogen reduction capabilities of freeze/thaw sludge conditioning. Water Res 28:2393–2398
85. HellstrÖm D, KvarnstrÖm E (1997) Natural sludge dewatering. 1. Combination of freezing, thawing, and drying as dewatering methods. J Cold Reg Eng 11:1–14
86. Kowalska E, Chmura K, Bien J (1978) Ultrasonics in the dehydration process of sludge. Ultrasonics 16:183–185
87. Bien J (1988) Ultrasonic preparation of sludges to improve dewatering. Filtr Sep 25:425–426
88. Sakai Y, Kurakata S, Takahashi F (1991) Magnetic forced sedimentation of flocs in activated sludge supplemented with ferromagnetic powder of iron oxide. J Ferment Bioeng 71:208–210
89. Sakai Y, Terakado T, Takahashi F (1994) A sewage treatment process using highly condensed activated sludge with an apparatus for magnetic separation. J Ferment Bioeng 78:120–122
90. Sakai Y, Miama T, Takahashi F (1997) Simultaneous removal of organic and nitrogen compounds in intermittently aerated activated sludge process using magnetic separation. Water Res 31:2113–2116
91. Bien JB, Kempa ES, Bien JD (1997) Influence of ultrasonic field on structure and parameters of sewage sludge for dewatering process. Water Sci Technol 36:287–291
92. Bien J, Wolny L (1997) Changes of some sewage sludge parameters prepared with an ultrasonic field. Water Sci Technol 36:101–106
93. Singh BP (1999) Ultrasonically assisted rapid solid-liquid separation of fine clean coal particles. Miner Eng 12:437–443
94. Chu CP, Chang B-V, Liao GS, Jean DS, Lee DJ (2001) Observations on changes in ultrasonically treated waste-activated sludge. Water Res 35:1038–1046

95. Hattori S, Watanabe M, Endo T, Togii H, Sasaki K (2001) Effects of an external magnetic field on the sedimentation of activated sludge. World J Microbiol Biotechnol 17:279–285
96. Hattori S, Watanabe M, Osono H, Togii H, Sasaki K (2001) Effects of an external magnetic field on the flock size and sedimentation of activated sludge. World J Microbiol Biotechnol 17:833–838
97. Hattori S, Watanabe M, Sasaki K, Yasuharu H (2002) Magnetization of activated sludge by an external magnetic field. Biotechnol Lett 24:65–69
98. Menéndez JA, Domínguez A, Inguanzo M, Pis JJ (2004) Microwave pyrolysis of sewage sludge: analysis of the gas fraction. J Anal Appl Pyrolysis 71:657–667
99. Yin X, Han P, Lu X, Wang Y (2004) A review on the dewaterability of bio-sludge and ultrasound pretreatment. Ultrason Sonochem 11:337–348
100. Wojciechowska E (2005) Application of microwaves for sewage sludge conditioning. Water Res 39:4749–4754
101. Dewil R, Baeyens J, Goutvrind R (2006) Ultrasonic treatment of waste activated sludge. Environ Prog 25:121–128
102. Yin X, Lu X, Han P, Wang Y (2006) Ultrasonic treatment on activated sewage sludge from petro-plant for reduction. Ultrasonics 44:e397–e399
103. Khanal SK, Grewell D, Sung S, Van Leeuwen J (2007) Ultrasound applications in wastewater sludge pretreatment: a review. Crit Rev Environ Sci Technol 37:277–313
104. KyllÖnen H, Pirkonen P, GrÖnroos A (2007) Ultrasonically assisted dewatering of waste activated sludge. In: Proceedings of the 19th international congress on acoustics, Madrid, Spain
105. Na S, Kim Y-U, Khim J (2007) Physiochemical properties of digested sewage sludge with ultrasonic treatment. Ultrason Sonochem 14:281–285
106. Feng X, Deng J, Lei H, Bai T, Fan Q, Li Z (2009) Dewaterability of waste activated sludge with ultrasound conditioning. Bioresour Technol 100:1074–1081
107. Huan L, Yiying J, Mahar RB, Zhiyu W, Yongfeng N (2009) Effects of ultrasonic disintegration on sludge microbial activity and dewaterability. J Hazard Mater 161:1421–1426
108. Yu Q, Lei H, Yu G, Feng X, Li Z, Wu Z (2009) Influence of microwave irradiation on sludge dewaterability. Chem Eng J 155:88–93
109. Yu Q, Lei H, Li Z, Li H, Chen K, Zhang X, Liang R (2010) Physical and chemical properties of waste-activated sludge after microwave treatment. Water Res 44:2841–2849
110. Saveyn H, Curvers D, Jacobsen R, Meeren PV (2010) Improved dewatering by freeze–thawing of predewatered sludge cakes. Asia Pac J Chem Eng 5:798–803
111. Pilli S, Bhunia P, Yan S, LeBlanc RJ, Tyagi RD, Surampalli RY (2011) Ultrasonic pretreatment of sludge: a review. Ultrason Sonochem 18:1–18
112. La Heij EJ (1994) An analysis of sludge filtration and expression. PhD dissertation thesis, TU Eindhoven, the Netherlands
113. La Heij EJ, Kerkhof PJAM, Kopinga K, Pel L (1996) Determining porosity profiles during filtration and expression of sewage sludge by NMR imaging. AIChE J 42:953–959
114. Reichmann B, Tomas J (2001) Expression behaviour of fine particle suspensions and the consolidated cake strength. Powder Technol 121:182–189
115. Falk T, Wallin ML (1987) Influence of energy in filter cake dewatering. Aufbereit-Tech 28:115–125
116. Bergins C, Berger S, Strauss K (1999) Mechanical/thermal dewatering–a perspective for an efficient solid/liquid separation. Ceram Forum Int 76:8–12
117. Bergins C, Berger S, Strauss K (1999) Dewatering of fossil fuel and suspension of ultrafine particles by mechanical/thermal dewatering. Chem Eng Technol 22:923–927
118. Bergins C, Berger S, Strauss K (1998) Process technology for mechanical/thermal dewatering I. Aufbereit-Tech 39:58–70
119. Artanto Y, Chaffee AL (2005) Dewatering low rank coals by mechanical thermal expression (MTE) and its influence on organic carbon and inorganic removal. Coal Prep 25:251–267

120. Wheeler RA, Hoadley AFA, Clayton SA (2009) Modelling the mechanical thermal expression behaviour of lignite. Fuel 88:1741–1751
121. Cussó Grau N (2009) Energy balance in thermal assisted mechanical dewatering (TAMD) process. Master thesis, Escola Tècnica Superior d'Enginyeria Industrial de Barcelona Mobilitat, Barcelona, Spain
122. Bergins C, Hulston J, Strauss K, Chaffee AL (2007) Mechanical/thermal dewatering of lignite. Part 3: Physical properties and pore structure of MTE product coals. Fuel 86:3–16
123. Bergins C (2003) Kinetics and mechanism during mechanical/thermal dewatering of lignite. Fuel 82:355–364
124. Korger V, Stahl W (1993) Vapour pressure dewatering-a new technique for combined mechanical and thermal dewatering of filter cakes. In: Proceedings of the 6th world filtration congress, Nagoya, pp 400–404
125. Korger V (1995) Dampfdruckentwässerung-Ein neues kombiniertes Entfeuchtungsverfahren, Fortschritt.-Ber. VDI Reihe 3, Nr. 403 Düsseldorf: VDI Verlag, Germany
126. Ruf J, Stahl W (1997) The hot filter press-the new technology of vapour pressure dewatering. In: Baumann R, Weisert L (eds) Advances in filtration and separation technology, vol 11. Advancing Filtration Solution, Minneapolis
127. Ruf J, Korger V, Stahl W (1997) Vapour pressure dewatering in a hot filter press. Aufbereit-Tech 38:411–416
128. Ruf J (1998) Die technische Umsetzung der Dampfdruckentwässerung in einer Heißen Filterpresse. PhD dissertation thesis, Institut für Mechanische Verfahrenstechnik und Mechanik, Karlsruhe, Germany
129. Couturier S, Valat M, Vaxelaire J, Puiggali JR (2004) Mechanical dewatering of residual sludge enhanced by a local thermal supply. In: Proceedings of the world filtration congress 9, New Orleans, Louisiana, USA
130. Couturier S, Valat M, Vaxelaire J, Peuchot C (2005) Thermal supply at the filter medium level: a solution to enhanced expression process. In: Proceedings of the 18th annual conference of AFS, Atlanta, Georgia, USA
131. Gerl S, Stahl W (1996) Improved dewatering of coal by steam pressure filtration. Coal Prep 17:137–146
132. Peuker UA, Stahl W (1999) Scale-up of steam pressure filtration. Chem Eng Process 38:611–619
133. Peuker UA, Stahl W (2002) Mechanical-thermal dewatering applying steam centrifugation. Chem Eng Technol 25:368–373
134. Peeters B (2010) Mechanical dewatering and thermal drying of sludge in a single apparatus. Dry Technol 28:454–459
135. Bongert W (1976) US Pat 3970552
136. Furedi P (1977) US Pat 4055491
137. Muralidhara HS, Ensminger D (1984) Acoustic dewatering and drying: state of the art review. In: Proceedings of the IV international dry technol symposium, Kyoto, Japan
138. Muralidhara HS, Ensminger D, Putnam A (1985) Acoustic dewatering and drying (low and high frequency): state of the art review. Dry Technol 3:529–566
139. Muralidhara HS, Parekh B, Senapati N (1985b) Solid liquid separation process for fine particle suspensions by an electric and ultrasonic field. US Pat 4561953
140. Bekker MC, Meyer JP, Pretorius L, Van Der Merwe DF (1997) Separation of solid-liquid suspensions with ultrasonic acoustic energy. Water Res 31:2543–2549
141. Muralidhara HS, Bread RB, Senapati N (1987) Mechanisms of ultrasonic agglomeration for dewatering colloid suspension. Filtr Sep 24:409–413
142. Wang F, Ji M, Lu S (2006) Influence of ultrasonic disintegration on the dewaterability of waste activated sludge. Environ Prog 25:257–260
143. Tuori T (1998) Enhancing filtration by electro-acoustic methods. PhD dissertation thesis, Lougborough University, UK

144. Gallego-Juárez JA, Elvira-Segura L, Rodríguez-Corral G (2003) A power ultrasonic technology for deliquoring. Ultrasonics 41:255–259
145. Swamy KM, Rao ARK, Narasimhan KS (1983) Acoustics aids dewatering. Ultrasonics 21:280–281
146. Svoboda J (1982) The influence of surface forces on magnetic separation. IEEE Trans Magn 18:862–865
147. Tsouris C, Scott TC (1995) Flocculation of paramagnetic particles in a magnetic field. J Colloid Interface Sci 171:319–330
148. Charles S (1988) Aggregation in magnetic fluids and magnetic fluid composites. Chem Eng Commun 67:145–180
149. Chaplin M (2011) Magnetic and electric effects on water. http://www.lsbu.ac.uk/water/magnetic.html. Accessed 25 Jan 2011
150. Tarleton ES, Wakeman RJ (2007) Solid/liquid separation: equipment selection and process design. Butterworth-Heinemann, Oxford
151. Watson JL, Gardner PL (1995) Multi-force dewatering for magnetic waste materials. Miner Eng 8:191–200
152. Fuchs B, Hoffman C, Keller K, Rey CM (2006a) PCT/US Pat 2005 005043
153. Fuchs B, Stolkarski M, Stahl W, Nirschi H (2006) Magnetic field enhanced cake filtration. Filtration 6:333–339
154. Jacob J, Chia LHL, Boey FYC (1995) Thermal and non-thermal interaction of microwave radiation with materials. J Mater Sci 30:5321–5327
155. Toreci I, Kennedy KJ, Droste RL (2009) Evaluation of continuous mesophilic anaerobic sludge digestion after high temperature microwave pretreatment. Water Res 43:1273–1284
156. Ponne CT, Bartels PV (1995) Interaction of electromagnetic energy with biological material relation to food processing. Radiat Phys Chem 45:591–607
157. Thostenson ET, Chou T-W (1999) Microwave processing: fundamentals and applications. Compos Part A Appl Sci Manuf 30:1055–1071
158. Kaatze U (1995) Fundamentals of microwaves. Radiat Phys Chem 45:539–548
159. Kaatze U (1995) Microwave dielectric properties of liquids. Radiat Phys Chem 45:549–566
160. Fung DYC, Cunningham FE (1980) Effect of microwaves on microorganisms in foods. J Food Prot 43:641–650
161. NiederwÖhrmeier B, BÖhm R, Strauch D (1985) Microwave treatment as an alternative pasteurization process for the disinfection of sewage sludge: experiences with the treatment of liquid manure. In: Havelaar AH, L'Hermite P, Strauch D (eds) Inactivation of microorganisms in sewage sludge by stabilization processes. Elsevier Applied Science, London
162. Woo I-S, Rhee I-K, Park H-D (2000) Differential damage in bacterial cells by microwave radiation on the basis of cell wall structure. Appl Environ Microbiol 66:2243–2247
163. Hong SM, Park JK, Lee YO (2004) Mechanisms of microwave irradiation involved in the destruction of fecal coliforms from biosolids. Water Res 38:1615–1625
164. Fu D, Cai M, Chao Allen C, Liu J (1999) Initial studies on sludge drying using microwave heating. In: Proceedings of the international conference on solid waste technology management. Widener University School Engineering, Chester, PA, USA
165. Barton WA, Miller SA, Veal CJ (1999) The electrodewatering of sewage sludges. Dry Technol 17:497–522
166. Yoshida H (1993) Practical aspects of dewatering enhanced by electro-osmosis. Dry Technol 11:787–814
167. Moulik SP (1971) Physical aspects of electrofiltration. Environ Sci Technol 5:771–776
168. Yukawa H, Kobayashi K, Tsukui Y, Yamano S, Iwata M (1976) Analysis of batch electrokinetic filtration. J Chem Eng Jpn 9:396–401
169. Yoshida H, Shinkawa T, Yukawa H (1985) Water content and electric potential distributions in gelatinous bentonite sludge with electroosmotic dewatering. J Chem Eng Jpn 18:337–342

170. Krishnaswamy P, Bahnii JR (1986) Electrically coupled flow through fine particle cakes. Filtr Sep 23:289–293
171. Muralidhara HS, Senapati N, Beard RB (1986) A novel electroacoustic separation process for fine particle suspensions. In: Muralidhara HS (ed) Advances in solid-liquid separation. Battelle, Columbus
172. Gopalakrishnan S, Mujumdar AS, Weber ME (1996) Optimal off-time in interrupted electro-osmotic dewatering. Sep Technol 6:197–200
173. Xian-shu D, Xiao-jie H, Su-ling Y, Wei-peng R, Zhi-zhong W (2009) Vacuum filter and direct current electro-osmosis dewatering of fine coal slurry. In: Proceedings of the 6th international conference on mining science and technology, Procedia Earth Planet Sci vol 1, pp 685–693
174. Heath LW, Dmirel T (1984) Pressurized electro-osmotic dewatering. Eng found conf flocculation sediment consol. Sea Island, Georgia
175. Kondoh S, Hiraoka M (1990) Commercialization of pressurized electroosmotic dehydrator (PED). Water Sci Technol 22:259–268
176. Kondoh S, Hiraoka M (1993) Studies on the improving dewatering method of sewage sludge by the pressurized electro-osmotic dehydrator with injection of polyaluminum chloride. In: Proceedings of the 6th world filtration congress, Nagoya
177. Buijs PJ, Van Diemen AJG, Stein HN (1994) Efficient dewatering of waterworks sludge by electroosmosis. Colloids Surf A Physiochem Eng Asp 85:29–36
178. Gazbar S, Abadie JM, Colin F (1994) Combined action of electro-osmotic drainage and mechanical compression on sludge dewatering. Water Sci Technol 30:169–175
179. Chen H, Mujumdar AS, Raghavan GSV (1996) Laboratory experiments on electroosmotic dewatering of vegetable sludge and mine tailings. Dry Technol 14:435–445
180. Larue O, Mouroko-Mitoulou T, Vorobiev E (2001) Pressurized electroosmotic dewatering in filter cycle. Dry Technol 19:2363–2377
181. Lee JK, Shin HS, Park CJ, Lee CG, Lee JE, Kim YW (2002) Performance evaluation of electro-dewatering system for sewage sludges. Korean J Chem Eng 19:41–45
182. Raats MHM, Van Diemen AJG, Laven J, Stein HN (2002) Full scale electrokinetic dewatering of waste sludge. Colloids Surf A Physicochem Eng Asp 210:231–241
183. Tuan PA, Jurate V, Mika S (2008) Electro-dewatering of sludge under pressure and non-pressure conditions. Environ Technol 29:1075–1084
184. Kobayashi K, Hakoda M, Hosoda Y, Iwata M, Yukawa H (1979) Electroosmotic flow through particle beds and electroosmotic pressure distribution. J Chem Eng Jpn 12:492–494
185. Yoshida H, Iwata M, Igami H, Murase T (1991) Combined operation of electroosmotic dewatering and mechanical expression. J Chem Eng Jpn 24:399–401
186. Iwata M, Igami H, Murase T (1991) Analysis of electroosmotic dewatering. J Chem Eng Jpn 24:45–50
187. Iwata M, Igami H, Murase T, Yoshida H (1991) Combined operation of electroosmotic dewatering and mechanical expression. J Chem Eng Jpn 24:399–401
188. Raghavan GSV, Lightfoot DG (1993) Combined field dewatering of seaweed. In: Proceedings of the American Society of Agricultural Engineers, Paper no. 936030. Summer Meeting, Spokane, WA
189. Lightfoot DG, Raghavan GSV (1994) Combined fields dewatering of seaweed (*Nereocystis luetkeana*). Trans Am Soc Agric Eng 37:899–906
190. Orsat V, Raghavan GSV, Norris ER (1996) Food processing waste dewatering by electro-osmosis. Can Agric Eng 38:63–67
191. Saveyn H, Pauwels G, Timmerman R, Meeren PV (2005) Effect of polyelectrolyte conditioning on the enhanced dewatering of activated sludge by application of an electric field during the expression phase. Water Res 39:3012–3020
192. Saveyn H, Meeren PV, Pauwels G, Timmerman R (2006) Bench-and pilot-scale sludge electrodewatering in a diaphragm filter press. Water Sci Technol 54:53–60

193. Mahmoud A, Olivier J, Vaxelaire J, Hoadley AFA (2011) Electro-dewatering of wastewater sludge: influence of the operating conditions and their interaction effects. Water Res 45 (9):2795–2810
194. Rabie HR, Mujumdar AS, Weber ME (1994) Electroosmotic dewatering of bentonite in thin beds. Sep Technol 4:180–182
195. Vijh AK (1999) The Significance of current observed during combined field and pressure electroosmotic dewatering of clays. Dry Technol 17:555–563
196. Vijh AK (1999) Electroosmotic dewatering (EOD) of clays and suspensions: components of voltage in an electroosmotic cell. Dry Technol 17:565–574
197. Weber K (2002) Untersuchungen zum Einfluss eines elektrischen Feldes auf die kuchenbildende Pressfiltration, VDI Verlag GmbH
198. Larue O, Vorobiev E (2004) Sedimentation and water electrolysis effects in electrofiltration of kaolin suspension. AIChE J 50:3120–3133
199. Smollen M, Kafaar A (1994) Electroosmotically enhanced sludge dewatering: pilot-plant study. Water Sci Technol 30:159–168
200. Snyman HG, Forssman P, Kafaar A, Smollen M (2000) The feasibility of electro-osmotic belt filter dewatering technology at pilot scale. Water Sci Technol 41:137–144
201. Hwang S, Min KS (2003) Improved sludge dewatering by addition of electro-osmosis to belt filter press. J Environ Eng Sci 2:149–153
202. Miller SA, Murphy A, Veal CJ, Young M (1998) Improved filtration of sewage sludges using electrodewatering. End of Project Report, CSIRO Investigation Report N° ET/IR140
203. Lockhart NC (1983) Electroosmotic dewatering of clays. I. Influence of voltage. Colloids Surf 6:229–238
204. Lockhart NC (1983) Electroosmotic dewatering of clays. II. Influence of salt, acid, and flocculants. Colloids Surf 6:239–251
205. Lockhart NC (1983) Electro-osmotic dewatering of fine tailings from mineral processing. Int J Miner Process 10:131–140
206. Rabie HR, Mujumdar AS, Weber ME (1994) Interrupted electro osmotic dewatering of clay suspensions. Sep Technol 4:38–46
207. Yoshida H, Kitajyo K, Nakayama M (1999) Electro osmotic dewatering under A.C. electric field with periodic reversals of electrode polarity. Dry Technol 17:539–554
208. Yoshida H (2000) Electro-osmotic dewatering under intermittent power application by rectification of AC electric field. J Chem Eng Jpn 33:134–140
209. Saveyn H, Meeren PV, Hofmann R, Stahl W (2005) Modelling two-sided electrofiltration of quartz suspensions: importance of electrochemical reactions. Chem Eng Sci 60:6768–6779
210. Kondo S, Sano S (1994) Method of electroosmotically dehydrating sludge. US Pat 5279718
211. Lightfoot DG, Raghavan GSV (1995) Combined fields dewatering of seaweed with a roller press. Appl Eng Agric 11:291–295
212. Friehmelt V, Gidarakos E (1996) Electrokinetic dewatering of effluent sludge as a continuous process. Aufbereit-Tech/Miner Proc 37:133–141
213. Watanabe S, Yamada Y, Hobo Y, Hayashi N, Sumi T, Dykes D, Touchard G (2001) A new method of analysing electro-osmosis using FET models. J Electrost 51–52:455–462
214. Vijh A (2002) Electroosmotic dewatering by a new method using a gate electrode: field effect transistor (EFT) model or simply a multistage dewatering. Dry Technol 20:705–710
215. Weber K, Stahl W (2003) Influence of an electric field on filtration in a filter press. Chem Eng Technol 26:44–48
216. Ho MY, Chen G (2001) Enhanced electro-osmotic dewatering of fine particle suspension using a rotating anode. Ind Eng Chem Res 40:1859–1863
217. Zhou J, Liu Z, She P, Ding F (2001) Water removal from sludge in a horizontal electric field. Dry Technol 19:627–638
218. Yoshida H, Okada M (2006) Influence of electric field application with decreasing one sided area of electrodes on electro-osmotic dewatering. Dry Technol 24:1313–1316

219. De-Gang M, Shu-Ting Z (2007) The electro-dewatering of sludge using adsorptive material. In: Proceedings of the 5th Asia-Pacific dry conference, Guohua Chen, Hong Kong, China
220. Yu X, Zhang S, Xu H, Zheng L, Lü X, Ma D (2010) Influence of filter cloth on the cathode on the electroosmotic dewatering of activated sludge. Chin J Chem Eng 18:562–568
221. Lai CKB (2001) Salinity effect on biological sludge dewatering. PhD dissertation thesis, University of Science and Technology, Hong Kong
222. Chu CP, Lee DJ, Liu Z, Jin WH (2004) Morphology of sludge cake at electroosmosis dewatering. Sep Sci Technol 39:1331–1346
223. Esmaeily A, Elektorowicz M et al (2006) Dewatering and coliform inactivation in biosolid using electrokinetic phenomena. J Environ Eng Sci 5:197–202
224. Banerjee S, Law SE (1997) Elecroosmotically enhanced drying of biomass. IEEE Ind Appl Soc 3:1859–1866
225. Laursen S, Jensen JB (1993) Electroosmosis in filter cakes of activated-sludge. Water Res 27:777–783
226. Gazbar S (1993) Evaluation et amélioration des performances des procèdes de déshydratation mécanique des boues résiduaires. PhD dissertation thesis, Nancy, France
227. Weng CH, Yuan C (2002) Enhancement of sludge dewatering: application of electrokinetic technique. J Chin Inst Environ Eng 12:235–243
228. Yuan C, Weng CH (2003) Sludge dewatering by electrokinetic technique: effect of processing time and potential gradient. Adv Environ Res 7:727–732
229. Olivier J, Vaxelaire J, Ginisty P (2004) Gravity drainage of activated sludge: from laboratory experiments to industrial process. J Chem Technol Biotechnol 79:461–467
230. Gingerich I, Neufeld RD, Thomas TA (1999) Electroosmotically enhanced sludge pressure filtration. Water Environ Res 71:267–276
231. Condie D, Szemes F, Veal C (1996) Electrodewatering of wastewater treatment plant sludges. In: Proceedings of the 7th world filtration congress, Budapest
232. Perry RH, Green DW (1997) Perry's chemical engineers' handbook. McGraw-Hill, New York
233. Mujumdar AS (2007) Handbook of industrial drying. CRC Press, Boca Raton
234. Friehmelt V, Gidarakos E, Laser U (1995) Dewatering of sludges using electrokinetic effects. Aufbereit-Tech 36:267–276
235. Citeau M, Larue O, Vorobiev E (2011) Influence of salt, pH and polyelectrolyte on the pressure electro-dewatering of sewage sludge. Water Res 45:2167–2180
236. Laursen S (1993) Electrokinetics in clays and filter cakes of activated-sludge. Water Sci Technol 28:181–188
237. Mikkelsen LH (2003) Applications and limitations of the colloid titration method for measuring activated sludge surface charges. Water Res 37:2458–2466
238. Acar YB, Alshawabkeh AN (1993) Principles of electrokinetic remediation. Environ Sci Technol 27:2638–2647
239. Acar YB, Hamed JT, Alshawabkeh AN, Gale RJ (1994) Removal of cadmium (II) from saturated kaolinite by the application of electrical current. Geotechnique 44:239–254
240. Acar YB, Zappi M (1995) Infrastructural needs in waste containment and environmental restoration. J Infrastruct Syst-ASCE 1:82–91
241. Yang GCC, Lin SL (1998) Removal of lead from a silt loam soil by electrokinetic remediation. J Hazard Mater 58:285–299
242. Altin A, Degirmenci M (2005) Lead (II) removal from natural soils by enhanced electrokinetic remediation. Sci Total Environ 337:1–10
243. Mahmoud A, Muhr L, Vasiluk S, Aleynikoff A, Lapicque F (2003) Investigation of transport phenomena in a hybrid ion exchange-electrodialysis system for the removal of copper ions. J Appl Electrochem 33:875–884
244. Mahmoud A, Muhr L, Grévillot G, Lapicque F (2007) Experimental tests and modelling of an electrodeionization cell for the treatment of dilute copper solutions. Can J Chem Eng 85:171–179

245. Monzie I, Muhr L, Lapicque F, Grévillot G (2005) Mass transfer investigations in electrodeionization processes using the microcolumn technique. Chem Eng Sci 60:1389–1399
246. Tuan PA, Sillanpää M (2010) Fractionation of macro and trace metals due to off-time interrupted electrodewatering. Dry Technol 28:762–772
247. Yang L, Nakhla G, Bassi A (2005) Electro-kinetic dewatering of oily sludges. J Hazard Mater 25:130–140
248. Huang J, Elektorowics M, Oleszkiewics JA (2008) Dewatering and disinfection of aerobic and anaerobic sludge using an electrokinetic (EK) system. Water Sci Technol 57:231–236
249. Saveyn H, Curvers D, Pel L et al (2006) In situ determination of solidosity profiles during activated sludge electrodewatering. Water Sci Technol 40:2135–2142
250. Miller SA, Sacchetta C, Veal C (1997) Electrodewatering of waste activated sewage sludge. In: Proceedings of the 17th AWWA federal convention, Melbourne
251. Yamaguchi M, Ari T, Matsushita H (1986) Dewatering of water and wastewater sludges with an electro-osmotic dewatering system. Water Waste 28:36–41
252. Yukawa H, Yoshida H (1986) Recent development of electroosmotic dewatering. In: Mujumdar AS (ed) Drying of solids recent international developments. Willey Eastern, New Delhi
253. Kondoh S, Suwa T, Sano S, Muroi O (1991) Compressive and electro-osmotic dehydrator. US Pat 5034111
254. ELODE Electro-Osmosis Dehydrator: www.elode.com.sg
255. Muralidhara HS, Chauhan SP (1988) Electro-acoustic dewatering (EAD) a novel approach for food processing and recovery. Sep Sci Technol 23:2143–2158
256. Muralidhara HS, Hampel H, Kroening H (1988) Dewatering of Hamburg's dredged material by electroacoustic dewatering. Dry Technol 6:535–546
257. Muralidhara HS, Parekh BK, Senapati N (1988b) Solid-liquid separation process for fine particle suspensions by an electric and ultrasonic field. US Pat 4747920
258. Ensminger D (1988) Acoustic and electroacoustic methods of dewatering and drying. Dry Technol 6:473–499
259. Ensminger D (1989) Electroacoustic dewatering of sludge. In: Proceedings of the american filtration society's annual technical conference on filtration and separation. Pittsburgh, PA
260. Chauhan SP, Kim BC et al (1989) Scale-up of electroacoustic dewatering (EAD) process for food products. In: Proceedings of the summer national AIChE meeting, Philadelphia, Pennsylvania
261. Heikkinen J, Tuori T, Wakeman RJ et al (2000) Development of doliquoring method enhanced by electric and acoustic fields. In: Proceedings of the 8th world filtration congress, Brighton
262. Abu-Orf M, Muller CD, Park C, Novak JT (2004) Innovative technologies to reduce water content of dewatered municipal residuals. J Residuals Sci Technol 1:83–91
263. Chauhan SP, Johnson HW (1990) Scale-up of electroacoustic dewatering of sewage sludges. In: Muralidhara HS (ed) Solid/liquid separation: waste management and productivity enhancement. Battelle, Columbus
264. Scholes ON (2005) Mechanical thermal expression of lignite: directional dewatering and permeability characteristics. PhD dissertation. Monash University, Clayton, Australia
265. EMICO Water Technologies-Electro-dewatering: www.ewt-cinetik.com

Chapter 10
Biosorption of Heavy Metals: Recent Trends and Challenges

Ravindra Kumar Gautam, Mahesh Chandra Chattopadhyaya, and Sanjay K. Sharma

10.1 Introduction

Rapid industrialisation and urbanisation coupled with increased population growth has lead to an increased disposal of heavy metals into the environment. Environmentalists are primarily concerned with the presence of heavy metals due to their toxicity and impact on human health, environment, and ecosystems. US Environmental Protection Agency has classified heavy metals as priority pollutants. Heavy metal pollution in the aqueous solutions has become a serious threat today and of great environmental concern as they are non-biodegradable and thus persistent in the nature. They accumulate in living cells of biological systems throughout the food chain which has humans at its top, multiplying the threat. Thus, it is necessary to control presence of heavy metals in the aqueous environment.

These potentially toxic and relatively accessible metals to the biological systems have attracted the keen attention of the scientists and researchers. These toxic metals occur in very small quantities in the Earth's crust and hence are called "rare metals" [1]. These are further arbitrarily subdivided on the basis of their densities. Those having densities less than 5 g/cm^3 are designated as "light metals," while those having densities more than 5 g/cm^3 are designated as "heavy metals". Thus, metals like Hg, Cd, Cr, Cu, Ni, Fe, and Pb are generally known as toxic heavy metals. Unexpectedly, even metalloids like Se, As, and Sb are also considered under this category, thereby making the term heavy metals misleading. Among the

R.K. Gautam • M.C. Chattopadhyaya (✉)
Environmental Chemistry Research Laboratory, University of Allahabad, Allahabad, U. P. 211 002, India
e-mail: ravindragautam1987@gmail.com; mcc46@rediffmail.com

S.K. Sharma
Department of Chemistry, JECRC University, Jaipur, India
e-mail: drsanjay1973@gmail.com

S.K. Sharma and R. Sanghi (eds.), *Wastewater Reuse and Management*,
DOI 10.1007/978-94-007-4942-9_10,

metallic species, Pb, Cd, Hg, Cr, Ni, and As have drawn greater attention for their environmental impacts as they can enter the human beings via food chain and often exceed the toxic levels before they produce visible toxic effects. World Health Organization has, recently, recognised the health hazards of toxic metals in food chain even at low concentrations. Environmental Protection Agency's "Top 20 Hazardous Substance Priority List" has ranked mercury as 6th, cadmium 7th, chromium 8th, and nickel 13th, while quoting arsenic as the king of poisons.

Various methods exist for the removal of heavy metal ions from wastewater which include chemical precipitation, coagulation, membrane technology, electrolytic reduction, ion exchange, and adsorption. Ion exchange is the most widely used method for heavy metal removal from aqueous streams. During removal, recovery, or processing of metals, ion exchange acts as a concentrator of metals. Coagulation–flocculation can also be employed to treat wastewater laden with heavy metals wherein the coagulation process destabilises colloidal particles by adding a chemical agent and results in sedimentation. Membrane filtration has received considerable attention for the treatment of inorganic effluent since it is capable of removing not only suspended solid and organic compounds but also inorganic contaminants such as heavy metals.

Among all the treatment processes mentioned, adsorption using sorbents is one of the most popular and effective processes for the removal of heavy metals from wastewater. The adsorption process offers flexibility in design and operation and in many cases produces treated effluent suitable for reuse, free of colour and odour. In addition, because adsorption is sometimes reversible, the regeneration of the adsorbent with resultant economy of operation may be possible.

Activated carbon adsorbents are used widely in the removal of organic contaminants and to a lesser extent heavy metal contaminants in product purification and pollution control. In spite of its prolific use, activated carbon remains an expensive material, and the higher the quality of activated carbon, the higher its cost. Also, in practice, activated carbon is employed more frequently for adsorption of organic compounds rather than heavy metal ions.

10.2 Sources of Heavy Metals

Various industries generate heavy metal containing wastewater includes the tanning, battery, glassware, ceramics, electroplating, fertiliser, mining, paints, and photographic industries. These wastewaters contain heavy metals such as chromium, lead, cadmium, arsenic, copper, iron, manganese, nickel, mercury, and cobalt, among others. The amount and the number of metals present in any wastewater are related directly to the operations carried out in an industry. For example, tanneries discharge chromium in wastewater; copper, chromium, zinc, and cadmium are widely generated from metal plating; the production of electrical equipment and mining, smelting, and fossil fuel combustion contribute to mercury pollution; and lead is generated from a number of industrial and mining sources. In

Table 10.1 Significant anthropogenic sources of heavy metal in the environment

Industry	Metals	Pollution arising	References
Electroplating	Cr, Ni, Zn, Cu	Liquid effluents from plating processes	[2] [3]
Batteries	Pb, Sb, Zn, Cd, Ni, Hg	Waste battery fluid, contamination of soil and groundwater	[4]
Paints and pigments	Pb, Cr, As, Ti, Ba, Zn	Aqueous waste from manufacture, old paint deterioration, and soil pollution	[5] [6]
Landfill leachate	Zn, Cu, Cd, Pb, Ni, Cr, Hg	Landfill leachate, contamination of ground and surface water	[7] [8]
Electronics	Pb, Cd, Hg, Pt, Au, Cr, As, Ni, Mn	Aqueous and solid metallic waste from manufacturing and recycling process	[9]
Metalliferous mining	Cd, Cu, Ni, Cr, Co, Zn	Acid mine drainage, tailings, slag heaps	[10]
Fertilisers	Cd, Cr, Mo, Pb, U, V, Zn	Run-off, surface and groundwater contamination, plant bioaccumulation	[11] [12]
Manures sewage sludge	Zn, Cu, Ni, Pb, Cd, Cr, As, Hg	Land spreading threat to ground and surface water	[11] [13]
Specialist alloys and steels	Pb, Mo, Ni, Cu, Cd, As, Te, U, Zn	Manufacture, disposal, and recycling of metals. Tailings and slag heaps	[14] [15]
Paper and pulp	Zn, Cu, Cd, Pb, Ni, Fe, Mn	Wastewater effluents	[16]

most wastewaters, the concentration of heavy metals present is much larger than the safe permissible limits, and therefore, they need to be removed. Table 10.1 summarises the anthropogenic sources of heavy metals in the environment.

10.3 Toxicity to Human Health and on Ecosystems

It becomes a matter of alarm when metals accumulate in a locality to concentrations approaching or exceeding those that can harm organisms, including human beings. Some heavy metals are essential for biological systems in small amounts but if ingested in larger amounts can disrupt physiological processes. Heavy metals, such as Cd, Hg, Pb, and Sn, may serve no particular biological function and can be toxic at relatively low concentrations. Heavy metals tend to bioaccumulate in the food chain, and their toxicity to organism ranges from chronic to acute and from sublethal to deadly.

At high exposure levels, lead causes encephalopathy, cognitive impairment, behavioural disturbances, kidney damage, anaemia, and toxicity to the reproductive system. Chromium is widely recognised to exert toxic effects in its hexavalent form to biological systems. Human exposure to Cr(VI) compounds is associated with a higher frequency of respiratory cancers. Cadmium is associated with nephrotoxic effects particularly at high exposure levels; long-term exposure may cause bone damage as well. High concentrations of mercury can lead to neurobehavioural disorders and developmental disabilities including dyslexia, attention deficit hyperactivity disorder, and intellectual retardation. Excessive copper concentrations can lead to weakness, lethargy, and anorexia, as well as damage to the gastrointestinal tract.

Due to their detrimental effects on human beings and biodiversity, it becomes necessary to remediate the heavy metals containing wastewater and industrial effluents. Biosorption offers a green approach for the treatment of heavy metals containing wastewater for management and sustainable development.

10.4 Biosorption

Biosorption has been defined as the property of certain biomolecules to bind and concentrate selected ions or other molecules from aqueous solutions [17]. As opposed to a much more complex phenomenon of bioaccumulation based on active metabolic transport, biosorption by dead biomass is passive and based mainly on the "affinity" between the sorbent and sorbate. Biosorption of heavy metals has become a popular environmentally driven research topic; it represents only one particular type of a concentration-removal aspect of the sorption process. The methodology of studying biosorption is based on an interdisciplinary approach to it, whereby the phenomenon can be studied, examined, and analysed from different angles and perspectives—by chemists, chemical engineers, as well as by environmentalists.

In recent years, research on the mechanisms of biosorption has intensified since biomass can be employed to sequester heavy metals from industrial effluents or to recover precious metals from waste aqueous solutions. Of the many types of biosorbents (i.e., fungi, bacteria, yeasts, weeds, and industrial and agricultural wastes) recently investigated for their ability to sequester heavy metals, these low-cost biomass has proven to be highly effective as well as reliable and predictable in the removal of, for example, Pb^{2+}, Cu^{2+}, Cd^{2+}, Cr^{6+}, and Zn^{2+} from aqueous solutions.

There has been an intense research effort aimed at characterising the metal-binding properties of various forms of biomass. These have included fresh and saltwater algae, bacteria, fungi, and yeasts and industrial and agricultural waste/by-products. Metal sequestration by diverse group of biosorbents has been reviewed by several authors worldwide. Uptake of metals by different low-cost biosorbents is summarised in Table 10.2.

Jha et al. [39] studied chitosan for Cd(II) removal and reported an adsorption capacity of 5.93 mg/g in the pH range of 4.0–8.3. It was observed that the presence of ethylenediaminetetraacetic acid (EDTA) significantly decreased the cadmium removal. The adsorption of some metal ions on chitosan was also investigated by McKay et al. [40] who found the adsorption capacity of chitosan for removal of Hg^{2+}, Cu^{2+}, Ni^{2+}, and Zn^{2+} as 815, 222, 164, and 75 mg/g, respectively. Udaybhaskar et al. [41] have studied the sorption of Cr(VI) onto chitosan and reported a highest adsorption capacity of 273 mg/g. Recently, Gupta et al. [42] have reported the removal of Pb^{2+}, Cu^{2+}, and Ni^{2+} using hydroxyapatite/chitosan composite as an adsorbent.

Table 10.2 Recently reported biosorption studies of biomass materials for heavy metals and metalloids

Biomass material	Metal	Biomass type	Biosorption capacity (mg/g)	pH	Reaction time (h)	Adsorbent concentration (g/L)	References
Garcinia cambogia	As (III)	Plants	704.11	6	0.5	5	[18]
Waste crab shells	As (V)	Animal	8.3	2.51	24	2.0	[19]
Saccharomyces cerevisiae (brewer's yeast) (from brewery)	Hg	Fungi (yeast)	133.3	6.0	–	2.0	[20]
Penicillium oxalicum var. Armeniaca (residue from fermentation industry)	Hg	Fungi	269.3	5–6.2	72	0.1, 0.2	[21]
Tolypocladium sp. (residue from fermentation industry)	Hg	Fungi	161.0	7	72	0.1, 0.2	[21]
Cystoseira baccata	Hg	Algae	329	6.0	4	2.5	[22]
Spirulina sp. (commercially available)	Cd	Bacteria	99.5	7	0.5	1	[23]
Saccharomyces cerevisiae (waste brewer's yeast)	Cd	Fungi (yeast)	15.4	4	3	2	[24]
Baker's yeast (lab cultured)	Cd	Fungi (yeast)	11.63	–	0.5	1	[25]
Phomopsis sp. (lab cultured)	Cd	Fungi	29 ± 1	6.0	24	2.0	[26]
Ulva onoi	Cd	Algae	61.9	7.8	12	1	[27]
Ulva onoi (soaked in 0.1 *M* NaOH for 1 h)	Cd	Algae	90.7	7.8	12	1	[27]
Gelidium sesquipedale	Cd	Algae	18.0 ± 0.2	5.3	1	2	[28]
Olive pomace	Cd	Plants	5.5 ± 0.1	5	–	10	[29]
Azolla filiculoides	Cd	Plants	111–132	5.5	10	2	[30]
Phragmites australis shoot	Cd	Plants	10.2 ± 0.3	–	3	1	[31]
Gloeocapsa gelatinosa	Pb	Bacteria	256.41	4	0.5	0.1	[32]
Saccharomyces cerevisiae (waste brewer's yeast)	Pb	Fungi (yeast)	85.6	4	3	2	[24]
Penicillium chrysogenum (lab cultured)	Pb	Fungi	204 ± 3	5.5	6	1	[33]
Penicillium oxalicum var. Armeniaca (residue from fermentation industry)	Pb	Fungi	47.4	5	72	0.1, 0.2, 0.3	[21]
Sargassum sp.	Pb	Algae	303	5	6	1	[34]

(continued)

Table 10.2 (continued)

Biomass material	Metal	Biomass type	Biosorption capacity (mg/g)	pH	Reaction time (h)	Adsorbent concentration (g/L)	References
Azolla filiculoides	Pb	Plants	264–297	5.5	10	2	[30]
Phragmites australis shoot	Pb	Plants	17.2	–	3	1	[31]
Saccharomyces cerevisiae (waste brewer's yeast)	Ni	Fungi (yeast)	6.34	4	3	2	[24]
Penicillium chrysogenum (lab cultured)	Ni	Fungi	55 ± 1	5.5	6	1	[33]
Sargassum sp.	Ni	Algae	71.6	5	6	1	[34]
Azolla filiculoides	Ni	Plants	70.3–80.1	5.5 ± 0.2	10	2	[30]
Phragmites australis shoot	Ni	Plants	7.92 ± 0.06	–	3	1	[31]
Phomopsis sp. (lab cultured)	Zn	Fungi	10.3 ± 0.3	5	24	2	[26]
Ulva onoi	Zn	Algae	74.6	7.8	12	1	[27]
Azolla filiculoides	Zn	Plants	64.1–84.4	5.5 ± 0.2	10	2	[30]
Phragmites australis shoot	Zn	Plants	5.75 ± 0.07	–	3	1	[31]
Spirulina sp.	Cr(III)	Bacteria	185	7	0.5	1	[23]
Saccharomyces cerevisiae (waste brewer's yeast)	Cr(III)	Fungi (yeast)	12.8	4	18	2	[24]
Bengal gram (*Cicer arientinum*) seed husk	Cr(VI)	Plants	91.64	2.0	5	10	[35]
Soybean hulls	Cr(VI)	Plants	58 ±2	3	24	10	[36]
Sugarcane bagasse	Cr(VI)	Plants	103 ±9	3	24	10	[36]
Corn stover	Cr(VI)	Plants	84 ± 10	3	24	10	[36]
Spirulina sp.	Cu	Bacteria	196	7	0.5	1	[23]
Sphaerotilus natans	Cu	Bacteria	60	6	0.5	3	[37]
Aspergillus niger	Cu	Fungi	26	6	24	2	[38]
Sargassum sp.	Cu	Algae	87.1	5	2	1	[34]
Olive pomace	Cu	Plants	10 ± 1	5	–	10	[29]
Phragmites australis shoot	Cu	Plants	9.91 ± 0.06	–	3	1	[31]

In addition to several natural materials, a number of agricultural wastes/by-products have been investigated as adsorbents for the removal of pollutants from aqueous solutions by a number of workers. Namasivayam and Periasamy [43] converted peanut hull into an adsorbent peanut hull carbon (PHC) by treating it with concentrated sulphuric acid, then carbonising in air and further treating with 1% sodium bicarbonate overnight. The authors observed much superior adsorption capacity (109.89 mg/g) for the bicarbonate-treated peanut hull compared to a commercial granular activated carbon (12.38 mg/g), which was suggested to be due to higher porosity and the moderate ion exchange capacity of the bicarbonate-treated peanut hull compared to the activated carbon.

Sun and Shi [44] have studied the feasibility of utilising sunflower stalks, a renewable agricultural waste available at low cost. The adsorption was found to be physical with an adsorption capacity of 29.3, 30.73, 42.18, and 25.07 mg/g for Cu^{2+}, Zn^{2+}, Cd^{2+}, and Cr^{3+}, respectively. Parab et al. [45] tested the efficiency of the coir pith as low-cost adsorbent and studied the adsorption efficiency of the material for the removal of Co(II), Cr(III), and Ni(II) using a batch adsorption technique from synthetic nuclear power plant coolant water. The researcher noted that even though the adsorptive capacity of coir pith is not excessively high for cobalt, chromium, and nickel, the low cost of the material together with its adsorptive ability could offer a promising procedure for the treatment of industrial wastewaters.

Raw bagasse is a waste biomaterial from sugar mills and has been used as a low-cost adsorbent for the removal of chromium and nickel from aqueous solution by Rao et al. [46]. The efficiency of the biosorbent for the removal of Cr(VI) at pH 6 and Ni(II) at pH 8, with a sorbate concentration 100 mg/L, contact time 90 min, adsorbent particle size 75 μm, and adsorbent dose 3.5 g/L, was found to be 91% and 85.6%, respectively.

10.5 Modelling of Biosorption: Isotherm and Kinctic Models

Assessment of a solid–liquid sorption system is usually based on two types of investigations: equilibrium batch sorption tests and dynamic continuous-flow sorption studies. Equilibrium isotherm model equations such as Langmuir and Freundlich are used to describe experimental adsorption data in batch mode. It is important to find best-fit isotherm to evaluate the efficacy of the prepared adsorbent to develop suitable industrial adsorption system designs.

10.5.1 Equilibrium Modelling of Biosorption

Equilibrium isotherm models are usually classified into the empirical equations and the mechanistic models. The mechanistic models are based on mechanism of metal ion biosorption, which are able not only to represent but also to explain and predict the experimental behaviour. Some empirical models for single solute systems are listed in Table 10.3. The Langmuir model (based on monolayer adsorption of solute) and the Freundlich model (developed for heterogeneous surfaces) are the most widely accepted and used in literatures. The Brunauer–Emmett–Teller (BET) model describes the multilayer adsorption at the biosorbent surface and assumes that the Langmuir isotherm applies to each layer. These models can provide information of metal uptake capacity and difference in metal uptake between various species.

10.5.2 Freundlich Isotherm

Freundlich isotherm is an empirical equation. This equation is one among the most widely used isotherms for the explanation of adsorption equilibrium. Freundlich isotherm is capable of describing the adsorption of heavy metals on a wide variety of biosorbents. This equation has the following form:

$$q_e = K_F \, C_e^{1/n}$$

It can also be expressed in the linearised logarithmic form

$$\log q_e = \log K_F + \frac{1}{n} \log C_e$$

The plot of log q_e versus log C_e has a slope with the value of $1/n$ and an intercept magnitude of log K_F. Log K_F is equivalent to log q_e when C_e equals unity. Though, in other case when $1/n \neq 1$, the K_F value depends on the units upon which q_e and C_e are expressed. On average, a favourable adsorption tends to have Freundlich constant n between 1 and 10. Larger value of n (smaller value of $1/n$) implies stronger interaction between biosorbent and heavy metal, while $1/n$ equal to 1 indicates linear adsorption leading to identical adsorption energies for all sites [47].

10.5.3 Langmuir Isotherm

Langmuir is well-known used model for describing heavy metal sorption onto biosorbent. Langmuir equation relates the coverage of molecules on a solid surface to concentration of a medium above the solid surface at a fixed temperature. This isotherm based on three assumptions, namely, adsorption is limited to monolayer coverage, all surface sites are alike and only can accommodate one adsorbed atom, and the ability of a molecule to be adsorbed on a given site is independent of its neighbouring sites occupancy [47]. By applying these assumptions, and a kinetic

Table 10.3 Summary of widely used isotherms for biosorption systems with their advantages and disadvantages

Isotherm	Functional form	Linear form	Plot	Advantages	Disadvantages
Langmuir	$q_e = \frac{q_m K_L C_e}{1+K_L C_e}$	$\frac{C_e}{q_e} = \frac{1}{K_L q_m} + \frac{1}{q_m} C_e$	$\frac{C_e}{q_e}\ VS \cdot C_e$	Has Henry law and finite saturation limit so valid over a wide range of concentration	Based on monolayer assumption
Freundlich	$q_e = K_F\ C_e^{\frac{1}{n}}$	$\ln q_e = \ln K_F + \frac{1}{n} \ln C_e$	$\ln q_e\ VS \cdot \ln C_e$	Simple expression and has parameter for surface heterogeneity	Does not have Henry law and no saturation limit, not structured, not applicable over wide range of concentration
Temkin	$q_e = \frac{RT}{b} \ln(K_T C_e)$	$q_e = \frac{RT}{b} \ln K_T + \frac{RT}{b} \ln C_e$	$q_e\ VS \cdot \ln C_e$	Simple expression	Same as Freundlich. It does not have correct Henry law limit and finite saturation limit, not applicable over wide range of concentration
Redlich–Peterson (R–P)	$q_e = \frac{K_{RP}\ C_e}{1+a_{RP}\ C_e^{\beta}}$	$\ln\left[\left(\frac{K_{RP}\ C_e}{q_e}\right) - 1\right] = \ln a_{RP} + \beta \ln C_e$	$\ln\left[\left(\frac{K_{RP}\ C_e}{q_e}\right) - 1\right]\ VS \cdot \ln C_e$	Approaches Freundlich at high concentration	No special advantages

principle (rate of adsorption and desorption from the surface is equal), the Langmuir equation can be written in the following form:

$$q_e = q_{\max} \frac{K_L C_e}{1 + K_L C_e}$$

The linear form of this equation is often written as

$$\frac{C_e}{q_e} = \frac{1}{q_{\max}} C_e + \frac{1}{K_L q_{\max}}$$

Within the Langmuir model, the saturation capacity $q_{\max}$ is supposed to coincide with saturation of a fixed number of identical surface sites, and as such, it should logically be independent of temperature.

Sometimes, these empirical models do not reflect any mechanisms of sorbate uptake and hardly have a meaningful physical interpretation for biosorption. Some author have pointed out that the results from the empirical models cannot be extrapolated, and no predictive conclusions can be drawn for systems operating under different conditions [48]. Both simple basic models (Langmuir and Freundlich models) also do not incorporate the effects of any external variable environmental factors, although they are capable of describing many biosorption isotherms in most cases. The mechanistic conclusions from the good fit of the models alone should be avoided. Moreover, biosorption isotherms may exhibit an irregular pattern due to the complex nature of both the biosorbents and its varied multiple active sites, as well as the complex solution chemistry of some metallic compounds.

10.5.4 Other Two-Parameter Isotherms

A particular model might be invalid in a particular circumstance, and in most cases, more than one model can explain the biosorption mechanism. Several models initially developed for gas phase adsorption can be implemented to correlate heavy metals biosorption processes. Some of these equations contain two fitting parameters (Temkin isotherm, Flory–Huggins, and Dubinin–Raduskevich equations), whereas others can have more than two parameters (Redlich–Paterson and Sips isotherms) [47].

10.5.5 Kinetic Studies in Biosorption of Heavy Metals

Adsorption equilibria studies are important to conclude the efficacy of adsorption. In spite of this, it is also necessary to identify the adsorption mechanism type in a given system. On the purpose of investigating the mechanism of biosorption and its potential rate-controlling steps that include mass transport and chemical reaction processes, kinetic models have been exploited to test the experimental data. In addition, information on the kinetics of metal uptake is required to select the optimum condition for full-scale batch metal removal processes.

Several adsorption kinetic models have been established to understand the adsorption kinetics and rate-limiting step. These include pseudo-first- and pseudo-second-order rate models, Weber and Morris sorption kinetic model, Adam–Bohart–Thomas relation, first-order reversible reaction model, external mass transfer model, first-order equation of Bhattacharya and Venkobachar, Elovich's model, and Ritchie's equation. The pseudo-first- and pseudo-second-order kinetic models are the most well-liked models to study the biosorption kinetics of heavy metals and quantify the extent of uptake in biosorption kinetics.

10.5.5.1 The Pseudo-First-Order Kinetic

The Lagergren first-order rate expression based on solid capacity is generally expressed as follows:

$$\frac{\mathrm{d}q}{\mathrm{d}t} = k_1(q_e - q)$$

Integration of the above equation with the boundary conditions, $t = 0$, $q = 0$, and $t = t$, $q = q$, gives

$$\ln(q_e - q) = \ln q_e - k_1 t$$

This equation can be written in the linear form:

$$q = q_e(1 - \exp(-k_1 t))$$

Kinetic adsorptions by numerous biological materials have been studied using the pseudo-first-order kinetic model [47].

10.5.5.2 The Pseudo-Second-Order Kinetic

Predicting the rate of adsorption for a given system is among the most important factors in adsorption system design, as the system's kinetics determine adsorbate residence time and the reactor dimensions [47]. As previously noted that although various factors govern the adsorption capacity, that is, initial heavy metals concentration, temperature, pH of solution, biosorbent particle size, and heavy metals nature, a kinetic model is only concerned with the effect of observable parameters on the overall rate.

Pseudo-second-order model is derived on the basis of the sorption capacity of the solid phase, expressed as

$$\frac{dq}{dt} = k_2(q_e - q)^2$$

Integration of above equation with the boundary conditions $t = 0, q = 0$, and at $t = t, q = q$ results in

$$\frac{1}{q_e - q} = \frac{1}{q_e} + k_2 t$$

This equation can be stated in the linear form as

$$\frac{t}{q} = \frac{t}{q_e} + \frac{1}{k_2 q_e^2}$$

The pseudo-second-order rate constants can be determined experimentally by plotting *t*/*q* against *t*. As such, in comparison to pseudo-first-order kinetic, this model is considered more appropriate to represent the kinetic data in biosorption systems [47].

Pseudo-first- and pseudo-second-order rate expressions have been and still in wide use for studying the biosorption of heavy metals from aqueous solutions. In chemisorption process, the pseudo-second-order is superior to pseudo-first-order model as it takes into account the interaction of adsorbent–adsorbate through their valency forces [47].

10.5.5.3 The Weber and Morris Sorption Kinetic Model

The Weber and Morris (WM) sorption kinetic model was initially employed by Pavasant et al. [49] to describe their biosorption experimental data. This model has the following form:

$$q = K_{WM}\sqrt{t}$$

In their investigation, the sorption process by *C. lentillifera* biomass for Cu(II), Cd(II), Pb(II), and Zn(II) was regulated by two main mechanisms, that is, intraparticle diffusion and external mass transfer. The intraparticle diffusion can be estimated with

$$D = \frac{\pi}{8640}\left(\frac{d_p K_{WM}}{q_e}\right)^2$$

The external mass transfer process was determined by

$$\frac{dq}{dt} = K'_L A\left(C - C_s^i\right)$$

They observed that the external mass transfer coefficients can be ordered from high to low values as Cu(II) > Pb(II) > Zn(II) > Cd(II), while the intraparticle diffusion coefficients were as follows: Cd(II) > Zn(II) > Cu(II) > Pb(II).

10.6 Thermodynamic Analysis

The Langmuir isotherm constant can be used in biosorption to estimate the thermodynamic parameters Gibbs free energy (ΔG), change in enthalpy (ΔH), and change in entropy (ΔS). Negative ΔG indicates the spontaneity of the adsorption process. ΔH is used to identify the nature of adsorption. A positive value of ΔS indicates increased randomness of adsorbate molecules on the solid surface than in solution.

The free energy of adsorption (ΔG) can be related with the Langmuir equilibrium constant by the following expression [50]:

$$\Delta G = -RT \ln K_L$$

Enthalpy and entropy changes are also related to the Langmuir equilibrium constant by the following expression:

$$\ln K_L = \frac{\Delta S}{R} - \frac{\Delta H}{RT}$$

Thus, a plot of $\ln K_L$ versus $1/T$ should be a straight line. ΔH and ΔS values could be obtained from the slope and intercept of this plot.

10.7 Advantages of Biosorption

There are several advantages in using low-cost biosorbents for removal and recovery of heavy metals from aqueous solutions. Some of them we have summarised below:

1. The metal removal rates from aqueous solution by biosorption are generally faster than those by other metabolically mediated processes. This property of biosorption would make it the most effective in treatment of voluminous water bodies containing low concentrations of metals, such as the final wastewater

treatment process, to meet regulation standards of metals before being discharged to the environment.

2. The costs for formulating biosorbents and operating the treatment systems are generally low, especially waste bacteria and fungi from various fermentation industries, agricultural by-products, and the biomass that exists abundantly in nature that is easily available can be utilised as cheap raw materials of biosorbents. It is not necessary to feed any substrates or energy for propagation of the raw biomass of biosorbents if these microbial wastes and abundant biomass are to be utilised. They are, however, often required to receive some modifications before introduction into the real wastewater treatment systems, and they will take additional costs according to the reagents and equipments used for the modifications.
3. It is generally easy to recover metals from metal-loaded biosorbents and regenerate the biosorbents for further adsorption–desorption cycles. Dilute acid and alkaline solutions, salt solutions, and chelating agents such as ethylenediaminetetraacetic acid (EDTA) solution are reported to be effective in eluting metals from the metal-laden biosorbents. Elution of the adsorbed metals with a smaller volume of eluent results in concentration of the metals from the original water bodies, and the metals in the eluate can be recovered and reused.
4. Performances of biosorbents can be improved through various kinds of physical and chemical pretreatments. The main purposes of modification of raw biosorbent materials are to (a) shape and impart mechanical strength and resistance to chemical and microbial degradation, and (b) improve biosorption selectivity for target metals. Pretreatments of raw biomass by physical or chemical methods have been also found to improve the metal adsorption performances of the biosorbents, although this is not always the case.
5. Biosorption treatment can be operated at a wide range of scales. A packed bed column reactor configuration is considered to be the most effective mode of the biosorption process, and this system can be scaled up to the treatment of a large volume of acid mine drainage and industrial wastewater and scaled down to the removal of As from contaminated well water.
6. Biosorbents are basically biodegradable, and the exhausted biosorbents that consist mainly of degradable organic matter could be converted into energy fuels through fermentation processes, thermal energy by combustion, or composts as fertiliser, if the complete (if not, sufficient) removal of adsorbed metals from the biosorbents is secured.

10.8 Disadvantages of Biosorption

Biosorption of metal ions is strongly affected by the properties of the water to be treated, such as pH, ionic strength, coexisting ions, and suspended solids. Optimal pH for biosorption of metal cations is generally around neutral, and their adsorbed amounts would decrease as solution pH is lowered because of diminishing of negatively charged sites on the biosorbents due to protonation. On the other hand, acidic pH favours biosorption of metal anions. Mine drainage and industrial wastewater usually contain more than one heavy metal species, and these metal ions would compete with each other for the binding sites on biosorbents. Not only heavy metals but also Ca and Mg present in hard water and Al and Fe in most of industrial effluents and mine drainage may strongly obstruct biosorption of the target heavy metals [51]. Thus, metal adsorption performance of biosorbents should be carefully evaluated in not only a single-metal solution system but also multimetal solution systems prior to their practical applications [52]. This evaluation would clarify at the same time the selectivity of the biosorbents in metal binding. The adsorption of metal cations on biosorbents would be also inhibited by inorganic and organic anions present in the solution due to complexation with the metal cations.

10.9 Biosorbents and Their Future Applications

In the development of new separation and purification technologies, such as removal of heavy metals, biosorption can play a key enabling role. A breakthrough in sorbent development is needed to solve the critical problem of removal and recovery of metal ions. However, to remove the poisoning of the metal ions from aqueous solutions, there is need for the further development of newer biosorbents of highest removal capacity.

In nature, biosorbents are biodegradable. This may be a serious challenge for long-term storage and application in sorption processes as sorbent recycling is necessary for cost-efficient use of these biosorbents for environmental cleanup. Cellulose exhibits excellent chemical stability and mechanical strength. Its fibres exhibit regions in which the molecules are densely and orderly packed, and where extensive intermolecular hydrogen bonding is observed [53]. These regions are referred to as crystalline domains, and they impart strength, swelling resistance, and stability to the fibre [54]. The detailed structural studies of biosorbents can play vital role in cost-effective modification of the biosorbent for enhancing the removal of heavy metals from aqueous solutions.

References

1. Srivastava S, Goyal P (2010) Novel biomaterials, decontamination of toxic metals from wastewater. Springer, Heidelberg
2. Castelblanque J, Salimbeni F (2004) NF and RO membranes for the recovery and reuse of water and concentrated metallic salts from waste water produced in the electroplating process. Desalination 167:65–73
3. Alvarez-Ayuso E, Garcia-Sanchez A, Querol X (2003) Purification of metal electroplating waste waters using zeolites. Water Res 37:4855–4862

4. Naiya TK, Bhattacharya AK, Mandal S, Das SK (2009) The sorption of lead(II) ions on rice husk ash. J Hazard Mater 163:1254–1264
5. Davis AP, Burns M (1999) Evaluation of lead concentration in runoff from painted structures. Water Res 33:2949–2958
6. Monken A (2000) Water pollution control for paint booths. Met Finish 98:464–471
7. Kjeldson P, Barlaz MA, Rooker AP, Baun A, Ledin A, Christensen TH (2002) Present and long term composition of MSW landfill leachate: a review. Crit Rev Environ Sci Technol 32:297–336
8. Fernandez Y, Maranon E, Castrillon L, Vazquez I (2005) Removal of Cd and Zn from inorganic industrial waste leachate by ion exchange. J Hazard Mater 126:169–175
9. Veglio F, Quaresima R, Fornari P, Ubaldini S (2003) Recovery of valuable metals from electronic and galvanic industrial wastes by leaching and electrowinning. Waste Manag 23:245–252
10. Aswathanarayana U (2003) Mineral resources management and the environment. Routledge, Leiden
11. Nicholson FA, Smith SR, Alloway BJ, Carlton-Smith C, Chambers BJ (2003) An inventory of heavy metals inputs to agricultural soils in England and Wales. Sci Total Environ 311:205–219
12. Otero N, Vitoria L, Soler A, Canals A (2005) Fertiliser characterisation: major, trace and rare earth elements. Appl Geochem 20:1473–1488
13. Walter I, Martinez F, Cala V (2006) Heavy metal speciation and phytotoxic effects of three representative sewage sludges for agricultural uses. Environ Pollut 139:507–514
14. Rule KL, Comber SDW, Ross D, Thornton A, Makropoulos CK, Rautiu R (2006) Diffuse sources of heavy metals entering an urban wastewater catchment. Chemosphere 63:64–72
15. Cheng SP (2003) Heavy metal pollution in China: origin, pattern and control. Environ Sci Pollut Res Int 10:192–198
16. Hakeem AS, Bhatnagar S (2010) Heavy metal reduction of pulp and paper mill effluent by indigenous microbes. Asian J Exp Biol Sci 1:201–203
17. Volesky B (2007) Biosorption and me. Water Res 41:4017–4029
18. Kamala CT, Chu KH, Chary NS, Pandey PK, Ramesh SL, Sastry ARK, Chandra Sekhar K (2005) Removal of arsenic(III) from aqueous solutions using fresh and immobilized plant biomass. Water Res 39:2815–2826
19. Niu CH, Volesky B, Cleiman D (2007) Biosorption of arsenic(V) with acid washed crab shells. Water Res 41:2473–2478
20. Yavuz H, Denizli A, Gungunes H, Safarikova M, Safarik I (2006) Biosorption of mercury on magnetically modified yeast cells. Sep Purif Technol 52:253–260
21. Svecova L, Spanelova M, Kubal M, Guibal E (2006) Cadmium, lead and mercury biosorption on waste fungal biomass issued from fermentation industry. I. Equilibrium studies. Sep Purif Technol 52:142–153
22. Herrero R, Lodeiro P, Rey-Castro C, Vilarino T, de Vicente MES (2005) Removal of inorganic mercury from aqueous solutions by biomass of the marine macroalga *Cystoseira baccata*. Water Res 39:3199–3210
23. Chojnacka K, Chojnacki A, Gorecka H (2005) Biosorption of Cr^{3+}, Cd^{2+} and Cu^{2+} ions by blue-green algae *Spirulina* sp.: kinetics, equilibrium and the mechanism of the process. Chemosphere 59:75–84
24. Chen C, Wang J (2007) Influence of metal ionic characteristics on their biosorption capacity by *Saccharomyces cerevisiae*. Appl Microbiol Biotechnol 74:911–917
25. Yu J, Tong M, Sun X, Li B (2007) Cystine-modified biomass for Cd(II) and Pb(II) biosorption. J Hazard Mater 143:277–284
26. Saiano F, Ciofalo M, Cacciola SO, Ramirez S (2005) Metal ion adsorption by *Phomopsis* sp. biomaterial in laboratory experiments and real wastewater treatment. Water Res 39: 2273–2280
27. Suzuki Y, Kametani T, Maruyama T (2005) Removal of heavy metals from aqueous solution by nonliving *Ulva* seaweed as biosorbent. Water Res 39:1803–1808

28. Vilar VJP, Botelho CMS, Boaventura RAR (2006) Equilibrium and kinetic modelling of Cd (II) biosorption by algae *Gelidium* and agar extraction algal waste. Water Res 40:291–302
29. Pagnanelli F, Mainelli S, De Angelis S, Toro L (2005) Biosorption of protons and heavy metals onto *Olive pomace*: modeling of competition effects. Water Res 39:1639–1651
30. Rakhshaee R, Khosravi M, Taghi Ganji M (2006) Kinetic modeling and thermodynamic study to remove Pb(II), Cd(II), Ni(II) and Zn(II) from aqueous solution using dead and living *Azolla filiculoides*. J Hazard Mater 134:120–129
31. Southichak B, Nakano K, Nomura M, Chiba N, Nishimura O (2006) *Phragmites australis*: a novel biosorbent for the removal of heavy metals from aqueous solution. Water Res 40: 2295–2302
32. Raungsomboon S, Chidthaisong A, Bunnag B, Inthorn D, Harvey NW (2006) Production, composition and Pb2+ adsorption characteristics of capsular polysaccharides extracted from a cyanobacterium *Gloeocapsa gelatinosa*. Water Res 40:3759–3766
33. Deng S, Ting YP (2005) Characterization of PEI-modified biomass and biosorption of Cu(II), Pb(II) and Ni(II). Water Res 39:2167–2177
34. Chen JP, Yang L (2005) Chemical modification of *Sargassum* sp. for prevention of organic leaching and enhancement of uptake during metal biosorption. Ind Eng Chem Res 44: 9931–9942
35. Ahalya N, Kanamadi RD, Ramachandra TV (2005) Biosorption of chromium(VI) from aqueous solutions by the husk of Bengal gram (*Cicer arientinum*). Electron J Biotechnol 8: 258–264
36. Wartelle LH, Marshall WE (2005) Chromate ion adsorption by agricultural byproducts modified with dimethyloldihydroxyethylene urea and choline chloride. Water Res 39: 2869–2876
37. Beolchini F, Pagnanelli F, Toro L, Veglio F (2006) Ionic strength effect on copper biosorption by *Sphaerotilus natans*: equilibrium study and dynamic modeling in membrane reactor. Water Res 40:144–152
38. Mukhopadhyay M, Noronha SB, Suraishkumar GK (2007) Kinetics modeling for the biosorption of copper by pretreated *Aspergillus niger* biomass. Bioresour Technol 98:1781–1787
39. Jha IN, Iyengar L, Rao AVSP (1998) Removal of cadmium using chitosan. J Environ Eng ASCE 114:962
40. McKay G, Blair HS, Findon A (1989) Equilibrium studies for the sorption of metal ions onto chitosan. Indian J Chem 28:356–360
41. Udaybhaskar P, Iyengar L, Rao A (1990) Hexavalent chromium interaction with chitosan. J Appl Polym Sci 39:739–747
42. Gupta N, Kushwaha AK, Chattopadhyaya MC (2012) Adsorptive removal of Pb^{2+}, Co^{2+} and Ni^{2+} by hydroxyapatite/chitosan composite from aqueous solution. J Taiwan Inst Chem Eng 43:125–131
43. Namasivayam C, Periasamy K (1993) Bicarbonate-treated peanut hull carbon for mercury (II) removal from aqueous solution. Water Res 27:1663–1668
44. Sun G, Shi W (1998) Sunflower stalks as adsorbents for the removal of metal ions from wastewater. Ind Eng Chem Res 37:1324–1328
45. Parab H, Joshi S, Shenoy N, Lali A, Sarma US, Sudersanan M (2006) Determination of kinetic and equilibrium parameters of the batch adsorption of Co (II), Cr (III) and Ni (II) onto coir pith. Process Biochem 41:609–615
46. Rao M, Parwate AV, Bhole AG (2002) Removal of Cr^{6+} and Ni^{2+} from aqueous solution using bagasse and fly ash. Waste Manag 22:821–830
47. Febrianto J, Kosasih AN, Sunarso J, Ju YH, Indraswati N, Ismadji S (2009) Equilibrium and kinetic studies in adsorption of heavy metals using biosorbent: a summary of recent studies. J Hazard Mater 162:616–645
48. Wang J, Chen C (2009) Biosorbents for heavy metals removal and their future. Biotechnol Adv 27:195–226

49. Pasavant P, Apiratikul R, Sungkhum V, Suthiparinyanont P, Wattanachira S, Marhaba TF (2006) Biosorption of Cu^{2+}, Cd^{2+}, Pb^{2+}, and Zn^{2+} using dried marine green macroalga *Caulerpa lentillifera*. Bioresour Technol 97:2321–2329
50. Nethaji S, Sivasamy A, Thennarasu G, Saravanan S (2010) Adsorption of Malachite Green dye onto activated carbon derived from *Borassus aethiopum* flower biomass. J Hazard Mater 181:271–280
51. Lee HS, Suh JH, Kim BI, Yoon T (2004) Effect of aluminum in two-metal biosorption by an algal biosorbent. Miner Eng 17:487–493
52. Volesky B, Holan ZR (1995) Biosorption of heavy metals. Biotechnol Prog 11:235–250
53. O'Connell DW, Birkinshaw C, O'Dwyer TF (2008) Heavy metal adsorbents prepared from the modification of cellulose: a review. Bioresour Technol 99:6709–6724
54. Sionkowska A, Kaczmarek H, Vicini S, Pedemonte E, Wisniewski M (2002) The influence of camphorquinone on the photochemical stability of cellulose. Polym Degrad Stab 78:175–182

Chapter 11
Water Desalination by Solar Energy

K. Sampathkumar, T.V. Arjunan, and P. Senthilkumar

11.1 Introduction

The availability of water on the surface of the Earth has made life, as we know it, possible. Water is one of the most ubiquitous resources and can be either consumed directly or used in applications such as irrigation, industrialisation and power generation. Increases in population, urbanisation, industrialisation and modern agriculture have put a tremendous amount of stress on bodies of freshwater, threatening the health of mankind. The quality of water is vastly deteriorating due to improper waste disposal, water management and environmental views. The World Health Organisation reported that all of the water on earth can be classified as 97.5% salt water and 2.5% freshwater, 70% of which is frozen in the polar icecaps. The remaining 30% of freshwater is mostly present as soil moisture. In the end, less than 1% of the world's freshwater (about 0.007% of all water on earth) is readily accessible for direct human use [1]. Globally, 3.575 million people die each year from water-related diseases and 98% of the deaths occur in developing countries. Therefore, the importance of supplying drinkable water to remote areas can hardly be overstressed.

K. Sampathkumar (✉)
Department of Mechanical Engineering, Tamilnadu College of Engineering,
Coimbatore 641 659, Tamilnadu, India
e-mail: ksktce@gmail.com

T.V. Arjunan
Department of Mechanical Engineering, Coimbatore Institute of Engineering and Technology,
Coimbatore 641 109, Tamilnadu, India

P. Senthilkumar
Department of Mechanical Engineering, KSR College of Engineering,
Tiruchengode 637 215, Tamilnadu, India

S.K. Sharma and R. Sanghi (eds.), *Wastewater Reuse and Management*,
DOI 10.1007/978-94-007-4942-9_11,

Solar energy is one of the cleanest energy sources and helps reduce the greenhouse effect. Solar distillation is a promising solution for converting brackish water into drinkable water, which can be supplied to relatively small communities and remote areas where the solar energy is abundant and access to other conventional energy sources is limited. Solar stills operate similarly to the natural hydrological cycle and are mainly classified as either passive or active. If the water in the basin is heated by only solar radiation, the system is classified as a passive solar still. If additional heat energy is fed into the basin in addition to the solar radiation, the system is an active solar still. Single basin single slope solar stills are the most popular still configuration due to their simple design, ease of fabrication and low operational costs. However, the productivity of this system is limited and needs to be increased before it can be deployed in remote areas. Therefore, studies on water purification techniques using solar energy resources are becoming important in the present scenario.

11.2 Supply and Demand for Water

Water, one of nature's most important gifts to mankind, is an essential element to good health. Natural water sources, especially rivers, canals, ponds, wells, etc. are being exploited, mistreated and contaminated, making drinking from them unsafe. Over the years, the canal water has become highly polluted with the presence of physical and chemical impurities, as well as harmful bacteria. The World Health Organisation (WHO) reported that more than 1.1 billion people across the globe draw water from unsafe sources [2].

Water supply depends on several factors in the water cycle, including the rates of rainfall, evaporation, the use of water by plants (transpiration), river and groundwater flows. It is estimated that less than 1% of all freshwater is available for people to use (the remainder is locked up in ice sheets and glaciers). Globally, around 12,500 km^3 of water are considered available for human use on an annual basis. This amounts to about 6,600 m^3 per person/year.

Each year, about four million people die due to waterborne diseases, including two million children who die of diarrhoea. Nearly 70–80% of these people live in rural areas. Based on the population growth, projections of development and climate change, it has been estimated that the proportion of the world's population living in countries with significant water problems will increase from approximately 34% in 1994 to 63% in 2025, including large areas of Africa, Asia and Latin America. This will impact their lives and livelihood. India accounts for 2.45% of land area and 4% of water resources of the world but represents 16% of the world population. With the present population growth rate (1.9% per year), the population is expected to cross the 1.5 billion mark by 2050.

11.3 Water Desalination Techniques

One of the techniques used by people for water purification in the world is desalination. Many systems were invented for desalination from time to time, and it has been accepted by people in olden days. Major desalination techniques like vapour compression distillation, reverse osmosis and electrolysis consume electricity as input energy. But in the recent years, most of the countries in the world have been significantly affected by energy crisis because of heavy dependency on conventional energy sources (coal power plants, fossil fuels, etc.). All these problems could be solved only through efficient and effective utilisation of renewable energy resources such as solar, wind, biomass, tidal and geothermal energy, etc. Safe drinking water from available water should be made using the technologies with limited financial resources and practices that minimise ecological disturbances [3].

Many options are available to distillate the brackish/saline water. Among the non-conventional methods to purify the polluted water, the most prominent method is the 'Solar Distillation'. This is an oldest method to get the potable water from brackish or saline water by utilising the solar energy which is the most abundant natural resource in the world. Solar desalination is applicable in all areas receiving 200–350 sunny days a year. Solar distillation of brackish water is a one of the good options to obtain freshwater for drinking, hospital and laboratories, etc., as it needs simple technology and low maintenance due to which it can be used anywhere with lesser number of problems. There are no subsequent costs for delivering water to the end user and no energy costs as the conventional one. The technology involved in distillation of saline or brackish water using solar energy is relatively simple, and semi-skilled/unskilled operators can carry out its operation and maintenance.

Nature itself provides most of the required freshwater through hydrological cycle. The essential features of hydrological cycle are summarised as the production of vapours above the surface of the liquids, then transport of vapours by winds, the cooling of air-vapour mixture, condensation and precipitation. This natural process is copied on a small scale in basin type solar stills. Solar distillation process has been developed in different countries all over the world like Australia, Greece, Spain, West Indies and India. There are various categories of solar stills, namely, basin solar stills, solar still coupled with flat plate collector, multiple condensing cover still, wick type solar still, vertical solar still, etc., of which more than 90% of all functioning stills are of the basin type. The details about classification and functions of various solar distillation systems are explained in the subsequent sections.

11.4 Solar Distillation

The solar still consists of an airtight space in which evaporation and condensation take place simultaneously. It is usually constructed out of concrete/cement, fibre-reinforced plastic (FRP), glass-reinforced plastic (GRP) or galvanised iron (GI)

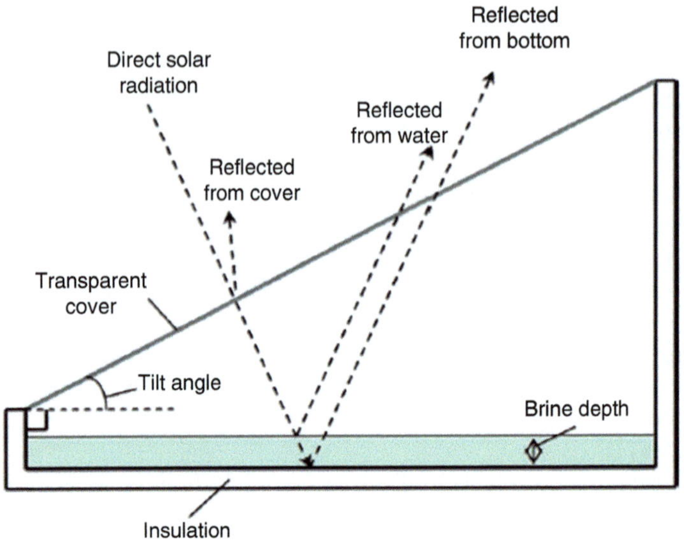

Fig. 11.1 Schematic of a single slope solar still (Reproduced from [4] with kind permission of © Elsevier BV)

sheet with a top cover of transparent material like glass or plastic, etc. The inner surface of the rectangular base is blackened to efficiently absorb incident solar radiation. The provision is made at lower end of the glass cover to collect the distillate output. The brackish or saline water is fed into the basin of the solar still for purification in a desired depth. The bottom and sides of the still is insulated properly to minimise the thermal losses to atmosphere. The top condensing cover is normally made of plane glass. The glass cover permits the short wave length solar radiation to enter into the solar still, which is mostly absorbed by blackened surface and further convected to water mass. The significant principles of this system which have been applied to solar still design are (Fig. 11.1):

1. The production of vapour above the water surface
2. The transport of this vapour by air
3. The cooling of the air-vapour mixture
4. Condensation and
5. Precipitation

The incident solar radiation on the glass cover is partially reflected and absorbed by the glass cover, and the rest is transmitted into the airtight enclosure of distillation unit. The transmitted radiation comes into contact of the water surface and part of its gets reflected and absorbed in the water mass. Finally, the solar radiation reaches the blackened surface where it is mostly absorbed. The blackened surface of the still usually known as basin liner releases the maximum thermal energy to the water mass through convection, and a very small amount of thermal energy is lost to atmosphere by conduction. Consequently, the water mass gets heated, leading to an increased temperature of water and glass cover. The heat transfer from water surface to the glass cover takes place mainly by three modes of heat transfer, namely, convection, evaporation and radiation. The evaporated water gets condensed over the inner surface of the glass cover after releasing the latent heat. The condensed water trickles under gravity into the channels provided at the lower end of the glass cover. The distillate is collected from the distillate channel to an appropriate jar through the plastic or rubber pipe for the end use. The thermal energy received by the inner surface of glass cover is lost to the ambient by radiation and convection from outer surface of glass cover.

11.5 History of Solar Distillation

Solar distillation has been used in many countries for various applications since many centuries. As early as in the fourth century B.C., Aristotle described a method to evaporate impure water and then condense it for potable use. However, historically, the earliest document work on solar distillation was reported by Mouchot in 1969, an Arab alchemist in the sixteenth century. The historical solar still apparatus consists of wide earthen pots and it is exposed to the intense heat of the solar rays to evaporate water and collect the condensate into vessel placed underneath as reported by Nebbia and Menozzi [5].

A Swedish engineer, C.Wilson, in 1872 was the first who designed conventional solar still for supplying freshwater to a nitrate-mining community in Chile, which become quite popular and was in operation for more than 40 years. During the 1950s, interest in solar distillation was revived, and in virtually all cases, the objective was to develop large centralised distillation plants. In California, the goal was to develop plants capable of producing 3,775 cubic metres of water per day. However, after about 10 years, researchers concluded that large solar distillation plants were too expensive to compete with conventional fuel-fired ones. Therefore, research shifted to smaller solar distillation plants. In 1970, 38 plants were built in 14 countries, with capacities ranging from a few hundred litres to around 30,000 l of water per day. Of these, about one third have since been dismantled or abandoned due to material failures. None in this size range is reported to have been built in the last 7 years [6]. Despite the growing discouragement over community-size plants, McCracken Solar Company in California continued its efforts to market solar stills for residential use. Worldwide interest in small residential units is growing, and now that the price of oil is ten times what it was in the 1960s, interest in the larger units may be revived.

11.6 Classification of Solar Stills

The solar distillation systems are mainly classified as passive solar still and active solar still. The numerous parameters are affecting the performance of the still such as water depth in the basin, material of the basin, wind velocity, solar radiation, ambient temperature and inclination angle. The productivity of any type of solar still will be a dependent factor by the temperature difference between the water in the basin and inner surface glass cover.

In a passive solar still, the solar radiation is received directly by the basin water and is the only source of energy for raising the water temperature and consequently the evaporation leading to a lower productivity. This is the main drawback of a passive solar still. The conventional passive solar still is the most economical solar still to provide drinking water for domestic applications at decentralised level. This is due to simplicity in design and fabrication, easy to handling and low cost of water per kg. Further, due to low operation and maintenance cost, it is most suitable for the rural areas of remote region. Later, in order to overcome the lower yield, many active solar stills have been developed. In active solar still, additional thermal energy is fed into the basin water using solar collectors or any other heat source to increase the basin water temperature and in turn improve its productivity. The classification solar distillation is given in Fig. 11.2. The different designs of passive and active solar stills are discussed as follows.

11.7 Passive Solar Stills

11.7.1 Single Slope Solar Still

The simplest structure of a solar still is single slope solar still and it is shown in Fig. 11.3. It contains a basin having a certain depth of saline water and a cover transparent to solar radiation, yet blocks the long wavelengths radiation emitted by the interior surface of the solar still. A sloped cover, which provides a cool surface for condensation of water vapour, facilitates an easy flow of the water droplets into condensate trough. The base of the still is blackened on the interior surface to maximise absorption of solar radiation and insulated on the exterior surface to minimise heat losses.

Tiwari et al. [7] presented the performance of various designs of solar stills and inferred that the single slope solar still gives better yield than double slope solar still in winter. Madani and Zaki [8] conducted the experiments in single slope solar still with porous basin and found that the daily yield is in the range of 2.5–5 kg/m^3. Lowering the cover temperature helps in increasing the productivity. The cooler inner glass surface increases the rate of condensation. The glass cover temperature

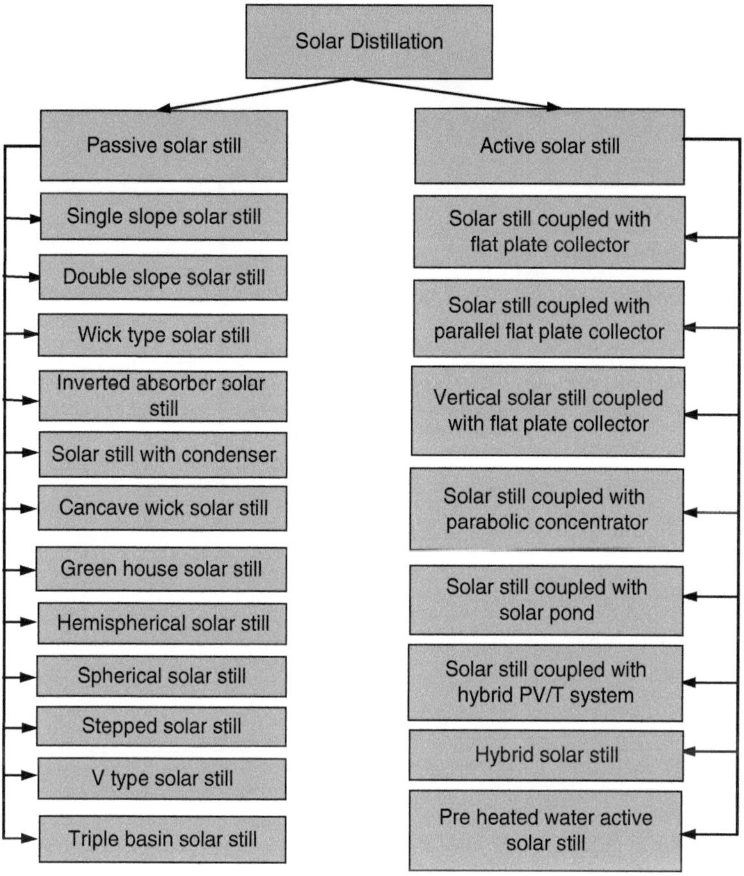

Fig. 11.2 Classification of solar distillation system

is reduced by a film of cooling water continuously flowing over the glass [9]. Toure and Meukam [10] studied the single basin solar still theoretically and experimentally. The results showed that the maximum yield was 5 l/m^2-day for a sunny day in Abidjan and also found that the wind speed has little effect on the total yield. Kalidasa Murugavel et al. [11] reviewed the progress of solar still effectiveness of the single basin passive solar still. They suggested that the orientation of the glass cover depends on the latitude and surface heating of water mass. Velmurugan et al. [12] conducted the experiments on fin type passive solar still and found the yield was increased by 52%. Khalifa and Hamood [4] presented the performance correlations for basin type solar stills and found that increasing the brine depth decreases the productivity of the basin type solar still. The correlation developed illustrates that the still productivity could be influenced by the brine depth alone by up to 33% for depth ranging from 1 to 10 cm.

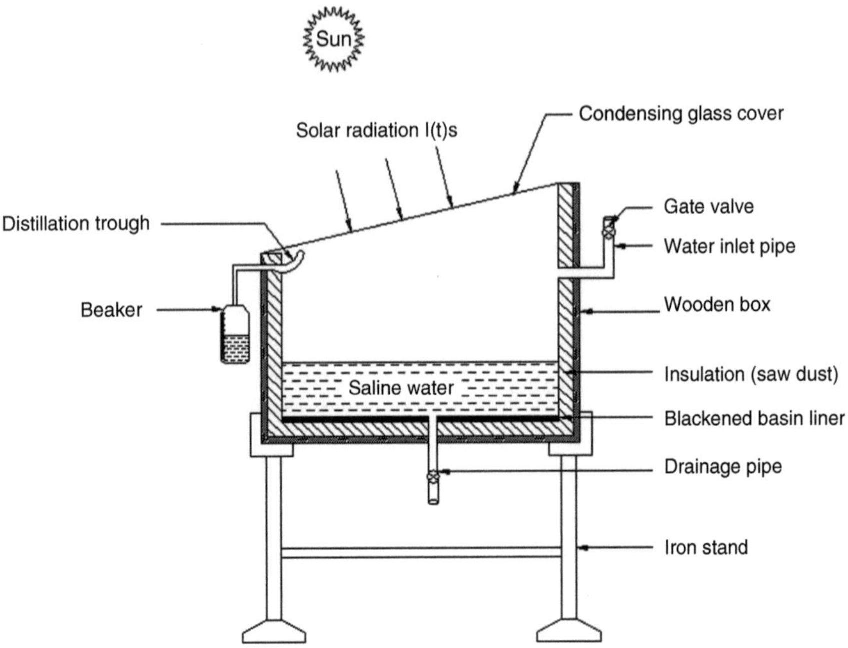

Fig. 11.3 Schematic of a single slope solar still

11.7.2 Double Slope Solar Still

The cross-sectional view of a double slope passive solar still is shown in Fig. 11.4, and it is placed normally in east west direction to receive the maximum solar radiation. The material normally used for construction of single and double slope solar still are galvanised iron sheet, fibre-reinforced plastic (FRP), glass-reinforced plastic (GRP) and concrete. The material required for construction of double slope solar still per square metre is less than two single slope solar stills due to absence of the two vertical walls.

The principle of solar still operation remains same as single slope solar still and distillates yield collected in both ends of double slope solar still. Dwivedi and Tiwari [13] found that the annual yield from single slope passive solar still is higher than double slope passive solar still. Kalidasa Murugavel and Srithar [14] experimentally studied the basin type double slope solar still with different wick materials like black cotton cloth, light jute cloth, waste cotton pieces, coir mate, aluminium fins and sponge sheet. The study revealed that the aluminium fins covered with cotton cloth gave the maximum yield of 3.58 kg/day.

11.7.3 Wick Type Solar Still

A conventional basin type solar still has some disadvantages: (1) the horizontal surface of water intercepts lesser solar radiation than a tilted surface, and (2) the

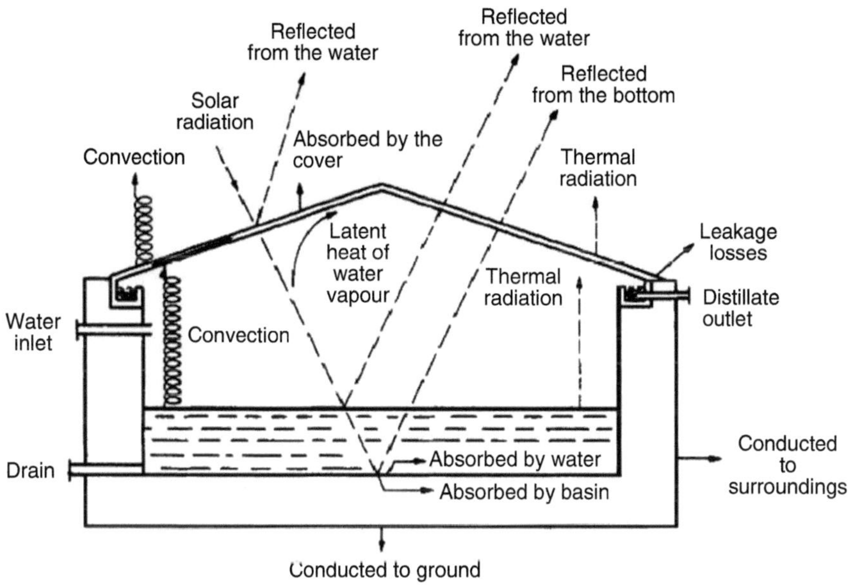

Fig. 11.4 Schematic of a double slope solar still

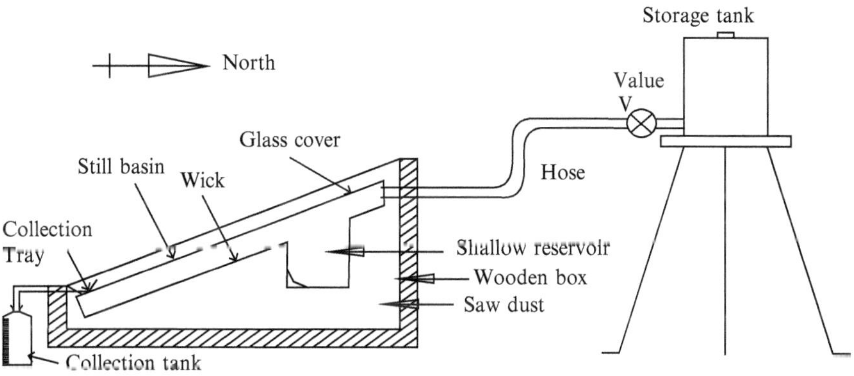

Fig. 11.5 Schematic of a wick type solar still (Reprinted from [15] with kind permission of © Elsevier Science)

output of basin type solar is also limited by the large thermal capacity of the water in the basin. A multi wick solar still (Fig. 11.5) is the best alternative for eliminating the above mentioned points. In which blackened wet jute cloth forms, the liquid surface which can be oriented to intercept maximum solar radiation and a smaller mass of water will be heated to higher temperature and will evaporate rapidly. The wet surface is created by a series of jute cloth pieces of increasing length separated by thin polythene sheets; these pieces are arranged along an incline and the upper edges are dipped in a saline water tank. Suction by the capillary action of the cloth fibre provides a surface of the liquid, and the arrangement ensures that all the surface irradiated by the sun is wet at all times; the portion of a piece of cloth, covered by the polythene sheet does not suffer evaporation and hence the exposed portion of the piece retains wetness [16, 17].

Sodha et al. [16, 17] observed that, overall efficiency of multiple wick solar still is 4% higher than the basin type still. Their results also show that the still costs less than half of the cost of a basin type still of same area and provide a higher yield of distillate.

Reddy et al. [18] showed that a multiple solar still with a condenser arrangement gives 15–25% higher than the non condenser type still. The excess vapour can be condensed on the additional surface and reduce the heat loads on the glass cover, and reduces glass cover temperature which in turn enhances evaporation rate. This concept was implemented by Tiwari et al. [19] on multi wick solar still. The authors concluded that the double condensing multiple wicks solar still gives nearly 20% higher yield than the simple wick solar still, and under cloudy and low intensity conditions, both stills show almost a similar performance. Minasian and Al-Karaghouli [20] connected a wick type solar still with a conventional basin type solar still. The result shows that the total yearly amounts of distilled water indicate that the wick basin type could produce 85% more than basin type and 43% more than wick type solar still.

11.7.4 Inverted Absorber Solar Still

The schematic diagram of inverted absorber solar still is shown in Fig. 11.6. The solar radiation, after transmission through the glass cover g_1, is reflected back to the inverted absorber of a solar still. The absorbed solar radiation is partially convected to the water mass above the inverted absorber; while the rest of the radiation is lost to the atmosphere through the glass covers g_1 and g_2. Now, the water gets heated. There are radiative, convective and evaporative heat losses from the water mass to the condensing cover. The evaporated water is condensed on the inner surface of the condensing cover, releasing its latent heat. The condensed water trickles down the condensing surface under gravity and is finally collected through drainage provided at the lower end [21].

Tiwari and Suneja [21] presented an analysis of an inverted absorber solar still. Their results show that the inverted absorber solar still gives about double the output of the conventional solar still; also the authors observed that the evaporative heat loss is a strong function of the operating temperature. Sunej et al. [22] also observed that an inverted absorber solar still gives a higher output than the conventional double effect solar still. The overall daily yield in the case of the inverted triple effect absorber solar still is 30% higher than the conventional triple effect solar still [23–25]. Suneja and Tiwari [26] optimised the number of effect on the multi basin solar still. They found that the yield from inverted absorber solar stills increases as the number of basins increased and reaches an optimum

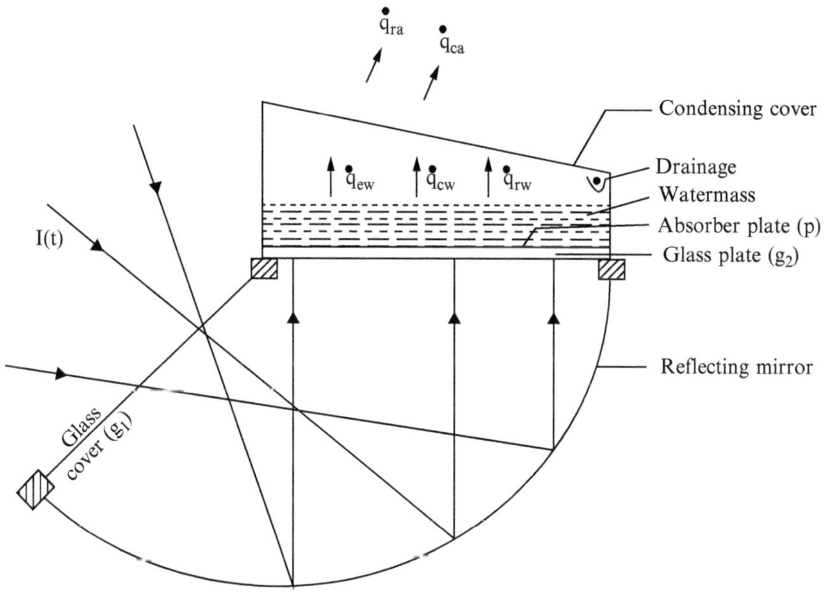

Fig. 11.6 Schematic of a inverted absorber solar still (Reprinted from [21] with kind permission of © Pergamon)

value when the number of basin is seven. The operation and maintenance expenses of inverted double basin solar still are small compared with the conventional solar system [23].

11.7.5 Solar Still with Condenser

The performance of a parallel double glass solar still with separate condenser was studied by El Bahi and Inan [27, 28]. In the system, a solar still designed and built utilises direct and reflected solar radiation incident on a parallel double glass cover to evaporate saline water.

Water vapour purges from the evaporator and diffuses to an integrated condenser due to pressure difference that exists because of the volume ration and temperature difference between the evaporator and condenser (Fig. 11.7).The study revealed that the efficiency was increased from 48% to more than 70% when the condenser was cooled down. El-Bahi and Inan [27] investigated the effect of adding an outside passive condenser to a single basin type solar still with minimum inclination and found that the daily yield of 7 kg/m^2 and efficiency of 75% were obtained during the summer months. Also found that the solar still was operated without a condenser, the yield was decreased to 70%.

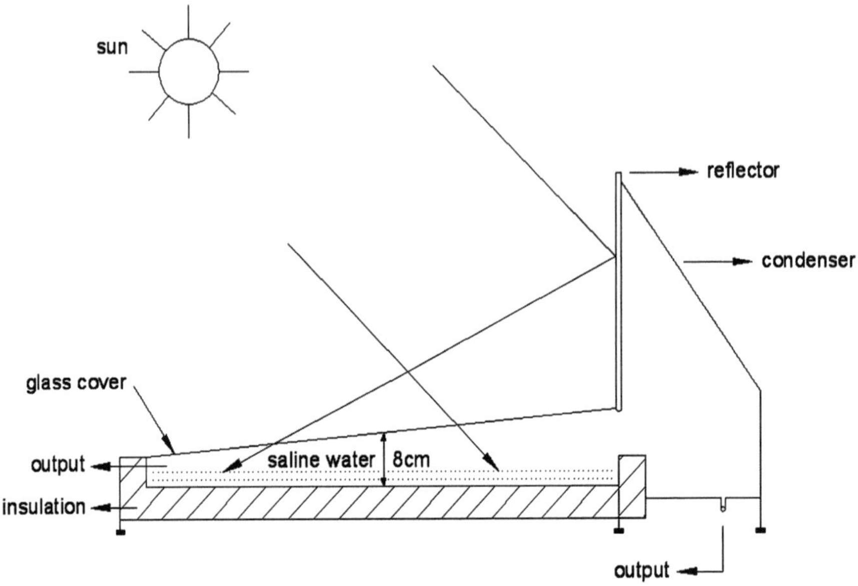

Fig. 11.7 Schematic of a solar still with condenser (Reprinted from [27, 28] with kind permission of © Elsevier BV)

11.7.6 Concave Wick Solar Still

The daily yield of basin type solar still mainly depends on the evaporative area and condensing surfaces. Increasing the surface area or decreasing the cover temperature will enhance the distillate output. The new design of solar still with wicked concave surface is shown in Fig. 11.8, and it will increase the absorption and evaporation areas and in turn increase the yield. Also, this concave still design reduces the shading effect than the conventional type because all sides are glass. The wick surface takes the same shape of the concave surface and is black painted to absorb maximum solar radiation. Glass cover is used at four sides of the solar still with ordinary window glass and tilted at an angle of 45° to horizontal surface. The distillate is collected by a galvanized iron channel fixed on the sides at the lower end of the glass cover.

The channels were inclined such that no accumulation of water was observed. The experimental study revealed that the average distillate productivity of the concave wick solar still during the 24 h time is about 4 kg/m^2 and the thermal efficiency was reached about 45% [29].

11.7.7 Greenhouse Type Solar Still

A schematic diagram of a solar still-cum greenhouse, indicating the solar energy absorbed by the roof and wall is shown in Fig. 11.9. The absorbed solar radiation by

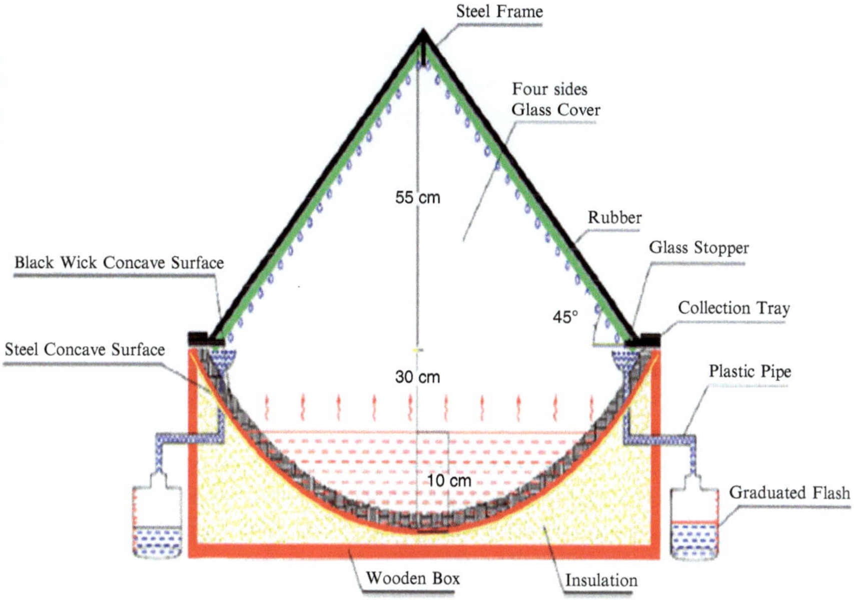

Fig. 11.8 Schematic of a concave wick solar still (Reprinted from [29] with kind permission of © Elsevier Science)

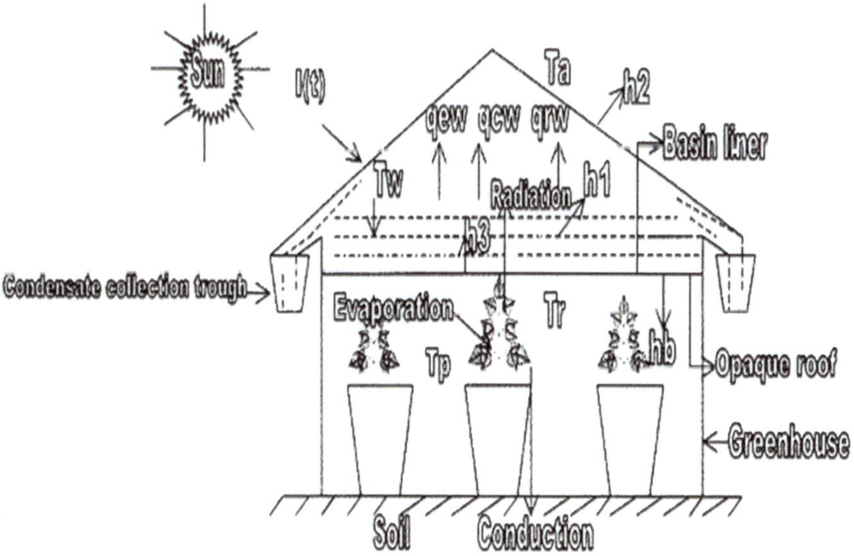

Fig. 11.9 A schematic of a solar still-cum-greenhouse (Reprinted from [30] with kind permission of © Elsevier BV)

the basin liner is partially transferred to water mass, and the rest is transferred to a room by conduction, convection and radiation. The stored energy in the water causes evaporation, and it then condenses and ultimately comes out in the form of freshwater to be used for irrigation purposes. The remaining transmitted solar energy is absorbed by the plants and the floor, after reflection. The absorbed energy is transferred to enclosed air by convection and radiation from the plant and floor; hence, room air is heated. A part of the energy absorbed by the floor is also conducted into the ground. If crops can be grown in strong sunlight without the debilitating effects of high temperatures, then high crop yields can be achieved with a significantly reduced consumption of water [30]. Sodha et al. [31] conducted experimental studies of roof type solar still and concluded that, in summer, the daily heat flux in the room gets reduced by 40% and in winter, mixed with black dye to obtain water, the daily heat flux in the room increased by a factor of two; productivity also increased.

Srivastava et al. [30] presented the performance evaluation of distillation cum greenhouse for a warm and humid climatic condition. Their result shows that (1) there is a marginal difference between the values of plant, and room temperature, water, transparent cover and basin liner temperatures also show a marginal difference; and (2) distillate output can be used as irrigation, and greenhouse will provide the desired temperature to plant in a warm and humid climate in coastal regions.

11.7.8 *Hemispherical Solar Still*

A schematic diagram of the hemispherical (dome-shaped) solar still is shown in Fig. 11.10 [32]. The still mainly consists of the circular basin (tray) and absorber plate carrying the saline water, the hemispherical cover, the conical shaped distillate collector, the distillate output plastic container and the mobile support structure (trolley).

The basin contains the absorber aluminium plate, and a hole was drilled into the tray to provide accessibility of saline water into the basin during initial filling, and the bottom section of the basin was insulated to reduce thermal losses to the surroundings. The hemispherical cover in the top of the solar still unit was made of a transparent plastic. For sealing purpose, rubber ring gasket was used and placed between the cover and the collector support structure where they were tightened and held in place using C-type clamps placed at equal distances around the periphery of the still. A new transportable hemispherical solar still was designed, fabricated and experimentally tested by Basel I Ismail [32] during daytime for 6 days under outdoors of Dhahran climatic conditions. It was found that over the hours of testing in daytime, the daily distillate yield produced from the still ranged from approximately 2.8 to 5.7 kg/m^2. The daily efficiency of the solar still reached as high as 33%.

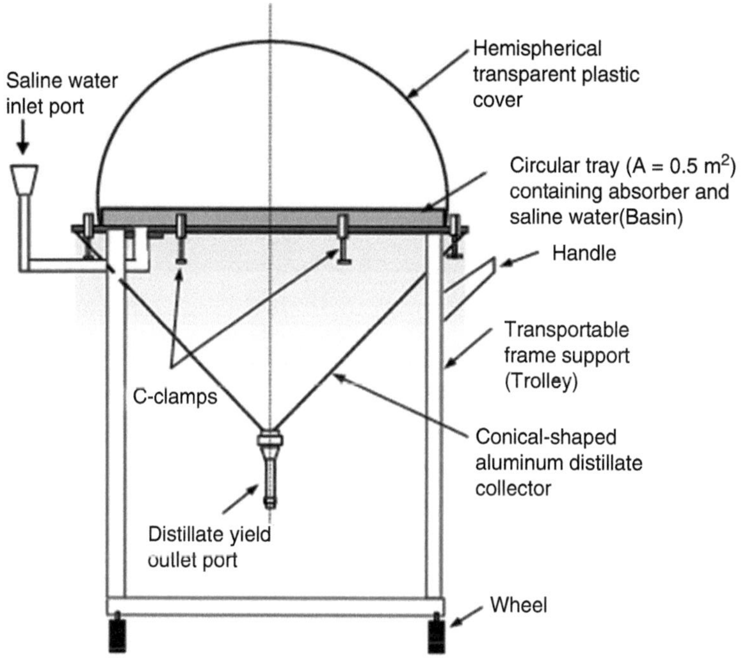

Fig. 11.10 Schematic view of hemispherical solar still (Reprinted from [32] with kind permission of © Pergamon)

11.7.9 Spherical Solar Still

The spherical solar still (Fig. 11.11) consists of a blackened metallic plate placed horizontally at the centre of a transparent envelope, which is spherical in shape and is usually made of glass. Naresh K Dhiman [33] presented a mathematical model to predict the thermal performance of a spherical solar still. The results show that spherical solar still has efficiency 30% more than the conventional solar still.

11.7.10 Stepped Solar Still

The schematic view of stepped solar still is shown in Fig. 11.12. In stepped solar still absorber, the plate is looking like a stepped structure. Normally, GI material is used for construction of stepped size absorber plate due to its low cost. The depth of the individual tray may be constant or varied based on the requirement. The absorber plate is placed in a wooden box with insulation material of saw dust. The absorber plate is painted in black to receive maximum solar radiation. One main advantage of this model is variation in water depth in trays. This leads to decrease of water depth and in turn increases the evaporation rate. Velmurugan et al. [34–36] experimentally

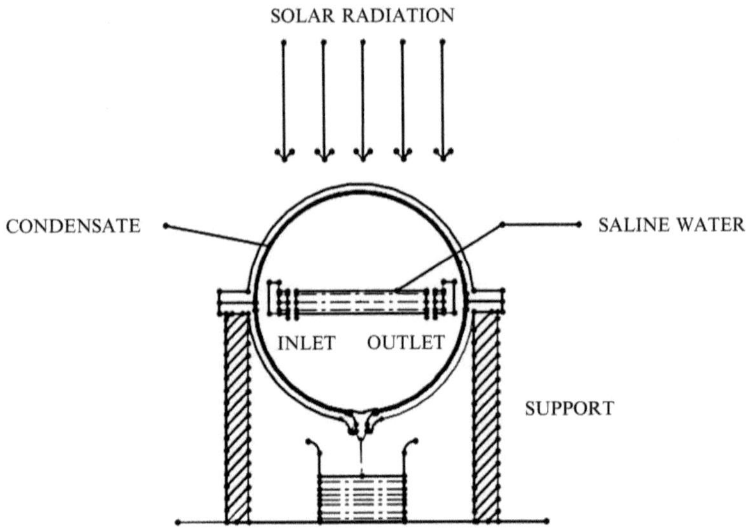

Fig. 11.11 Schematic of a spherical still (Reprinted from [33] with kind permission of © Elsevier BV)

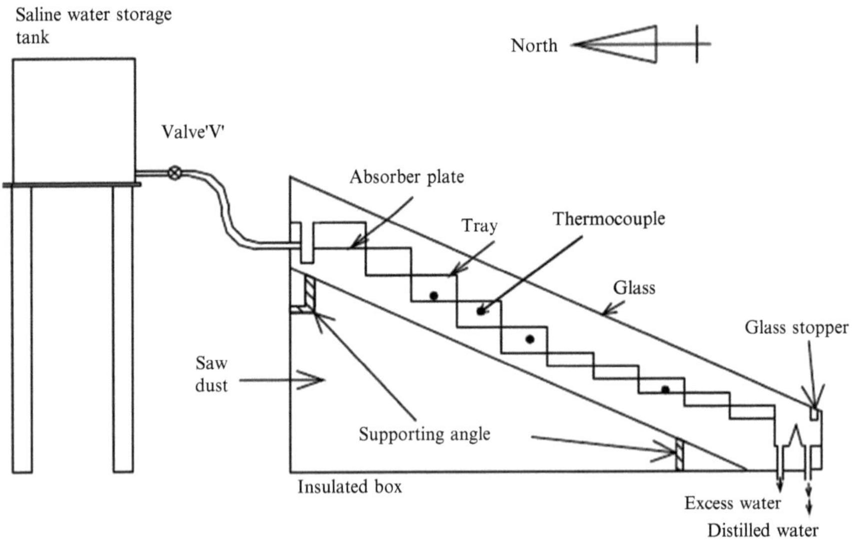

Fig. 11.12 Schematic view of stepped solar still (Reprinted from [34–36] with kind permission of © Elsevier Science)

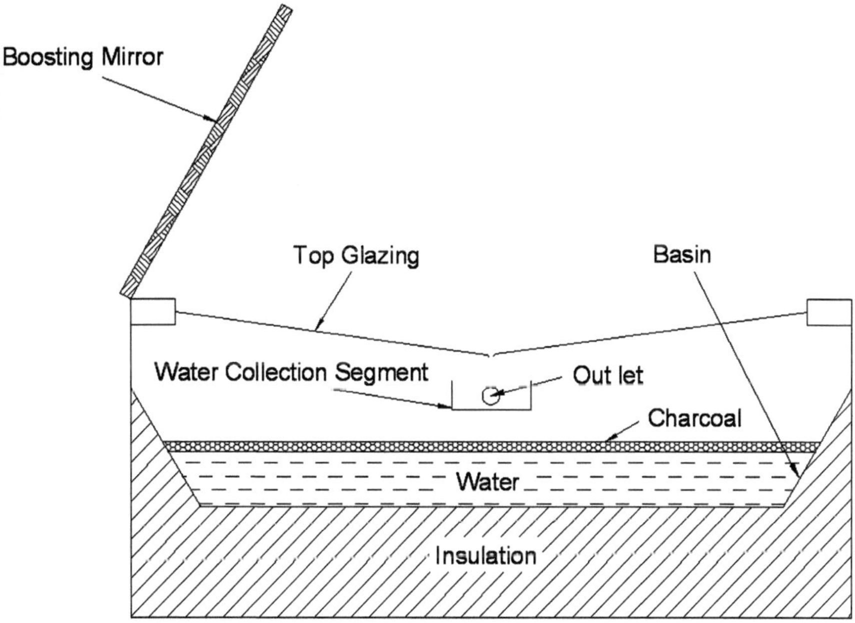

Fig. 11.13 Schematic view of 'V' type solar still (Reprinted from [37] with kind permission of © Elsevier BV)

investigated the stepped solar still with various energy-absorbing materials. They found that the production rate was increased by 53.3, 68 and 65% when fins, sponge and pebbles are used in the stepped solar still, respectively.

11.7.11 'V' Type Solar Still

The schematic view of the 'V' type solar still is shown in Fig. 11.13. Water collection through the single channel is the more prominent feature of the 'V' type solar still. The main advantage of this type of still is that the distilled water collection is directed towards the central water collection channel [37]. A water basin is made up of mild steel and water condensate collector is provided. The ordinary window glass with an inwards slope of 22° is maintained towards the centre of the still. This shape is maintained on both sides of the solar still to make it 'V' shaped. The two glass covers are jointly in the middle by using chemical adhesive without any air leakage.

A boosting mirror also provided in one side to increase the solar radiation. Selvakumar et al. [37] conducted the experiments in 'V' type solar still and they found that the efficiency of solar still is 24.47%.

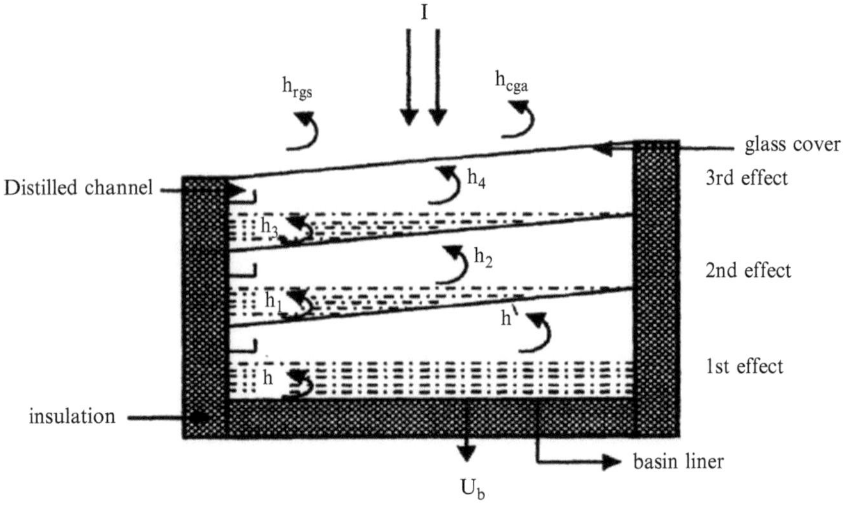

Fig. 11.14 Schematic of a triple basin solar still (Reprinted from [38] with kind permission of © Elsevier BV)

11.7.12 Triple Basin Solar Still

A triple basin solar still was experimentally studied by El-Sebaii [38] and the schematic is shown in Fig. 11.14. In the triple basin solar still, two glass sheets are fixed in between the basin liner and the glass cover of the single basin still. These glass sheets serve as the base of extra shallow depths of saline water, and the whole assembly behaves as three simple basin solar stills are placed one above the other. The water in the middle and upper basins makes use of the latent heat from condensation released at the inner surfaces of the glass covers of the lower and middle basins, respectively; consequently, evaporation occurs in the three effects.

The total productivity of the system is a sum of the yields of the individual basins. The study revealed that the daily total yield of the still was found to be 12.635 kg/m^2 on a typical summer day.

11.8 Active Solar Still

The performance of a solar still could neither be predicted nor improved by some of the uncontrollable parameters like intensity of solar radiation, ambient temperature and wind velocity. But, there are certain parameters such as depth of water, glass cover angle, fabrication materials, temperature of water in the basin and insulation thickness, which affect the performance of the solar still that could be modified for improving the performance. The temperature difference between water in the basin and condensing glass cover also has a direct effect in the performance of the still.

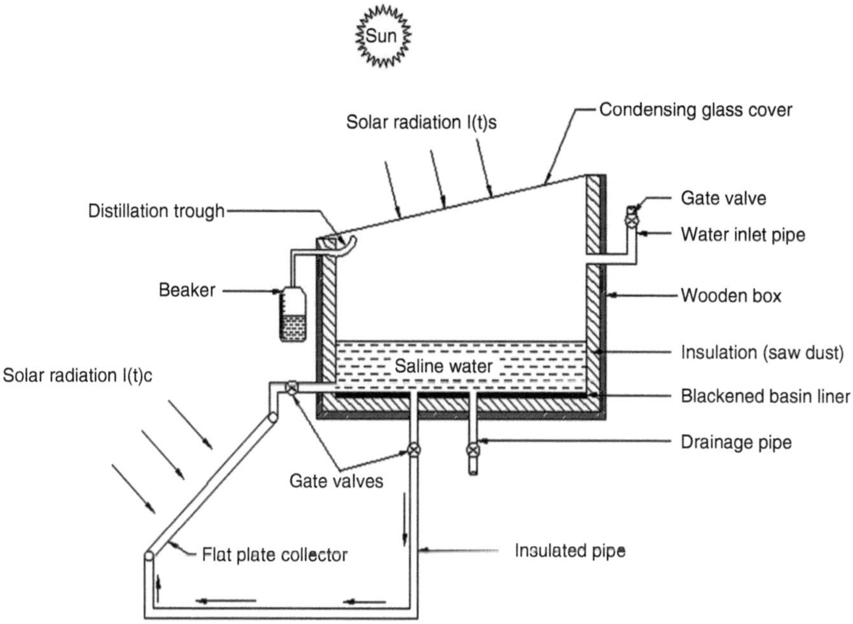

Fig. 11.15 Schematic of an active solar still integrated with a flat plate collector

The increased temperature of the water in basin can increase the temperature difference between the evaporating and condensing surfaces. To achieve better evaporation and condensation rate, the temperature of water in the basin could be raised by feeding thermal energy from some external sources. The detailed review of active solar distillation system was presented by Sampathkumar et al. [39].

11.8.1 Solar Still Coupled with Flat Plate Collector

The solar still coupled with flat plate collector is working as high temperature distillation method. The solar still coupled with flat plate collector (FPC) works either in forced circulation mode or natural circulation mode. In forced circulation mode, a pump is used for supplying water. In natural circulation mode, water flows due to the difference in the density of water.

The flat plate collector gives an additional thermal energy to the basin of the solar still. A pump is used to circulate the water from the basin via flat plate collector to the basin. Many researchers have been carried out in this method and the first being reported by Rai and Tiwari [40]. They found that the daily distillate production of a coupled single basin still is 24% higher than that of an uncoupled one using forced circulation mode. A schematic diagram of an active solar still integrated with a flat plate collector under forced circulation mode is shown in Fig. 11.15.

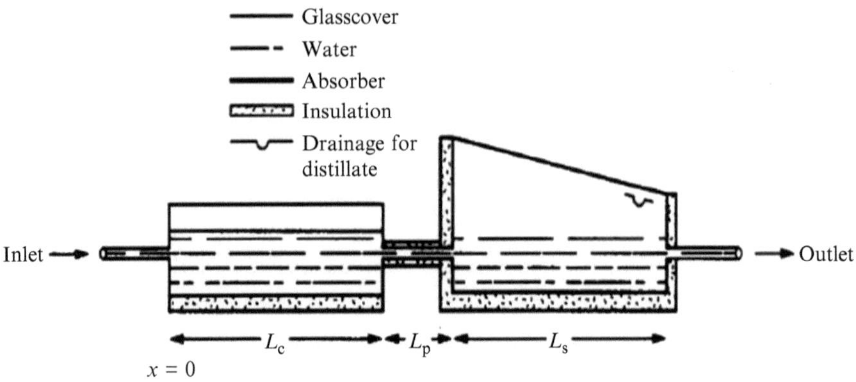

Fig. 11.16 Schematic diagram of a solar still integrated with a parallel flat plate water collector (Reprinted from [44] with kind permission of © Pergamon)

The bottom insulation is an important design parameter of the active solar still and for drinking purposes, the conventional solar still will give better performance because the efficiency of the system reduces with the increase in the effective area as reported by Tiwari and Dhiman [41]. Their experimental study showed that there was only 12% rise in yield of the system if the length of the heat exchanger is varied from 6.0 to 12.0 m and the overall efficiency of the system varied from 15 to 19%. Sanjeev Kumar and Tiwari [42] observed that temperature of water and thermal efficiency decreased with an increase in basin area due to the large storage capacity of the water mass in the basin and depth of water respectively. Yield increased with increase in the number of collectors, as expected, owing to increased heat transfer from the collector panel into the basin, and the optimum number of collectors for maximum yield is 8 m^2 since; beyond that, the increase in gain will be lower than the thermal loss. Singh and Tiwari [43] found that the annual yield is at its maximum when the condensing glass cover inclination is equal to the latitude of the place and the optimum collector inclination for a flat plate collector is 28.58°, for a condensing glass cover inclination of 18.58° for New Delhi climatic condition.

11.8.2 Solar Still Coupled with Parallel Flat Plate Collector

Yadav and Prasad [44] experimentally studied the solar still integrated with parallel flat plate collector. The schematic diagram of a solar still integrated with a parallel flat plate solar energy collector is shown in Fig. 11.16. The collector essentially consists of a parallel flat plate placed over the insulation with an air gap through which the water will flow below the absorber.

There is a glass sheet over the absorber and the whole assembly is enclosed in a wooden box. The top of the plate (absorber) is blackened by blackboard paint before the glass cover is placed over the absorber. The collector outlet is connected

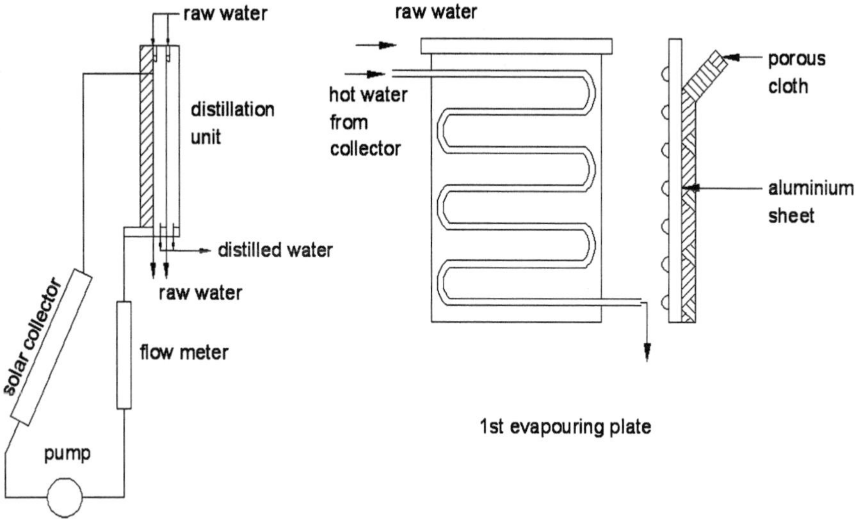

Fig. 11.17 Schematic sketch of the multiple effects still with a flat plate collector [45] with kind permission of © Elsevier Ltd

to the still by a pipe covered with insulation. The circulation of water between the collector and the still can be made either via a pump (forced circulation system) or by placing the collector over a supporting structure at such a height as to provide adequate head for natural circulation of water (thermosiphon) in the system. The results show that a significant rise in the distillate output is observed when the still is coupled with the collector and this system can be preferred as cost-effective compared to the flat plate collector.

11.8.3 Vertical Solar Still Coupled with Flat Plate Collector

Kiatsiriroat et al. [45] analysed the multiple effect of vertical solar still coupled with flat plate solar collector. The schematic sketch is shown in Fig. 11.17. The distillation unit consists of 'n' parallel vertical plates. The first plate is insulated on its front side and the last plate is exposed to ambient. Each plate in the enclosure is covered with wetted cloth on one side. The cloth is extended into a feed trough along the upper edge of each plate. Feedwater in the trough is then drawn onto the plate surface by capillary. Excess water moves down the plate and is conducted out of the still. The last plate is cooled by air or water. The authors found that the distillation output increases slightly when the plate number is over 5, and it increased by about 34 and 15% when the evaporating plate numbers are 1 and 6, respectively.

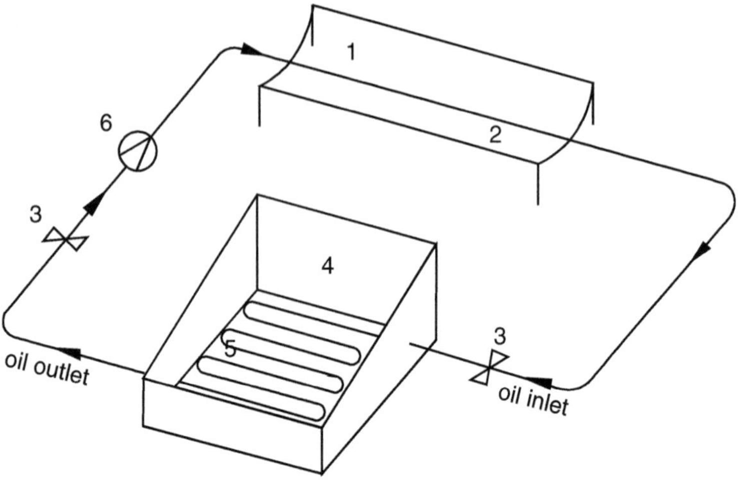

Fig. 11.18 Solar still coupled with parabolic concentrator. *1* Parabolic through, *2* Oil pipeline, *3* Valves, *4* Solar still, *5* Oil heat exchanger, *6* Pump (Reprinted from [44] with kind permission of © Elsevier BV)

11.8.4 Solar Still Coupled with Parabolic Concentrator

The schematic diagram of the solar still coupled with parabolic concentrator is shown in Fig. 11.18. The parabolic-shaped concentrator or solar collector concentrates the incident solar radiation on large surface and it focuses on to a small absorber or receiver area. The performance of concentrators is much affected by the sun tracking mechanism. The tracking mechanism should move the collectors throughout the day to keep them focused on the sunrays to achieve the higher efficiency. These types of solar collectors reach higher temperature compared to flat plate collectors owing to reduced heat loss area [46].

The various types of concentrators were used over the years based on the applications. To achieve higher yield, the contractor is coupled with solar still by means of increasing water temperature in the basin. The water or oil will be supplied to trough receiver pipe by natural circulation mode or forced circulation mode. They found that the fresh water productivity was increased by an average of 18%. Singh et al. [47] found an analytical expression for water temperature of an active solar still with flat plate collectors and parabolic concentrator through natural circulation mode. The results show that the efficiency of the system with concentrator is higher than parabolic collector as the evaporative heat transfer coefficient is higher in concentrator.

11.8.5 Solar Still Coupled with Solar Pond

Solar pond is an artificially constructed pond in which significant temperature rises are caused to occur in the lower regions by preventing convection. Solar ponds are used for collection and storage of solar energy, and it is used for various thermal applications, and this detailed review of solar pond has been done by Velmurugan and Srithar [48]. Velmurgan and Srithar [49] theoretically and experimentally analysed the mini solar pond assisted solar still with sponge cube. The results show that average increase in productivity when a pond is integrated with a still is 27.6% and when pond and sponge are integrated with a still, is 57.8%.

Velmurugan et al. [34–36] studied the augmentation of saline streams in solar stills integrated with a mini solar pond. Industrial effluent was used as feed for fin type single basin solar still and stepped solar still. A mini solar pond connected to the stills to enhance the productivity and tested individually. The schematic diagram of experimental setup is shown in Fig. 11.19. The results show that maximum productivity of 100% was obtained when the fin type solar still was integrated with pebble and sponge. The productivity increases with increase in solar intensity and water-glass temperature difference, and decreases with increase in wind velocity. Velmurugan et al. [34–36] experimentally investigated the possibility of enhancing the productivity of the solar stills by connecting a mini solar pond, stepped solar still and a single basin solar still in series. Pebbles, baffle plates, fins and sponges are used in the stepped solar still for productivity augmentation. Their finding shows that maximum productivity of 78% occurred when fins and sponges were used in the stepped solar still and also found that the productivity during night also improved when pebbles were used in the solar stills.

El-Sebaii et al. [50] experimentally studied to improve the productivity of the single effect solar stills, a single slope single basin solar still integrated with a shallow solar pond (SSP). They found that the annual average values of daily productivity and efficiency of the still with SSP were higher than those obtained without the SSP by 52.36 and 43.80%, respectively.

11.8.6 Solar Still Coupled with Hybrid PV/T System

The problem encountered with normal PV cells is that most of the solar radiation that is absorbed by a solar cell is not converted into electricity. The excess energy which goes unabsorbed by the solar cell increases the temperature of the photovoltaic cell and reduces the efficiency. Natural or forced circulation of a fluid cooling medium reduces the cell temperature. Cooling is often applied for concentrating photovoltaic systems, in which the irradiance on the cell surface is high. An alternative to ordinary photovoltaic modules is to use photovoltaic-thermal (PV/T) modules, which are photovoltaic modules coupled to heat extraction devices. Hence, these systems, in addition to converting sunlight into electricity, collect the residual thermal energy and delivers both heat and electricity in usable forms.

Kumar and Tiwari [51] conducted an experimental study of hybrid photovoltaic/thermal (PV/T) active solar still and found that the yield increased by more than 3.5 times than the passive solar still. The schematic of a hybrid (PV/T) active solar still is shown in Fig. 11.20. Also found was that the maximum daily yield

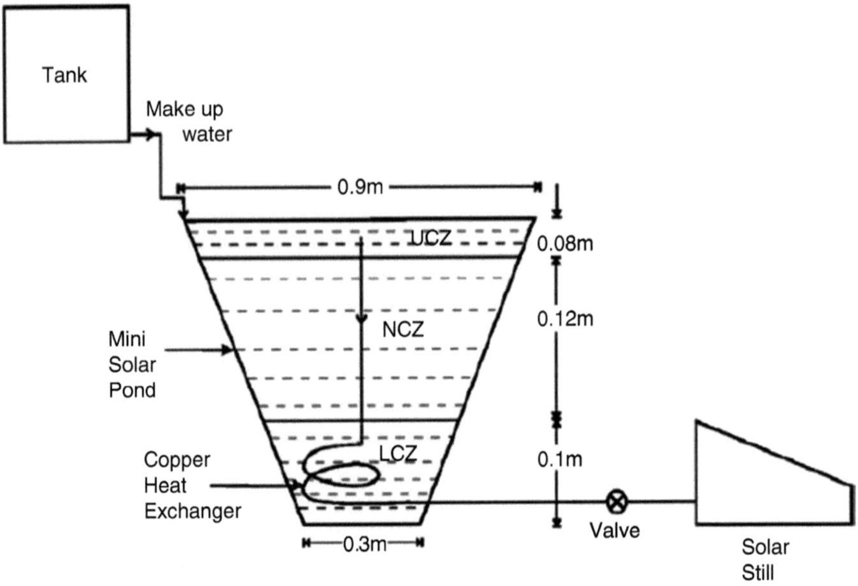

Fig. 11.19 Schematic of the mini solar pond integrated with single basin solar still (Reprinted from [49] with kind permission of © Elsevier BV)

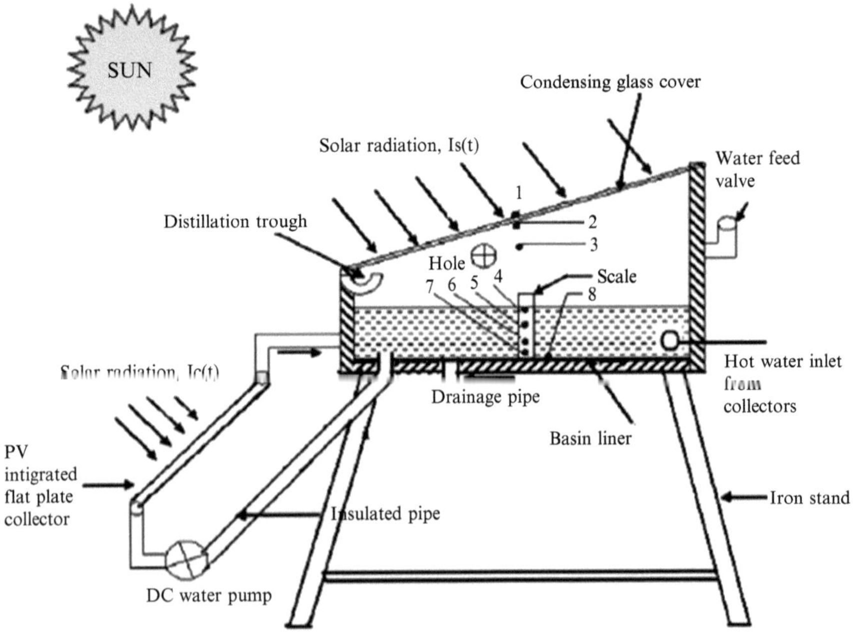

Fig. 11.20 Schematic of a hybrid (PV/T) active solar still ([51] with kind permission of © John Wiley & Sons Ltd)

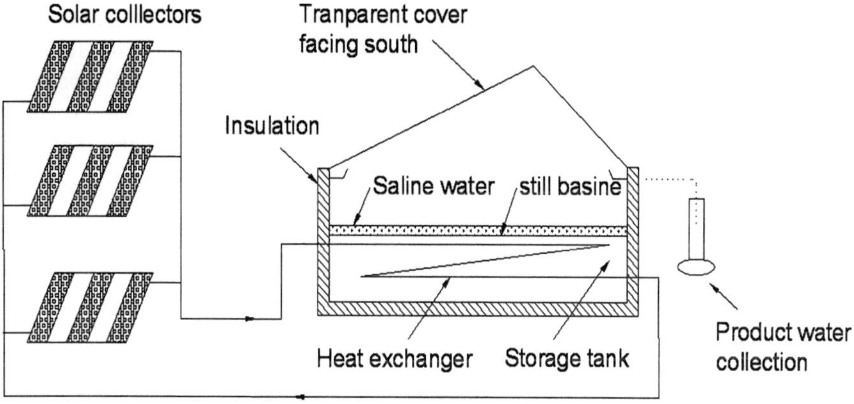

Fig. 11.21 Schematic diagram of hybrid solar distillation system (Reprinted from [52] with kind permission of © Elsevier BV)

of 2.26 and 7.22 kg were obtained from passive and hybrid active solar still. The hybrid active solar still also provides higher electrical and overall thermal efficiency, which is about 20% higher than the passive solar still.

11.8.7 Hybrid Solar Distillation System

The hybrid solar still can produce the desalinated and hot water from the same system. These types of designs have more advantages over the other type of systems. Voropoulos et al. [52] experimentally investigated the hybrid solar still coupled with solar collectors (Fig. 11.21). The results show that (1) the productivity of the coupled system is about double that of the still only; (2) significant raises in distilled water productivity have been obtained not only during the day but mainly during night operation of the system, reaching triples the solar only system productivity; and (3) the continuous heating of basin water from tank water results in higher production rates in all operation periods as a result of significantly higher differences between water and cover temperatures, mainly at night. Voropoulos et al. [53] studied the energy behaviour of hybrid solar still and concluded that the developed method can be a valuable tool for the system optimization, used during its design and also for evaluation of an existing solar distillation installation through short-term testing.

Mathioulakis and Belessiotis [54] investigated the possibilities of using optimization of a simple solar still through its incorporation in a multi-source and multi-use environment, and observed that the design of such systems depends on the available heat sources and/or expected consumption of hot water usage.

Voropoulos et al. [55] conducted experimental study of a hybrid solar desalination and water heating system. The results show that the output of a conventional solar still can be significantly increased if it is coupled with a solar collector field and hot water storage tank. The distilled water production was gradually reduced when the increase in delivered energy through hot water draws-off. Ben Bacha et al. [56] developed a mathematical model to give the ability to estimate the expected performance of the system under given climatic conditions, allowing the choice of the proper design solutions in relation to the desired usage.

11.8.8 Preheated Water Active Solar Still

In this method, preheated water is used to increase the water temperature in the basin. The waste hot water is available from various sources like paper industries, chemical industries, thermal power plants and food processing industries, and the same may be utilised for solar distillation plant to increase the productivity. The hot water will be supplied directly to the basin or through heat exchangers. Proctor [57] proposed the technique of using waste heat in a solar still and predicted that productivity increased 3.2 times compared with ordinary still. Sodha et al. [16–17] presented the experimental results on utilisation of waste hot water for distillation. In that test, two modes were studied: (1) flowing waste hot water from thermal power plants at constant rate through the solar still and (2) feeding waste hot water obtained from thermal power plants once a day.

Their results showed that length of solar still, depth of water in basin, inlet water temperature and solar radiation are the parameters which affect the performance of the still, and the still fed with hot water at constant rates gives higher yield in comparison to a still with hot water filled only once in a day. Ashok Kumar and Tiwari [58] investigated the use of hot water in double slope solar still through heat exchanger (Fig. 11.22).

The authors observed that the evaporative heat transfer coefficient depends strongly on temperature and advised to use the waste hot water with either higher temperature or during off sunshine hours. Yadav [59] analysed the performance of double basin solar still coupled to a heat exchanger. The author observed that the efficiency of a double basin solar still coupled to a heat exchanger is significantly less, as compared to that without heat exchanger.

11.9 Conclusion

In finale, solar stills are cost-effective when compared to all other available desalinations systems in use today. The main advantage of the solar still is easy to construct and operate even in remote areas. The only disadvantage of the solar still is very low yield and it may be improved by increasing the basin area. Passive solar stills are more suitable for freshwater supply in rural areas as compared to active solar stills. Although solar distillation at present cannot compete with

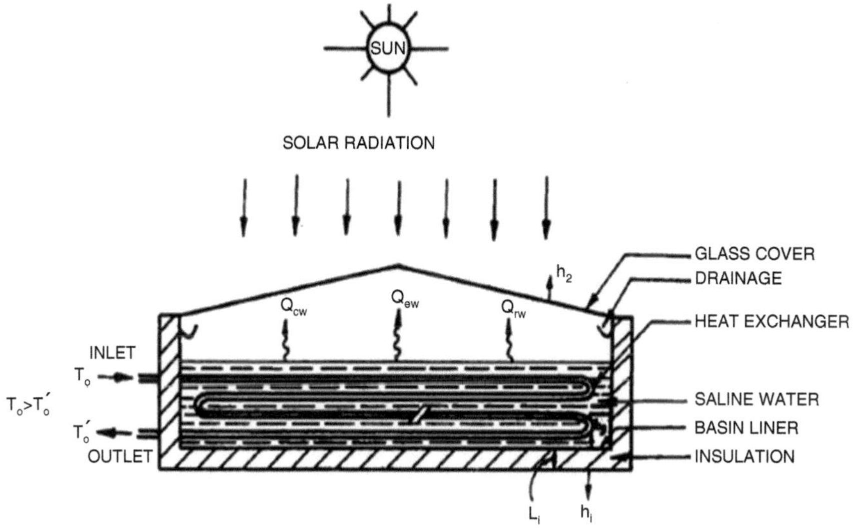

Fig. 11.22 Schematic diagram of double slope single basin solar still with heat exchanger (Reprinted from [58] with kind permission of © Pergamon)

oil-fired desalination in large central plants, it has potential to become a viable technology in near future, when oil supplies stops. However, producing freshwater from saline water with solar stills is much useful for remote areas where no other economically viable method for obtaining freshwater supply is available.

References

1. Vinothkuumar K, Kasturibai R (2008) Performance study on solar still with enhanced condensation. Desalination 230:51–61
2. www.wateraid.org, Drinking water and sanitation status in India (2009)
3. Tiwari GN, Tiwari AK (2008) Solar distillation practice for water desalination systems. Anamaya Publishers, New Delhi
4. Khalifa AJN, Hamood AM (2009) Performance correlation for basin type solar stills. Desalination 249:24–28
5. Nebbia G, Mennozi G (1966) A short history of water desalination. In: Proceedings of international symposium. Milano, New York, pp 129–172
6. Arjunan TV, Aybar HŞ, Nedunchezhian N (2009) Status of solar desalination in India. Renewable and Sustainable Energy Reviews 13:2408–2418
7. Tiwari GN, Mukherjee K, Ashok KR, Yadav YP (1986) Comparison of various designs of solar stills. Desalination 60:191–202
8. Madani AA, Zaki GM (1995) Yield of stills with porous basin. Appl Energy 52:273–281
9. Abu-Hijleh BAK (1996) Enhanced solar still performance using water film cooling of the glass cover. Desalination 107:235–244
10. Toure S, Meukam P (1997) A numerical model and experimental investigation for a solar still in climatic conditions in Abidjan. Renew Energy 11:319–330

11. Kalidasa Murugavel K, Chockalingam KnKSK, Srithar K (2008) Progress in improving the effectiveness of the single basin passive solar still. Desalination 220:677–686
12. Velmurugan V, Deenadayalan CK, Vinod H, Srithar K (2008) Desalination of effluent using fin type solar still. Energy 33:1719–1727
13. Dwivedi VK, Tiwari GN (2008) Annual energy and exergy analysis of single and double slope passive solar stills. Trends Appl Sci Res 3:225–241
14. Kalidasa Murugavel K, Srithar K (2011) Performance study on basin type double slope solar still with different wick materials and minimum mass of water. Renew Energy 36:612–620
15. Velmurugan V, Srithar K (2011) Performance of solar stills based on various factors affecting the productivity. Renew Sustain Energy Rev 15:1294–1304
16. Sodha MS, Kumar A, Tiwari GN, Tyagi RC (1981) Simple multiple wick solar still: analysis and performance. Solar Energy 26:127–131
17. Sodha MS, Kumar A, Tiwari GN (1981) Utilization of waste hot water for distillation. Desalination 37:325–342
18. Reddy MS, Navin Chandra DJ, Sehgal HK, Sabberwal SP, Bhargava AK, Jither Chandra DS (1983) Performance of a multiple-wick solar still with condenser. Appl Energy 13:15–21
19. Tiwari GN, Sharma SB, Sodha MS (1984) Performance of a double condensing multiple wick solar still. Energy Convers Manage 24:155–159
20. Minasian AN, Al-Karaghouli AA (1995) An improved solar still: the wick basin type. Energy Convers Manage 36:213–217
21. Tiwari GN, Suneja S (1998) Performance evaluation of an inverted absorber solar still. Energy Convers Manage 39:173–180
22. Suneja S, Tiwari GN, Rai SN (1997) Parametric study of an inverted absorber double-effect solar distillation system. Desalination 109:177–186
23. Suneja S, Tiwari GN (1999) Effect of water depth on the performance of an inverted absorber double basin solar still. Energy Convers Manage 40:1885–1897
24. Suneja S, Tiwari GN (1999) Effect of water flow on internal heat transfer solar distillation. Energy Convers Manage 40:509–518
25. Suneja S, Tiwari GN (1999) Parametric study of an inverted absorber triple effect solar still. Energy Convers Manage 40:1871–1884
26. Suneja S, Tiwari GN (1998) Optimization of number of effects for higher yield from an inverted absorber solar still using the Runge-Kutta method. Desalination 120:197–209
27. El-Bahi A, Inan D (1999) Analysis of a parallel double glass solar still with separate condenser. Renew Energy 17:509–521
28. El-Bahi A, Inan D (1999) A solar still with minimum inclination, coupled to an outside condenser. Desalination 123:509–521
29. Kabeel AE (2009) Performance of solar still with a concave wick evaporation surface. Energy 34:1504–1509
30. Srivastava NSL, Din M, Tiwari GN (2000) Performance evaluation of distillation-cum-greenhouse for a warm and humid climate. Desalination 128:67–80
31. Sodha MS, Kumar A, Srivastava A, Tiwari GN (1980) Thermal performance of "still on roof" system. Energy Convers Manage 20:181–190
32. Ismail BI (2009) Design and performance of a transportable hemispherical solar still. Renew Energy 34:145–150
33. Dhiman NK (1988) Transient analysis of a spherical solar still. Desalination 69:47–55
34. Velmurugan V, Mandlin J, Stalin B, Srithar K (2009) Augmentation of saline streams in solar stills integrating with a mini solar pond. Desalination 249:143–149
35. Velmurugan V, Naveenkumar KJ, Haq N, Srithar K (2009) Performance analysis in stepped solar stills for effluent desalination. Energy 34:1179–1186
36. Velmurugan V, Pandiarajan S, Guruparan P, Subramanian H, Prabaharan CD, Srithar K (2009) Integrated performance of stepped and single basin solar stills with mini solar pond. Desalination 249:902–909

37. Selvakumar B, Sanjaykumar, Jayaprakash R (2008) Performance analysis of a "V" type solar still using a charcoal absorber and a boosting mirror. Desalination 229:217–230
38. El-Sebaii AA (2005) Thermal performance of a triple basin solar still. Desalination 174:23–37
39. Sampathkumar K, Arjunan TV, Pitchandi P, Senthilkumar P (2010) Active solar distillation – a detailed review. Renew Sustain Energy Rev 14:1503–1526
40. Rai SN, Tiwari GN (1983) Single basin solar still coupled with flat plate collector. Energy Convers Manage 23:145–149
41. Tiwari GN, Dhiman NK (1991) Performance study of a high temperature distillation system. Energy Convers Manage 32:283–291
42. Kumar S, Tiwari GN (1998) Optimization of collector and basin areas for a higher yield for active solar stills. Desalination 116:1–9
43. Singh HN, Tiwari GN (2004) Monthly performance of passive and active solar stills for different Indian climatic condition. Desalination 168:145–150
44. Yadav YP, Prasad AS (1995) Performance analysis of a high temperature solar distillation system. Energy Convers Manage 36:365–374
45. Kiatsiriroat T, Bhattacharya SC, Wibulswas P (1987) Performance analysis of multiple effect vertical solar still with a flat plate solar collector. Solar Wind Technol 4:451–457
46. Abdel Rehim ZS, Lasheen A (2007) Experimental and theoretical study of a solar desalination system located in Cairo, Egypt. Desalination 217:52–64
47. Singh SK, Bhatnagar VP, Tiwari GN (1996) Design parameters for concentrator assisted solar distillation system. Energy Convers Manage 37:247–252
48. Velmurugan V, Srithar K (2008) Prospects and scopes of solar pond: a detailed review. Renew Sustain Energy Rev 12:2253–2263
49. Velmurugan V, Srithar K (2007) Solar stills integrated with a mini solar pond – analytical simulation and experimental validation. Desalination 216:232–241
50. El-Sebaii AA, Ramadan MRI, Aboul Enein S, Slaem N (2008) Thermal performance of a single basin solar still integrated with a shallow solar pond. Energy Convers Manage 49:2839–2848
51. Kumar S, Tiwari A (2008) An experimental study of hybrid photovoltaic thermal (PV/T) active solar still. Int J Energy Res 32:847–858
52. Voropoulos K, Mathioulakis E, Belessiotis V (2001) Experimental investigation of a solar still coupled with solar collectors. Desalination 138:103–110
53. Voropoulos K, Mathioulakis E, Belessiotis V (2003) Solar stills coupled with solar collectors and storage tank-analytical simulation and experimental validation of energy behavior. Solar Energy 75:199–205
54. Mathioulakis E, Belessiotis V (2003) Integration of solar still in a multi-source, multi-use environment. Solar Energy 75:403–411
55. Voropoulos K, Mathioulakis E, Belessiotis V (2004) A hybrid solar desalination and water heating system. Desalination 164:189–195
56. Ben Bacha H, Dammak T, Ben Abdalah AA, Maalej AY, Ben Dhia H (2007) Desalination unit coupled with solar collectors and a storage tank: modelling and simulation. Desalination 206:341–352
57. Proctor D (1973) The use of waste heat in a solar still. Solar Energy 14:433–449
58. Kumar A, Tiwari GN (1990) Use of waste hot water in double slope solar still through heat exchanger. Energy Convers Manage 30:81–89
59. Yadav YP (1993) Performance analysis of a solar still coupled to a heat exchanger. Desalination 91:135–144

Chapter 12
AOPs Application on Dyes Removal

Tuty Emilia Agustina

12.1 Introduction

Approximately 10,000 different dyes and pigments are manufactured worldwide with a total annual market of more than 7×10^5 tonnes per year [1]. All the dyes absorb light in the visible region. Untreated dye effluent such as textile effluents is highly colored. The color is aesthetically unpleasant, and it also reduces light penetration into water decreasing the efficiency of photosynthesis in aquatic plants, thereby having undesirable impact on their growth. In addition, some of the dyes might be toxic to some organisms. Nowadays, the color of the effluent released into receiving waters has become a serious environmental problem.

On the other hand, advanced oxidation processes (AOPs) have shown great potential in treating pollutants at low and high concentration and have been utilized for treatment of water, wastewater, odors, and sludges.

Developments in chemical treatment have made a range of AOPs suitable for water and wastewater applications. Table 12.1 lists those AOPs that have been expanded. The list highlights the range and variety of processes developed or under development that could have application in water and wastewater treatment. A number of these processes are commercially available and in some instances widely used. Processes such as the chemical combinations of H_2O_2, O_3, and UV, Fenton's reagent, wet air oxidation, supercritical oxidation, and electron beam have all been used at full scale. Other processes such as photocatalysis and ultrasound have been evaluated at pilot scale. While the list comprises individual processes, much research have been undertaken using a combination of the processes, such as UV, ultrasound, and ozone together, which offer significant kinetic and performance benefits over each of the processes alone.

T.E. Agustina (✉)
Chemical Engineering Department, Sriwijaya University, Palembang, Indonesia
e-mail: tuty_agustina@unsri.ac.id

S.K. Sharma and R. Sanghi (eds.), *Wastewater Reuse and Management*,
DOI 10.1007/978-94-007-4942-9_12, © Springer Science+Business Media Dordrecht 2013

Table 12.1 AOPs evaluated for water and wastewater treatment [2]

Catalysis	Pulsed plasma
Electrochemical	Supercritical water oxidation
Fenton's reagent	Ultrasound
Ferrate	UV
Ionizing radiation	UV/H_2O_2
Microwave	$UV/H_2O_2/O_3$
Photo-Fenton's reagent	Vacuum UV
Photocatalysis	Wet air oxidation

Advanced oxidation formed by a variety of chemicals (O_3, Fenton's reaction, Fenton/O_3, modified Fenton, or Fenton-like oxidation), photochemical (UV/O_3, UV/H_2O_2, photo-Fenton, modified Photo-Fenton's oxidation, photocatalytic ozonation) as well as near-UV-vis or solar light-induced TiO_2-mediated heterogeneous photocatalytic oxidation system appear to have the ability to completely decolorize and at least partially mineralize textile industry dyes in short reaction periods without producing any exhaust gases or harmful sludge [3, 4]. Besides, no toxic and/or carcinogenic by-products have usually been observed when applying AOPs for textile dyes decolorization [2]. In this chapter, the application of AOPs on the treatment of dyes used in textile industry will be discussed.

12.2 Definition and Classification

A dye is a colored substance that has an affinity to the substrate to which it is being applied. The dye is generally applied in an aqueous solution and may require a mordant, a substance used to set dyes on fabrics or tissues section by forming a complex with the dye which then attaches to the fabric or tissue, to improve the fastness of the dye on the fiber. Dyes appear to be colored because they absorb some wavelengths of light more than others.

Dyes are used primarily to instruct color in textile. It is also used in application of leather, paints, cosmetic, and food industries. Many natural dyes have been largely replaced by synthetic dyes that were developed at the end of the nineteenth century. Dyes should be safe, with no toxicity, carcinogenicity, or allergenicity. However, the most frequently reported causes of unexpected side effects of garments are textile dyes.

There are several ways for classification of dyes. It should be noted that each class of dye has a very unique chemistry, structure, and particular way of bonding. Dyestuff can be classified according to their origin (natural/synthetic), chemical and/or physical properties, or characteristics related to the application process. Another categorization is based on the application classes. A systematic classification of dyes according to chemical structure is the color index.

Dyes are indicated in the color index (CI) with two systems. A numeric one with five numbers, corresponds to the CI number, for example, CI 11110. The second system is a CI name, indicating the chemical category, the color, and identification number, for example, CI disperse red 1 for the previous molecule. However, the CI does not contain all information about dyes, and some textile dyes have no CI number.

According to their chemical structures and the CI system, dyes can be classified into 17 groups: nitro dyes, triphenylmethane derivatives, xanthenes, acridine derivatives, quinoline derivatives, azines, anthraquinones, indigoid dyes, phthalocyanines dyes, oxidation bases, insoluble azo dye precursors, and azo dyes (classes XII–XVII). In practice, dyes are classified into different application classes: disperse, acid, basic, direct, vat, fiber reactive, sulfur, premetallic, solvent dyes, and naphthols [5].

In accordance with US EPA, by using general dye chemistry as the basis for classification, textile dyestuffs are grouped into 14 categories or classes: (1) acid dyes, (2) direct (substantive) dyes, (3) azoic dyes, (4) disperse dyes, (5) sulfur dyes, (6) fiber reactive dyes, (7) basic dyes, (8) oxidation dyes, (9) mordant (chrome) dyes, (10) developed dyes, (11) vat dyes, (12) pigments, (13) optical/fluorescent brighteners, and (14) solvent dyes [6].

Six of the above dye classes (1–6), selected by EPA's Office of Toxic Substances, were the focus of intensive data collection efforts in an attempt to describe each class, its chemistry, and chemical and physical properties as fully as possible. Information collected for the remaining eight classes (7–14) was similar, but not as detailed. Selection of these classes (1–6) was based in part on a projection of new dye development activities and concerns about the health and environmental effects of these dyes. The following paragraphs briefly identify these six classes of dyes and describe their use by the textile dyeing industry.

Acid dyes are typically used to dye acrylics, wool, nylon, and nylon/cotton blends. They are called acid dyes because they are normally applied to the nitrogenous fibers of fabrics in organic or inorganic acid solutions. Chemical reactions between the dye and fiber form an insoluble color molecule on the fiber. The three most commercially important acid dyes are azo, anthraquinone, and triarylmethane. These dyes are generally applied as a liquid at elevated temperatures of greater than 39 °C (102 °F). Although acid dyes may be used for transfer printing, this use is not commercially important. In general, these dyes have poor wet fastness. Molecular weights range from 200 to 900, and the higher the molecular weight, the poorer the leveling (uniformity of dye uptake) property.

Direct dyes are applied in an aqueous bath containing ionic salts and electrolytes. These dyes are normally used to dye cotton and other cellulosic fibers; they bond to fibers by electrostatic forces. These dyes are highly soluble even in cold water; most have solubility's in water from 8 to 40 g l^{-1}. A few have solubility up to 80 g l^{-1}.

Azoic dyes are applied by combining two soluble components impregnated in the fiber to form an insoluble color molecule. These dye components, sold as paste-like dispersions and powders, are used chiefly for cellulosic, especially cotton. Dye bath temperatures of 16–27 °C (60–80 °F) are generally used to the shade.

Disperse dyes are colloidal and have very low water solubility. Most of these dyes are used for polyester, nylon, acetate, and triacetate fibers. They are usually applied from a dye bath as dispersions by direct colloidal absorption. Dye bath conditions (temperature, use of carrier) are varied based on the degree of difficulty encountered by the dyes in penetrating the fiber being dyed. They are sometimes applied dry at high temperatures by means of a sublimation process followed by colloidal absorption. High temperature sublimes the dye and, once it is inside the fiber, the dye condenses to a solid colloidal state and is absorbed on the fiber.

Table 12.2 Classification of dyestuffs, their associated fibers, and fixation rates [7]

Dyestuff	Characteristics	Attachment mechanism and associated fibers
Acid	Anionic, highly water soluble, poor wet fastness	Ionic bond, nylon, wool
Metal (Cr, Co, Cu) complexed acid	Anionic, low water solubility, good wet fastness	Ionic bond, nylon, wool
Direct	Anionic, highly water solubility, poor wet fastness	Ionic bond, cotton, viscose
Basic and cationic	Cationic, highly water soluble	Ionic bond, acrylics
Disperse	Colloidal dispersion, very low water solubility, good wet fastness	Colloidal impregnation and adsorption, polyester, nylon, acrylic, cellulose acetate
Reactive	Anionic, highly water soluble, good wet fastness	Covalent bonds, cotton, viscose, wool
Sulfur	Colloidal after reaction in fiber, insoluble	Dye precipitated *in situ* in fiber, cotton, viscose
Vat	Colloidal after reaction in fiber, insoluble	Dye precipitated *in situ* in fiber, cotton, viscose
Azoic	Colloidal after reaction in fiber, insoluble	Dye precipitated *in situ* in fiber, cotton, viscose

Sulfur dyes are used primarily for cotton and rayon. The application of sulfur dyes requires carefully planned transformations between the water-soluble reduced state of the dye and the insoluble oxidized form. Sulfur dyes can be applied in both batch and continuous processes; continuous applications are preferred because of the lower volume of dye required. These dyes generally have a poor resistance to chlorine. In general, sulfur blacks are the most commercially important colors and are used where good color fastness is more important than shade brightness. Sulfur dyes are not applicable to wool or silk because the fibers are chemically damaged by the dyeing process.

Fiber reactive dyes derive their name from the fact that they form covalent bonds with the fiber molecules to be dyed. Molecules of fiber reactive dyes are much smaller than the complex molecules of direct dyes. Fiber reactive dyes are unique in that they become an integral part of the textile fiber that is dyed. Although more expensive than direct dyes, advantages of reactive dyes are excellent shade reproducibility and good leveling properties. These dyes also have outstanding wet fastness. These dyes can be subdivided into either "hot" or "cold" dyeing groups, based on the temperature of application. Although silk and nylons can be dyed with fiber reactive dyes, the chief fibers dyed are cellulosic and wool. These dyes are also popular for printing textiles, since even the brightest colors are wet fast. In addition, the fact that the dye shades are very reproducible and wash fast indicates a continued increase in importance of fiber reactive dyes in the future. The fiber reactive dye class is expected to experience the largest percentage increase in production of any of the 14 dye classes.

General characteristic of major types of textile dyes together with their fixation degree for different dye and fiber combinations are presented in Table 12.2.

12.3 Dyes as Environmental Problem

In the production of dyes and pigments used in textiles and other industries, dyes are synthesized in a reactor, filtered, dried, and blended with other additives to produce the final product. The synthesis step involves reactions such as sulfonation, halogenation, amination, diazotization, and coupling followed by separation processes which may include distillation, precipitation, and crystallization. In general, organic compounds such as naphthalene are reacted with an acid or an alkali along with an intermediate (such as a nitrating or a sulfonating compound) and a solvent to form a dye mixture. The dye is then separated from the mixture and purified. On completion of the manufacture of actual color, finishing operations including drying, grinding, and standardization are performed and these are important for maintaining consistent product quality.

Liquid effluents resulting from equipment cleaning after batch operation can contain toxic organic residues. Cooling waters are normally recirculated. Wastewater generation rates are of the order of 1–700 l kg^{-1} of product except for vat dyes. The wastewater generation rate for vat dyes can be of the order of 8,000 (l kg^{-1}) of product. The biochemical oxygen demand (BOD_5) and chemical oxygen demand (COD) levels of reactive and azo dyes can be of the order of 25 and 80 kg per kg of product, respectively. Values for other dyes are, for example, 6 kg of BOD_5 together with 25 kg of COD, 6 kg of suspended solids, and 30 g of oil and grease per kg of product. Examples of wastes considered toxic include wastewater treatment sludge, spent acids, and process residues from the manufacture of chrome yellow and orange pigments, molybdate orange pigments, zinc yellow pigments, chrome and chrome oxide green pigments, iron blue pigments, and azo dyes [8].

Although not generally toxic, dyes damage the transparency of water already at low concentrations and may thus influence photosynthetic activity in water bodies due to reduced light penetration. Most commercially used dyes are of synthetic origin and are recalcitrant to microbial degradation due to their complex aromatic molecular composition. There are some side effects of the use of synthetic dyes such as allergic contact dermatitis, carcinogenesis, and mutagenesis [5] .

Accumulation of dyes in certain forms of aquatic life is suspected of leading to toxic and carcinogenic degradation products. The primary concern, even though not legally adopted in many countries, is the striking visibility of dye effluent, which renders it aesthetically unacceptable to the general public. Additionally, textile effluents usually contain secondary contaminants, such as dispersing agents, salts, emulsifiers, leveling agents, and in some cases, heavy metals. Depending on their respective concentrations, they have a pronounced effect on legally restricted quality parameters like pH, COD, TOC, and conductivity.

Many dye houses discharge their colored effluents directly to the main sewer or municipal wastewater treatment plants, while others provide pretreatment or on-site treatment facilities prior to discharge. Over 700,000 t of approximately 10,000 types of dyes and pigments are produced annually worldwide. From this amount, about 20% are discharged as industrial effluents during the textile dyeing and finishing processes without previous treatment [9].

A wide range of methods has been developed for the removal of synthetic dyes from waters and wastewaters to decrease their impact on the environment. Research on textile effluent decolorization has often focused on reactive dyes at least for four reasons. First, reactive dyes represent an increasing market share, currently about 20–30% of the total market for dyes, because they are used to dye cotton which makes up about half of world's fiber consumption. Second, a large fraction, usually around 30%, of the applied reactive dyes is wasted because of dye hydrolysis in the alkaline dye bath. As a result, dye house effluents typically contain 0.6–0.8 g l^{-1} of dye [10]. Third, these dyes have low fixation rates, which results in highly colored spent dye-baths whose treatment is inadequate in conventional wastewater treatment plants [11]. Fourth, of special concern is the reactive dyeing process, where an average of 10 times more water is consumed for preparation, dyeing, washing, and rinsing stages than for dyeing with other dye types. Reactive dyes have been identified as the most environmental problematic compounds in textile dye effluents [10, 12]. Thus, the treatment of reactive dyestuff will be focused in the forthcoming section.

12.4 AOPs Application for Removal of Dyes

The traditional treatment technologies applied in dye wastewaters treatment, such as coagulation/flocculation, membrane separation, or activated carbon adsorption, only do a phase transfer of the pollutant, and biological treatment is not a suitable solution to the problem due to biological resistance of some dyes [13]. Therefore, it is necessary to find an effective treatment technology that leads to complete destruction of the dye molecules, in terms of limited water resources management and the need for nature preservation. Among these treatments, advanced oxidation processes (AOPs) are a powerful alternative to conventional treatment methods for such wastewater decontamination [14]. These treatments are based on the in situ generation of highly reactive hydroxyl radicals (HO•). In the sake of simplicity, the AOPs' function is divided by non-UV and UV-assisted in the following discussion.

12.4.1 Non-UV-assisted AOPs

The generation of free radicals is possible in the absence of light by several methods. The most popular methods are including ozonation, Fenton's reaction, or combination of both methods. Also the ultrasound and ozone (sonolysis ozonation) method will be discussed in brief.

12.4.1.1 Ozonation (O_3)

Ozone can also be used for degradation of organic pollutants in aqueous solution. Ozone is a powerful oxidant (electrochemical oxidation potential of 2.07 vs. 2.8 V for hydroxyl radical), which is generally produced by an electric discharge method in the presence of air or oxygen. Ozonation, as an effective oxidation process, has found application in the decolorization of synthetic dyes.

The reaction of ozone with a substrate may involve both direct, selective reaction of ozone with the substrate (typically electrophilic addition reactions), and that of the substrate with $HO^{\bullet}$, the relative proportions of which will depend on various organic and inorganic species being present or added. The net reaction for OH-initiated ozone decomposition is as shown in equation below [15]:

$$3\,O_3 + HO^- \rightarrow 2\,HO^{\bullet} + 4\,O_2 \tag{12.1}$$

In the application of ozone on the decolorization study of reactive dyes, the ozone technique was applied to reduce color content caused by two commercial reactive dyes, namely, Reactive Blue 19 (RB19) and Reactive Orange 13 (RO13). Results show that the ozone dosage and pH were important factors, which controlled the effectiveness of decolorization. Results reveal that pH value of 3 favored the decolorization of RB19, and the decolorization of RO13 was enhanced at pH value of 10. This was attributed to the fact that ozone molecules and hydroxyl radicals are the predominant oxidants under low and high pH conditions, respectively. For oxidation of dyes, molecular ozone was more selective toward certain dye structures, while hydroxyl radical toward others. This implies that pH value could influence the rate of decolorization. The degree of decolorization was primarily proportional to the ozone dosage [16].

Other researchers also reported the ozonation study on the Reactive Blue 19 dye. The ozonation was found highly efficient in the decolorization of the dye solutions. The pH of unbuffered solution does not significantly affect the removal efficiencies of color. Inorganic dye degradation by-products such as sulfate, nitrate, and formate were detected and monitored during ozonation by ion chromatography which is an evidence of oxidation and cleavage of the dye chemical structure. The presence of sodium chloride and sulfate salts in the dye solution slightly reduces the decolorization efficiency, while sodium carbonate and bicarbonate salts slightly increase the rate of decolorization. The COD and TOC reduction was about 55 and 17%, respectively, for 90 min ozonation of 800 mg l^{-1} dye solution. A rapid decrease of pH during the ozonation process gave evidence for production of acidic by-products [17]. These organic acids have been confirmed by Fanchiang and Tseng [18] in their study on the degradation of anthraquinone dye C.I. Reactive Blue 19 (RB19) in aqueous solution by ozonation. The identification by LC–MS and GC–MS analyses indicated that some organic acids (e.g., phthalic acids) and 1,3-indanone could be the primary degradation products, respectively. In addition, the results of UV/VIS and FTIR spectra showed that the anthraquinone structures, nitrogen linkages, and amino groups of RB-19 were destroyed under direct ozone reaction. The detoxification accompanied biodegradability enhancement revealed by BOD/COD ratio increasing from 0.15 to 0.33 after 10 min of ozonation.

Based on the analyses of decolorization efficiency and power consumption, the study about degradation of C.I. Reactive Red 2 (RR2) using ozone-based systems suggests that the ozonation systems are acceptable for decolorizing RR 2 at pH 10. Results showed that the ozonation was a very effective method for reactive dye decolorization.

12.4.1.2 Fenton's Reagent (H_2O_2/Fe^{2+})

Among AOPs, Fenton process is very promising since the systems achieves high reaction yields, offers a cost-effective source of hydroxyl radicals and it is easy to operate and maintain [19]. Fenton's reagent oxidation is a homogeneous catalytic oxidation process using a mixture of hydrogen peroxide and ferrous ions. Hydrogen peroxide is added to an aqueous system in an acidic condition. The overall reaction is [13]:

$$Fe^{2+} + H_2O_2 \rightarrow Fe^{3+} + HO^{\bullet} + HO^{-} \tag{12.2}$$

Fenton's reagent is an attractive oxidative system for wastewater treatment due to the fact that iron is very abundant and nontoxic and hydrogen peroxide is easy to handle and environmentally safe. The main advantage of the Fenton's reagent is its simplicity. The chemicals are readily available at moderate cost and there is no need for special equipment [20]. This process could be a good option to treat and eliminate textile dyes. The process appears to have capacity to completely decolorize and partially mineralize textile industry dyes in short reaction time, as reported by some studies [19, 21–23].

Degradation of dyes in aqueous solutions by the Fenton process was investigated by Xu et al. [24]. The reactive dyes studied include Reactive Brilliant Red K-2BP, Reactive Brilliant Red K-2 G, and Reactive Yellow KD-3 G. The dyes were decolorized 98, 100, and 96%, respectively. Their TOC values were decreased 76, 72, and 79%, respectively.

The oxidative degradation of a reactive dye, C. I. Reactive Red 2 by hydroxyl radicals generated by Fenton's reagent has been carried out in aqueous acidic media. Effects of different parameters like initial concentration of dye, H_2O_2, Fe^{2+}, pH of the solution, reaction temperature, and added electrolytes on the oxidation process have been studied. The results indicate that 1.63×10^{-4} mol dm^{-3} dye can be most effectively degraded at a dye: Fe^{2+}: H_2O_2 molar ratio of 1:0.22:8.13 at pH $\pm$ 2.7 and at 299 K. The addition of excess 2-propanol or t-butyl alcohol, well-known scavengers of hydroxyl radicals, almost stopped the degradation of the dye indicating the absence of any possible reductive pathways in the degradation [25].

Lucas and Peres [13] examined decolorization of Reactive Black 5 (RB5) in aqueous solution using Fenton's reagent. Batch experiments were carried out to investigate the process's optimal operational conditions: pH, H_2O_2 dosage, Fe^{2+} dosage, RB5 concentration, and optimal $[Fe^{2+}]_0/[H_2O_2]_0/[RB5]_0$ ratio, to obtain the best results at low cost, making AOP competitive with other processes. The optimal conditions found were a ratio $[H_2O_2]_0/[RB5]_0$ of 4.9:1, a ratio $[H_2O_2]_0/[Fe^{2+}]_0$ of 9.6:1, and pH = 3.0. A color degradation of 97.5% and TOC removal of 21.6% were achieved for optimal conditions.

The oxidative decolorization and mineralization of three reactive dyes in separately prepared aqueous solutions C.I. Reactive Yellow 3 (RY3), C.I. Reactive Blue 2 (RB2), and C.I. Reactive Violet 2 (RV2) by using Fenton process have been investigated. The effects of H_2O_2, Fe^{2+} concentrations, and Fe^{2+}/H_2O_2 molar ratios have been studied. Optimal operational conditions for the efficient degradation of all three dye solutions (100 mg l^{-1}) were found to be $Fe^{2+}/H_2O_2 = 0.5$ mM/20 mM at pH 3 and T = 296 K. By using this process, the high levels of mineralization (more than 78%) and decolorization (more than 95%) were achieved [10].

12.4.1.3 Fenton-Ozonation ($H_2O_2/Fe^{2+}/O_3$)

Among the AOPs, treatment with ozone [26–28] or Fenton-type [21, 24, 25] processes have yielded very good results. The combined Fenton-type process with ozone logically leads to produce more hydroxyl radicals and consequently compose more efficient system.

The oxidative degradation of azo dye Reactive Orange 113 aqueous solution by H_2O_2/Fe^{2+} and ozone was investigated by Gutowska et al. [28]. The full decolorization was attained during the first minutes of oxidation process by H_2O_2/Fe^{2+} and ozone at room temperature. However, the complete mineralization of aqueous solutions of Reactive Orange 113 dye cannot be achieved even under the optimal reaction conditions.

The decolorization of C.I. Reactive Red 2 using ozone, Fenton-like and ozone/Fenton-like hybrid systems was studied by Wu [29]. Decolorization rate constants of ozone/Fenton-like hybrid ($O_3/H_2O_2/Fe^{3+}$) is bigger than ozone-hydrogen peroxide (O_3/H_2O_2) systems at pH 7. In the study, based on an appropriate hydrogen peroxide dosage of 1,000 ppm and to prevent the generation of a large quantity of sludge, the Fe^{3+} dosage was set at 25 ppm in all experiments in which Fe^{3+} was used, in order to maintain the hydrogen peroxide: catalyst ratios of 40:1 as recommended for optimal Fenton process.

12.4.1.4 Sonolysis Ozonation

When a liquid is sonolyzed, gas molecules entrapped by micro-bubbles can be fragmented into radical species. Hydroxyl radicals generated by the combination of ozone with ultrasound can be represented as follows [30]:

$$O_3 +))) \rightarrow O_2(g) + O(^3P)(g) \tag{12.3}$$

$$O(^3P)(g) + H_2O \rightarrow 2HO^\bullet \tag{12.4}$$

Ince and Tezlani [31] combined sonolysis and ozonation for hydrolyzed Reactive Black 5 (RB5) degradation. They found that the joint action of ultrasound (520 kHz) and ozone induced a synergistic effect on both the decolorization of the dye and the overall degradation process. It was found during lab-scale treatment of a reactive dye bath effluent with ozone, that the rate of color degradation achieved in conjunction with sonolysis was twice as fast as that accomplished by ozonation alone. The synergism observed for the overall degradation (mineralization) of the dye reflects the potential of the US/ozone combination treatment to minimize the survival time of the reaction intermediates. Its effects can be attributed to the mechanical effects of cavitations, which increased the mass transfer and decomposition of ozone, thus enhancing both the rate of direct reactions and that of $HO^{\bullet}$-mediated oxidation of intermediate products.

More recent research conducted by Zhiqiao He et al. [32] on the decolorization Reactive Black 5 by using combination of sonolysis and ozonation has been reported. The experimental results indicate that the rate of decolorization is influenced by the initial concentration of dye, temperature, and pH. Increasing the initial concentration of RB5 led to a decreasing decolorization rate. Under the experimental conditions, the decolorization rate increased with an increase in temperature. The optimum pH for the reaction was 11, and both lower and higher pH decreased the decolorization rate. Based on the decolorization rate constants obtained at different temperatures within the range 287–338 K and the Arrhenius equation, the apparent activation energy (Ea) was calculated to be 11.2 kJ mol^{-1}. This indicated that the reaction has little dependence on temperature. The color decay was considerably faster than the decrease in total organic carbon (TOC), which was attributed to the ease of chromophore destruction. Hence, the efficiency of decolorization was 84% compared with 4% of TOC removal after 5 min reaction. Additionally, muconic acid, (2Z)-pent-2-enedioic acid, and maleic acid were identified as main oxidation products by gas chromatography coupled with mass spectrometry (GC–MS) after 150 min of reaction.

Degradation of C.I. Reactive Yellow 84 (RY84) in aqueous solution by sonolytic ozonation was measured under various conditions [33]. As the initial concentration of RY84 decreased, the degradation rate increased. The impact of US density on mineralization was slight, and increasing ozone dose had a positive effect on the mineralization of RY84. TOC removals observed were 9, 42, and 56% with US, O_3, and US/O_3, respectively. The better performance of the combined system was attributed to enhanced mass transfer of ozone to accelerate its direct reactions with RY84 and the generation of excess radical species upon its thermal decomposition in collapsing cavity bubbles. These results illustrate that ozonation combined with sonolysis for removal of TOC is more efficient than ozonation alone or ultrasonic irradiation alone without considering the operating costs. With the initial pH value at 10.0, the ozone dose at 4.5 g h^{-1}, the energy density of ultrasound at 176 W l^{-1}, and the initial concentration of RY84 at 100 mg l^{-1}, the extent of mineralization measured as TOC removal were maximized.

Mineralization of an anthraquinone dye, C.I. Reactive Blue 19 (RB19), by single ozonation, single sonolysis, and ozonation combined with sonolysis (O_3/US) was carried out in a laboratory-scale experiment [34]. O_3/US treatment rendered a more effective result than the other two methods. The experiments showed that the mineralization rate of O_3/US was affected by initial dye concentration, pH, O_3 dose, and US energy density. Ultrasonic irradiation alone yielded negligible results, but O_3/UStreatment exhibited a higher mineralization rate compared with ozonation alone. GC/MS and IC analysis allowed the identification of intermediates and products. The element C in RB19 could be degraded to an organic acid with a smaller molecular weight, indicating that sonolytic ozonation is a highly effective way to mineralize RB19 in aqueous solution.

12.4.2 *UV-Assisted AOPs*

The principal mechanism of AOPs' function is the generation of very powerful and nonselective oxidizing agent, free hydroxyl radical ($HO^{\bullet}$), to destroy hazardous pollutants. Consequently, a combination of two or more AOPs enhances the radical generation, which finally leads to higher oxidation rates.

12.4.2.1 Photolysis of Ozone (UV/O_3)

Additional free hydroxyl radicals can be produced in the aqueous media from ozone by pH modification or can be introduced by combining ozone either with hydrogen peroxide or UV irradiation from a high-pressure mercury lamp [35].

The process features direct molecular ozonation, ozone decomposition (free radical chemistry) and a photolytic reaction (direct photolysis of aqueous ozone producing hydrogen peroxide and indirect photolysis of hydrogen peroxide formed *in situ*) [36]. The principal reaction is given in the following equation [15]:

$$O_3 + \frac{1}{2}H_2O \xrightarrow{hv} H_2O_2 \xrightarrow{hv} 2\,HO^{\bullet} \tag{12.5}$$

Degradation of C.I. Reactive Red 2 (RR2) using ozone based systems at various pH was studied by Wu and Ng [37]. The decolorization efficiency of C.I. Reactive Red 2 (RR2) is generally higher in UV/ozone-based systems than in ozone-based systems. After 30 min of reaction, decolorization efficiency for pH 4, pH 7, and pH 10 were 60, 78, and 97, respectively. Hence, UV/O_3 and O_3 systems are acceptable for decolorizing RR2 at pH 10.

12.4.2.2 Photolysis of Hydrogen Peroxide (UV/H_2O_2)

Principally, the cleavage of hydrogen peroxide by UV-C radiation is the most direct method for $HO^{\bullet}$ generation. In this process, hydrogen peroxide is activated by UV-C irradiation (λ 254 nm) to form $HO^{\bullet}$ (Eq. 12.6) [15]:

$$H_2O_2 + hv \rightarrow 2\,HO^{\bullet} \tag{12.6}$$

Mechanism and a kinetic model for the decolorization of the diazo dye Reactive Black 5 (RB5) by UV radiation and hydrogen peroxide (UV/H_2O_2) was developed

[38]. The study is based on experimental results as well as on a reaction kinetic mechanism of the photolysis of H_2O_2. This mechanism depends on known chemical reactions. The experiments were carried out in a batch reactor and any applied UV intensity. For higher hydrogen peroxide concentration and at low applied intensities, the reaction is only controlled by the intensity. For lower hydrogen peroxide concentration and higher intensity, the reaction rate depends on hydrogen peroxide concentration only. The model is able to predict the decolorization rate as a function of dye concentration, hydrogen peroxide concentration, and intensity in very good agreement with experimental results.

Decolorization of vinylsulphone reactive dyes by using H_2O_2–based processes was conducted by Kurbus et al. [39]. In their study, six vinylsulphone reactive dyes were used to determine the effectiveness of three advanced oxidation processes (H_2O_2/UV, H_2O_2/O_3, and H_2O_2/Fe^{2+}). The efficiency of decolorization was measured by decolorization time, absorbance, COD, and TOC, and the three methods of decolorization are compared. The H_2O_2/ Fe^{2+} process achieved the fastest decolorization. The H_2O_2/UV decolorization process is faster than the H_2O_2/O_3. The selected dyes were efficiently decolorized by all of the three advanced oxidation processes, their absorbance values after decolorization being zero or approximately zero. The COD values obtained after decolorization with H_2O_2/UV is the lowest. The TOC values obtained in all of the three decolorization processes were still high. Although decolorization with H_2O_2/Fe^{2+} was the fastest and the costs of such a process are low, it is not really suitable for environmental reasons, namely, the formed sludge needs to be removed. Decolorization with H_2O_2/UV is an environment-friendly procedure while the COD removal is the highest. It causes no sludge and no additional water pollution because of the physical activation of hydrogen peroxide, but investment and operating costs are high due the UV lamp utilization.

12.4.2.3 Photo-Fenton's Reagent (UV/H_2O_2/Fe^{2+})

In the generally accepted mechanism of Fenton's reaction, hydroxyl radicals ($HO^{\bullet}$) are produced by interaction of H_2O_2 with ferrous salts:

$$\mathrm{Fe(II)} + H_2O_2 \rightarrow \mathrm{Fe(III)} + HO^{\bullet} + OH^{-} \quad (12.7)$$

The rate of degradation of organic pollutants by Fenton's reaction can be increased if an irradiation source is present. The positive effect of irradiation on the degradation rate is due to the photoreduction of Fe(III) to Fe(II) ions, a step that produces new $HO^{\bullet}$ radicals and regenerates Fe(II) ions that can further react with more H_2O_2 molecules. The photoreduction of Fe(III) follows the equation:

$$FeOH^{2+} + h\nu \rightarrow Fe^{2+} + HO^{\bullet} \quad (12.8)$$

$Fe(OH)^{2+}$ being the dominant Fe(III) species in solution at pH 2–3 [40].

The oxidative decolorization of Reactive Black 5 (RB5) in aqueous solution has been studied by using Fenton (H_2O_2/Fe^{2+}) and photo-Fenton (H_2O_2/Fe^{2}+/UV) processes. This investigation reveals that both methods can remove the color of RB5. Batch experiments were carried out to investigate the process's optimal operational conditions: pH, H_2O_2 dosage, Fe^{2+} dosage, RB5 concentration, and optimal $[Fe^{2+}]_0/[H_2O_2]_0/[RB5]_0$ ratio, to obtain the bests results at low cost and to make AOP process's competitive with other processes. The optimal conditions found were a ratio $[H_2O_2]_0/[RB5]_0$ of 4.9:1, a ratio $[H_2O_2]_0/[Fe^{2+}]_0$ of 9.6:1, and pH = 3.0. The decolorization experiments indicate that RB5 can be effectively decolorized using Fenton and photo-Fenton processes with a little difference between the two processes, 97.5 and 98.1%, respectively, for optimal conditions. This small difference in dye decolorization is not similar to TOC removal: with photo-Fenton process there is a significant increment (46.4% TOC removal) relatively to Fenton process (only 21.6% TOC removal). Although the decolorization is very similar in both processes, the employment of the UV lamp benefits the azo dye degradation. It is possible to conclude that the UV lamp, though has little power (15 W), is very useful in Fenton process to increase the dye mineralization [13].

The benefits of using combined solar radiation with Fenton process have been reported in more recent publication. For instance, the oxidation of C.I. Reactive Blue 4 (RB4) by photo-Fenton process mediated by ferrioxalate was investigated under artificial and solar irradiation. The RB4 degradation in acidic medium (pH 2.5) was evaluated by the decrease in total organic carbon (TOC) content and color, measured by the decrease in chromophore absorption band (600 nm). The influence of ferrioxalate and H_2O_2 concentrations on the dye degradation was studied and best results were obtained using 1.0 mM ferrioxalate and 10 mM of hydrogen peroxide. Under these experimental conditions, 80% of TOC and 100% of color removal were obtained for a 0.1 mM RB4 dye in 35 min of solar irradiation.

Concentrations of RB4 up to 0.1 mM can be photodegraded resulting in 80% mineralization and 100% color removal under solar irradiation at summer clear sky days in 35 min exposition in the presence of 1.0 mM ferrioxalate and 10 mM H_2O_2. However, when studying the influence of initial FeOx and H_2O_2 concentrations, it was observed that 0.7 mM of FeOx and 7.0 mM H_2O_2 were sufficient to obtain similar photodegradation. It is possible to conclude that the photodegradation of the RB4 dye under solar irradiation is highly dependent on the irradiation intensity. The yellow coloration characteristic of Fe (III) species was observed for both solar and artificial irradiation conditions, but the residue generated is easily removed after rest for 4 h, when 100% decolorization is obtained. These results indicate that visible irradiation can effectively accelerate the generation of hydroxyl radicals, due to the regeneration of Fe^{2+}. The rate of the reaction is diminished on cloudy days. However, the lower cost intrinsic to solar energy utilization can compensate for small differences in photodegradation rates, especially in tropical regions where solar energy is abundant. The homogeneous photo-Fenton system presents the advantage that larger reactors are not necessary since the results presented here were obtained with small exposition areas of about 57 cm^2 [9]. In this research, the highest value of dye mineralization percentage in the term of TOC removal was obtained by using Fenton application under solar irradiation.

Another research describing the solar exploration using a compound parabolic collector (CPC) applied to degradation of RB4 solutions was examined by Duran et al. [41]. The efficiency of photocatalytic degradation was determined from the analysis of color removal, total organic carbon (TOC), and chemical oxygen demand (COD). In their experiments, the solar photo-Fenton–ferrioxalate process increases degradation rate of RB4 since ferrioxalate complexes absorb strongly and a higher portion of the solar spectrum can be used. Although the addition of oxalic acid increases operating costs, it improves the process and it also helps to reduce pH in solution, decreasing charges derived from this operation. Under the optimum conditions, ($[H_2O_2] = 120$ ppm (in two additions), $[Fe(II)] = 7$ ppm, $[(COOH)_2] = 10$ ppm, pH 2.5), color and COD were completely removed whereas TOC was reduced up to 66%.

The highest TOC removal on the treatment of RB4 dye has been achieved in the study of photodegradation of RB4 solutions under ferrioxalate-assisted UV/solar photo-Fenton system with continuous addition of H_2O_2 and air injection, which was conducted by Montegado et al. [42]. In their research, the effect of the continuous addition of hydrogen peroxide and air injection on the degradation of RB4 solutions was examined. The reaction was carried out in a CPC solar pilot plant reactor. It was shown that the degree of dye solution mineralization was enhanced because the scavenger effect of H_2O_2 was minimized. Air bubbling had a negative effect because oxygen reacted with oxalyl radical anions, diminishing the amount of generated Fe(II), and consequently, the concentration of hydroxyl radicals produced under irradiation of oxalate in the presence of peroxide was lower. In addition, this system permits the use of lower concentrations of ferrous ion, reducing the costs of the later iron removal process. Under the optimal conditions selected in the system, TOC removal was increased from 61 to 82% with a shorter reaction time as compared with a peroxide dose at the beginning of the reaction. The efficiency of the mineralization of RB4 solutions was enhanced when the solar CPC reactor worked together with an artificial UV-A/C pilot plant, increasing TOC removal up to 95%. Artificial UV lamps can be used either to improve the process or as an alternative to solar CPC on cloudy days. The authors explain that the degree of mineralization of dye solutions was enhanced when hydrogen peroxide was continuously dosed because the scavenger effect of H_2O_2 was minimized. Air bubbling had a negative effect because less Fe(II) and $HO^{\bullet}$ radicals were generated. Oxalic acid improved RB4 solution mineralization significantly, because ferrioxalate complexes absorb strongly and a higher portion of the solar spectrum was used.

12.4.2.4 TiO_2-Based Photocatalysis

The combination of UV irradiation with photocatalysts such as TiO_2 is one of such methods which have described significant attention in recent years, due to its effectiveness in mineralization of organic compounds, including removal of dye pollutants from residual textile treatment waters [22, 43–45].

The photocatalytic degradation of three commercial reactive textile dyes with different structure has been investigated using TiO_2 (Degussa P25) photocatalyst in aqueous solution under solar irradiation. The experimental results indicate that TiO_2 (Degussa P25) is the best catalyst in comparison with other commercial photocatalysts such as, TiO_2 (Merck), ZnO, ZrO_2, WO_3, and CdS. Among the catalysts, TiO_2 (Degussa P25) is found to be an effective one because of its multiphase purity, large band gap, and noncorrosive nature [46]. Additionally, the same catalyst can be used for a number of cycles which will reduce the cost of operation. Though the UV irradiation can efficiently degrade the dyes, naturally abundant solar irradiation is also very effective in the mineralization of dyes.

It has been proven that the presence of catalysts enhances the rate of photodecomposition. Most investigations of the photo-assisted decomposition of dyes have utilized TiO_2 as a model. Interest in this photocatalytic process mainly was because of its low cost, combined with the high photocatalytic reactivity, chemical and biological stability, nontoxicity, insolubility, and ready availability. A further important advantage is the fact that the process can be powered by natural sunlight [47–51], thus reducing significantly the electrical power requirements and operating costs [46, 52].

The following mechanism is the widely accepted to describe the photodegradation of organic compounds by UV/TiO_2 [53]. Photogenerated holes are generated as a result of TiO_2 particles having been irradiated with UV light (Eq. 12.9). Hydroxyl radicals are produced by the oxidation of $HO^{\bullet}$ or H_2O by these photogenerated holes (Eqs. 12.10 and 12.11) which are principally responsible for the destruction of organic species. Oxygen is used as an efficient electron trap (Eq. 12.12), preventing the recombination of electrons and photogenerated holes. Eqs. (12.13), (12.14), and (12.15) represent the other reactions of UV/TiO_2. If oxygen is limited, rapid recombination of photoproduced electrons and holes in TiO_2 lowers the efficiency of the photocatalytic reactions [54]; consequently, the efficacy of such a system has limited practical application.

$$TiO_2 + hv \rightarrow TiO_2(e_{CB\_} + h_{VB+}) \tag{12.9}$$

$$TiO_2(h_{VB+}) + OH^- \rightarrow TiO_2 + HO^{\bullet} \tag{12.10}$$

$$TiO_2(h_{VB+}) + H_2O \rightarrow TiO_2 + H^+ + HO^{\bullet} \tag{12.11}$$

$$TiO_2(e_{CB^-}) + O_2 \rightarrow TiO_2 + O_2^{-\bullet} \tag{12.12}$$

$$O_2^{-\bullet} + H^+ \rightarrow HO_2^{\bullet} \tag{12.13}$$

$$TiO_2(e^{\bullet}{}_{CB^-}) + H^+ + O_2{}^{-\bullet} \rightarrow TiO_2 + HO_2{}^- \tag{12.14}$$

$$HO_2{}^- + H^+ \rightarrow H_2O_2 \tag{12.15}$$

Among the reactions, the availability of electron trapping in aqueous solution is important to inhibit the recombination of electrons and photogenerated holes. Adding the electron acceptor (hydrogen peroxide, persulfate, and copper ions) and hydroxyl radical trap (ethanol) was attempted by Wu [55]. In the study, the effects of dye concentration, pH, $Na_2S_2O_8$ dose, and addition of $Na_2S_2O_8$, H_2O_2, Cu^{2+}, and ethanol on the decolorization of C.I. Reactive Red 198 in UV/TiO_2-based systems was examined. Decolorization rate declined as dye concentration increased. The k values for the UV/TiO_2, UV/TiO_2/$Na_2S_2O_8$, and UV/TiO_2/H_2O_2 systems increased with decreasing pH. Decolorization rate increased as the concentration of persulfate ions added increased. The persulfate ions trapped photogenerated conduction band electrons, preventing the recombination of the electrons with positive holes and, simultaneously, generated sulfate free radicals. H_2O_2 trapped photogenerated electrons and inhibited the recombination of photogenerated electron/hole pairs. Adding Cu^{2+} suppressed the recombination of photogenerated electron/hole pairs and generated additional persulfate free radicals in the UV/TiO_2/$Na_2S_2O_8$ system, thereby accelerating decolorization. The UV/TiO_2/H_2O_2/Cu system resembles the photo-Fenton-like process since the induced photo-Fenton-like reactions acted as an additional source of hydroxyl radicals. This investigation suggests that the primary decolorization pathway involves hydroxyl radicals and that direct oxidation by photogenerated holes is significant in the UV/TiO_2 system. During decolorization, some Cl^-, NO_3^-, and SO_4^{2-} ions were released and, even when decolorization was complete, the TOC was reduced by only 50%.

A synergistic effect of photocatalysis and ozonation on textile effluent treatment was also reported by Agustina et al. [56]. The decolorization was almost complete and the highest TOC reduction was achieved. It was shown that the photocatalytic oxidation in the presence of ozone is a process, which is qualitatively and quantitatively different from either the well-known photocatalytic oxidation with oxygen or ozonation without photocatalyst.

The main problem of AOPs lies in the high cost of reagents such as ozone, hydrogen peroxide, or energy source such as ultraviolet light. For the photocatalytic ozonation methods, the energy demand of the UV/TiO_2/O_3 could be considerably decreased by the use of solar irradiation and in situ electrochemical O_3 generation. However, the use of solar radiation as an energy source can reduce the costs.

12.5 Conclusions

It is clear from this chapter and many of the available literatures that AOPs have proven performance in removing dyes associated with the textile industry activities. Especially in terms of color degradation, the complete color removal can be accomplished even in single AOPs application. But in terms of mineralization, different TOC removal can be reached by single or more than one AOPs function.

Under non-UV-assisted AOPs, ozonation is a very effective method in the dyes treatment. But this method needs to utilize ozone generator which is associated to high operational cost. Fenton systems offer the cost-effective, as high color degradation can be obtained by using a small concentration of reagent. Moreover, low chemical dose leading to lower residuals and, where applicable, no residual processes for Fenton-based meaning. But still the Fenton methods cannot reach a satisfied TOC removal (only 80%); even when Fenton method is combined with ozonation, the complete TOC removal cannot be obtained. In sonolysis ozonation process, this system is better than ultrasound or ozonation alone in the dyes treatment. Besides, the TOC reduction is not so high. However, the experimental results show that sonolysis ozonation as a developing technology deserves further study and potentially will be an important and effective technology for removal of dyes in wastewater treatment.

UV-assisted AOPs systems emerge to have the ability to completely decolorize and at least partially mineralize textile industry dyes in short reaction periods. The highest TOC removal obtains of 95% has been reported in the application of UV/Fenton system. The UV-based systems need the source of UV radiation for the processes which make the investment and operating costs are high due the UV lamp utilization. The TiO_2-based photocatalysis systems propose a very potential treatment as the TiO_2 catalyst can be powered by natural sunlight. So the UV employment is replaced by solar energy. Solar energy will appear as an alternative cost-effective light source because of its great quantity and nonhazardous nature.

References

1. Doble M, Kumar A (2005) Degradation of dyes. In: Biotreatment of industrial effluents Butterworth-Heinemann, Burlington, pp 111–122
2. Parsons S (2004) Advanced oxidation processes for water and wastewater treatment. IWA Publishing, London
3. Kang SF, Liao CH, Po ST (2000) Decolorization of textile wastewater by photo-Fenton oxidation technology. Chemosphere 41(8):1287–1294
4. Kusic HN, Koprivanac N, Srsan L (2006) Azo dye degradation using Fenton type processes assisted by UV irradiation. A kinetic study. J Photochem Photobiol A 181(2–3):195–202
5. Le Coz CJ (2005) Dyes. In: Encyclopedia of toxicology. Elsevier, California, pp. 104–114
6. Kulkarni SV, Blackwell CD, Blackard AL, Stackhouse CW, Alexander MW (1985) Textile dyes and dyeing equipment classification properties and environmental aspects. United States Environmental Protection Agency, Research Triangle Park, North Carolina
7. Arslan-Alaton I (2004) Advanced oxidation of textile industry dyes. In: Parsons S (ed) Advanced oxidation processes for water and wastewater treatment. IWA Publishing, London, pp 302–328
8. Environmental guidelines for dye manufacturing. In: Multilateral investment guarantee agency (pp. 363–366)
9. Carneiro PA, Pupo Nogueira RF, Zanoni MVB (2007) Homogeneous photodegradation of C.I. Reactive Blue 4 using a photo-Fenton process under artificial and solar irradiation. Dyes Pigment 74:127–132

10. Papic CS, Vujevi CD, Koprivanac N, Sinko D (2009) Decolourization and mineralization of commercial reactive dyes by using homogeneous and heterogeneous Fenton and UV/Fenton processes. J Hazard Mater 164:1137–1145
11. Easton JR (1995) The dye maker's view. In: Cooper P (ed) Colour in dyehouse effluent the society of dyers and coulorists. Alden Press, Oxford, pp 6–21
12. Garcia-Montano J, Ruiz N, Munoz I, Domenech X, Garcia-Hortal JA, Torrades F, Peral J (2006) Environmental assessment of different photo-Fenton approaches for commercial reactive dye removal. J Hazard Mater A 138:218–225
13. Lucas MS, Peres JA (2006) Decolorization of the azo dye Reactive Black 5 by Fenton and photo-Fenton oxidation. Dyes Pigment 71:236–244
14. Munter R (2001) Advanced oxidation processes-current status and prospects. Proc Estonian Acad Sci Chem 50:59–80
15. Arslan-Alaton I (2003) A review of the effects of dye-assisting chemicals on advanced oxidation of reactive dyes in wastewater. Coloration Technol 119(6):345–353
16. Chen TY, Kao CM, Hong A, Lin CE, Liang SH (2009) Application of ozone on the decolorization of reactive dyes Orange 13 and Blue 19. Desalination 249(3):1238–1242
17. Tehrani-Bagha AR, Mahmoodi NM, Menger FM (2010) Degradation of a persistent organic dye from colored textile wastewater by ozonation. Desalination 260(1–3):34–38
18. Fanchiang JM, Tseng DH (2009) Degradation of anthraquinone dye C.I. Reactive Blue 19 in aqueous solution by ozonation. Chemosphere 77(2):214–221
19. Neamtu M, Yediler A, Siminiceanu I, Kettrup A (2003) Oxidation of commercial reactive azo dye aqueous solutions by the photo Fenton and Fenton like processes. J Photochem Photobiol A 161:87–93
20. Gogate PR, Pandit AB (2004) A review of imperative technologies for wastewater treatment oxidation technologies at ambient conditions. Adv Environ Res 8:501–551
21. Meric S, Kaptan D, Olmez T (2004) Color and COD removal from wastewater containing Reactive Black 5 using Fenton's oxidation process. Chemosphere 54:435–441
22. Arslan I, Balcioglu IA, Bahnemann DW (2000) Advanced chemical oxidation of reactive dyes in simulated dyehouse effluents by ferrioxalate Fenton/UVA and TiO_2/UV-A processes. Dyes Pigments 47:207–218
23. Feng J, Hu X, Yue PL, Zhu HY, Lu GQ (2003) Discoloration and mineralization of Reactive Red HE-3B by heterogeneous photo-Fenton reaction. Water Res 37:3776–3784
24. Xu XR, Li HB, Wang WH, Gu JD (2004) Degradation of dyes in aqueous solutions by the Fenton process. Chemosphere 57(7):595–600
25. Dutta K, Bhattacharjee S, Chaudhurib B, Mukhopadhyay S (2002) Chemical oxidation of C. I. Reactive Red 2 using Fenton-like reactions. J Environ Monit 4:754–760
26. Wu C, Chang C (2006) Decolorization of Procion Red MX-5B by advanced oxidation processes comparative studies of the homogeneous and heterogeneous systems. J Hazard Mater 128:265–272
27. Wu C, Chang C, Kuo C (2008) Decolorization of Procion Red MX-5B in electrocoagulation (EC) UV/TiO_2 and ozone related systems. Dyes Pigments 76(1):187–194
28. Gutowska A, Kaluzna-Czaplinska J, Józwiak WK (2007) Degradation mechanism of Reactive Orange 113 dye by H2O2/Fe2+ and ozone in aqueous solution. Dyes Pigments 74(1):41–46
29. Wu CH (2008) Decolorization of C.I. Reactive Red 2 in O_3 Fenton-like and O_3 Fenton-like hybrid systems. Dyes Pigments 77(1):24–30
30. Tezcanli Güyer G, Ince NH (2008) Individual and combined effects of ultrasound ozone and UV irradiation a case study with textile dyes. Ultrasonics 42(1–9):603–609
31. Ince NH, Tezcanlí G (2001) Reactive dyestuff degradation by combined sonolysis and ozonation. Dyes Pigments 49(3):145–153
32. He Z, Song S, Zhou H, Ying H, Chen J (2007) C.I. Reactive Black 5 decolorization by combined sonolysis and ozonation. Ultrason Sonochem 14(3):298–304

33. He Z, Song S, Xia M, Qiu J, Ying H, Lü B, Jiang Y, Chen J (2007) Mineralization of C.I. Reactive Yellow 84 in aqueous solution by sonolytic ozonation. Chemosphere 69(2):191–199
34. He Z, Lin L, Song S, Xia M, Xu L, Ying H, Chen J (2008) Mineralization of C.I. Reactive Blue 19 by ozonation combined with sonolysis: performance optimization and degradation mechanism. Sep Purif Technol 62(2):376–381
35. Chiron S, Fernandez Alba A, Rodriguez A, Garcia-Calvo E (2000) Review paper Pesticide chemical oxidation state of the art. Water Res 34(2):366–377
36. Legrini O, Olivero E, Braun AM (1993) Chem Rev 93:671
37. Wu CH, Ng HY (2008) Degradation of CI Reactive Red 2 (RR2) using ozone based systems. Comparisons of decolorization efficiency and power consumption. J Hazard Mater 152(1): 120–127
38. El Mohey, Dein A, Libra JA, Wiesmann U (2003) Mechanism and kinetic model for the decolorization of the azo dye Reactive Black 5 by hydrogen peroxide and UV radiation. Chemosphere 52(6):1069–1077
39. Kurbus T, Le Marechal AM, Voncina DB (2003) Comparison of H_2O_2/UV, H_2O_2/O_3 and H_2O_2/Fe^{2+} processes for the decolorisation of vinylsulphone reactive dyes. Dyes Pigments 58 (3):245–252
40. Torrades F, GarcíaMontaño J, Antonio J, GarcíaHortal JA, Domènech X, Peral J (2004) Decolorization and mineralization of commercial reactive dyes under solar light assisted photo Fenton conditions. Sol Energy 77(5):573–581
41. Durán A, Monteagudo JM, Amores EE (2008) Solar photo-Fenton degradation of Reactive Blue 4 in a CPC reactor. Appl Catal Environ 80:42–50
42. Monteagudo JM, Durán A, Aguirre M, Martín IS (2010) Photodegradation of Reactive Blue 4 solutions under ferrioxalate assisted UV/solar photo Fenton system with continuous addition of H_2O_2 and air injection. Chem Eng J 162(2):702–709
43. Hachem C, Bocquillon F, Zahraa O, Bouchy M (2001) Decolourization of textile industry wastewater by the photocatalytic degradation process. Dyes Pigments 49(2):117–125
44. Saquib M, Muneer M (2003) TiO_2-mediated photocatalytic degradation of a triphenylmethane dye (gentian violet) in aqueous suspensions. Dyes Pigments 56(1):37–49
45. Saquib M, Muneer M (2003) Titanium dioxide mediated photocatalyzed degradation of a textile dye derivative acid orange 8 in aqueous suspensions. Desalination 155(3):255–263
46. Neppolian B, Choi HC, Sakthivel S, Arabindoo B, Murugesan V (2002) Solar/UV-induced photocatalytic degradation of three commercial textile dyes. J Hazard Mater B89:303–317
47. Saquib M, Muneer M (2002) Semiconductor mediated photocatalysed degradation of an anthraquinone dye Remazol Brilliant Blue R under sunlight and artificial light source. Dyes Pigments 53(3):237–249
48. Augugliaro V, Baiocchi C, Bianco Prevot A, García López E, Loddo V, Malato S, Marcí G, Palmisano L, Pazzi M, Pramauro E (2002) Azo-dyes photocatalytic degradation in aqueous suspension of TiO_2 under solar irradiation. Chemosphere 49(10):1223–1230
49. Stylidi M, Kondarides DI, Verykios XE (2003) Pathways of solar light-induced photocatalytic degradation of azo dyes in aqueous TiO_2 suspensions. Appl Catal Environ 40(4):271–286
50. Stylidi M, Kondarides DI, Verykios XE (2004) Visible light-induced photocatalytic degradation of Acid Orange 7 in aqueous TiO_2 suspensions. Appl Catal Environ 47(3):189–201
51. Chen Y, Wang K, Lou L (2004) Photodegradation of dye pollutants on silica gel supported TiO_2 particles under visible light irradiation. J Photochem Photobiol A Chem 163(1–2): 281–287
52. Gonçalves MST, Pinto EMS, Nkeonye P, Oliveira Campos AMF (2005) Degradation of C.I. Reactive Orange 4 and its simulated dyebath wastewater by heterogeneous photocatalysis. Dyes Pigments 64(2):135–139
53. Konstantinou IK, Albanis TA (2004) TiO_2-assisted photocatalytic degradation of azo dyes in aqueous solution kinetic and mechanistic investigations a review. Appl Catal Environ 49:1–14

54. Serpone N, Maruthamuthu P, Pichat P, Pelizzetti E, Hidaka H (1995) Exploiting the interparticle electron transfer process in the photocatalysed oxidation of phenol 2-chlorophenol and pentachlorophenol chemical evidence for electron and hole transfer between coupled semiconductors. J Photochem Photobiol A Chem 85:247–255
55. Wu CH (2008) Effects of operational parameters on the decolorization of C.I. Reactive Red 198 in UV/TiO_2-based systems. Dyes Pigments 77:31–38
56. Agustina TE, Ang HM, Vareek VK (2005) A review of synergistic effect of photocatalysis and ozonation on wastewater treatment. J Photochem Photobiol C Photochem Rev 6:264–273

Chapter 13
Constructed Wetlands for Reducing Pesticide Inputs into Surface Water and Groundwater

Luis Alfredo Ortega-Clemente, Víctor M. Luna-Pabello, and Héctor M. Poggi-Varaldo

13.1 Introduction

Industrial development, intensive farming, and population growth have led to the emergence of emissions and wastes that constitute a growing threat to the environment and the human health. Agriculture produces the food for the world, mostly by intensive practices that are directly or indirectly related to the use and misuse of agrochemicals, such as pesticides for pest control. Pesticide residues and their transformation metabolites are eventually released into the environment and have been related to negative effects on flora, fauna, soil quality, water and groundwater quality, and human health [1].

Due to the increase in the water and groundwater contamination caused by modern agricultural practices, remediation or treatment alternatives are required in order to improve the quality of polluted streams and aquifers [2]. In this regard, the design and application of wetlands for remediation of effluents have received increasing attention from the scientific and engineering communities. Wetlands are transitional areas between the terrestrial and aquatic environments and serve as a dynamic link between the two. The water that moves up and down the moisture

L.A. Ortega-Clemente (✉)
Boca del Río Institute of Technology (ITBoca), Division of Graduate Studies and Research,
Carr. Veracruz-Cordoba km. 12, Boca del Río, P.O. Box 94290, Veracruz, México
e-mail: alfclemen2002@yahoo.com.mx

V.M. Luna-Pabello
Laboratorio de microbiología Experimental, Departamento de Biología,
Facultad de Química, Universidad Nacional Autónoma de México, Ciudad Universitaria,
P.O. Box 04510, México, D.F, México
e-mail: victormlp@yahoo.com

H.M. Poggi-Varaldo
CINVESTAV del IPN, México, Department Biotechnology and Bioengineering,
Environmental Biotechnology R&D Group, P.O. Box 14.740, CP 07000 México, D.F, México
e-mail: hectorpoggi2001@gmail.com

S.K. Sharma and R. Sanghi (eds.), *Wastewater Reuse and Management*,
DOI 10.1007/978-94-007-4942-9_13,

gradient takes a variety of physical and chemical constituents into solution either as detritus or as sediment; these in turn are transformed and transported to the surrounding landscape. Due to these characteristics, wetlands may occur or are formed in water catchment areas that can receive a great amount of both the natural and agricultural runoff from the surrounding area. Because of that, wetlands may receive considerable inputs of pesticides, unintentionally or by design, among other agrochemicals and pollutants.

In this chapter, the current knowledge on the use of constructed wetlands (CWs) to reduce pesticide inputs into surface water and groundwater and their effectiveness when applied in practice are discussed. Because of their simplicity in design, and relatively low costs of investment and operation, CWs seem to be a viable and economic alternative technology for the treatment of effluents contaminated with pesticides, compared with conventional treatment systems.

13.1.1 Pesticides

Any substance or mixture of substances that is intended to control any pest, including vectors of human and animal diseases, as well as unwanted species that cause damage or interfere with agricultural production, livestock, and forestry are called pesticides [3, 4]. The intensive use of pesticides began with the massive development of agriculture and agribusiness, especially after World War II. As other xenobiotics, pesticides often have mutagenic, teratogenic, and carcinogenic effects on animals and humans. Different pesticides may distribute into water or accumulate mainly in sediments, with a varying degree of persistence [5–8].

Pesticides are used worldwide to control a variety of pests. Substantial amounts (0.1–5%) of pesticides might be lost from the application fields into the surface waters via runoff and drainage [3, 4], as well as volatilization [9–11]. Of all the pesticides on the market, between 55 and 60%, are herbicides; 400 organic compounds have been developed and reported so far.

The use of pesticide agrochemicals requires a rational selection and management to prevent toxicities to nontarget organisms [12, 13]. Of the environmental compartments into which pesticides can partition, contamination of fresh water sources often presents the highest risk because of its ubiquitous use by organisms and high potential to direct exposure [13]. Organochlorine pesticides (OCPs) have been widely used, globally, since 1940s because of their high effectiveness in controlling pests and diseases [14, 15]. Organochlorine pesticides (OCPs) were the first synthetic organic pesticides to be used in agriculture. DDT and gamma-hexachlorocyclohexane (γ-HCH) are probably the best known of this type of pesticides. The production and intensive agricultural and industrial use of OCP have caused the widespread contamination of the environment [15, 16]. Organochlorine pesticides and polychlorinated biphenyls (PCBs) are very persistent compounds, and their persistence in the environment is measured in years. As a result of their persistence and high capacity for accumulation in living organisms [15, 17–19], the use of the majority of OC compounds has been banned in developed countries.

The contamination of water bodies with pesticides can pose a significant threat to aquatic ecosystems and water resources. However, the risk for the aquatic community or the human health can often be substantially reduced by appropriate measures. Pesticides can enter river systems as point sources, that is, in some places along the river, or as diffuse sources, which are inputs along the whole water flow [20]. Examples of point sources of pesticides are sewage plants, sewer overflows, and losses due to bad management practices from farmers. As diffuse input pathways runoff, drain flow, drift, deposition, and a contribution through groundwater can be distinguished.

Pesticides are being detected in water bodies of small rural communities on an increasingly frequent basis. The use of phytosanitary products in agricultural activities is the major cause of pesticide contamination in water due to their being continuously discharged into aquatic environmental via surface runoff [21–23].

The pesticide physical removal reactions involve sedimentation, flocculation, absorption, coprecipitation, and precipitation; the pesticide movements take place in water, sediments, suspended matters, and plants [24, 25]. The pesticide can be transported from one component to another, for example, from water to sediment or biota, or by the suspension of the material or vice versa. Moreover, the chemical removal of pesticides involves a series of anion and cation exchange reactions, oxidation-reduction. It may also include UV radiation, especially for surface-flow systems, where some organic pollutant molecules undergo photolytic decomposition due to exposure to the UV wavelengths of sunlight [25, 26].

13.2 Pathways of Entry of Pesticides in Soil, Surface Water, and Groundwater

13.2.1 Pesticides in Soil and Surface Water

Pesticides can move in ecosystems according to their chemical properties such as soil organic carbon/water distribution coefficient, K_{oc}, half-live time ($t_{1/2}$), vapor pressure, etc. [4, 27]; the environmental conditions; and their application mode.

The toxic effects of substances applied in agroecosystems operate over a wide area of influence, because the culture is connected with the environment. In this way, natural ecosystems are part of the pesticides used on crops, and the fate of them is influenced by the affinity for water, soil, air, or biota [1]. Pesticides react with biomolecules such as DNA, proteins, and other metabolites, and their activity leads to toxic effects, including physiological changes, nutritional changes, and even death as a result of altering the metabolic pathways. They can also enter the natural trophic chains via irrigation with contaminated water used to wash food and containers. In addition to this, bioaccumulation and biomagnifications (Fig. 13.1) can lead to hazardous concentration in living organisms [1, 26]. In view of these facts, the attention paid by the scientific community to establish degradation mechanisms and techniques for treating wastewater that contains pesticides has grown considerably in the last decades [26, 28–30].

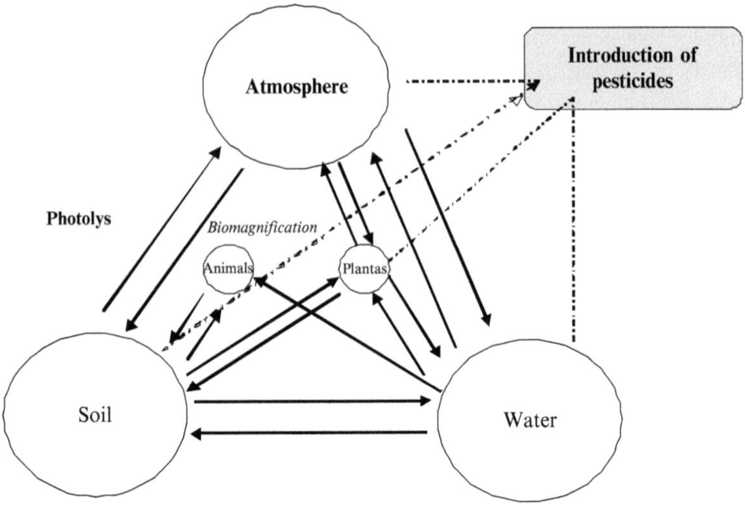

Fig. 13.1 Outline of the different routes a pesticide can take once it is scattered on the ground

In the soil, pesticides have different mechanisms by which they can move or detain; those mechanisms may act alone or in combination, as they depend on other variables, such as humidity, temperature, organic matter, clay type, pH, and ion exchange land, among others. Once the pesticides are in the soil, they can follow different routes or pathways, which can range from disposal by chemical degradation to photochemical and microbial decomposition or volatilization. Figure 13.1 shows some of the possible interactions of pesticides sprayed over the soil.

When a pesticide has been distributed on the soil, its degradation and its fate will depend on a number of both biotic and abiotic processes, such as hydrolysis, photo degradation, transformation by plant roots, and soil microorganisms. The area where the most reactive pollutant present is transferred is the root zone (or rhizosphere). This is where the physicochemical and biological processes [31–34] that occur are induced by the interaction of plants, microorganisms, soil, and pollutants (Fig. 13.2). The biological removal mechanisms may include antimicrobial activity of root exudates. From this moment on, several possible interactions of pesticides with different components of soil and environment begin. It is possible that this is an exchange form, over time, being absorbed and desorbed by the different soil compartments. This exchange form can cause the transformation into other compounds which may be more dangerous than the parent compound. Furthermore, pesticides and their transformation products may end up in leachates that contaminate groundwater and surface water (Fig. 13.3) [32, 33, 35].

The processes that could be greater importance in the removal of pesticides are the processes of sorption/desorption and biodegradation. Both are strongly influenced by the presence of a sediment layer [20, 36].

The fate of pesticides in surface water is determined by their sorption behavior. The effect of physical transport during the sorption process can directly or indirectly influence the degradation of pesticides. The chemical reactivity of a sorbed pesticide in sediment is significantly different from that of it in solution [20, 36].

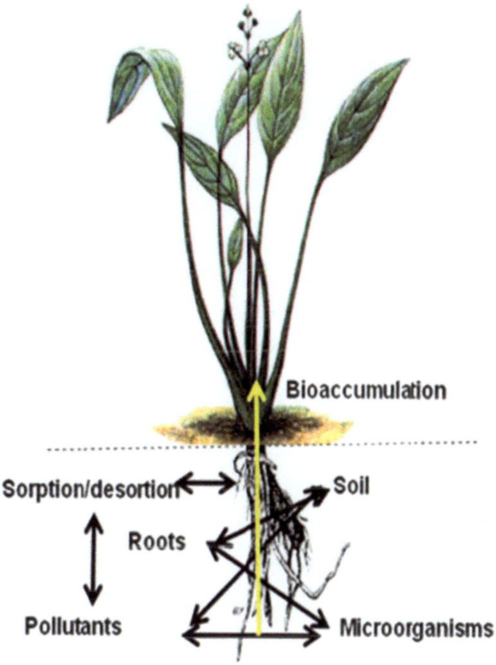

Fig. 13.2 Possible interactions among plant, pollutants, soil, and microorganisms present in the root zone

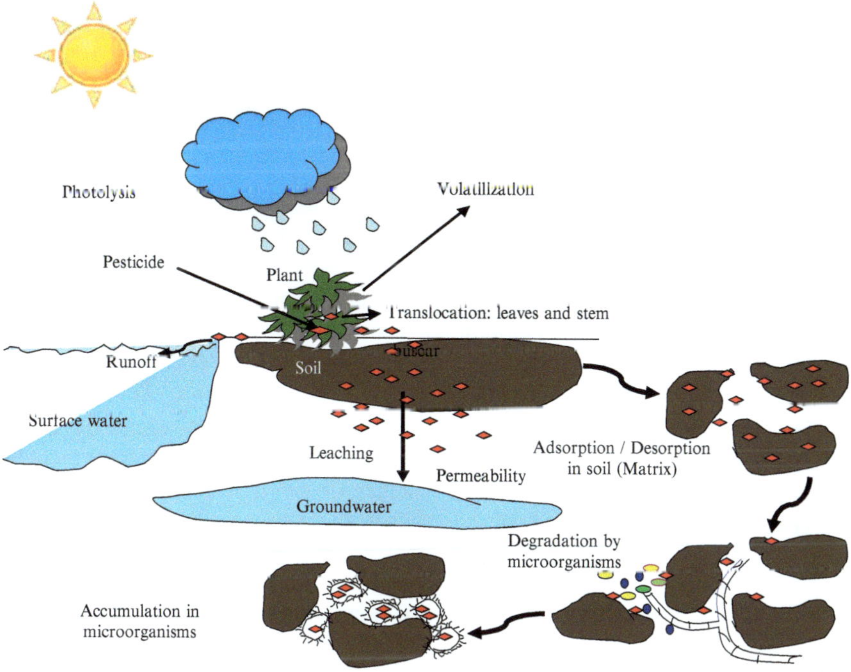

Fig. 13.3 Factors involved in the distribution and degradation mechanisms of pesticides in soil

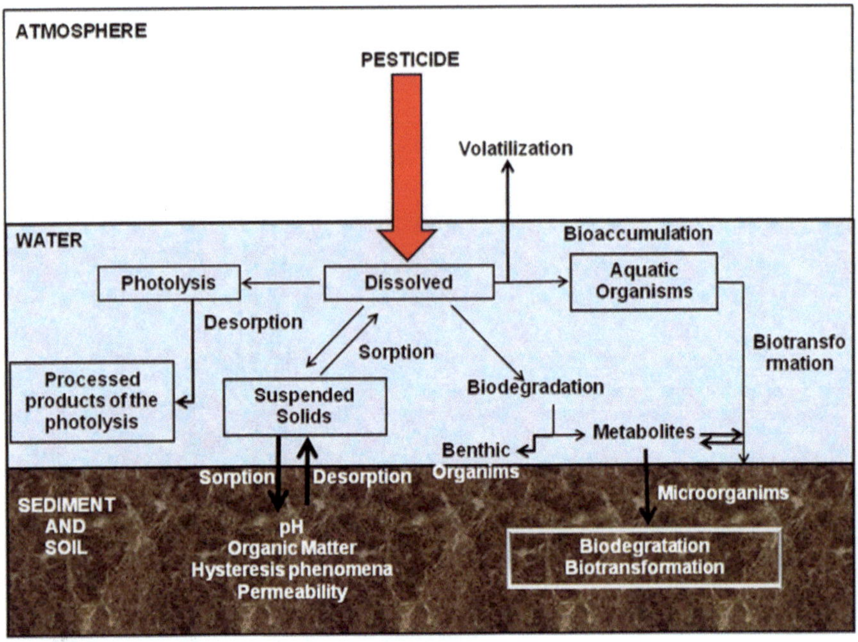

Fig. 13.4 Processes of pesticides in surface water

For instance, the chemical and bioavailability of the sorbed chemical are lower than those in solution. Hysteresis or irreversibility of the sorption/desorption process plays a distinctive role on pollutant availability.

13.2.2 Pesticides in Groundwater

In a rural setting, there are potential contaminant sources associated with human activity that may affect groundwater quality. These can be point sources, where potential contaminants are concentrated or stored in one spot, such as the pesticide storage. There are also nonpoint sources, where potential contaminants are spread over a wider area, such as applied nutrients or pesticides on cultural fields [21, 25, 37].

The vulnerability of groundwater to contamination may vary across the place [20]. The vulnerability of each aquifer is unique, and determining factors include the type of formation, its depth, and whether or not it is protected by a low permeability (Figs. 13.3 and 13.4).

Various physical, chemical, and biological processes determine the environmental fate of a pesticide. The rates and importance of each of these processes are, in turn, affected by different factors [20, 38–41]. Table 13.1 shows the influence of these factors on the specific groundwater vulnerability to pesticides. The following paragraphs discuss the different points in more details.

Table 13.1 Factors affecting groundwater vulnerability to contamination by pesticides

Factor	Examples	Reference
Subsoil	Thickness, degradation sites	[39]
		[40]
Pesticide properties	Sorption, degradation	[39]
		[24]
		[38]
Climate	Timing of first rainfall, temperature, potential evapotranspiration	[20]
		[36]
		[25]
Groundwater	Groundwater flow, dilution	[25]
		[35]
Soil and crop properties	Organic matter content, texture, structure, plant uptake	[38]
		[42]

13.3 Constructed Wetland and Wastewater Treatment

13.3.1 Wetlands

Pollution impacts should be reduced or eliminated, so it is crucial to develop technologies for the treatment and remediation of contaminated water. Several existing treatment technologies may be suitable for this goal; however, either their high costs or technology complexity, or both, preclude their successful application. In this regard, CWs appear as a viable alternative technology and economic treatment of effluents contaminated with pesticides, compared with conventional treatment systems [43–46]. CWs have featured characteristic of appropriate technology, that is, low cost and simple operation that render them very well fitted for application in underdeveloped countries where capital and skilled operators are scarce.

Wetlands, commonly known as biological filters, have emerged as an alternative for solving a wide range of environmental and water quality issues [47–50]. A great deal of research on the use of wetlands to treat a variety of wastewater to remove pollutants from water in order to reduce the possible impact on humans and the rest of the ecosystem have been performed in the past decades [51].

The characteristic of a natural wetland is the presence of water during extended periods of time enough to alter the soil microorganisms and their flora communities. The depths of these typical areas are less than 0.60 m where emergent plants grow as rushes of water, *Typha* "totora," and *Lemna gibba* (duckweed gibbous) "water lentil" (Fig. 13.5), contributing the reduction of pollutants through aerobic degradation processes [52, 53].

Wetlands have been used for pollution control for centuries [54–56], although the scientifically based use of wetlands for wastewater treatment in constructed wetlands began in the 1950s and 1960s, with investigations by Seidel and Kickuth in Germany [47, 57–60].

Fig. 13.5 Aquatic plants in natural wetlands

13.3.2 Constructed Wetlands Systems (CWs)

The CWs seek to imitate the properties of natural wetlands in an environment that can be manipulated and controlled [51, 61]. CWs are defined as engineered wetlands that utilize natural processes involving wetland vegetation, soil, and their associated microbial assemblages to assist, at least partially, in treating wastewater or other polluted water sources [62]. The number of constructed treatment wetlands receiving wastewater from municipal, industrial, agricultural, and storm water sources has increased to more than 20,000 across the world [63].

A CW is an artificial system of wastewater treatment that consists of a shallow pond or channel, no more than 0.60 m, which in most cases has aquatic plants planted to treat wastewater. The CWs have advantages over alternative treatment systems because they require little or no energy to operate. If there are sufficient areas of land available near the installation of the CWs for aquatic cultivation, they can be an effective low-cost alternative. CWs provide habitat for wildlife and are aesthetically pleasing to the eye. Furthermore, for their ecological value, wetlands have important roles, such as preventing floods, stabilizing the shoreline, and recharging underground aquifers to stabilize local conditions, particularly rainfall and temperature [64, 65].

The CWs are man-made copies of natural wetlands that optimally exploit the biogeochemical cycles that normally occur in these systems for the purpose of wastewater treatment. Different types can be distinguished, based on water flow characteristics and plant species. Systems with aboveground flow are referred to as surface-flow (SF) CWs and the ones with belowground flow as subsurface-flow (SSF) CWs. The latter ones can be further subdivided according to the water flow direction, that is, horizontal or vertical flow [66]. Commonly used plants in all types of systems are helophytes, like reed and cattails. Some SF systems also make use of free-floating macrophytes, like duckweed and water hyacinth; floating-leaved bottom-rooted macrophytes, like lotus; or submersed macrophytes, like waterweed, but these are less frequently used and therefore not further discussed in this chapter. For a further overview on system types and pollutant removal mechanisms, the relatively recent work of Kadlec and Wallace is recommended [67].

Different CWs do not necessarily function as stand-alone treatment plants but can be combined with each other or with other low-tech or high-tech wastewater treatment units, in order to exploit the specific advantages of the different systems. CWs are indeed increasingly used as a tertiary treatment step after activated sludge, rotating biological contactors, UASB, etc.

Constructed wetland systems are considered as low-cost alternatives for wastewater treatment, especially suitable for underdeveloped countries. They also have low operation and maintenance requirements. Treatment of wastewater in constructed wetland systems includes biological and biochemical processes [68].

13.3.2.1 Surface-Flow CWs (SF-CWs)

Surface-flow constructed wetlands (SF-CWs) are most similar to natural environments due to the permanent standing water and conditions favorable for wetland plant species [69]. The SF-CWs generally have a soil bottom, emergent vegetation, and a water surface above the substrate [70]. The design may or may not include areas of open water in addition to the vegetated areas. These systems typically consist of ponds or canals, with some sort of barrier to prevent underground seepage, soil or other convenient means to handle the emergent vegetation, and water depth in a relatively low (0.1–0.6 m) usually less than 0.4 m [62]. The low water depth, low velocity flow, and the presence of plant stems and litter regulate water flow. The treatment occurs when the flow of water slowly passes through the stem and root of the emergent vegetation (Fig. 13.6).

13.3.2.2 Subsurface-Flow CWs

The technology of wastewater treatment by means of constructed wetlands with horizontal subsurface flow (SSF-CWs) was started in Germany and by Reinhold Kickuth in the 1970s [46, 58, 71–73]. In SSF-CWs systems, water passes either horizontally (horizontal subsurface-flow constructed wetlands HSSF-CW) or vertically (vertical subsurface-flow constructed wetland VSSF-CW) through the porous filter material (usually sand and/or gravel). SSF-CWs consist of a bed, which helps the growth of plants. In SSF-CWs, wastewater flows underneath and through the plant rooting media, and water level is maintained below the top of the substratum. The subsurface-flow systems are reported as causing fewer problems arising from odors, insects, or public exposure, which are thus suitable to treat landfill

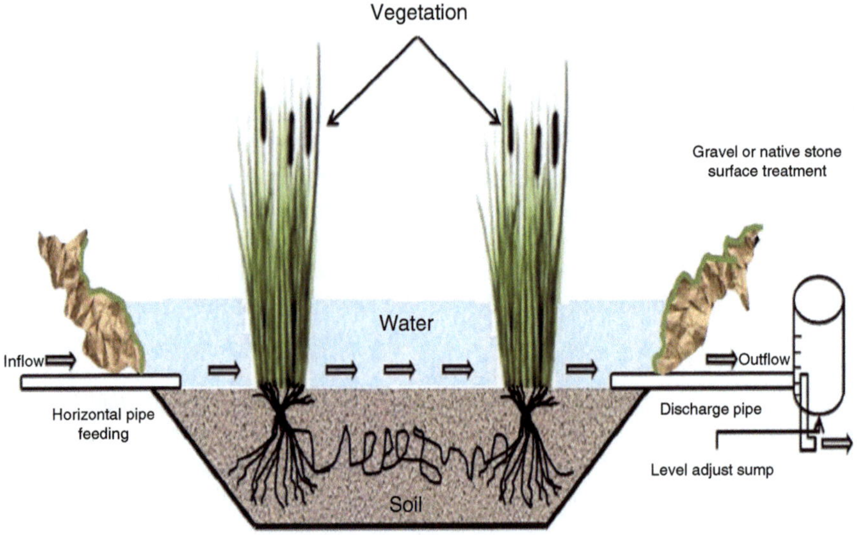

Fig. 13.6 Schematic representation of a surface-flow constructed wetland

leachate [124, 74–79]. The wastewater is treated by microbiological degradation and by physical–chemical processes in a SSF-CWs. Anaerobic and aerobic processes take place within the pores of the filter media (sand or gravel).

Figure 13.7 shows a typical HSSF-CW; the wastewater is fed in at the inlet and flows slowly through the porous medium under the surface of the bed in a more or less horizontal path, until it reaches the outlet zone where it is collected before leaving via level control arrangement at the outlet [46]. During this passage, the wastewater will come into contact with a network of aerobic, anoxic, and anaerobic zones. The aerobic zones occur around roots and rhizomes that leak oxygen into the substrate [80, 81]. Major design parameters, removal mechanisms, and treatment performance have been reviewed by several authors [52, 62, 63, 67, 81–84]. The general design of SSF-CWs is based on a hydraulically isolated filter bed that can be vegetated with different plants [70].

In the case of HSSF-CW, mechanically pretreated wastewater is fed in at the inlet and passes slowly through the filtration substrate below the filter surface horizontally until the outlet zone. Due to a long retention time of the wastewater, BOD_5, COD, and TSS removal is possible to a high degree, normally higher than 80–90%. Also, nitrogen reduction is not significant, but fully nitrification is limited due to lack of oxygen that is characteristic for this kind of systems [70, 79].

In the VSSF-CWs, the wastewater is fed on the whole surface area through a distribution system and passes the filter in a more or less vertical path. The dosing of wastewater on the bed is performed intermittently [70, 79]. Due to the mostly aerobic conditions in a VSSF-CW system, oxygen-requiring nitrifying bacteria is favored, and nitrification can be achieved in these systems. However, denitrification may not take place to a large extent. As the constructed wetlands age, the rate of

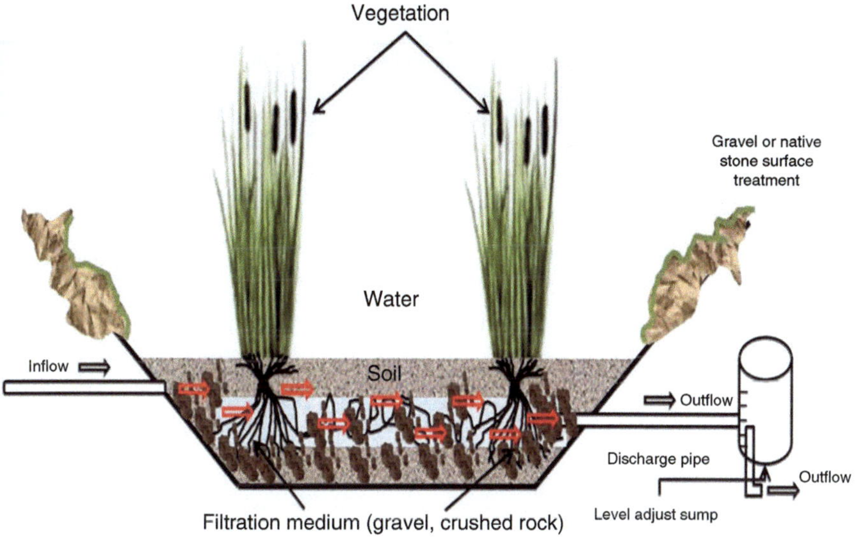

Fig. 13.7 Schematic representation of a subsurface-flow constructed wetland

organic removal increases. This is mainly due to an increase in microbial population. Total nitrogen and phosphorus removal in constructed wetland systems can be as high as 98–99%, respectively [79, 85].

SF-CW systems are less numerous than subsurface-flow systems, although free-water surface wetlands are one of the oldest designs in Europe [83]. In order to enhance wastewater purification efficiency, hybrid systems have been introduced, which consist of various types of constructed wetlands arranged in sequence [70].

Table 13.2 summarizes some papers published in recent years in high impact journals on subjects related to the use of constructed wetlands for industrial and agricultural wastewater treatment (primarily wastewater contaminated with pesticides).

13.4 Removal Processes of Pesticides in the Environment

Once a pesticide is introduced into the environment, whether through an application, a disposal, or a spill, the pathway it takes is influenced by many processes. These processes determine the persistence and movement, if any, of the pesticide and its ultimate fate.

Significant research efforts have been dedicated to understanding the fate and transport of pesticides in the environment, and the relationship among the fate and transport of the pesticide and specific environmental parameters, such as organic carbon and pH (Fig. 13.4) in soils, are generally understood at least qualitatively [25].

Table 13.2 Examples of the use of constructed wetlands for treatment of contaminated wastewater

Wastewater	Country	Reference
Agricultural runoff (pesticide)	China	[86]
	Argentina	[87, 88]
	India	[89]
	Africa	[90, 91]
	Brazil	[92]
Tannery	Portugal	[93]
	Turkey	[94]
	USA	[93, 95, 96]
	México	[97]
Food	Slovenia	[98]
	Netherlands	[99]
	USA	[100, 101]
	Italy	[102]

Three hydrological factors affect the depuration capacity of the constructed wetlands: the hydro period [103, 104], the residence time of water into the wetland [65, 105, 106], and the origin and contents of the feeding water. Pesticides either in solution or adsorbed to the solid phase undergo a chemical degradation by oxidation or photolysis induced or catalyzed by soil components. The abiotic degradation is often incomplete and leads to intermediate substrates for biological reactions. Degradation is presented by changes in molecular structure and the formation of metabolites due to the action of chemical, photochemical, and biological processes [105].

13.5 Processes That Reduce Pesticides in Constructed Wetlands

The contamination of water bodies with agricultural pesticides can pose a significant threat to aquatic ecosystems and drinking water resources [12, 62, 90]. However, the risk for the aquatic community or to human health can often be substantially reduced by appropriate measures. Mitigation of pesticide inputs into water bodies includes both reduction of diffuse-source (runoff and erosion, tile drainage, spray drift, leaching to groundwater) and of point-source inputs (mainly farmyard runoff), which in some regions of Europe (e.g., Western Germany, Sweden) have been shown to make a highly significant contribution to the observed pesticide loads in rivers [12, 13, 107–109].

The natural and artificial water bodies, such as CWs, are highly likely to receive contaminated water, with and without treatment, resulting in the detriment of their quality and thus the loss of native biota. Constructed wetlands action on pesticides is twofold: either as a sink due to storage, transformation, and elimination; or as a source as compounds that may be transferred to water surface and groundwater, due to plant interception and temporary storage due to sorption into sediment, soil, or suspended matter [25]. In the environment, pesticides are partitioned into liquid, solid, and gaseous phases. Pesticides presence in solid phase is due to adsorption phenomena that control the distribution in the other phases, while their most mobile portion is located in liquid and gaseous phases. This portion is available for microbial degradation and for vertical or lateral transfer related to ground and surface water contamination. Generally, the solid phase retention minimizes the pesticide mobility risk but makes pesticide disappearance (i.e., degradation) more difficult [25].

Most studies on CWs have been unable to distinguish bioattenuation from physical and chemical removal of contaminants, since removal is commonly determined based on differences in input and output, considering the wetlands as a black box. In particular, hydraulic retention time (HRT) of pesticides, substrate (granulometry, porosity, uniformity), and hydraulic efficiency in CWs are the main parameters to be considered for the performance of the biological treatment. Contrary to most of plant treatments, where flow rates are in a narrow range of values, in the case of CWs, flow rates are often close to zero in a few hours after a storm event, while they are very high during storm events with lower hydraulic retention times than the time needed for biological treatment [20, 36]. Therefore, pesticide storage in CWs along with close contact between pesticides and plants (and/or microorganisms) should be improved at the same time as biological removal processes are designed. Macrophytes play a role in biological removal process through their ability to extract metals and/or organic compounds [25]. The hydraulic design (hydraulic retention time, HRT) and the use of adsorbing materials can be useful to increase the pesticides residence time and the contact between pesticides and biocatalysts. Pesticide fluxes can be reduced by 50–80% when increasing ten times the retention time. This, in turn, leads to CW lengths that are much longer than those for municipal wastewater treatment [25, 26].

On the whole, global efficiency is generally calculated by the ratio of inlet/outlet concentration without describing all the processes involved. This black box approach does not allow the highlighting of main processes but provides evidence of efficiency.

The use of vegetated wetlands for accelerating pesticide removal from agricultural runoff is gaining acceptance as a best management practice. CWs are promising tools for mitigating pesticide inputs via runoff/erosion and drift into surface waters, but their effectiveness still has to be demonstrated for weakly and moderately sorbing compounds [12]. However, the bioaccumulated pesticide in the plant can accelerate its movement trophic chain. Poissant et al. [4] in recent studies assessed the pesticides field losses in extensive agricultural catchments to survey their potential interaction with downstream wetlands; this work presents innovative approaches for pesticides fate and field losses studies in agricultural wetlands catchments. Furthermore, their research reports organochlorinated, organophosphorus, aryloxyacid pesticides, and other group of pesticides. The metabolites as well as the active ingredients of pesticides were detected in the wetland and its catchment. Overall, the pesticides lost from agricultural field toward the streams were $<1\%$ of the quantity applied. On the other hand, Rose et al. [13] modeled the rates of dissipation processes and evaluated the influence of aquatic plants on these rates, in order to formulate some general recommendations regarding the use of aquatic plants in constructed wetlands for polishing pesticide-contaminated agricultural runoff.

There is also evidence that aquatic vegetation can accelerate pesticide removal compared to open water systems [109, 110]. This reportedly occurs because of the increased capacity for plant/biofilm sorption and subsequent immobilization (Table 13.3), breakdown, or uptake of pesticides [13, 110, 115–117]. However, up to date, there has been limited effort in quantifying the effect of aquatic vegetation on other pesticide dissipation pathways from aquatic systems [13, 116]. This is despite the pathways leading to the dissipation of organic contaminants, such as pesticides from aquatic systems having been discussed in detail [118].

Most of these studies suggest that the CWs are very effective in reducing pesticide inputs in surface waters; however, CWs need large extensions of land. The largest wetland investigation in relation to pesticide removal was 134 m long and 36 m wide [117].

However, smaller and less demanding wetland areas (e.g., 50 m long and 1.5 m wide, [119]) have also been found to be very effective in removing pesticides from water passing through the wetland [25].

Publications revised and in some way related to mitigation of pesticide inputs into water bodies are discussed in the context of this study. However, not all were useful for quantitative evaluation, because no quantitative estimates of the reduction of pesticides or surface water concentration by the respective mitigation measures were reported (Table 13.3).

Finally, some limitations of CWs into their application for remediation of waters polluted with pesticides are worth mentioning: CWs may not be suitable for high concentrations of xenobiotics; wetlands should be carefully managed in order to avoid an over contamination by pesticides of the wetland site, and associated negative effects (i.e., wetlands that require remediation); provisions for effluent storage and recycling in order to increase hydraulic retention times should be provided with consequent increases in cost and operation complexity.

13.6 Conclusions

The contamination of water bodies with pesticides can pose a significant threat to aquatic ecosystems and water resources. However, the risk to the aquatic community or the human health can often be substantially reduced by appropriate measures. Previous research shows that plants can accelerate pesticide removal from aquatic systems by several mechanisms. In this regard, constructed wetlands

Table 13.3 Reduction efficiency of pesticides using constructed wetlands to treat surface water and groundwater

Wetland type	Water	Pesticide	Condition	Reducing (%)	Remarks	Reference
SF-CWs[a]	Surface water	Endosulfan Fluometuron	Operation: 250 h non-vegetated, vegetated pond, 58 m^3 in each pond, pH 8.0, initial concentration: endosulfan 2.2 μg L^{-1} and fluometuron 5.6 μg L^{-1}	40 (Open pond) 55 (vegetated pond) 80 (in both systems)	Plants accelerate pesticide dissipation from aquatic systems by increasing sedimentation	[13]
SSF-CWs[b]	Surface water	Atrazine	Synthetic wastewater, under different salinity, operation was continued for 30 days, initial concentration 0.1 mg L^{-1}, pH 6.54–6.9	44	Atrazine half-life in SSFCW was approximately 17.5 days, concentration of 0.1 mg L^{-1}, atrazine decreased to 0.053, 0.054, and 0.056 mg L^{-1}	[111]
SF-CWs	Surface water	Pyrethroid insecticides: Lambda-cyhalothrin and cyfluthrin	Initial concentration: 0.12 kg lambda-cyhalothrin ha^{-1}; 0.57 kg cyfluthrin ha^{-1}. pH 7. Operation: 55 days	49 and 76% lambda-cyhalothrin and cyfluthrin, respectively	The study's measured lambda-cyhalothrin and cyfluthrin masses were associated with vegetation	[112]
HSSF-CW[c]	Surface water	simazine, alachlor, chlorpyrifos, pentachlorobenzene, pentachlorophenol, endosulfan, diuron, and lindane	Average water depth of 0.3 m, a surface area of 55 m^2, an average granular size of 3.5 mm, and an HLR[d] at the time of the experiments of 36 mm day^{-1}.	>90% (lindane, pentachlorophenol, endosulfan, pentachlorobenzene); 80–90% (alachlor and chlorpyrifos); 20% (mecoprop and simazine)	Pollutants that was recalcitrant to elimination (0%), namely, clofibric and diuron	[113]

(continued)

Table 13.3 (continued)

Wetland type	Water	Pesticide	Condition	Reducing (%)	Remarks	Reference
			Operation: 450 h, concentration: 0.1 to 100 μg ml^{-1}, pH 7.5–7.7, HRT[e] 5.4, 6.2 days			
SSF-CWs	Surface water	Atrazine	Soil: pH 6.03, 2.13% organic carbon. Flow low <250 L min^{-1}, medium 250–500 L min^{-1}, high >500 L min^{-1}. 2–4 week intervals. concentration of 400 μg L^{-1} atrazine (5 experiments)	7-d period ranged from 18 to 24% in 1998 (experiments 1–3) and 16–17% in 1999 (experiments 4 and 5)	The assay indicated the existence in the wetland of a low-density population of microorganisms with the potential to mineralize atrazine's ethyl side chain. An increased density of wetland plants did not appear to affect treatment	[114]
Ws[f] SF-CWs	Groundwater/ surface water	Atrazine	HRT: 4 and 5 days. Concentration of 2.55 mg atrazine kg^{-1} wet weight sediment and 2.8 mg L^{-1} for water samples	60–80% mineralization of atrazine (1 month)	Atrazine mineralization derived from the first-order rate expression. Half-lives were in the order of 5–7 days in the most active samples	[17]

[a] Surface-flow constructed wetlands
[b] Subsurface-flow constructed wetlands
[c] Horizontal subsurface-flow constructed wetlands
[d] Hydraulic loading rate
[e] Hydraulic retention time
[f] Natural wetlands

(CWs) where plants grow have been used to treat waters polluted with both herbicides and insecticides under various conditions.

One of the main mechanisms of pesticide removal in CWs is sorption onto plants, support media, and sediments; yet, CWs effectiveness still has to be demonstrated for weakly and moderately sorbing compounds. The hydraulic design (hydraulic retention time, HRT) and the use of adsorbing materials can be useful to increase the pesticides residence time and the contact between pesticides and biocatalysts. Pesticide fluxes can be reduced by 50–80% when increasing ten times the retention time. This, in turn, leads to CW lengths that are much longer than those for municipal wastewater treatment. CWs are a viable and economic alternative technology for the treatment of effluents contaminated with pesticides, compared to conventional treatment systems, due to simple design and operation and relatively low costs of investment and operation. Yet, there are some limitations of CWs, *inter alia*, they are not suitable for high concentrations of xenobiotics; they should be carefully managed in order to avoid an over contamination by pesticides of the wetland site and associated negative effects (i.e., wetlands that require remediation).

Acknowledgments The authors wish to thank PAPIME (PE205706) and PAPIIT (IN107209) for their financial support given to this research work in wetlands, as well as to DGAPA-UNAM for the Postdoctoral Fellowship awarded to Dr. Ortega-Clemente.

References

1. Calamari D, Barg U (1993) Hazard assessment of agricultural chemical by simple simulation models. Prevention of water pollution by agricultural and related activities. In: Proceedings of the FAO expert consultation, Santiago, Chile, 20–23 Oct 1992 (pp 207–222). Water report 1. FAO, Rome
2. Álvarez-Fajardo E, Ksoll W, Luna Pabello V, Miranda-Ríos M, Ramírez-Carrillo H (2002) Remoción de *Escherichia coli* y ortofosfatos en aguas residuales parcialmente tratadas empleando diferentes materiales pétreos. Una estrategia para la protección de la vida acuática silvestre. En Memorias del XXVIII Congreso Interamericano de Ingeniería Sanitaria y Ambiental. Cancún, México.
3. Leu C, Singer H, Stamm C, Muller SR, Schwarzenbach RP (2004) Variability of herbicide losses from 13 fields to surface water within a small catchment after a controlled herbicide application. Environ Sci Technol 38(14):3835–3841
4. Poissant L, Beauvais C, Lafrance P, Deblois C (2008) Pesticides in fluvial wetlands catchments under intensive agricultural activities. Sci Total Environ 404:182–195
5. Boone MD, James SM (2003) Interactions of an insecticide, herbicide, and natural stressors in amphibian community mesocosms. Ecol Appl 13(3):829–841
6. De la Vega-Salazar MY, Martinez-Tabche L, Macias-Garcia C (1997) Bioaccumulation of methyl parathion and its toxicology in several species of the freshwater community in Ignacio Ramirez Dam in Mexico. Ecotoxicol Environ Saf 38:53–62
7. Peterson HG, Boutin C, Martin PA, Freemark KE, Ruecker NJ, Moody MJ (1994) Aquatic phyto-toxicity of 23 pesticides applied at expected environmental concentrations. Aquat Toxicol 28(3/4):275–292
8. Zaga A, Little EE, Rabeni CF, Ellersieck MR (1998) Photoenhanced toxicity of a carbamate insecticide to early fife stage anuran amphibians. Environ Toxicol Chem 17(12):2543–2553

9. Aulagnier F, Poissant L (2005) Some pesticides occurrence in air and precipitation in Québec, Canada. Environ Sci Technol 39:2960–2967
10. Garmounma M, Poissant L (2004) Occurrence, temperature and seasonal trends of a- and g-HCH in air (Quebec, Canada). Atmos Environ 38:369–382
11. Poissant L, Koprivnjak JF (1996) Fate and atmospheric concentration of α- and γ-hexachlorocyclohexane in Québec, Canada. Environ Sci Technol 30:845–851
12. Reichenberger S, Bach M, Skitschak A, Frede HG (2007) Mitigation strategies to reduce pesticide inputs into ground- and surface water and their effectiveness: a review. Sci Total Environ 384:1–35
13. Rose MT, Crossan AN, Kennedy IR (2008) The effect of vegetation on pesticide dissipation from ponded treatment wetlands: quantification using a simple model. Chemosphere 72:999–1005
14. Chen S, Lili S, Zhengjun S, Qiuhui H (2007) Determination of organochlorine pesticide residues in rice and human and fish fat by simplified two-dimensional gas chromatography. Food Chem 104:1315–1319
15. Kalyoncu L, Arca I, Aktunmsek A (2009) Some organochlorine pesticide residues in fish species in Konya, Turkey. Chemosphere 74:885–889
16. Ayas Z, Ekmekci G, Ozmen M, Yerli SV (2007) Histopathological changes in the livers and kidneys of fish in Sariyar Reservoir, Turkey. Environ Toxicol Pharmacol 23:242–249
17. Andersson PL, Berg AH, Bjerselius R, Norrgren L, Olsen H, Olsson PE, Orn S, Tysklind M (2001) Bioaccumulation of selected PCBs in zebrafish, three-spined stickleback and Arctic char three different routes of exposure. Arch Environ Contam Toxicol 40:519–530
18. Stefanelli P, Muccio AD, Ferrara F, Barbini DA, Generali T, Pelosi P, Amendola G, Vanni F, Muccio SD, Ausili A (2004) Estimation of intake of organochlorine pesticides and chlorobiphenyls through edible fishes from the Italian Adriatic Sea during 1997. Food Control 15:27–38
19. Zaranko TD, Griffiths RW, Kaushik NK (1997) Biomagnification of polychlorinated biphenyls through a riverine food web. Environ Toxicol Chem 16:1463–1471
20. Holvoet KMA, Seuntjens P, Vanrolleghem PA (2007) Monitoring and modeling pesticide fate in surface waters at the catchment scale. Review. Ecol Model 209:53–64
21. He ZL, Yanga XE, Stoffellab PJ (2005) Trace elements in agroecosystems and impacts on the environment. J Trace Elem Med Biol 19:125–140
22. Kloeppel H, Koerdel W, Stein B (1997) Herbicide transport by surface runoff and herbicide retention in a filter strip rainfall and runoff simulation studies. Chemosphere 35:129–141
23. Olette R, Couderchet M, Biagianti S, Eullaffroy P (2008) Toxicity and removal of pesticides by selected aquatic plants. Chemosphere 70:1414–1421
24. Gao JP, Maguhn J, Spitzauer P, Kettrup A (1998) Sorption of pesticides in the sediment of the Teufelsweiher pond (Southern Germany). I: equilibrium assessments, effect of organic carbon content and pH. Water Res 32(5):1662–1672
25. Gregoire C, Elsaesser D, Huguenot D, Lange J, Lebeau T, Merli A, Mose R, Passeport E, Payraudeau S, Schutz T, Schulz R, Tapia-Padilla G, Tournebize J, Trevisan M, Wanko A (2008) Mitigation of agricultural nonpoint source pesticide pollution in artificial wetland ecosystems. Review. Environ Chem Lett. doi:10.1007/s10311-008-0167-9
26. Araña J, Garriga I, Cabo C, Fernández-Rodríguez C, Herrera-Melián JA, Ortega-Méndez JA, Doña-Rodríguez JM, Pérez-Peña J (2008) Combining TiO_2-photocatalysis and wetland reactors for the efficient treatment of pesticides. Chemosphere 71:788–794
27. Wania F, Haugen JE, Lei YD, Mackay D (1998) Temperature dependence of atmospheric concentrations of semivolatile organic compounds. Environ Sci Technol 32:1013–1021
28. Cheng S, Vidakovic-Cifrek Z, Grosse W, Karrenbrock F (2002) Xenobiotics removal from polluted water by a multifunctional constructed wetland. Chemosphere 48:415–418
29. Essam T, Hamid Z, Aly-Amin M, El Tayeb O, Mattiasson B, Guieysse B (2006) Sequential UV-biological degradation of chlorophenols. Chemosphere 63:277–284

30. Essam T, Aly-Amin M, El Tayeb O, Mattiasson B, Guieysse B (2007) Sequential photochemical–biological degradation of chlorophenols. Chemosphere 66:2201–2209
31. Chung AKC, Tam NFY, Wong MH (2008) Nitrogen and phosphate mass balance in a sub-surface flow constructed wetland for treating municipal wastewater. Ecol Eng 32:81–90
32. Daniels R (2001) Enter the root-zone: green technology for leather manufacturer. World Leather 14(4):63–67
33. Daniels R (2001) Enter the root-zone: green technology for the leather manufacturer, part 3. World Leather 14(6):85–88
34. Stottmeister U, Wießner A, Kuschk P, Kappelmeyer U, Kastner M, Bederski O, Muller RA, Moormann H (2003) Effects of plants and microorganisms in constructed wetlands for wastewater treatment. Biotechnol Adv 22:97–117
35. Zbytniewski R, Buszewski B (2002) Sorption of pesticides in soil and compost. Pol J Environ Stud 11(2):179–184
36. Warren N, Allan IJ, Carter JE, House WA, Parker A (2003) Pesticides and other micro-organic contaminants in freshwater sedimentary environments. A review. Appl Geochem 18:159–194
37. Harris BC, Bonner JS, Autenrieth RL (1999) Nutrient dynamics in marsh sediments contaminated by an oil spill following a flood. Environ Technol 20:795–811
38. Ahmad R, Nelson PN, Kookana RS (2006) The molecular composition of soil organic matter as determined by 13C NMR and elemental analyses and correlation with pesticide sorption. Eur J Soil Sci 57:883–893
39. Arias-Estévez M, López-Periago E, Martínez-Carballo E, Simal-Gándara J, Mejuto JC, García-Río L (2008) The mobility and degradation of pesticides in soils and the pollution of groundwater resources. Agric Ecosyst Environ 123:247–260
40. Guo L, Jury WA, Wagenet RJ, Flury M (2000) Dependence of pesticide degradation on sorption: nonequilibrium model and application to soil reactors. J Contam Hydrol 43 (1):45–62
41. He Q, Mankin KR (2002) Performance variations of COD and nitrogen removal by vegetated submerged bed wetlands. J Am Water Resour Assoc 38:1679–1689
42. Spark KM, Swift RS (2002) Effect of soil composition and dissolved organic matter on pesticide sorption. Sci Total Environ 298(1–3):147–161
43. Brix H (1993) Wastewater treatment in constructed wetland: system design, removal process and treatment performance. En constructed wetlands for water quality improvement (pp 9–22). Editado por G. Moshiri, CRC Press Inc Boca de Raton, FL. U.S.A.
44. Brix H (1994) Use of constructed wetlands in water pollution control: historical development, present status, and future perspectives. Water Sci Technol 30:209–223
45. Miranda-Ríos M, Luna-Pabello VM (2001) Estado del arte y perspectivas de aplicación de los humedales artificiales de flujo horizontal en México. Facultad de Química, UNAM. México, D.F. 121 p
46. Vymazal J (2009) The use constructed wetlands with horizontal sub-surface flow for various types of wastewater. Ecol Eng 35:1–17
47. Bavor HJ, Roser DJ, Adcock PW (1995) Challenges for the development of advanced constructed wetlands technology. Water Sci Technol 32(3):13–20
48. Greenway M, Simpson JS (1996) Artificial wetlands for wastewater treatment, water reuse and wildlife in Queens Land, Australia. Water Sci Technol 33(10–11):221–229
49. Greenway M (1997) Nutrient content of wetland plants in constructed wetlands receiving municipal effluent in tropical Australia. Water Sci Technol 35(5):135–142
50. Nitisoravut S, Klomjek P (2005) Inhibition kinetics of salt-affected wetlands for municipal wastewater treatment. Water Res 39:4413–4419
51. Sheoran AS, Sheoran V (2006) Heavy metal removal mechanism of acid mine drainage in wetlands: a critical review. Miner Eng 19:105–116

52. Calheiros CSC, Duque AF, Moura A, Henriques IS, Correia A, Rangel AOSS, Castro PML (2009) Substrate effect on bacterial communities from constructed wetlands planted with *Typha latifolia* treating industrial wastewater. Ecol Eng. doi:10.1016/j.ecoleng.2008.11.010
53. Tchobanoglous G, Schroeder E (1985) Water quality. Addison Wesley Pub Co, Reading
54. Brix H, Arias CA (2005) The use of vertical flow constructed wetlands for on-site treatment of domestic wastewater: new Danish guidelines. Ecol Eng 25:491–500
55. Mitsch WJ, Jørgensen SE (1989) Ecological engineering: an introduction to ecotechnology. Wiley, New York, p 472
56. Mitsch WJ, Jørgensen SE (2004) Ecological engineering and ecosystem restoration. Wiley, New York, p 411
57. Gray S, Kinross J, Read P, Marland A (2000) The nutrient assimilative capacity of Mareal as a substrate for waste treatment. Water Res 34(8):2183–2190
58. Kickuth R (1970) Ökochemische Leistungen höherer Pflanzen. Naturwissenschaften 57:55–61
59. Kickuth R (1984) Das Wurzelraumverfahren in der Praxis. Landschaft Stadt 16:145–153
60. Seidel K (1966) Reinigung von Gewassern durch hohere Pflanzen. Naturwiss 53:289–297
61. Griffin P (2003) Ten years experience of treating all flows from combined sewerage systems using package plant and constructed wetland combinations. Water Sci Technol 48 (11–12):93–99
62. Kadlec RH, Knight RL (1996) Treatment wetlands. Lewis Publishers/CRC Press, Boca Raton, 893
63. Vymazal J, Kröpfelová L (2008) Wastewater treatment in constructed wetlands with horizontal sub-surface flow. Springer, Heidelberg/Berlin, p 566
64. Llagas-Chafloque WA, Guadalupe-Gómez E (2006) Diseño de humedales artificiales para el tratamiento de aguas residuales en la UNMSM. Rev Inst Inv 15(17):85–96
65. Rousseau DPL, Vanrolleghem PA, De Pauw N (2004) Model-based design of horizontal subsurface flow constructed treatment wetlands: a review. Water Res 38:1484–1493
66. Rousseau DPL, Lesage E, Story A, Vanrollegheme PA, De Pauw N (2008) Constructed wetlands for water reclamation. Desalination 218:181–189
67. Kadlec RH, Wallace SD (2008) Treatment wetlands, 2nd edn. CRC Press, Boca Raton
68. Wittgren HB, Maehlum T (1997) Wastewater treatment wetlands in cold climates. Water Sci Technol 35(5):45–53
69. Scholz M, Lee BH (2005) Constructed wetlands: a review. Int J Environ Stud 62:421–447
70. Truu M, Juhanson J, Truu J (2009) Microbial biomass, activity and community composition in constructed wetlands. Sci Total Environ. Stoten-11035; No. of Pages 14
71. Kickuth R (1977) Degradation and incorporation of nutrients from rural wastewaters by plant hydrosphere under limnic conditions. In: Utilization of manure by land spreading (pp 335–343). Comm. Europ. Commun., EUR 5672e, London
72. Kickuth R (1978) Elimination gelöster Laststoffe durch Röhrichtbestände. Arbeiten des Deutschen Fischereiverbandes 25:57–70
73. Kickuth R (1981) Abwasserreinigung in Mosaikmatritzen aus aeroben and anaeroben Teilbezirken. In: Moser F (ed) Grundlagen der Abwasserreinigung. Verlag Oldenburg, Munchen/Wien, pp 639–665
74. Del Bubba M, Lepri L, Cincinelli A, Griffini O, Tabani F (2000) Linear alkylbenzene-sulfonates (LAS) removal in a pilot submerged horizontal flow constructed wetland. In: Proceedings of the 7th international conference on wetland systems for water pollution control (pp 919–925). University of Florida, Gainesville.
75. Huang Y, Latorre A, Barcelo D, Garcia J, Aguirre P, Mujeriego R, Bayona JM (2004) Factors affecting linear alkylbenzene sulfonates removal in subsurface flow constructed wetlands. Environ Sci Technol 38:2657–2663
76. Orozco CE, Cruzi AM, Rodríguez MA, Pohlan AJ (2006) Humedal subsuperficial de flujo vertical como sistema de depuración terciaria en el proceso de beneficiado de café. Higiene y Sanidad Ambiental 6:190–196

77. Paing J, Voisin J (2005) Vertical flow constructed wetlands for municipal wastewater and septage treatment in French rural area. Water Sci Technol 51:145–155
78. Tanner CC, Kadlec RH (2003) Oxygen flux implications of observed nitrogen removal rates in subsurface-flow treatment wetlands. Water Sci Technol 48(5):191–198
79. Yalcuk A, Ugurlu A (2009) Comparison of horizontal and vertical constructed wetland systems for landfill leachate treatment. Bioresour Technol 100:2521–2526
80. Brix H (1987) Treatment of wastewater in the rhizosphere of wetland plants the root zone method. Water Sci Technol 19:107–118
81. Cooper PF, Job GD, Green MB, Shutes RBE (1996) Reed beds and constructed wetlands for wastewater treatment. WRc Publications, Medmenham/Marlow
82. Kadlec RH, Knight RL, Vymazal J, Brix H, Cooper P, Haberl R (2000) Constructed wetlands for pollution control: processes, performance, design and operation. IWA specialist group on use of macrophytes in water pollution control. Scientific and technical report no. 8. IWA Publishing, London
83. Vymazal J (2005) Horizontal sub-surface flow and hybrid constructed wetlands systems for wastewater treatment. Ecol Eng 25:478–490
84. Vymazal J, Sven EJ, Brian F (1998) Constructed wetlands, sub-surface flow. Encyclopedia of ecology. Academic, Oxford, pp 748–764
85. Picard C, Fraser HL, Steer D (2005) The interacting effects of temperature and plant community type on nutrient removal in wetland microcosms. Bioresour Technol 96:1039–1047
86. Gao J, Liu L, Liu X, Lu J, Zhou H, Huang S, Wang Z, Spear PA (2008) Occurrence and distribution of organochlorine pesticides lindane, p, p′-DDT, and heptachlor epoxide in surface water of China. Environ Int 34:1097–1103
87. Simonit S, Cattaneo F, Perrings C (2005) Modelling the hydrological externalities of agriculture in wetlands: the case of rice in Esteros del Iberà. Argent Ecol Mod 186:123–141
88. Miglioranzaa KSB, Morenoa JEA, Morenoa VJ (2004) Organochlorine pesticides sequestered in the aquatic macrophyte *Schoenoplectus californicus* (C.A. Meyer) Soják from a shallow lake in Argentina. Water Res 38:1765–1772
89. Sarkar SK, Bhattacharya BD, Bhattacharya A, Chatterjee M, Alam A, Satpathy KK, Jonathan MP (2008) Occurrence, distribution and possible sources of organochlorine pesticide residues in tropical coastal environment of India: an overview. Environ Int 34:1062–1071
90. Dabrowski JM, Schulz R (2003) Predicted and measured levels of azinphos-methyl in the Lourens river, South Africa: comparison of runoff and spray drift. Environ Toxicol Chem 22:494–500
91. Dabrowski JM, Bollen A, Bennett ER, Schulz R (2005) Pesticide interception by emergent aquatic macrophytes: potential to mitigate spray-drift input in agricultural streams. Agric Ecosyst Environ 111:340–348
92. Laabs V, Wehrhan A, Pinto A, Dores E, Amelung W (2007) Pesticide fate in tropical wetlands of Brazil: an aquatic microcosm study under semi-field conditions. Chemosphere 67:975–989
93. Calheiros CSC, Rangel AOSS, Castro PML (2007) Constructed wetland systems vegetated with different plants applied to the treatment of tannery wastewater. Water Res 41 (8):1790–1798
94. Kucuk OS, Sengul F, Kapdan IK (2003) Removal of ammonia from tannery effluents in a reed bed constructed wetland. Water Sci Technol 48(11–12):179–186
95. Dotro G, Fitch M, Larsen D, Palazolo P (2006) Treatment of chromium-bearing wastewaters from tannery operations with constructed wetlands. In: Proceedings of the 10th international conference on wetland systems for water pollution control, MAOTDR 2006 (pp 1725–1733). Lisbon, Portugal.
96. Calheiros CSC, Rangel AOSS, Castro PML (2008) Evaluation of different substrates to support the growth of Typha latifolia in constructed wetlands treating tannery wastewater over long-term operation. Bioresour Technol 99(15):6866–6877

97. Aguilar JRP, Cabriales JJP, Vega MM (2008) Identification and characterization of sulfur-oxidizing bacteria in an artificial wetland that treats wastewater from a tannery. Int J Phytoremediation 10:359–370
98. Vrhovsek D, Kukanja V, Bulc T (1996) Constructed wetland (CW) for industrial waste water treatment. Water Res 30:2287–2292
99. De Zeeuw W, Heijnen G, De Vries J (1990) Reed bed treatment as a wastewater (post) treatment alternative in the potato starch industry. In: Cooper PF, Findlater BC (eds) Constructed wetlands in water pollution control. Pergamon Press, Oxford, pp 551–553
100. White KD (1994) Enhancement of nitrogen removal in subsurface-flow constructed wetlands by employing a 2-stage configuration, an unsaturated zone, and recirculation. In: Proceedings of the 4th international conference on wetland systems for water pollution control, ICWS'94 Secretariat, Guangzhou, P.R. China (pp 219–229)
101. Wallace S (2002) Treatment of cheese-processing waste using subsurface-flow wetlands. In: Nehring KW, Brauning SE (eds) Wetlands and remediation II. Battelle, Columbus, pp 197–203
102. Mantovi P, Piccinni S, Lina F, Marmiroli M, Marmiroli N (2007) Treating wastewaters from cheese productions in H-SSF constructed wetlands. In: Borin M, Bacelle S (eds) Proceedings of the international conference on multi functions of wetland systems (pp 72–73). P.A.N. s.r. l., Padova, Italy
103. Bojcevska H, Tonderski K (2007) Impact of loads, season, and plant species on the performance of a tropical constructed wetland polishing effluent from sugar factory stabilization ponds. Ecol Eng 29:66–76
104. Prochaska CA, Zouboulis AI, Eskridge KM (2007) Performance of pilot-scale vertical-flow constructed wetlands, as affected by season, substrate, hydraulic load and frequency of application of simulated urban sewage. Ecol Eng 31:57–66
105. Holland JF, Martin JF, Granata T, Bouchard V, Quigley M, Brown L (2004) Effects of wetland depth and flow rate on residence time distribution characteristics. Ecol Eng 23:189–203
106. Kjellin J, Worman A, Johansson H, Lindahl A (2007) Controlling factors for water residence time and flow patterns in Ekeby treatment wetland, Sweden. Adv Water Resour 30:838–850
107. Jaeken P, Debaer C (2005) Risk of water contamination by plant protection products (PPP) during pre- and post treatment operations. Annu Rev Agric Eng 4:93–114
108. Kreuger J, Nilsson E (2001) Catchment scale risk-mitigation experiences key issues for reducing pesticide transport to surface waters. In: Pesticide behaviour in soil and water, vol. 78. BCPC symposium proceedings (pp 319–324)
109. Rose MT, Sanchez-Bayo F, Crossan AN, Kennedy IR (2006) Pesticide removal from cotton farm tailwater by a pilot-scale ponded wetland. Chemosphere 63:1849–1858
110. Schulz R, Moore MT, Bennett ER, Farris JL, Smith S, Cooper CM (2003) Methyl parathion toxicity in vegetated and nonvegetated wetland mesocosms. Environ Toxicol Chem 22:1262–1268
111. Lin T, Wen Y, Jiang L, Li J, Yang S, Zhou Q (2008) Study of atrazine degradation in subsurface flow constructed wetland under different salinity. Chemosphere 72:122–128
112. Moore MT, Cooper CM, Smith S, Cullum RF, Knight SS, Locke MA, Bennett ER (2009) Mitigation of two pyrethroid insecticides in a Mississippi Delta constructed wetland. Environ Pollut 157:250–256
113. Matamoros V, Puigagut J, García J, Bayona JM (2007) Behavior of selected priority organic pollutants in horizontal subsurface flow constructed wetlands: a preliminary screening. Chemosphere 69:1374–1380
114. Runes HB, Jenkins JJ, Moore JA, Bottomley PJ, Wilson BD (2003) Treatment of atrazine in nursery irrigation runoff by a constructed wetland. Water Res 37:539–550
115. Bouldin JL, Farris JL, Moore MT, Smith S, Stephens WW, Cooper CM (2005) Evaluated fate and effects of atrazine and lambda-cyhalothrin in vegetated and unvegetated microcosms. Environ Toxicol 20:487–498

116. Schulz R (2004) Field studies on exposure, effects, and risk mitigation of aquatic nonpoint-source insecticide pollution: a review. J Environ Qual 33:419–448
117. Schulz R, Peall SKC (2001) Effectiveness of a constructed wetland for retention of nonpoint-source pesticide pollution in the Lourens River Catchment, South Africa. Environ Sci Technol 35:422–426
118. Warren CS, Mackay D, Bahadur NP, Boocock DGB (2002) A suite of multisegment fugacity models describing the fate of organic contaminants in aquatic systems: application to the Rihand Reservoir, India. Water Res 36:4341–4355
119. Moore MT, Rodgers JH Jr, Smith S Jr, Cooper CM (2001) Mitigation of metolachlor-associated agricultural runoff using constructed wetlands in Mississippi, USA. Agric Ecosyst Environ 84:169–176
120. Anderson KL, Kevin A, Wheeler KA, Robinson JB, Tuovinen OH (2002) Atrazine mineralization potential in two wetlands. Water Res 36:4785–4794
121. FAO (1993) Pesticide residue in food. Evaluations part I residues FAO. Plant production and protection paper. 124,351
122. Kantawanichkul S, Wara-Aswapati S (2005) LAS removal by a horizontal flow constructed wetland in tropical climate. In: Vymazal J (ed) Natural and constructed wetlands. Nutrients, metals and management. Backhuys Publishers, Leiden, pp 261–270
123. Knight RL, Kadlec RH, Harry M (1999) The use of treatment wetlands for petroleum industry effluents. Environ Sci Technol 33:973–980
124. Yang L, Hu CC (2005) Treatments of oil-refinery and steel-mill wastewaters by mesocosm constructed wetland systems. Water Sci Technol 51(9):157–164

Part III
Risk Assessment

Chapter 14
Tracing Nitrogen Sources and Cycle in Freshwater Through Nitrogen and Oxygen Isotopic Research

Zi-Xiang Chen, Xue-Bin Yin, Guang Liu, and Gui-Jian Liu

14.1 Introduction

Freshwater systems are vulnerable to pollution, due to excess nitrogen nutrient input, which poses a severe threat to drinking water supplies and promotes eutrophication and seasonal hypoxia in surface and groundwater around the world [1, 2]. High levels of nutrient input also increase hazards to the aquatic system, such as heavy metals and toxic bacteria, leading to further deterioration in water quality [3]. Nitrate is the dominant form of nitrogen contamination in freshwater systems worldwide. Rising nitrate concentrations, above normal levels, have been recorded in many areas of the world in recent years. The overapplication of fertilizers and poorly treated human sewage and animal waste in hydrologic systems make an important contribution to these high nitrate levels. High nitrate levels create potential health risks, especially to infants, who can contract methemoglobinemia, also termed blue-baby disease [4, 5]. Hence, a limit on nitrate concentration (10 mg/L) in drinking water has been set by WHO (World Health Organization) and USEPA (United States Environmental Protection Agency).

In order to provide better management and improvement of water quality and to make effective remediation plans for heavily polluted areas, it is important to identify the sources of nitrates and map the nitrogen cycle. Traditional analysis of

Z.-X. Chen • X.-B. Yin (✉) • G. Liu
School of Earth and Space Science, University of Science and Technology of China (USTC), Hefei, Anhui 230026, China

Advanced Lab for Ecological Safety and Human Health, Suzhou Institute of USTC, Suzhou, Jiangsu 215123, China
e-mail: xbyin@ustc.edu.cn

G.-J. Liu
School of Earth and Space Science, University of Science and Technology of China (USTC), Hefei, Anhui 230026, China

S.K. Sharma and R. Sanghi (eds.), *Wastewater Reuse and Management*,
DOI 10.1007/978-94-007-4942-9_14,

nitrate sources is conducted by determining nitrate concentrations in freshwater systems and potential nitrate sources of water supplies. However, there are many deficiencies in this method, such as the possibility of multiple nitrate sources or other elements impacting on nitrate distributions in surface and groundwater, the presence of mixing point and nonpoint sources along a flow path, and potential occurrence of some biogeochemical processes which would control changes in nitrate and other chemical concentrations. Given these complex possibilities, the introduction of stable isotopic composition in nitrate seems crucial, because stable isotopes can provide a direct means of nitrate source identification, and there are distinctive nitrogen and oxygen isotopic compositions present among different sources. Recently, the technique of using nitrogen and oxygen isotopic compositions to identify nitrate sources has been extensively used in many aquatic systems worldwide [6–8]. The method of using single nitrogen isotope ($\delta^{15}N$) to elevate nitrate sources and processes has been successfully practiced in a variety of research over the past several decades [1]. Since the early 1990s, the combined use of oxygen isotopes in nitrate ($\delta^{18}O$) has been employed [6]. Recently, many isotopic approaches have combined use of nitrate isotopes with multi-isotope ratios, such as strontium(Sr), boron(B), and $\delta^{18}O$-H_2O, with proven success in tracing nitrate sources in freshwater [9, 5].

This chapter will focus on using nitrogen and oxygen isotopic compositions of nitrate to identify the sources of dissolved nitrate and assess the process of nitrogen cycle in freshwater systems. More details of applications of this technique will be discussed in the sections on surface and groundwater. It is a kind of interdisciplinary approach, and many interacting factors in freshwater systems will be also incorporated into and discussed in this chapter.

14.1.1 Nitrogen Isotopes in Nitrate

Nitrogen has two naturally occurring stable isotopes, ^{14}N and ^{15}N, present as 99.64 and 0.36% in all nitrogen atoms, respectively (Faure 1986). There are many different kinds of nitrogen compounds, because of the multivalence of nitrogen, ranging from +5 (e.g., NO_3^-) to −3 (e.g., NH_4^+), results in a wide natural range of isotopic compositions. Nitrogen isotopic ratios are generally reported using δ notation relative to the nitrogen isotopic ratio of air (Peterson and Fry 1987). Stable isotopic ratios are expressed using the delta per mil (‰) notation relative to international reference standards:

$$\delta_{\text{sample}}(‰) = \left[\left(R_{\text{sample}} - R_{\text{standard}}\right)/R_{\text{standard}}\right] \times 1{,}000 \qquad (14.1)$$

R is the ratio of ^{15}N to ^{14}N of the sample and standard. Analytical precision of 0.3 ‰ or better is accepted. Some reference materials with distinct $\delta^{15}N$ values have been widely used as standards for the comparison of different researchers' laboratory, such as standard reference materials N-1(IAEA), N-2 (IAEA), and N-3(IAEA), whose reported $\delta^{15}N$ values are +0.45, +20.35, and +22.3‰, respectively [7, 10].

14.1.2 Oxygen Isotopes in Nitrate

The stable oxygen isotopic composition of nitrate has been extensively used as a marker to identify nitrate sources and map the nitrogen cycle in freshwater systems. There are three kinds of oxygen isotopic compositions, such as ^{16}O (99.759‰), ^{17}O (0.037‰), and ^{18}O (0.204‰) [11]. According to Eq. 14.1, stable oxygen isotopic compositions are also expressed in terms of $^{18}O/^{16}O$ ratios of the sample and standard, and the $\delta^{18}O$ values of nitrate are reported in the delta per mil (‰) relative to the Vienna Standard Mean Ocean Water (VSMOW). Analytical precision of 0.3 ‰ or better is also accepted. Positive δ_{sample} values indicate enrichment in the heavy isotope, whereas negative δ_{sample} values indicate depletion in the heavy isotope. In theory, two of the oxygen atoms in nitrate are thought to be derived from ambient environmental H_2O and one from atmospheric air [12]. The larger range of $\delta^{18}O$ values suggests that oxygen isotopes in nitrate are fractionated from their source compositions during atmospheric processes.

14.2 Methods

The technique of using the $\delta^{15}N$ and $\delta^{18}O$ values of nitrate to trace nitrogen sources has been proven effective; however, the analytic techniques for the determination of $\delta^{15}N$ and $\delta^{18}O$ values must be accurate, inexpensive, and rapid. Below is a brief description of some recent methods to solve the problem of concentrating the dissolved nitrate in an aquatic system and analyzing the natural abundance of nitrogen and oxygen isotopic compositions.

The first approach is the "$AgNO_3$ technique," also known as the "ion-exchange method," which was proposed by Chang et al. [13] and Silva et al. [14]. Their work provides a detailed, practical description of both field and laboratory methods of concentrating and purifying dissolved nitrate in freshwater samples, then converting the dissolved nitrate to $AgNO_3$ crystals, before determining $\delta^{15}N$ and $\delta^{18}O$ values. The specific experimental procedures can be briefly stated as follows. To determine the $\delta^{15}N$ nitrate value, the first step is to pass water samples through anion exchange resin columns to concentrate nitrate. Then, nitrate is eluted by hydrochloric acid (HCl). The next step uses silver oxide to neutralize the eluted solution, which is then passed through filters to remove AgCl. To determine the nitrate $\delta^{18}O$ value, on the base of the experimental steps to determine the $\delta^{15}N$ value, all non-nitrate oxygen-bearing anions, such as SO_4^{2-}, PO_4^{3-}, and CO_3^{2-} in silver nitrate solutions, must be removed by $BaCl_2$, and the precipitate must be filtered out. After this step is complete, the filtered $AgNO_3$ solution is passed through anion exchange resin columns to remove excess Ba^{2+} and re-neutralized with Ag_2O. Finally, low-blank-value active carbon is used to adsorb some other potential interference factors. Both the resulting treated $AgNO_3$ solutions can be freeze dried [14] or oven thermal dried [15] to produce the required $AgNO_3$ salts. Traditional analysis of $\delta^{15}N$ and $\delta^{18}O$ values in the resulting $AgNO_3$ salts uses the sealed tube combustion method, converting $AgNO_3$ salts to N_2 and CO_2 gas for IRMS (Isotope Ratio Mass Spectrometer) determination to obtain the $\delta^{15}N$ and $\delta^{18}O$ values independently [10]. Although this method is practical and useful, the pretreatment apparatus is too cumbersome, and the amount of $AgNO_3$ salts required is rather high. Recently, a more efficient and precise method of determination of $\delta^{15}N$ and $\delta^{18}O$ values in $AgNO_3$ salts has been used extensively. Both $\delta^{15}N$ and $\delta^{18}O$ values can be determined by TC/EA-IRMS (thermal conversion/element analyzer-IRMS). The essential principle of the method is that $AgNO_3$ salts set in aluminum and silver tubes are combusted at 1,400 °C via EA and then converted to N_2 and CO gas for IRMS. The advantages of this method include its precision and sensitivity and the fact that minor isotope fractionation occurs throughout the experimental procedure. The drawbacks of this method are that the volume of water sample required is very large and that the necessary chemical reagents are very expensive. Overall, this technique is very suitable for freshwater samples and is widely used [2, 5, 15].

The second approach is the "bacterial denitrification method," which has also been successfully applied by many research groups [16, 17]. As the name implies, the essence of this method is the use of denitrifying bacteria which is featured with lacking activity of catalyzing N_2O reduction to N_2 to convert nitrate to N_2O, followed by IRMS analysis to simultaneously determine $\delta^{15}N$ and $\delta^{18}O$ values. Briefly, the specific experimental procedures can be generalized as follows. Denitrifying bacteria are firstly cultured in amended tryptic soy broth (TSB) for about 6–10 days and then divided into centrifuge tubes of 40 mL aliquots. After the centrifugation is complete, the supernatant is decanted, and approximately 4 mL TSB is added to the remaining materials in the tubes to obtain a tenfold level of denitrifying bacteria. These tubes are reserved and then vortexed to confirm the bacteria culture is homogenized. 2*2 mL aliquots are transferred into 20-mL headspace vials, and then, Teflon-backed silicone septa are used to seal these vials. Before the injection of samples into the bacterial culture, the headspace vials are purged with nitrogen gas for about 3 h to ensure there is no N_2O and that the environment is fully anaerobic. Approximately 100 nmol of dissolved nitrate samples are added to the headspace vials and cultured for approximately 12 h to ensure absolute conversion from NO_3^- to N_2O. After conversion, to prevent further denitrifying bacterial activity and remove the CO_2 gas, which can impair determination of N_2O, about 0.1 mL of 10N NaOH is added to the headspace vials. $\delta^{15}N$ and $\delta^{18}O$ values are determined simultaneously by analysis of the prepared N_2O via IRMS. There are two different methods of completing the measurement process, online and offline. The online method uses an auto-sampler injection system to extract N_2O online, and Nafion is employed as an additional water trap for the IRMS analysis. The offline method uses a vacuum line to freeze N_2O out offline and then conducts IRMS analysis [18]. The significant advantages of using this method are the reduction in labor and costs and the low number of samples. Whereas, there are also some disadvantages in using this method, for example, the bacterial growth time is very long (approximately 12 days) and all the experimental procedures must be conducted in a sterile environment.

Recently, a new method, called the "cadmium reduction method" or "azide method" has been proposed [19]. This approach uses the chemical reduction method to complete the conversion from NO_3^- to N_2O in order to avoid the disadvantages of the "bacterial denitrification method." A two-step reduction method is employed, firstly using cadmium reduction to convert NO_3^- to NO_2^- and then using sodium azide to convert NO_2^- to N_2O for IRMS analysis to determine $\delta^{15}N$ and $\delta^{18}O$ values, in the same way as employed in the "bacterial denitrification method." The first step of this method has been described by many water chemistry researchers in the last 30–50 years [20, 21]. The traditional method adds 1 g spongy cadmium (pH = 9) which has first been shaken overnight at 120 rpm to approximately 50-mL water sample in Teflon-sealed batches to complete NO_3^- to NO_2^- conversion. The concentration of sodium chloride in the freshwater samples should be increased to 0.5 mol/L. The samples are then centrifuged to separate the spongy cadmium to produce NO_2^-. In recent years, a new method for this first reduction step has been developed by Schiliman and Teplyakov [22]. To fully achieve NO_3^- to NO_2^- conversion and simultaneously remove some factors causing any isotopic fractionation, the water samples are passed through a copperized granular cadmium-filled column at a rate of 5 mL/min for several minutes and then transferred into 60-mL vials and sealed with Teflon-lined septa. The vials are purged with nitrogen gas or helium gas to ensure any N_2O gas is removed. After NO_2^- is produced, in a subsequent reduction step, 2 mL 1:1 sodium azide (2 mol/L)/acetic acid solution (20%) is added to the samples to complete the final NO_2^- to N_2O and N_2 conversion, while 1 mL of 6 mol/L NaOH is added to the samples to remove the HN_3. Finally, the pure N_2O and N_2 gas are subjected to IRMS analysis to determine $\delta^{15}N$ and $\delta^{18}O$ values. This technique successfully avoids the drawbacks of the "bacterial denitrification method." However, the major disadvantage of this method is the toxicity of the chemical reduction reagents employed.

The three methods discussed above are representative of the methods of determining $\delta^{15}N$ and $\delta^{18}O$ values of dissolved nitrate in freshwater samples in widespread use worldwide in recent years. There are some other methods which are not discussed in detail in this chapter. However, any experimental method has both advantages and disadvantages: these methods need to be further developed to enhance their benefits and minimize their disadvantages.

14.3 $\delta^{15}N$ Values of Nitrate Sources

There are different kinds of nitrate sources with distinctive ranges of $\delta^{15}N$ values in freshwater systems. Overall, the ranges of $\delta^{15}N$ values in terrestrial nitrate sources extend from almost −20% to +30‰. Although wider ranges of $\delta^{15}N$ values have

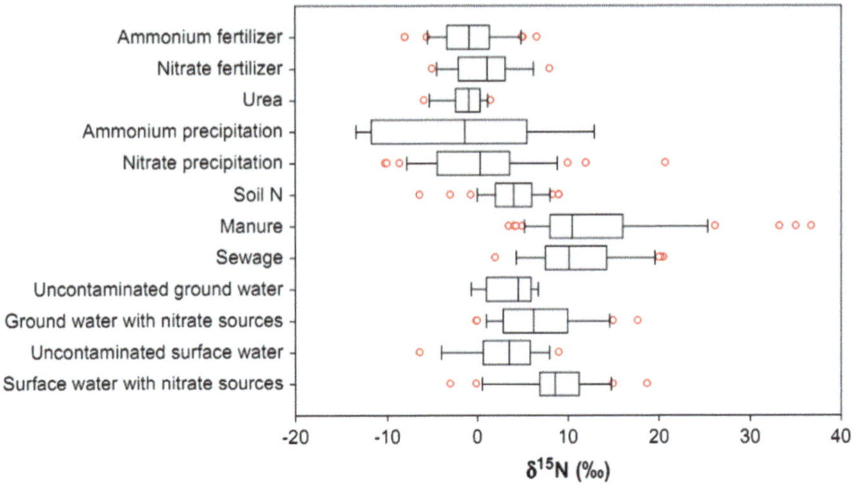

Fig. 14.1 Box plots of $\delta^{15}N$ values pf NO_3^- from various sources and sinks. *Box plots* illustrate the 25th, 50th and 75th percentiles; the *whiskers* indicate the 10th and 90th percentiles; and the *circles* represent outliers (Xue et al. [31])

been reported, some extremely high or low $\delta^{15}N$ values may derive in chief from some nitrate sources produced by unusual nitrogen transformation processes, which are not discussed in detail in this chapter. Precisely because of the distinctive ranges of $\delta^{15}N$ values in terrestrial nitrate sources, the isotopic differences among and changes in these nitrate sources have been used for some time to identify nitrate sources in freshwater systems. The first study using this technique was conducted by Kohl et al. [23], whose research used this method to assess fertilizer contributions to nitrate sources in a river system. The dominant sources of nitrate in freshwater systems are atmospheric deposition, N-bearing fertilizers, sewage and manure, and soil organic nitrogen. While the nitrogen isotopic compositions of some of these sources is distinctive, others are less distinct, with overlapping $\delta^{15}N$ values, for example, between N-bearing fertilizers and soil organic nitrogen. The specific classification of all nitrate sources with different nitrogen isotopic ratios is summarized in Fig. 14.1. A discussion of each of these different nitrate sources appears below.

14.3.1 Atmospheric Deposition

There is a large variation in the nitrogen isotopic composition of atmospheric deposition, which usually falls within the range of $\delta^{15}N$ values from −15% to +15‰ [4, 6, 24]. This large range of $\delta^{15}N$ values is controlled by complex chemical reactions in a complex atmospheric environment [6, 25]. The formation process of these atmospheric nitrate sources is also complex. Nitrate may be produced naturally by volatilization of ammonia from soils and animal waste, nitrification and denitrification may occur in soil and surface water, or production may result from atmospheric N_2 in thunderstorms. Nitrate may be produced by various anthropogenic source inputs, such as combustion of fossil fuels. Given this complex interaction of environment and impacting elements, few comprehensive studies of $\delta^{15}N$ values in atmospheric deposition sources have been conducted to date [25]. Many studies have demonstrated that seasonal changes in $\delta^{15}N$ values in atmospheric deposition occur. Specifically, lower $\delta^{15}N$ values in nitrate have been recorded in spring and summer and higher values in winter. There are two possible explanations for this phenomenon: one reason is that warm, moist environments favor the nitrification activities of N-bearing fertilizers; the other is that increased combustion of fossil fuel in colder seasons may result in increased production of enriched NO_x [1, 26].

The combined use of $\delta^{18}O$ and $\delta^{15}N$ values in nitrates could enable researchers to better distinguish between atmospheric and terrestrial nitrate sources [4, 27, 28]. A combination of $\delta^{18}O$ and $\delta^{15}N$ values can also be used to differentiate among anthropogenic nitrate sources. Furthermore, oxygen isotopic compositions nitrate ratios have been successfully used to identify N_2O produced from nitrification and denitrification activities [29]. More details will be provided in the following section.

14.3.2 N-Bearing Fertilizers

N-bearing fertilizers are also significant nitrate sources in freshwater systems, as to some extent agricultural land use occurs in the environs of most freshwater worldwide, and since large quantities of N-bearing fertilizers are used on crops, nitrate pollution caused by excessive N-bearing fertilizer input seems inevitable. In general, N-bearing fertilizers can be categorized as organic and inorganic fertilizers, such as ammonium fertilizer, ammonium nitrate, potassium nitrate, and urea. All these N-bearing fertilizers are artificial and produced from fixation of atmospheric N_2. Because few isotopic fractionations occur during the process of fixation of atmospheric N_2, small differences in nitrogen isotopic ratios can be found among these artificial fertilizers [30]. The range of $\delta^{15}N$ values of these artificial fertilizers generally extends from −6 to +6‰. The nitrogen isotopic ratio of organic fertilizers (such as cover crops, plant compost, liquid, and solid manure, with a wide range of δ15N value from +2 to +30‰) is mostly higher than inorganic fertilizers. This can be attributed to their different origins. The nitrogen isotopic ratios of the N-bearing fertilizers discussed here are different from those of soil organic nitrogen fertilizers.

14.3.3 Sewage and Manure

Sewage and manure play an important part in contributing nitrate pollutants to the freshwater systems; indeed, they are the dominant nitrate source in such systems. It has often been observed that the nitrogen isotopic ratios in sewage and manure are more highly enriched than other nitrate sources. The high enrichment in $\delta^{15}N$ value of these animal waste products is mainly due to the volatilization of ^{15}N-depleted ammonia; the remaining ammonia, containing high levels of $\delta^{15}N$ is subsequently oxidized into $\delta^{15}N$-enriched nitrate. As a result of this process, nitrate sources deriving from manure range from +5 to +25‰, while sewage ranges from +4 to +19‰ [31]. In practical studies, high levels of $\delta^{15}N$ value in sewage and manure can be taken to signify anthropogenic contamination sources.

14.3.4 Soil Organic Nitrogen

The nitrogen cycle is highly complex and results from continuous interaction among the atmosphere, soil, water, and the entire biosphere. Hence, there are many factors which impact on the nitrogen isotopic ratios of soil organic nitrogen. These include some natural factors, such as soil depth, vegetation type, climate characteristics, and site history, and some bacterial activities, such as mineralization and nitrification process. All these factors cause small variations in nitrogen isotopic ratios of between 0 and 8‰ in soil organic nitrogen. Nitrate is the dominant nitrogen form in soil and derives from different sources, such as soils fertilized with synthetic nitrate fertilizers or manure and soils polluted by surface and groundwater. This explains the overlap in nitrogen isotopic compositions between synthetic nitrate fertilizers and soil organic nitrogen and hence the deficiency of using single nitrogen isotopic ratios to identify nitrate sources. Because there are a large range of $\delta^{15}N$ values in these nitrate sources, the average $\delta^{15}N$ value of nitrate sources recorded in one place cannot be taken to represent other places: this is better explained [32] through the complexities of nitrate $d^{15}N$ data.

14.3.5 Nitrate Sources in Freshwater

During the past decade, nitrate pollution has accumulated and resulted in unacceptable levels of nitrate concentration in freshwater systems (surface water and groundwater) in many parts of the world. This has caused worldwide concern. Moreover, human activities are a major contributor to this worsening of surface and groundwater quality [4, 5]. Nitrate pollution in surface and groundwater is largely affected by land use around the water area. For example, if the dominant land use is agricultural, nitrate contamination mainly derives from the extensive use of fertilizers, animal manures, and some other N-containing organic matter, whereas in industrial areas, intensive land use and high levels of human activity will contribute the majority of nitrates found in surface and groundwater. Consequently, the nitrogen isotopic ratios of dissolved nitrates in surface and groundwater range from −4 to +15‰ [31].

14.4 $\delta^{18}O$ Values of Nitrate Sources

As seen in the discussion of nitrogen isotopic ratios in different nitrate sources in Sect. 1.3, the disadvantage of using single dissolved nitrate $\delta^{15}N$ value is the inability to identify nitrate sources or plot the nitrogen transformation process in freshwater systems due to overlaps in $\delta^{15}N$ value among such nitrate sources. Therefore, the $\delta^{18}O$ value of dissolved nitrate has been extensively used to provide an additional, comprehensive means of identifying nitrate sources in freshwater systems precisely [7, 33, 34]. In fact, the $\delta^{18}O$ value of nitrate was determined and studied as early as the 1980s [27, 35]. However, because the method used at that time was very labor-intensive and hazardous, determination of the $\delta^{18}O$ value of nitrate was not widely employed. With the development of techniques to determine $\delta^{15}N$ and $\delta^{18}O$ values in nitrates, dual-isotope research began to be used in many parts of the world. There are large differences in $\delta^{18}O$ value ranges between atmospheric nitrate deposition and microbially produced soil nitrate: the typical ranges of $\delta^{18}O$ value in atmospheric nitrate deposition and microbially produced soil nitrate are from +52.5 to +60.9‰ and from +0.8 to +5.8‰, respectively [28]. This distinction is helpful in distinguishing between atmospheric nitrate deposition and microbially produced soil nitrate. Moreover, the $\delta^{18}O$ value of nitrates also can be a good indicator of synthetic nitrate fertilizer in other nitrate sources [36]. The specific $\delta^{18}O$ value of different nitrate sources is shown in Fig. 14.2 [31]. Detailed $\delta^{18}O$ values of different nitrate sources (such as atmospheric nitrate deposition, synthetic fertilizer, and nitrification of N-bearing organic matter) will be discussed individually below.

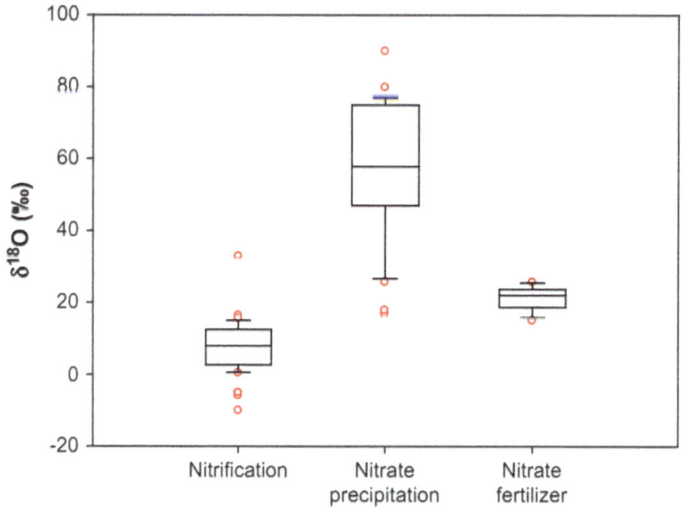

Fig. 14.2 Box plots of $\delta^{18}O$ values pf NO_3^- generated during nitrification, nitrate precipitation and nitrate fertilizer. *Box plots* illustrate the 25th, 50th and 75th percentiles; the *whiskers* indicate the 10th and 90th percentiles; and the *circles* represent outliers (Xue et al. [31])

14.4.1 Atmospheric Nitrate Deposition

A much larger range of $\delta^{18}O$ value in atmospheric nitrate deposition is observed, as a result of the complex atmospheric processes and considerable spatial or temporal variability in the atmosphere. Possible reasons for the larger range of $\delta^{18}O$ value may include the following factors: isotopic fractionation during the process of nitrate production in thunderstorms, incomplete combustion of fossil fuels in power plants and vehicle exhaust emissions, and photochemical reactions in the atmosphere. These processes also play an important role in the fractionation of nitrogen isotopic ratios in the atmosphere (as discussed in Sect. 14.3.1). There are different natural and anthropogenic nitrate sources in the atmosphere, which also affect the oxygen isotopic ratios in atmospheric nitrate deposition. The $\delta^{18}O$ value of the oxygen naturally occurring in the atmosphere is around +23‰. When natural or anthropogenic NO_x is oxidized to format nitrate from reactive oxygen in the atmosphere [6, 34], on the base of impacts factors to alert $\delta^{18}O$ value in atmospheric nitrate deposition, the oxygen isotopic ratio of atmospheric nitrate deposition exhibits a wide range, from +25 to +75‰ [31].

14.4.2 Synthetic Fertilizer

The $\delta^{18}O$ value of synthetic fertilizers is almost constant, and there are few factors which impact on the oxygen isotopic ratio. Since the 1980s, when the $\delta^{18}O$ value of synthetic fertilizers was determined by Amberger and Schmidt [27], in their study, many different kinds of anthropogenic nitrate fertilizer and synthetic nitrate which the total number of three oxygen atoms is totally derived from atmospheric oxygen were determined to obtain $\delta^{18}O$ value in synthetic fertilizers, and the $\delta^{18}O$ value typically ranges from +17 to +25‰, which is similar to the $\delta^{18}O$ value of atmospheric oxygen (23.5‰).

14.4.3 Nitrification of N-Bearing Organic Matters

Nitrification is the process of converting N-bearing organic matter from all nitrate sources (such as atmospheric deposition, N-bearing fertilizers, sewage and manure, and soil organic nitrogen) into nitrate by nitration bacteria. In theory, when nitrification activities occur, two oxygen atoms of the newly formed nitrate are derived from ambient water, and one oxygen atom is derived from atmospheric oxygen [6, 7, 28, 33, 37]. According to this theory, we would anticipate a range of $\delta^{18}O$ value in nitrification nitrates from -10 to +10‰, because the $\delta^{18}O$ value of ambient water ranges from -25 to +4‰, while that of atmospheric oxygen is approximately +23.5‰ [6, 28]. However, some studies have reported a maximum $\delta^{18}O$ value of nitrate from microbial activities of approximately +15‰ [6, 33], which is +5‰ higher than the expected $\delta^{18}O$ value (+10‰). Possible explanations for this discrepancy might be that the $\delta^{18}O$ value of ambient water could be isotopically fractionated to a high level during the evaporation and respiration process or isotopic fractionation might occur during the nitrification process. Some studies have suggested that the amount of atmospheric oxygen used by the nitration bacteria could exceed the theoretical level [6, 33]. These hypotheses provide convincing explanations as to explain why the $\delta^{18}O$ value from microbially produced nitrate is higher than the $\delta^{18}O$ value predicted in theory.

14.5 Factors Affecting Nitrogen and Oxygen Isotopic Compositions

It is very important to understand some factors affecting nitrogen and oxygen isotopic compositions when the technique of using $\delta^{15}N$ and $\delta^{18}O$ values is used to identify nitrate sources in freshwater systems. There are many kinds of nitrate sources derived from high-density urban and agricultural activities, which make important contributions of nitrate to surface and groundwater. Thus, nitrate is accumulated and nitrate pollution is caused. However, during the process of nitrate transformation, complex isotopic fractionations occur, caused by different kinds of nitrogen cycle processes, which ultimately alter the nitrogen and oxygen isotopic compositions in nitrates [38]. Hence, these multiple nitrogen cycle process should be studied in order to understand precisely how the $\delta^{15}N$ and $\delta^{18}O$ values in nitrate change. Over the past 10 years, many studies and reviews discussing this issue have been published [6, 31, 39–41].

14.5.1 Mixing

As this term implies, there are many kinds of nitrate sources with different nitrogen and oxygen isotopic compositions which contribute to surface and groundwater, and they are mixed together, which causes an alteration in $\delta^{15}N$ and $\delta^{18}O$ values in nitrate, and this deviation affects identification of nitrate sources. When mixing occurs, the $\delta^{15}N$ and $\delta^{18}O$ values of nitrate in water samples at a given site are not constant. In addition, before or after the mixing process, the initial nitrogen and oxygen isotopic compositions of these different nitrate sources may be altered by various isotopic fractionation activities in the field [6, 39]. Therefore, such mixing will cause inaccurate identification of nitrate sources in surface and groundwater [6], and in practical research, the types of mixing which occur locally should be considered so that nitrate sources in freshwater systems may be traced more effectively.

14.5.2 Ammonification

In ammonification, sometimes also termed mineralization, ammonium is produced from soil organic nitrogen matter (organic nitrogen → NH_4^+), and where ammonification occurs, there will be limited isotopic fractionation (about ±1‰) between soil organic matter and soil ammonium. Some researchers have hypothesized that the entire nitrate was produced from organic nitrogen matter by some reaction steps during the process of ammonification, one of which is rate limiting [42, 43]. However, in fact, isotopic fractionations are caused by the nitrification of ammonium (nitrification), not the conversion of organic N to ammonium (ammonification) [44, 45]. Confusion about these processes which control nitrogen and oxygen isotopic compositions should be avoided in experimental studies.

14.5.3 Ammonia Volatilization

Ammonia volatilization describes the loss of ammonia gas from soil to the atmosphere: during the process of ammonia volatilization, the nitrogen isotopic ratio of the residual ammonium in the soil will be higher than that of the ammonia gas in the atmosphere. The isotopic fractionation caused by ammonia volatilization comprises two steps. The first, which is reversible, is that equilibrium fractionations occur between ammonium and ammonia in solution and between aqueous and gaseous ammonia. The second irreversible step is that kinetic reactions result in isotopic fractionation, which is caused by the loss of $\delta^{15}N$ values of depleted ammonia in the atmosphere. Both these processes will cause an increase in $\delta^{15}N$ values. In ammonia volatilization, pH values in the environment are the dominant factor. For example, urea and manure are applied to fields for agricultural use, and urea is also the dominant N form in manure; hence, in this situation, the urea and ammonia fertilizer can be hydrolyzed to ammonia and then oxidized to form a nitrate by nitrification [1], as expressed in this equation: $CO(NH_2)_2 \rightarrow NH_3 \rightarrow \leftarrow NH_4^+ \rightarrow NO_3^-$. This reaction will cause a temporary increase in pH value, creating the perfect environment for the loss of ammonia gas by volatilization. Hence, the pH value in during ammonia loss will remain acid, while the residual ammonium will have a higher $\delta^{15}N$ value [6]. There are also natural factors which affect the degree of $\delta^{15}N$ value enrichment in the residual ammonium, such as the pH value and degree of moisture of the soil and the temperature.

14.5.4 Assimilation

Assimilation is the biological reaction in which N-bearing compounds, specifically ammonium, nitrite, and nitrate, are incorporated into organisms. The assimilation process can be briefly described as follows. NO_x compounds are first converted into ammonium by nitrate or nitrite reductases and finally assimilated into organic matter. The assimilation process, like other factors, can alter the nitrogen isotopic ratio and generally prefer the incorporation of ^{14}N over ^{15}N. Many studies have been conducted to seek to determine the nitrogen isotopic ratio and isotopic fractionation values [25, 46]. They concluded that fractionations caused by assimilation activities slightly impact on the isotopic composition of the residual fertilizer or soil organic matter. Fogel and Cifuentes [47] reported a wide range of fractionations in nitrate and ammonium assimilation by algae in aquatic areas, from −270 to 0‰ in field tests and laboratory simulations. The major reason for the large range of fractionations was believed to be interaction with some kinetic, equilibrium isotope, which impacts significantly on environmental conditions.

14.5.5 *Nitrification*

Nitrification is a multistep oxidation process by which ammonium is converted to various nitrogen oxides as intermediate species, such as NO_2^-, NO, and N_2O, and finally nitrate is produced. The entire process is also accompanied by distinct nitrogen isotopic fractionations, which induce depletion of the $\delta^{15}N$ value in the nitrate relative to in $\delta^{15}N$ value of the initial ammonium, while the $\delta^{15}N$ value in the residual ammonium is enriched [46, 48]. The whole nitrification process can be elaborated as a series of oxidation reactions, as shown below:

$$NH_4^+ + H_2O \rightarrow NH_2OH + 2[H] + H^+ \quad (14.2)$$

$$NH_2OH + O_2 \rightarrow NO_2^- + [H] + H^+ \quad (14.3)$$

$$NO_2^- + H_2O \longleftrightarrow NO_3^- + 2[H] \quad (14.4)$$

The reaction of Eq. 14.2 is oxidized by ammonia monooxygenase, Eq. 14.3 is oxidized by hydroxylamine oxidoreductase, and the final reaction described in Eq. 14.4 is oxidized by nitrite oxidoreductase. In this reaction, the oxygen from water has been incorporated, because this step is reversible, so it is evident that the oxygen atom can be exchanged between water and nitrate or nitrite [49]. The isotopic fractionations of $\delta^{15}N$ value during the entire nitrification process have been extensively discussed [6, 25]. The isotopic enrichment in $\delta^{15}N$ value of nitrate produced by nitrification ranges from −12 to −29‰ [6].

Studies of the source of the newly produced oxygen in these reactions have attracted considerable interest [6, 37, 50]. Researchers have concluded that two of the oxygen atoms in nitrate newly produced by nitrification derive from ambient water, and one oxygen atom derives from O_2 in the atmosphere, as previously stated. Hence, the $\delta^{18}O$ value of the newly formed nitrate is decided by the interaction between the ambient water and atmospheric O_2, not the oxygen isotopic composition of the organic matter which has been converted to ammonium.

As there are three reaction steps in the nitrification process, identification of the step which is rate determining seems crucial. This also directly impacts on the isotopic fractionation caused by nitrification. If, prior to these nitrification reactions, the ammonia is produced from organic matter, as discussed before, it is not the dominant factor in isotopic fractionation and also because the step of nitrite conversion to nitrate is generally rapid in natural systems. Therefore, the rate determining step must be the oxidation of ammonium to nitrite. The degree of isotopic fractionation caused by nitrification will certainly depend on the amount of ammonium available. If there is little ammonium available, the isotopic fractionation may exhibit a very small range. However, when there is a high level of ammonium, nitrification occurs frequently: in this case, the oxidation of ammonium must be the rate determining step and will probably cause considerable isotopic fractionation. In studies of surface and groundwater, the nitrification process could not be determined from the $\delta^{15}N$ and $\delta^{18}O$ values recorded in situ alone: in order to precisely identify the nitrification process involved, it was also necessary to consider water chemistry. This will be discussed in more detail below.

14.5.6 *Denitrification*

Denitrification is another multistep process by which nitrate is converted to N_2 by chemical or biological reduction reactions, also producing various nitrogen oxides as intermediate compounds, such as N_2O and NO. This process may alter the nitrogen and oxygen isotopic ratios in residual nitrate. Two factors cause microbial denitrification: one is an anaerobic or oxygen-limited environment; the other is the presence of some organic carbon [51, 52]. The process of microbial denitrification more commonly converts the isotopes ^{14}N and ^{16}O, which are lighter in composition, to N_2 and N_2O. This naturally causes enrichment of the heavier nitrogen and oxygen isotopes in the residual nitrate [4, 15, 33]. Thus, during the denitrification process, as the concentration of nitrate decreases, the $\delta^{15}N$ and $\delta^{18}O$ values will be enriched in the residual nitrate. This trend can allow us to identify whether denitrification occurs in the surface and groundwater studies. Moreover, the enrichment factors in nitrogen and oxygen isotopic ratios associated with denitrification have been determined by many research groups. For example, the enrichment factor in a nitrogen isotopic ratio ranges from −40 to −5‰ [53, 54], while the enrichment factor in an oxygen isotopic ratio ranges from −18 to −8‰ [15, 55, 56]. The enrichment factors for $\delta^{15}N$ and $\delta^{18}O$ values associated with denitrification can be also assessed by using a simplified equation as below:

$$\delta_s = \delta_{s0} + \in \ln(f), \tag{14.5}$$

where δ_s denotes delta values in the substrate, δ_{s0} denotes initial delta values in the substrate, $\in$ is the enrichment factor, and f denotes the substrate concentration divided by the initial substrate concentration [6]. This equation has been extensively used to provide useful information about the denitrification process [15, 57]. Some studies also show that if there is a linear relationship representing an enrichment factor of $\delta^{15}N$ value relative to $\delta^{18}O$ value from 1.3 to 2.1, then denitrification activities must have occurred [7, 15, 56]. There are several methods for determining the occurrence and degree of denitrification process, such as different kinds of enzyme-blockage methods (e.g., the acetylene-blockage method), methods of ^{15}N tracer [58], and some methods using natural isotope ratios. All these topics will also be discussed in detail in later studies which seek to identify nitrate sources.

14.6 Identifying Nitrogen Sources by Using Dual-Isotope Methods and Water Chemistry

Recently, the technique of using $\delta^{15}N$ and $\delta^{18}O$ values of dissolved nitrate, and combining the recorded $\delta^{15}N$ and $\delta^{18}O$ values with water chemistry characteristics to precisely identify nitrate sources, has been applied in many parts of the world [2, 5, 33]. In fact, in initial research using the isotopic tool to identify nitrate sources, a single stable nitrogen isotope was used; the differences in $\delta^{15}N$ value in nitrate can serve as a direct signal of different nitrate sources, especially in sewage and manure sources with a distinctively high level of $\delta^{15}N$ value relative to other nitrate sources. Much research has also been conducted using $\delta^{15}N$ values alone to identify nitrate sources in freshwater systems [2, 59]. It is certainly feasible to use the $\delta^{15}N$ value alone in a relatively simple environment. However, such use is not recommended if the hydrologic system of a freshwater environment is very complex, for example, when there are many isotopic factors, such as ammonia volatilization, nitrification, and denitrification, occurring in a freshwater system and impacting on $\delta^{15}N$ values. The existence of many kinds of point and nonpoint sources mixing in the freshwater system will also hinder the identification of nitrate sources if the $\delta^{15}N$ value is used alone. Hence, the combined use of nitrate $\delta^{18}O$ value with $\delta^{15}N$ value can successfully provide more useful and accurate clues when tracing nitrate sources and plotting the nitrogen cycle. The images in Figs. 14.1 and 14.2 cited from Xue et al. [31] comprehensively show the normal range $\delta^{15}N$ and $\delta^{18}O$ values of all nitrate sources and clearly suggest that good complementarity exists between $\delta^{15}N$ and $\delta^{18}O$ values when both $\delta^{15}N$ and $\delta^{18}O$ values are employed to identify nitrate sources. For example, where there is an overlap of $\delta^{15}N$ values of synthetic fertilizer and natural soil organic nitrogen, the $\delta^{18}O$ values can provide useful data to distinguish between these two nitrate sources. There have been many dual-isotope studies of groundwater and surface water, which have been demonstrably successful in using $\delta^{15}N$ and $\delta^{18}O$ values to identify nitrate sources in freshwater [2, 5].

The types of land use around the surface and groundwater to be studied play an important role in the better identification of nitrate sources. For example, a study of the $\delta^{15}N$ value range of different types of land use in a groundwater system was conducted by Choi et al. [60], who concluded that $\delta^{15}N$ values ranged from +4.5 to +8.5‰ in agriculture land using mineral fertilizer, from +8.7 to +17.6‰ in farmland using animal manure, and was higher than +10‰ in high-density areas. These different ranges of $\delta^{15}N$ value for different types of land use almost correspond to the range of $\delta^{15}N$ value of different nitrate sources, providing an additional tool to identify nitrate sources. Similar research has also been conducted into surface water [2, 5]. Areas surrounded by urban activity and agricultural land generally feature high nitrate concentrations of high $\delta^{15}N$ value, relative to other areas. Moreover, there is generally a positive correlation between the nitrate $\delta^{15}N$ value and nitrate concentration in these areas, which is fully attributable to anthropogenic source input [61]. As a result, a combination of land use data with determined $\delta^{15}N$ value has been successfully used to aid the identification of nitrate sources in freshwater systems.

The concentration and determined $\delta^{15}N$ and $\delta^{18}O$ values of nitrate have typically been usually combined to provide useful clues to identify nitrate sources and trace the nitrification or denitrification process in surface and groundwater [2, 6, 7, 60]. Earlier researchers have concluded that when sewage and manure are the dominant nitrate sources in a freshwater system, the concentration and nitrogen isotopic composition of dissolved nitrate will be high, and a positive correlation between the concentration and nitrogen isotopic composition of dissolved nitrate must exist. This conclusion only holds true for single nitrate contamination sources. However, the relationship between nitrate concentration and $\delta^{15}N$ value is also very important, in practical studies, when determining whether nitrification or denitrification has occurred. For example, a series of water samples are collected from one part of the surface water system: both the nitrate concentration and $\delta^{15}N$ value are determined, and there is a marked negative correlation between nitrate concentration and $\delta^{15}N$ value, which suggests that the nitrate distribution in this area is probably due to denitrification. This is just one method of determining denitrification: if there is a linear relationship of from 1.3 to 2.1 between the enrichment factor of the $\delta^{15}N$ value and that of the $\delta^{18}O$ value, this can also confirm the occurrence of denitrification. Both $\delta^{15}N$ and $\delta^{18}O$ values of nitrate can also be used to track the mixing process from different nitrate sources, because when the $\delta^{15}N$ values of some nitrate sources are indistinguishable, the $\delta^{18}O$ value can be used to provide better source resolution, as the oxygen isotopic composition of nitrate generally changes in tandem with nitrogen isotopic composition during the denitrification process.

The concentrations of anions and cations in freshwater are also very important. These have been combined with nitrate $\delta^{15}N$ value to provide some useful information when identifying nitrate sources, since anions and cations in freshwater may also be an indicator of natural sources [7]. In studies, the concentrations of major cations (Mg^{2+}, Ca^{2+}, K^+, Na^+) and anions (Cl^-, SO_4^{2-}, NO_3^-, NO_2^-, PO_4^{3-}) have been determined. In a freshwater system, if Ca^{2+} and Na^+ are the dominant cations, Cl^- and SO_4^{2-} are the dominant anions. Freshwater lakes such as Taihu Lake, one of five largest surface lakes in China, according to stoichiometric analysis, can be principally characterized as SO_4·Cl-Na·Ca types. Lang et al. [62] concluded that Na^+ and SO_4^{2-}, as well as Cl^-, principally derived from industrial and agricultural distribution as chloride, are not subject to the physical, chemical, and microbiological processes occurring in surface water. Therefore, determining NO_3^-/Cl^- levels can provide more information to help distinguish the effect of the N removal process by dilution from denitrification [52]. If there is a good positive correlation between NO_3^- and Cl^-, the mixing process may have an important effect on nitrate transportation.

In studies of surface and groundwater, there are multiple potential sources of nitrate in different ecosystems, the isotopic compositions of these nitrate sources are rarely constant, and the initial isotopic compositions are altered by different kinds of fractionating processes before or after mixing. Therefore, estimates of quantification of relative contributions from different nitrate sources are crucial. Deutsch et al. [63] successfully employed the dual-isotope technique to quantify three nitrate sources, namely, atmospheric deposition, groundwater nitrate sources, and nitrate sources from artificially drained agriculture soils, and quantified these three nitrate sources as 3, 11, and 86%, respectively, of the total input. Voss et al. [61] also successfully employed the dual-isotope technique to quantify three nitrate sources, namely, sewage input, atmospheric deposition, and nitrate sources from pristine soils, contributing to 12 Baltic rivers. The basis of these two successful pieces of quantification research was the use of a mass-balance mixing model [64]. Generally, if using $\delta^{15}N$ and $\delta^{18}O$ values to quantify three nitrate sources, the basic mass-balance mixing model can be expressed as follows:

$$\delta^{15}N_M = f_1\delta^{15}N_1 + f_2\delta^{15}N_2 + f_3\delta^{15}N_3 \tag{14.6}$$

$$\delta^{18}O_M = f_1\delta^{18}O_1 + f_2\delta^{18}O_2 + f_3\delta^{18}O_3 \tag{14.7}$$

$$f_1 + f_2 + f_3 = 1, \tag{14.8}$$

where $\delta^{15}N_M$ and $\delta^{18}O_M$ denote the isotopic ratios of mixed nitrate sources; the 1, 2, and 3 in the subscripts denote any of the three nitrate sources; and f denotes the proportion of nitrate input from each of the three nitrate sources. Given the $\delta^{15}N$ and $\delta^{18}O$ values, the contributions of three nitrate sources can be easily quantified. There are some disadvantages to this mixing model [65], the most significant of which is that the uncertainty of some nitrate sources has not been considered. For example, nitrate $\delta^{15}N$ and $\delta^{18}O$ values vary temporally and spatially. Moreover, some isotopic fractionations may also alter the isotopic composition of nitrate. A further important limitation is that this model can only be used for three nitrate sources, yet there may be some other nitrate sources which cannot be precisely quantified in studies. Despite its limitations, however, this mixing model is important and can be a useful tool in relatively simple studies. What is more, it is believed that a new, improved method of quantifying the contribution of each nitrate source in a freshwater system will be developed in the near future.

14.7 Combining Other Isotopic Tools to Identify Nitrate Sources

There are many kinds of isotopic tools that can provide additional clues when identifying and possibly quantifying anthropogenic nitrate sources in freshwater systems. Some of these will be briefly presented below.

14.7.1 Isotopic Ratios of Plants

Many studies have discussed the use of the $\delta^{15}N$ value of algae and terrestrial plants, with an analysis of the nitrate $\delta^{15}N$ value and specific land use, as an important tool in the identification of nitrate sources [66–69]. For example, Kohzu et al. [68] successfully used the $\delta^{15}N$ value of riparian macrophytes to determine the anthropogenic nitrogen input to an aquatic ecosystem through analysis of land use and comparison with the nitrate $\delta^{15}N$ value.

14.7.2 Water Isotopic Ratios

Known $\delta^{18}O$ and $\delta^{2}H$ water values can also provide valuable clues to help identify nitrate sources in freshwater systems [2, 5, 70]. Generally, as evaporation occurs and freshwater is mixed from different geographic water sources, the $\delta^{18}O$ and $\delta^{2}H$ values of surface and groundwater are different. For example, Li et al. [2] successfully used $\delta^{18}O$ and $\delta^{2}H$ water values to identify the water sources flowing into the Yangtze River to facilitate identification of nitrate sources, while Lee et al. [5] used $\delta^{18}O$ values to determine whether microbial nitrification occurred in the Han River.

14.7.3 Boron (B) Isotopic Ratios

Boron (B) isotopic ratios are also important tools when identifying nitrate sources in freshwater systems [71]. B (B) has two stable isotopes, ^{10}B and ^{11}B, with natural abundance of 20 and 80%, respectively. B is widely employed in manufacturing, agriculture, and cosmetic industry and is also easily soluble and hardly removed in aquatic solutions. The most important characteristic of B is that $\delta^{11}B$ is not altered by any transformation process (e.g., adsorption with clay minerals, denitrification, and mineral precipitation) [71, 72]. Hence, $\delta^{11}B$ is used as a conservative tracer of waste-water source and has been employed in many studies to identify sewage contamination sources in surface and groundwater ([73]; Vengosh 1998; [9]). These researchers reported a range of $\delta^{11}B$ values in sewage from −7.7 to +12.9‰. In addition, $\delta^{11}B$ value can be as a tracer of animal manure and mineral nitrogen fertilizer, which also contain B, and the recorded $\delta^{11}B$ values (from +6.9 to +42.1‰) in different kinds of animal manure are markedly higher than in mineral nitrogen fertilizer (from +8 to +17‰). Therefore, given the distinctive $\delta^{11}B$ values in these different contamination sources, a comparison of $\delta^{11}B$ value will allow for easy discrimination between them [74].

14.7.4 Strontium (Sr) Isotopic Ratios

Strontium (Sr) isotopic ratios play a significant role in identifying nitrate sources from anthropogenic contaminants in groundwater. Sr (Sr) has two stable isotopes, ^{87}Sr and ^{86}Sr, which can be used as valuable tools for distinguishing phosphate fertilizers derived from phosphorite or carbonatite [75]. Like boron (B) isotopic ratios, strontium (Sr) isotopic ratios are not easily affected by any isotopic fractionation process. This can be attributed to the lack of significant difference in mass between ^{87}Sr and ^{86}Sr. This feature of strontium (Sr) isotopes renders them important when determining whether mixing processes have occurred and assessing water–rock interactions in the aquifer. The only disadvantage of using the Sr (Sr) isotopic ratio to trace such interactions is that flux from water–rock interactions is generally larger than flux from anthropogenic origins, so that the isotope signal may be neglected. Vitoria et al. [75] successfully used some other isotopic tracers (N, B, and Sr isotope) to provide information useful in identifying nitrate sources in contaminated groundwater. More details are shown in the literature.

14.7.5 Lithium (Li) Isotopic Ratios

There are also some similarities between lithium (Li) isotopes and other isotopic tracers (B, Sr). Lithium also has two stable isotopes, ^{6}Li and ^{7}Li. $\delta^{7}Li$ value in natural materials generally exhibits variances. Bullen and Kendall [76] concluded that the $\delta^{7}Li$ value in minerals derived from marine water is heavier than in minerals derived from igneous and metamorphic rocks. Like boron, lithium has been widely used in industrial, agricultural, and pharmaceutical fields [77]. Therefore, the δ^7Li value can act as a marker of waste-water sources from anthropogenic origins. In studies using δ^7Li values, some affecting factors were identified, such as discharges associated with lithium from lithium-processing plants and accumulated lithium in aquatic systems. Some specific studies employing δ^7Li value can be seen in Bullen and Senior [78] and Bullen and Kendall [76].

14.7.6 $\delta^{18}O$ Value of Phosphate

Much similar as nitrogen, phosphorus (P) is needed for DNA, RNA, and energy transfer, considered as a primary nutrient supporting the growth of plants and algae in aquatic ecosystem [79, 80]. The need to reduce anthropogenic P inputs to aquatic ecosystem in order to protect the water resources due to eutrophication problems has been widely recognized. Till now, the traditional methods which use infield interviews and mathematic statistical tools to identify the sources of phosphorus inputs are much more unmanageable because of its labor cost and time waste. Recently, the oxygen isotopic ratios of phosphate have been used as a tool to study the sources and cycling of phosphorus pollutions [81–84]. Even though very few works have been conducted, it still presents a promising prospect for further applications on lakes and other waters. Phosphate can be assimilated by phytoplankton and altered to organic phosphorus compounds [85], as nitrogen and other nutrients are usually sufficient for algae growth (such as cyanobacteria) [79]. To understand the sources and cycling of phosphate plays a key role on controlling most eutrophication problems. Compared with nitrogen, carbon, and other elements, P has only one stable isotope (while two radioactive isotopes), which cannot be used as a stable isotopic tracer. However, it has been already found that most of P is strongly bound to oxygen in the environment, which has three stable isotopes. The P-O bond in phosphate is resistant to inorganic hydrolysis at the temperature and pH of most natural systems. Only if recycled by organism, phosphate will exchange oxygen with ambient water, causing isotopic fractionation [85, 86]. Thus, the isotopic ratios of oxygen bound to P can be an indicator which provides useful information of phosphate sources and cycling.

Oxygen isotopic composition of phosphate was firstly applied on marine apatite. Earliest researches found the oxygen isotope in biogenic apatite had a correlation with the seawater temperature, indicating an isotopic equilibrium of the phosphate biological cycling [87]. Later, several laboratory studies characterized this fractionation by algae culture experiment and demonstrated that the bacterial metabolic processes significantly changed the phosphate oxygen isotopic values, resulting in a trend closing to the equilibrium, even if the phosphate concentration is high [86]. And then, a seasonal variation in the oxygen isotopic composition of DIP in the San Francisco Bay estuarine was reported by McLaughlin in 2002–2004 [82]. The newest published work used the phosphate oxygen isotope to trace phosphorus sources, and cycling in Lake Erie was studied by Elsbury et al. [81]; in this study, the oxygen isotopic values of dissolved phosphate were largely out of equilibrium with ambient conditions, ranging from +10 to +17‰, and this conclusion also claimed that there might be a likely source released from sediments under reducing conditions created during anoxic events in the bottom of the central basin of Lake Erie.

14.8 Applications for Dual-Isotope Studies

Some studies using $\delta^{15}N$ and $\delta^{18}O$ values to identify nitrate sources and track nitrogen transformation in surface and groundwater are discussed in detail below: these provide a scientific guide to practical researchers. Two research groups' studies are selected for further study. The first is "Using $\delta^{15}N$ and $\delta^{18}O$ values to identify nitrate sources in Karst Ground Water, Guiyang, Southwest China," by Liu et al. [7]. The second is "Tracking the sources of nitrate in the Han River watershed in Korea, using $\delta^{15}N$-NO_3^- and $\delta^{18}O$-NO_3^- values," by Lee et al. [5]. The first study focused on groundwater, while the second focused on surface water.

14.8.1 Application in a Groundwater Study

Nitrate pollution in karstic groundwater is becoming a very serious problem in Guiyang, the capital city of Guizhou Province in southwest China. Hence, identification of the nitrate source, which could directly aid the remediation of groundwater in Guiyang, is a priority [7]. Both groundwater and surface water samples in Guiyang were collected during summer and winter. Major cations (Mg^{2+}, Ca^{2+}, K^+, and Na^+) and anions (Cl^-, SO_4^{2-}, NO_3^-) were determined, and $\delta^{15}N$ and $\delta^{18}O$ values of nitrate were determined by the "ion-exchange method" discussed in chapter 1.2 [14]. Our discussion will focus on four typical analysis figures (Figs. 1.3–1.6) from this research. As we have seen, the NO_3^-/Cl^- method can provide much useful information when tracking nitrate transformation. Figure 14.3 shows that there was no correlation between $NO_3^-$and Cl^- in winter, but a good positive correlation ($r^2 = 0.66$) in summer, which suggests that the mixing process probably controls nitrate transportation in summer. Figure 14.4 shows that a trend of similar Cl^- concentrations and NO_3^-/Cl^- molar ratios is seen in surface water in summer, which is probably due to the rapid response of groundwater to rain and surface water. And municipal input was found to contribute significant levels of nitrates to surface water in winter. Figure 14.5 shows that there was no negative correlation between $\delta^{15}N$ value and NO_3^- in any of the samples, but a slight positive correlation ($r^2 = 0.46$), which suggests that denitrification may not have occurred in the groundwater and that mixing possibly occurred in summer. Figure 14.6 shows that the nitrate in suburban groundwater mostly originated from

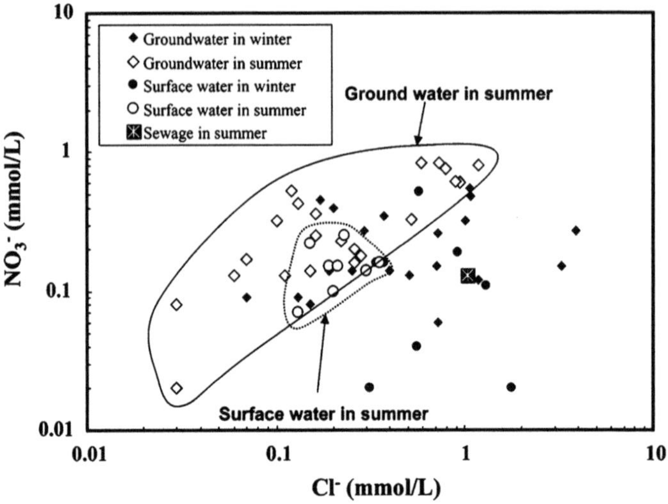

Fig. 14.3 Relationship between Cl^- and NO_3^- concentrations and groundwater. (Liu et al. [7])

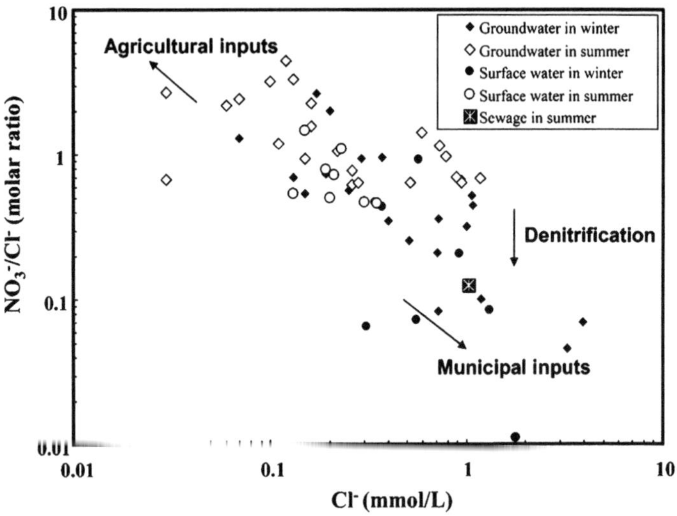

Fig. 14.4 Variations of NO_3^- /Cl^- molar ratios with Cl^- molar concentrations of groundwater, surface water, and sewage in both winter and summer seasons (Liu et al. [7])

some distinctive nitrate fertilizer, because of the low $\delta^{15}N$ and high $\delta^{18}O$ values. Meanwhile, denitrification may control nitrate transportation in some urban groundwater characterized by high $\delta^{15}N$ and $\delta^{18}O$ values. Similar studies of groundwater were conducted: this is merely one representative example.

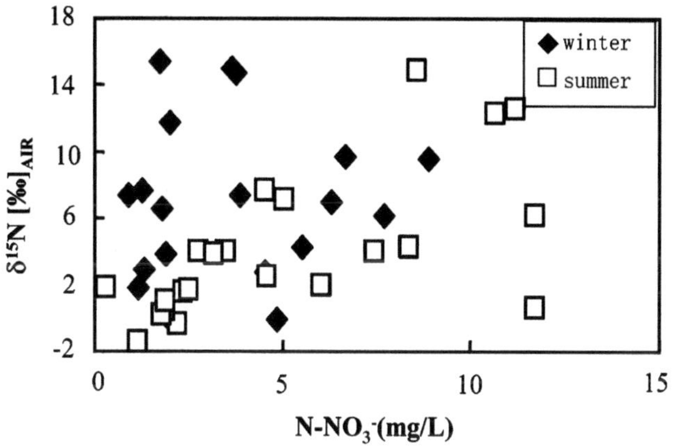

Fig. 14.5 Cross-plot of $\delta^{15}N$ nitrare versus N-NO_3^- concentrations in groundwater (Liu et al. [7])

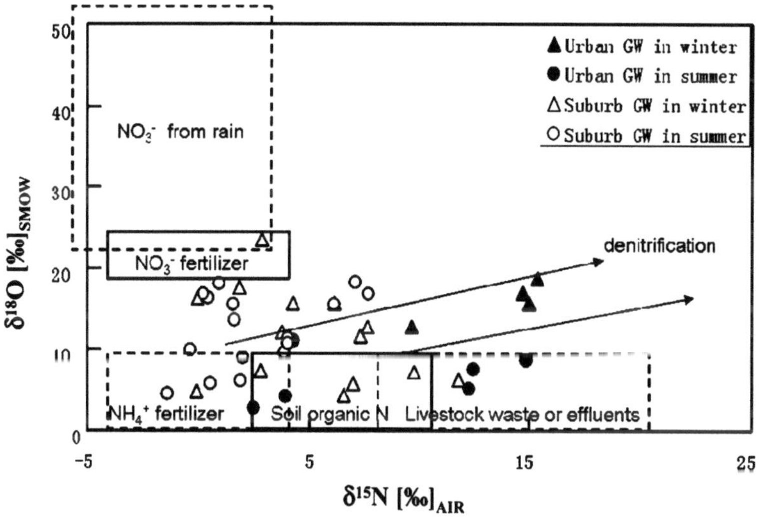

Fig. 14.6 Cross-plot of $\delta^{18}O$ nitrate versus $\delta^{15}N$ nitrate in groundwater (Liu et al. [7])

14.8.2 Application in a Surface Water Study

The dual-isotope technique has also been widely used in surface water systems. A study of the identification of nitrate sources in the Han River in Korea by Lee et al. [5] will provide an example of the application of this technique. In Lee et al.'s

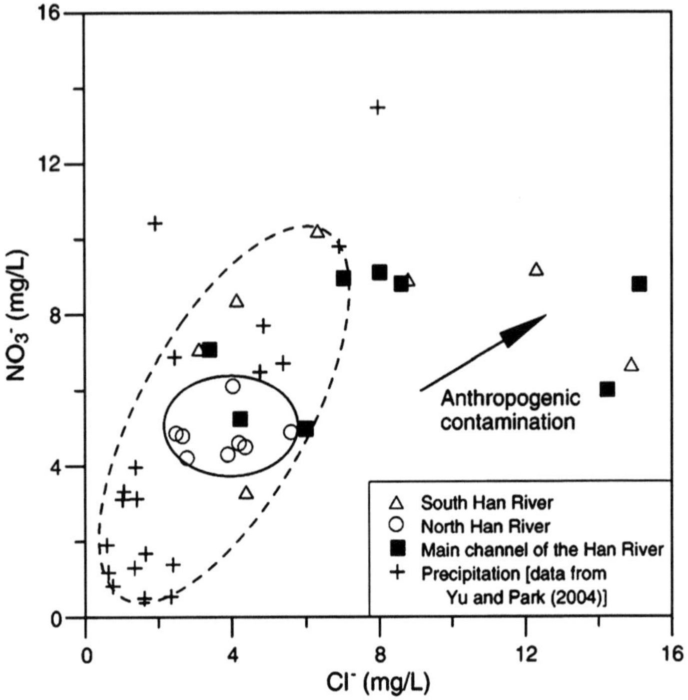

Fig. 14.7 Plot of NO_3^- versus Cl^- in the Han River (Lee et al. [5])

study, surface water samples were collected from the North Han River (NHR) and South Han River (SHR) in four different seasons. Major anions (Cl^-, SO_4^{2-}, NO_3^-, HCO_3^-) were determined. The $\delta^{15}N$ and $\delta^{18}O$ values of nitrate were determined by Silva et al.'s "ion-exchange method" (2002). Additional $\delta^{18}O$-H_2O values of the water samples were determined by IRMS. Our discussion will focus on four typical analysis figures (Figs. 14.7, 14.8, 14.9, and 14.10) from this research. Figure 14.7 shows that there were only slight seasonal and spatial variances in NHR, but widely scattered values recorded in SHR, which suggests that the sources of nitrate in these two tributaries are different. Figure 14.8 shows the nitrate mostly originated from sewage and manure, with limited contributions from soil organic matter or atmospheric deposition. Furthermore, the $\delta^{15}N$ value in summer was markedly lower than in other seasons, which suggests that dilution by rainwater in summer may have a major impact on nitrate sources. Figure 14.9 shows that neither atmospheric deposition nor synthetic fertilizer was the dominant nitrate source in the Han River: most of the nitrate was probably derived from microbial nitrification. Figure 14.10 illustrates a positive correlation between $\delta^{18}O$-H_2O and $\delta^{18}O$-NO_3^-, which

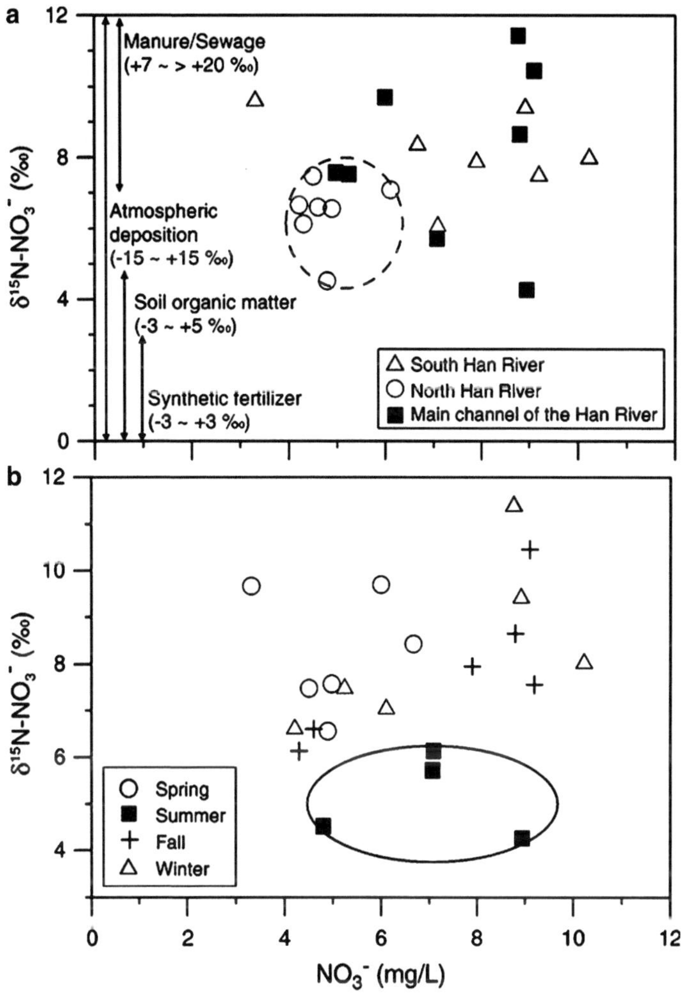

Fig. 14.8 Cross-plot versus showing the marked differences between two major tributaries of the Han River (**a**) and their seasonal patterns (**b**) (Lee et al. [5])

suggests that the $\delta^{18}O$- NO_3^- value is highly dependent on the $\delta^{18}O$-H_2O value in the river during nitrification: this conclusion further confirms the occurrence of microbial nitrification in the Han River. There was no obvious negative correlation between $\delta^{15}N$ value and NO_3^-; hence, denitrification may not be the dominant factor controlling nitrate transportation in the river. This study is also a successful example of the application of dual stable isotopes in the identification of nitrate sources in surface water.

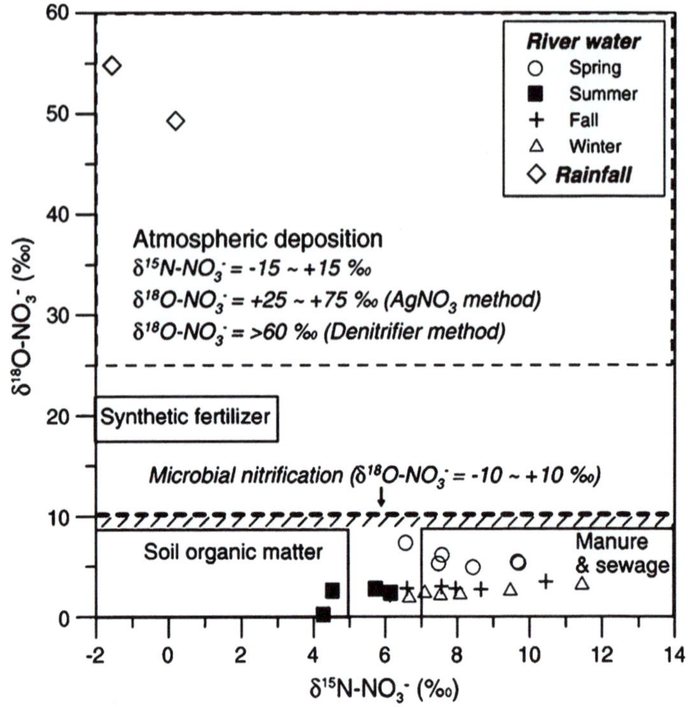

Fig. 14.9 Cross-plot of $\delta^{18}O\text{-}NO_3^-$ and $\delta^{15}N\text{-}NO_3^-$ in the Han River (Lee et al. [5])

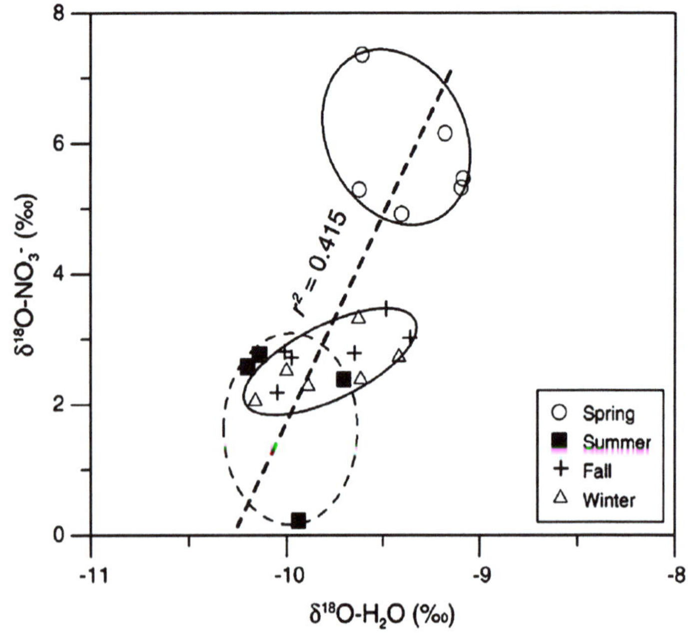

Fig. 14.10 Cross-plot of $\delta^{18}O\text{-}NO_3^-$ versus $\delta^{18}O\text{-}H_2O$ in the Han River (Lee et al. [5])

14.9 Summary

The technique of using nitrogen and oxygen isotopic compositions to trace nitrogen sources and map the nitrogen cycle has been extensively applied in freshwater research in the last few decades. Nitrate pollution in freshwater has caused concern worldwide: if nitrate sources could be identified, this major form of pollution could be combated. Thus, this technique provides much-needed scientific help with the remediation of freshwater. As the method of determining nitrogen and oxygen isotopic ratios has become more precise, many deficiencies in earlier methods of determination have been overcome. Moreover, each kind of nitrate source and some isotopic fractionation processes have also been well studied; all these factors play an important role in research into the use of isotopic tools in tracing nitrate sources. The combined use of dual isotopes and water chemistry is providing very detailed and useful information relating to the identification of nitrate sources and the tracking of nitrate transformation in freshwater systems. Recently, some other isotopic tools, which are not affected by biogeochemical processes, have been incorporated into dual-isotope research. These tools are better able to respond to the complexities of environments containing multiple potential nitrate sources coexisting in freshwater system, as well as the mixing of point and nonpoint sources in freshwater systems. Although the dual-isotope tool has been successfully used to identify nitrate sources in freshwater systems in some parts of the world, there are some problems yet to be solved, such as the precise quantification of different nitrate sources to the quantification of the extent of nitrate transformation. In conclusion, some current methods and tools need to be further improved, and there are still some issues to be resolved.

References

1. Heaton THE (1986) Isotopic studies of nitrogen pollution in the hydrosphere and atmosphere: a review. Chem Geol (Isot Geosci Sect) 59:87–102
2. Li SL, Liu CQ, Li J, Liu XL, Benjamin C, Wang B, Wang FS (2010) Assessment of the sources of nitrate in the Changjiang River, China using a nitrogen and oxygen isotopic approach. Environ Sci Technol 44(5):1573–1578
3. Wu CC et al (2009) Inter comparison of targeted observation guidance for tropical cyclones in the Northwestern Pacific. Mon Weather Rev 137:2471–2492
4. Kendall C, Elliott EM, Wankel SD (2007) Tracing anthropogenic inputs of nitrogen to ecosystems. In: Michener R, Lajtha K (eds) Stable isotopes in ecology and environmental science, 2nd edn. Blackwell, Oxford, UK. doi:10.1002/9780470691854.ch12
5. Lee KS, Bong YS, Lee DH, Kim YJ, Kim KJ (2008) Tracking the sources of nitrate in the Han River watershed in Korea, using δ^{15}N-NO_3^- and δ^{18}O-NO_3 values. Sci Total Environ 395 (2–3):117–124
6. Kendall C (1998) Tracing sources and cycling of nitrate in catchments. In: Kendall C, McDonnell JJ (eds) Isotope tracers in catchment hydrology. Elsevier, Amsterdam, pp 519–576

7. Liu CQ, Li SL, Lang YC, Xiao HY (2006) Using delta N-15 and delta O-18 values to identify nitrate sources in karst ground water, Guiyang, Southwest China. Environ Sci Technol 40 (22):6928–6933
8. Silva SR, Lee PB, Ebbert RW et al (2002) Forensic applications of nitrogen and oxygen isotopes of nitrate in an urban environment. Environ Forensic 3:125–130
9. Seiler RL (2005) Combined use of ^{15}N and ^{18}O of nitrate and ^{11}B to evaluate nitrate contamination in groundwater. Appl Geochem 20:1626–1636
10. Kendall C, Grim E (1990) Combustion tube method for measurement of nitrogen isotope ratios using calcium oxide for total removal of carbon dioxide and water. Anal Chem 62:526–529
11. Cook GA, Lauer CM (1968) Oxygen. In: Hampel CA (ed) The encyclopedia of the chemical elements. Reinhold Book Corporation, New York, pp 499–512, LCCN 68-29938
12. Anderson IC, Levine JS (1986) Relative rates of nitric oxide and nitrous oxide production by nitrifiers, denitrifiers, and nitrate respirers. Appl Environ Microbiol 51:938–945
13. Chang CCY, Langston J, Riggs M, Campbell DH, Silva SR, Kendall C (1999) A method for nitrate collection for ^{15}N and ^{18}O analysis from waters with low nitrate concentrations. Can J Fish Aquat Sci 56:1856–1864
14. Silva SR, Kendall C, Wilkison DH, Ziegler AC, Chang CCY, Avanzino RJ (2000) A new method for collection of nitrate from fresh water and the analysis of nitrogen and oxygen isotope ratios. J Hydrol 228:22–36
15. Fukada T, Hiscock KM, Dennis PF, Grischek T (2003) A dual isotope approach to identify denitrification in ground water at a river bank infiltration site. Water Res 37:3070–3078
16. Rock L, Ellert BH (2007) Nitrogen-15 and oxygen-18 natural abundance of potassium chloride extractable soil nitrate using the denitrifier method. Soil Sci Soc Am J 71:355–361
17. Sigman DM, Casciotti KL, Andreani M, Barford C, Galanter M, Böhlke JK (2001) A bacterial method for the nitrogen isotopic analysis of nitrate in seawater and freshwater. Anal Chem 73:4145–4153
18. Casciotti KL, Sigman DM, Galanter Hastings M, Bohlke JK, Hilkert A (2002) Measurement of the oxygen isotopic composition of nitrate in seawater and freshwater using the denitrifier method. Anal Chem 74:4905–4912
19. McIlvin MR, Altabet MA (2005) Chemical conversion of nitrate and nitrite to nitrous oxide for nitrogen and oxygen isotopic analysis in freshwater and seawater. Anal Chem 77:5589–5595
20. Margeson JH, Suggs JC, Midgett MR (1980) Reduction of nitrate to nitrite with cadmium. Anal Chem 52:1955–1957
21. Wood ED, Armstrong FAJ, Richards FA (1967) Determination of nitrate in seawater by cadmium–copper reduction to nitrite. J Mar Biol Assoc UK 47:23–31
22. Schilman B, Teplyakov N (2007) Detailed protocol for nitrate chemical reduction to nitrous oxide for delta^{15}N and delta^{18}O analysis of nitrate in fresh and marine waters. Annual report submitted to the Earth Science Research Administration, Ministry of National Infrastructures Jerusalem, December 2007 TR-GSI/15/2007
23. Kohl DH, Shearer GB, Commoner B (1971) Fertilizer nitrogen contribution to nitrate in surface water in a corn belt watershed. Science 174:1331–1334
24. Elliott EM, Kendall C, Burns DA et al (2006) Nitrate isotopes in precipitation to distinguish NO_x sources, atmospheric processes, and source areas in the United States. Eos (Trans. Am. Geophys. Union) 87(36)
25. Hübner H (1986) Isotope effects of nitrogen in the soil and biosphere. In: Fritz P, Fontes JC (eds) Handbook of environmental isotope geochemistry. Elsevier, Amsterdam, pp 361–425
26. Freyer HD (1991) Seasonal variation of 15N/14N ratios in atmospheric nitrate species. Tellus 43B:30–44
27. Amberger A, Schmidt HL (1987) Naturliche Isotopengehalte von Nitat als Indikatoren fur dessen Herkunft. Geochim Cosmochim Acta 51:2699–2705
28. Durka W, Schulze ED, Gebauer G, Voerkelius S (1994) Effects of forest decline on uptake and leaching of deposited nitrate determined from ^{15}N and ^{18}O measurements. Nature 372:765–767

29. Wahlen M, Yoshinari T (1985) Oxygen isotope ratios in N_2O from different environments. Nature 313:780–782
30. Flipse WJ, Bonner FT (1985) Nitrogen-isotope ratios of nitrate in ground water under fertilized fields, Long Island, New York. Ground Water 23:59–67
31. Xue DM, Botte J et al (2009) Present limitations and future prospects of stable isotope methods for nitrate source identification in surface- and groundwater. Water Res 43(5):1159–1170
32. Fogg GE, Rolston DE, Decker DL, Louie DT, Grismer ME (1998) Spatial variation in nitrogen isotope values beneath nitrate contamination sources. Ground Water 36:418–426
33. Mayer B, Boyer EW, Goodale C, Jaworski NA, Breemen NV, Howarth RW, Seitzinger S, Billen G, Lajtha K, Nadelhoffer K, Dam DV, Hetling LJ, Nosal M, Paustian K (2002) Sources of nitrate in rivers draining sixteen watersheds in the northeastern U.S.: isotopic constraints. Biogeochemistry 57(58):171–197
34. Pardo LH, Kendall C, Pett-Ridge J, Chang CCY (2004) Evaluating the source of stream water nitrate using ^{15}N and^{18}O in nitrate in two watersheds in New Hampshire, USA. Hydrol Process 18:2699–2712
35. Kendall C, Silva SR, Chang CCY, Burns DA, Campbell DH, Shanley JB (1996) Use of the Delta 18-O and Delta 15-N of nitrate to determine sources of nitrate in early spring runoff in forested catchments. In: Isotopes in water resources management. International atomic energy agency symposium vol. 1, Vienna (pp 167–176)
36. Wassenaar LI (1995) Evaluation of the origin and fate of nitrate in the Abbotsford Aquifer using the isotopes of ^{15}N and ^{18}O in NO_3^-. Appl Geochem 10:391–405
37. Hollocher TC (1984) Source of the oxygen atoms of nitrate in the oxidation of nitrite by Nitrobacter agilis and evidence against a P–O–N anhydride mechanism in oxidative phosphorylation. Arch Biochem Biophys 233:721–727
38. Kellman LM (2005) A study of tile drain nitrate-delta N-15 values as a tool for assessing nitrate sources in an agricultural region. Nutr Cycle Agroecosyst 71:131–137
39. Griffiths H (ed) (1998) Stable isotopes: integration of biological, ecological and geochemical processes. Bios Scientific, Herndon, 438 p
40. Handley LL, Scrimgeour CM (1997) Terrestrial plant ecology and ^{15}N natural abundance: the present limits to interpretation for uncultivated systems with original data from a Scottish old field. Adv Ecol Res 27:133–212
41. Högberg P (1997) ^{15}N natural abundance in soil-plant systems. New Phytol 137:179–203
42. Feigin A, Shearer G, Kohl DH, Commoner B (1974) The amount and nitrogen-15 content of nitrate in soil profiles from two central Illinois fields in a corn-soybean rotation. Soil Sci Soc Am Proc 38:465–471
43. Létolle R (1980) Nitrogen-15 in the natural environment. In: Fritz P, Fontes JC (eds) Handbook of environmental isotope geochemistry, vol 1. Elsevier, New York, pp 407–433
44. Glibert PM, Capone DG (1993) Mineralization and assimilation in aquatic, sediment, and wetland systems. In: Knowles R, Blackburn TH (eds) Nitrogen isotope techniques. Academic Press, San Diego, California, pp 243–272
45. Powlson DS, Barraclough D (1993) Mineralization and assimilation in soil-plant systems. In: Knowles R, Blackburn TH (eds) Nitrogen isotope techniques. Academic Press, San Diego, California, pp 209–242
46. Mariotti A, Germon JC, Hubert P, Kaiser P, Letolle R, Tardieux A, Tardieux P (1981) Experimental determination of nitrogen kinetic isotope fractionation: some principles: illustration for the denitrification and nitrification processes. Plant Soil 62:423–430
47. Fogel ML, Cifuentes LA (1993) Isotope fractionation during primary production. In: Engel MH, Macko SA (eds) Organic geochemistry. Plenum Press, New York, pp 73–98
48. Macko SA, Ostrom NE (1994) Molecular and pollution studies using stable isotope. In: Lajtha K, Michner R (eds) Stable isotopes in ecology and environmental science. Blackwell Scientific, Oxford, UK, pp 45–62

49. Kool DM, Wrage N, Oenema O, Dolfing J, Van Groenigen JW (2007) Oxygen exchange between (de)nitrification intermediates and H_2O and its implication for source determination of NO_3- and N_2O: a review. Rapid Commun Mass Spectrom 21:3569–3578
50. Andersson KK, Hooper AB (1983) O_2 and H_2O and each the source of one O in NO_2 produced from NH3 by Nitrosomonas: ^{15}N evidence. FEBS Lett 164:236–240
51. Knowles R (1982) Denitrification. Microbiol Rev 46:43–70
52. Koba K, Tokuchi N, Wada E et al (1997) Intermittent denitrification: the application of a ^{15}N natural abundance method to a forested ecosystem. Geochim Cosmochim Acta 61:5043–5050
53. Sebilo M, Billen G, Mayer B et al (2006) Assessing nitrification and denitrification in the Seine River and Estuary using chemical and isotopic techniques. Ecosystems 9:564–577
54. Smith RL, Howes BL, Duff JH (1991) Denitrification in nitrate-contaminated groundwater: occurrence in steep vertical geochemical gradients. Geochim Cosmochim Acta 55:1815–1825
55. Lehmann MF, Reichert P, Bernasconi SM, Barbieri A, McKenzie A (2003) Modelling nitrogen and oxygen isotope fractionation during denitrification in a lacustrine redox-transition zone. Geochim Cosmochim Acta 67:2529–2542
56. Mengis M, Schiff SL, Harris M, English MC, Aravena R, Elgood RJ, Maclean A (1999) Multiple geochemical and isotopic approaches for assessing ground water NO_3^- elimination in a riparian zone. Ground Water 37:448–457
57. Chen FJ, Jia GD, Chen JY (2009) Nitrate sources and watershed denitrification inferred from nitrate dual isotopes in the Beijing River, south China. Biogeochemistry 94(2):163–174
58. Mosier AR, Schimel DS (1993) Nitrification and denitrification. In: Knowles R, Blackburn TH (eds) Nitrogen isotope techniques. Academic Press, San Diego, California, pp 181–208
59. Town-Small A, McCarthy MJ, Brandes JA, Yang LY, Zhang L, Gardner WS (2007) Stable isotopic composition of nitrate in Lake Taihu, China, and major inflow rivers. Hydrobiologia 194(3):135–140
60. Choi WJ, Han GH, Lee SM, Lee GT, Yoon KS, Choi SM, Ro HM (2007) Impact of land-use types on nitrate concentration and $\delta^{15}N$ in unconfined ground water in rural areas of Korea. Agric Ecosyst Environ 120:259–268
61. Voss M, Deutsch B, Elmgren R, Humborg C, Kuuppo P, Pastuszak M, Rolff C, Schulte U (2006) Sources identification of nitrate by means of isotopic tracers in the Baltic Sea catchments. Biogeosciences 3:663–676
62. Lang YC, Liu CQ, Zhao ZQ, Li SL, Han GL (2006) Geochemistry of surface and ground water in Guiyang city, China: water/rock interaction and pollution in karst hydrological system. Appl Geochem 21:887–903
63. Deutsch B, Kahle P, Voss M (2006) Assessing the source of nitrate pollution in water using stable N and O isotopes. Agron Sustain Dev 26:263–267
64. Phillips DL, Koch PL (2002) Incorporating concentration dependence in stable isotope mixing models. Oecologia 130:114–125
65. Moore JW, Semmens BX (2008) Incorporating uncertainty and prior information into stable isotope mixing models. Ecol Lett 11:470–480
66. Elliott EM, Brush GS (2006) Sedimented organic nitrogen isotopes in fresh water wetlands record long-term changes in watershed nitrogen source and land use. Environ Sci Technol 26 (40):2910–2916
67. Harrington RR, Kennedy BP, Chamberlain CP et al (1998) ^{15} N enrichment in agricultural catchments: field patterns and applications to tracking Atlantic salmon (Salmo salar). Chem Geol 147:281–294
68. Kohzu A et al (2008) Use of stable nitrogen isotope signatures of riparian macrophytes as an indicator of anthropogenic N inputs to river ecosystems. Environ Sci Technol 42 (21):783–7841
69. Saurer M, Cherubini P, Ammann M et al (2004) First detection of nitrogen from NO_x in tree rings: a $^{15}N/^{14}N$ study near a motorway. Atmos Environ 38:2779–2787
70. Aravena R, Robertson WD (1998) Use of multiple isotope tracers to evaluate denitrification in ground water: study of nitrate from a large-flux septic system plume. Ground Water 36:975–982

71. Bassett RL (1990) A critical evaluation of the available measurements for the stable isotopes of boron. Appl Geochem 5:541–554
72. Vengosh A, Heumann KG, Juraske S, Kasher R (1994) Boron isotope application for tracing sources of contamination in groundwater. Environ Sci Technol 28:1968–1974
73. Eisenhut S, Heumann K, Vengosh A (1996) Determination of boron isotopic variations in aquatic systems with negative thermal ionization mass spectrometry as a tracer for anthropogenic influences. Fresenius J Anal Chem 345:903–909
74. Komor SC (1997) Boron contents and isotopic compositions of hog manure, selected fertilizers, and water in Minnesota. J Environ Qual 26:1212–1222
75. Vitoria L, Otero N, Soler A, Canals A (2004) Fertilizer characterization: isotopic data (N, S, O, C, and Sr). Environ Sci Technol 38(12):3254–3262
76. Bullen TD, Kendall C (1998) Tracing of weathering reactions and water flowpaths: a multi-isotope approach. In: Kendall C, McDonnell JJ (eds) Isotope tracers in catchment hydrology. Elsevier, Amsterdam, pp 611–646
77. Qi HP, Coplen TB, Wang QZ, Wang YH (1997) Unnatural isotopic composition of lithium reagents. Anal Chem 69:4076–4078
78. Bullen TD, Senior LA (1992) Lithogenic Sr and anthropogenic Li in an urbanized stream basin: isotopic tracers of surface water-groundwater interaction. Eos (Trans Am Geophys Union) 73:F130
79. Conley DJ, Paerl HW, Howarth RW, Boesch DF, Seitzinger SP, Havens KE, Lancelot C, Likens GE (2009) Controlling eutrophication: nitrogen and phosphorus. Science 323:1014–1015
80. Smith VH, Tilman GD, Nekola JC (1999) Eutrophication: impacts of excess nutrient inputs on freshwater, marine, and terrestrial ecosystems. Environ Pollut 100:179–196
81. Elsbury K, Paytan A, Kendall C, Young MB, McLaughlin K, Rollog M, Watson S (2009) Using oxygen isotopes of phosphate to trace phosphorus sources and cycling in Lake Erie. Environ Sci Technol 43:3108–3114
82. McLaughlin K, Kendall C, Silva S, Young M, Paytan A (2006) Phosphate oxygen isotope ratios as a tracer for sources and cycling of phosphate in North San Francisco Bay, California. J Geophys Res 111:G03003
83. McLaughlin K, Cade-Menun BJ, Paytan A (2006) The oxygen isotopic composition of phosphate in Elkhorn Slough, California: a tracer for phosphate sources. Estuar Coast Shelf Sci 70:499–506
84. Young MB, Mclaughlin K, Kendall C et al (2009) Characterizing the oxygen isotopic composition of phosphate sources to aquatic ecosystems. Environ Sci Technol 43:5190–5196
85. Paytan A, McLaughlin K (2007) The oceanic phosphorus cycle. Chem Rev 107:563–576
86. Blake RE, O'Neil JR, Surkov AV (2005) Biogeochemical cycling of phosphorus: Insights from oxygen isotope effects of phosphoenzymes. Am J Sci 305:596–620
87. Blake RE, O'Neil JR, Garcia GA (1997) Oxygen isotope systematics of biologically mediated reactions of phosphate: I. Microbial degradation of organ phosphorus compounds. Geochim Cosmochim Acta 61:4411–4422

Chapter 15
A Stepwise Approach to Assess the Fate of Nitrogen Species in Agricultural Lowlands

Micòl Mastrocicco, Nicolò Colombani, and Giuseppe Castaldelli

15.1 Introduction

During the twentieth century, human activities have greatly accelerated the rate at which nitrogen reaches ground and surface waters [1, 2]. Leaching and surface run-off from heavily exploited agricultural areas [3] and point pollution from sewage plants treating large urban loads [4] increase the flux of nutrients into water bodies, leading to nitrate accumulation in shallow groundwater aquifers [5, 6] and fuelling eutrophication processes of surface and coastal waters [7, 8]. On a broad scenario of increasing drinking water scarcity, groundwater nitrate contamination [9–11] is increasing health hazards such as methaemoglobinaemia or "blue child syndrome" [12], congenital malformations [13] and different forms of cancer [14].

At present, two main synergic causes are responsible for nitrogen pollution: (a) the large surplus of this element, particularly in highly productive agricultural areas [15, 16], and (b) the simplification of the landscape, which makes nitrogenous species, in particular nitrate, even more mobile among compartments. Intensive agricultural systems and improper agricultural practices are usually regarded as the major source of nitrate pollution [17–19]. Excessive application of nitrogen fertilizers has been demonstrated to have the potential to pollute not only soil but also groundwater [5, 6, 20, 21].

In order to counteract the degradation of water quality, efforts to reduce loading from watersheds are ongoing (Water Framework Directive, 2000/60/CE; Clear Waters Act, Section 319, USEPA) through the definition of best management

M. Mastrocicco (✉) • N. Colombani
Earth Sciences Department, University of Ferrara, Via G. Saragat 1, 44100 Ferrara, Italy
e-mail: mtm@unife.it; clo@unife.it

G. Castaldelli
Biology and Evolution Department, University of Ferrara, Via L. Borsari 46, 44100 Ferrara, Italy
e-mail: ctg@unife.it

S.K. Sharma and R. Sanghi (eds.), *Wastewater Reuse and Management*, DOI 10.1007/978-94-007-4942-9_15, © Springer Science+Business Media Dordrecht 2013 practices (BMP) in agriculture. Regulations have not become fully effective yet, partly due to the high intrinsic complexity of a soil matrix and furthermore for site specificity, which is determined by several terms as climate, cropping, fertilization type and timing, etc. [22]. A basic example is given by the difficulty to discriminate between the relative contribution of leaching and the surface run-off, on the loss of nutrients from agricultural land [3]. An increasing complexity is then found in aquifers, where nitrate concentrations are often found spatially and temporally variable [23–25] because of groundwater flow fluctuation and variation in denitrification rate [26, 27]. Nitrate attenuation from surface to groundwater systems may occur via bacterial heterotrophic denitrification, using NO_3^- as electron acceptor and a carbon (C) source as electron donor producing nitrogen gases [28, 29]. This process has been extensively studied in ecosystems [30] but has not been often referred to the complexity of agricultural practices [31]. For example, the quantity and quality of organic matter (its labile/refractory nature) its an underestimated term which may play a central role in regulating denitrification, thus buffering the risk of percolation to groundwater [32, 33].

The most widely used method to describe the above-mentioned processes, and determine water and agrochemicals losses from agricultural land, is the use of process-based mathematical models, which describe water movement, transport and transformations of dissolved species through the soil profile [34–37]. This approach extends to a wide scale of complexity, from the more simple empirical models (like vulnerability or risk indices), which use fewer and more accessible data (climatic data, topography and general soil physical properties) [38], to process-based simulation models.

Empirical models can be employed easily in large-scale application using geographical information systems and support efficiently the agricultural management using decision support systems [39–41]. The most popular vulnerability indices to describe the intrinsic vulnerability of the groundwater to contamination are DRASTIC index [42], SINTACS index [43], RISKE index [44], COP [45] and MERLIN [46].

Process-based simulation models to predict nitrate leaching at the watershed scale are more valid but more difficult to handle due to their complexity and data requirements. Carey and Loyd [47] used a numerical distributed transport model with a primitive GIS, in order to simulate nitrate concentrations in groundwater. Several process-based models combined with GIS, such as NLEAP model [48, 49], NITS-SHETRAN [50], AgriFlux-IDRISI GIS [51], DAISY MIKE SHE [52], GLEAMS [53], GIS NIT-1 [54] and MT3D-MODFLOW [27], have been used in order to predict the spatial and temporal distribution of nitrate leaching and to assess nitrate contamination in groundwater.

No matter if the first cited empirical approach for vulnerability assessment is more simple than process-based simulation models, data input are needed in both cases, and, for a correct application, these data must be collected following a hierarchical approach.

Hereafter, we present a stepwise approach, to quantify the fate of nitrogen species in agricultural lands, according to a scheme that can be increasingly implemented.

As a case study, we selected different alluvial soil types in the lowest Po Basin. This river plain is an intensively cultivated land, which has been fertilized with synthetic nitrogen, mainly urea and ammonium nitrate for decades. In Italy, the Po River valley is the largest and more intensively farmed alluvial plain and is heavily impacted by NO_3^- groundwater contamination [55–59] and surface water eutrophication [60, 61].

The procedure here shown is a hierarchical scheme that can be implemented in agricultural lands to quantify the fate of nitrogen species. As stepwise approach, we intend at first to perform a well-designed continuous monitoring of the field physical parameters, dedicated to properly estimate cumulative infiltration; besides, tracer tests have to be used to quantify the field velocity and the dispersion coefficient to determine the timescale of water mass transfer through the subsurface. Subsequently, tests on other relevant soil parameters should be performed at different scales in a laboratory controlled environment (e.g. batch test leaking, 1D columns elution and 3D tank experiments) in order to gain the necessary flow and transport parameters to select or discard which processes are occurring in the studied environments and thus to refine the conceptual model. Finally, a hierarchical numerical assessment via forward and inverse modelling, based on the achieved parameters, can be performed to quantify indirectly the rates at which nitrate is removed from the soil in a gaseous form, via biogeochemical reactions.

15.2 Water Balance

To assess nitrogen fate in agricultural lowlands, the water balance must be resolved. It is essential to quantify the groundwater recharge and consequently the input of nitrogen to the aquifer. To achieve this goal, all water inputs and outputs of the investigated system must be assessed. Usually, the balance equation is written as:

$$I - P + Ir - Et - R - \Delta S$$

where I is infiltration, P is precipitation, Ir irrigation water, Et evapotranspiration, R run-off and ΔS is the soil storage between two successive infiltration events.

Obviously, there are many different levels of characterization of each balance components: for instance, temporal and spatial discretization depends on the target of the research. Thus, rainfall events can be recorded on a monthly basis or every half an hour depending if the aim of the research is to quantify annual recharge trends or if a wet front propagation is being studied. In addition, different weather stations can be placed within the study domain to quantify the spatial variability of precipitation, solar radiation, humidity, etc. Another important variable of the water balance in agricultural lands is the evapotranspiration, although is more challenging to measure respect to precipitation. Several techniques have been developed to measure evapotranspiration: by weighting lysimeters [62], by field water balance equation [63] or micrometeorological methods [64]. All these techniques are threatened to be time consuming and expensive. Evapotranspiration can also be estimated from climatic data, linking evapotranspiration with one or more climatic variables [65]. The performance of various equations has been assessed under a variety of climate conditions [66, 67]. The Penman–Monteith equation is now considered the best approximation to estimate evapotranspiration over most of climates [68]. This equation has now become the United Nations Food and Agriculture Organization (FAO) standard equation to estimate evapotranspiration [66]. In addition to evapotranspiration, studies on water stress on plants and roots have been performed at various scales, from microscopic to field sites and ecosystems [69, 70]. Then, to correctly plan a site characterization for nitrogen leaching, the objective of the research must be clear at the beginning, i.e. to avoid data redundancy and to optimize the costs of the research. In particular, the water balance can be approached in a qualitative or a quantitative manner. The first approach is suitable for large-scale studies (e.g. watershed studies) where field data collection is usually not economically feasible, while the second one is usually employed at the plot scale by means of numerical modelling. The qualitative approach can provide useful data on the major component of the water balance but is not able to distinguish between sub-parameters and is usually descriptive of the system. The quantitative approach, via forward and inverse modelling, can provide insights on the water balance components that need further characterization. In this way, the numerical modelling can lead to more robust and consistent results although is more time consuming and needs a multidisciplinary approach. To deal with this complex task, usually more than one expert is needed, like an agronomist to study the transpiration, a hydrogeologist or a soil physicist to characterize the infiltration and run-off and an engineer to manage the numerical code. Although it must be keep in mind that there is not a clear distinction between the two approaches described above, but more realistically, there are many approaches with increasing level of complexity that can be enriched or simplified depending on the goal of the research.

15.3 Nitrogen Mass Balance

After the water budget is accomplished, a mass balance is needed to unravel the fate of nitrogen species. This is a very demanding task in terms of analytical costs, and it is time consuming. For these reasons, it can be accomplished at various levels of detail. Usually, to characterize the nitrogen inputs and outputs from agricultural lands, the following terms need to be quantified:

$$N_{\mathrm{sf}} + N_m + N_r + N_{\mathrm{dd}} = Ncrop_{\mathrm{up}} + D + Amm_v + N_l - \Delta S_N$$

where N_{sf} represents nitrogen inputs via synthetic fertilizers, N_m the manure, N_r the inputs from rain, N_{dd} dry deposition and $Ncrop_{\mathrm{up}}$ the nitrogen output from crop uptake, D denitrification, Amm_v ammonia volatilization, N_l stands for nitrogen leaking and finally ΔS_N represents the variation of the soil organic and mineral nitrogen between two successive sampling dates.

To compare topsoil N concentrations at the beginning and end of the study, soil samples have to be taken from the measurement area in each of the experimental fields. The sample size should be based on estimates of N variability in soil obtained from previous studies, or, if this information is not available, a regular sampling grid needs to be assessed. Soil samples should be taken in each field at the depth interested by cultivation. Each sample must be analysed separately for total N [71].

Wet N deposition can be measured at the field site by a rainfall collector. Rainfall is usually collected through a filter funnel into 3-l polythene bottles. The latter are usually held inside double-skinned metal cylinders designed to insulate the bottles and minimize evaporation. Samples are then analysed for ammonium and nitrate using standard methods [71].

To directly estimate N_2 emissions from denitrification activity at times of large emission, the acetylene (C_2H_2) inhibition technique can be used [72].

Measurements of ammonia volatilization can be carried out using the modified aerodynamic gradient method [73]. Wind speed and ammonia concentration profiles are made linear with respect to the logarithm of height using the relationship between zero plane displacement and crop height [74].

For the ammonia flux measurements, anemometers and duplicate passive samplers should be placed at 3 or 4 heights on a mast located in the centre of the field. These heights should be varied during the growing season to ensure that the bottom anemometer and samplers lie above the height of the growing crop.

After the harvest, the crop grain must be weighed. Grain subsamples have to be taken for analysis of dry matter (DM) and N concentrations, which is usually determined by near infrared reflectance [71]. Above-ground crop samples, cut at ground level, should be taken from each area just before the harvest in order to determine DM and N harvest indices. Plant uptake of N is then calculated by dividing the combine-derived values for grain N by the N harvest index.

15.4 Monitoring Grid at the Plot Scale

In the following paragraph, an example of monitoring grid for the water and nitrogen mass balance estimation is shown in details. First it is explained how the grid for the water balance has been implemented and subsequently how the grid for the nitrogen mass balance has been implemented.

15.4.1 An Example of Implementation of the Monitoring Grid for the Water and Nitrogen Mass Balance

To correctly quantify water budget components in lowland agricultural areas as above mentioned, it is necessary to monitor precipitations, soil water and groundwater. Meteorological stations recording rainfall, wind speed, solar radiation, temperature and humidity must be present and located near the field sites. Usually meteorological data are available online from meteorological regional services and from local web services. To validate the meteorological data, a rain gauge can be installed in each site to record daily rainfall. Then, the linear correlation of daily data between the rain gauges and the nearest weather stations must be plotted to test the extendibility of the data recorded by the online weather stations to the experimental field sites. The irrigation amount and time must also be estimated via flow metres attached to the irrigation system. The soil water monitoring is usually accomplished by using arrays of soil moisture probes vertically inserted into augered holes at different depths in each field site. A thermocouple is usually inserted adjacent to each soil moisture probe to compensate the readings for the actual soil temperature. Standard tensiometers (measurement range 0–8 m) can be installed at the same depths of soil moisture probes to monitor and correct any deviance of soil moisture probe readings from the real values.

To monitor the soil water content, total domain reflectometry (TDR) probes are usually employed as an indirect and relatively cheap method. Soil-water content measurements, via the gravimetric method [71], can be performed in soil core samples, collected at different time intervals and depths from randomly selected locations within the plot area. This direct method, based on weight loss after water removal at 105 °C for 24 h, is more time consuming than CTR, but it allows measures of soil water content and comparisons with soil water potential to define retention curves at various depths.

In the following figure is presented an example of nitrate profiles in two different sites: a loamy soil, fertilized with synthetic urea, and a sandy soil, previously amended with chicken manure; more details can be found in Mastrocicco et al. [75]. To define the site stratigraphy of each field site, triplicate core logs were drilled manually with an Eijkelkamp Agrisearch auger equipment down to 2 m below ground level (b.g.l.). The soil stratification was subdivided in distinct layers: usually the upper one interested by tillage, roots and weathering and a lower undisturbed layer. In the loamy soil site, the upper layer was 0.75 m thick and in the sandy soil site was 0.65 m, while the lower layer extended until 2 m b.g.l. in all sites. From collected core samples at 0.25, 0.50, 0.75, 1 and 2 m b.g.l., particle size curves were obtained using a settling tube for the sandy fraction and an X-ray Micromeritics Sedigraph 5100 for the finer one. Organic matter content was measured by loss of ignition method [71], while bulk density was determined gravimetrically.

A series of nested piezometers (2.5 cm inner diameter) screened from 1.5 to 4 m a.s.l., was installed near the soil moisture arrays to monitor the level and quality of groundwater. Monitoring started on 27 March 2008, and it is still going on. LTC M10 Levelogger Solinst data loggers were placed in piezometers to monitor hourly groundwater level, electrical conductivity and temperature. All the piezometers were sampled at variable intervals, via low-flow purging, for major ions and TOC determination. Two arrays of soil solution suction samplers were installed at 0.25, 0.50, 0.75 and 1 m b.g.l. in each field site to analyse soil water in the unsaturated zone. In addition to suction sampler, the unsaturated zone was sampled every 4 months by means of auger coring (from 0 to 2 m b.g.l.), and sediments were analysed for major anions and cations.

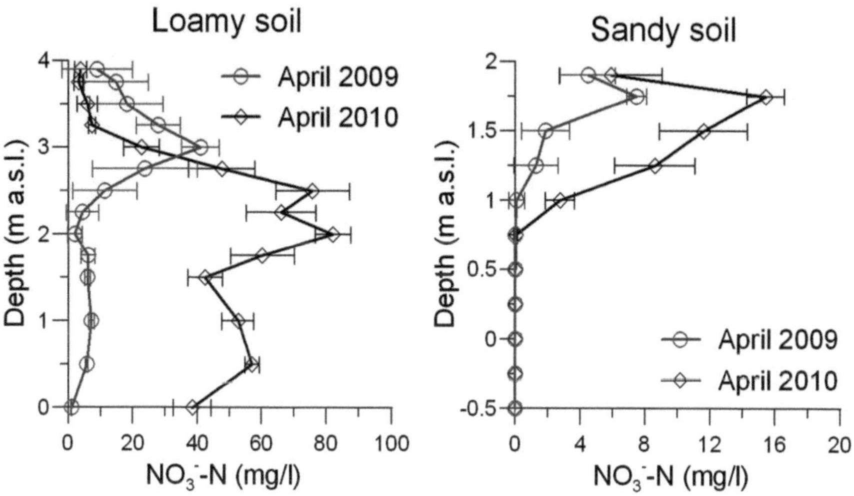

Fig. 15.1 NO_3^--N depth profiles in a loamy soil (*left*) and in a sandy soil amended with chicken manure (*right*) collected in April 2009 and April 2010. Concentrations measured in solids were corrected for bulk density and water content. *Error bars* show the standard deviation of three replicates

Unsaturated zone sediment analysis consisted of a batch with a sediment/water ratio of 1:10, using 10 g of air-dried sample dispersed in 100 ml of Milli-Q water (Millipore, USA). A biological inhibitor (1 g/l phenylmercuric acetate) was added to prevent microbial activity, and the solution was stirred for 1 h and then allowed to stand for 1 day. The insoluble residue was removed by filtration and analysed for major cations and anions. In-well parameters were determined with the HANNA Multi 340i instrument which includes a HIcell-31 pH combined electrode with a built-in temperature sensor for pH measurements, a CellOx 325 galvanic oxygen sensor for DO measurements, a combined AgCl-Pt electrode for Eh measurement, and a HIcell-21 electrode conductivity cell for EC measurements.

The major cations, anions and oxyanions (acetate and formate) were determined with isocratic dual-pump ion chromatography ICS-1000 Dionex, equipped with an AS9 HC 4 × 250-mm high-capacity column and an ASRS-ULTRA 4-mm self-suppressor for anions, and a CS12A 4- × 250-mm high-capacity column and a CSRS-ULTRA 4-mm self-suppressor for cations. An AS40 Dionex auto-sampler was employed to run the analyses; quality control (QC) samples were run every 10 samples. The standard deviation for all QC samples run was better than 4%. Charge balance errors in all analyses were less than 5% and predominantly less than 3%. Samples were filtered through 0.22-μm Dionex vial caps. Total organic carbon (TOC) and total inorganic carbon (TIC) were determined with a carbon analyzer (carbon analyzer Shimadzu TOC-V-CSM) after acidification with one drop of 2 M HCl to remove dissolved carbonate.

This monitoring strategy would allow gaining chemical profiles along the saturated and unsaturated zone. Repeated profiles in time would give information on the movement of nitrogen species and their impact on the aquifer. Since the most intense recharge in Northern Italy occurs during the spring time, Fig. 15.1 shows

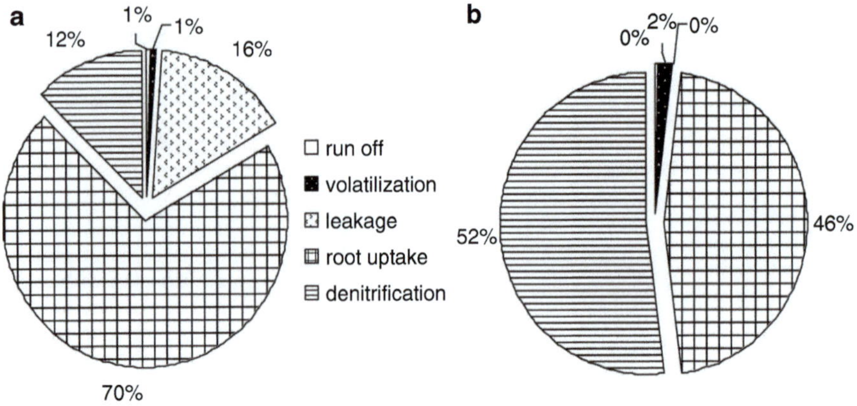

Fig. 15.2 NO_3^- balance cake plot for (**a**) loamy soil and (**b**) sandy soil

nitrate profiles in April 2009 and 2010 after recharge and before spring fertilization. In the loamy soil, NO_3^- concentration observed in April 2009 shows a peak at 3 m a.s.l. in correspondence of the water table, while in the saturated zone NO_3^- decreases rapidly to concentrations below 50 mg/l. In April 2010, NO_3^- profile shows very high concentration with a maximum peak below the water table. The elevated NO_3^- concentration was due to a combination of factors: (1) the water table was higher than in 2009, (2) the temperature was lower than in 2009 and (3) precipitation was less intense. The higher water table suggests that more NO_3^- trapped in the vadose zone could be dissolved in groundwater and the lower temperature may have decreased bacterial activity and denitrification rate. In addition, the lower rainfall contributed to concentrate NO_3^-.

A similar trend is visible for the sandy soil site, although NO_3^- profile appears quite different from the one in loamy soil. In fact, NO_3^- was found only in the upper part of the profile, while below 1 m a.s.l. NO_3^- was always below detection limits. This trend was explained by the higher availability of soil labile organic matter determined by chicken manure addition in the previous autumn, providing a source of organic substrates (fraction of organic carbon: 0.042), like acetate (34 mg/l), fuelling denitrification. On the other hand, in the silty loamy soil, there was a chronic deficiency of labile organic matter that could not prevent nitrogen leaking towards the shallow unconfined aquifer. This result emphasizes the importance of considering the role of labile organic matter in buffering NO_3^- excess via denitrification reactions.

NO_3^- leaching was observed in the loamy soil where the redox conditions remained oxidizing throughout the year, while in the sandy soil, residual content of organic matter from fertilization with chicken manure likely favoured reducing conditions and nitrate removal via denitrification, preventing its migration towards the saturated zone (Fig. 15.2).

Both water budget and nitrogen mass balance were calculated and used in arable lands of the Ferrara Province, in order to gain information on the risk of aquifer nitrogen contamination. In the following, a set of case study will be illustrated, spanning from the laboratory experiments, used to understand the leaching behaviour and the degradation potential, to the monitoring at provincial scale, for quantifying the pollution related to agricultural practices and soil types.

These results highlight the need to pay attention to the kind of manure used and the relative degradation kinetics which may also be heavily affected by farming type (organic or industrial farming). In fact, in industrial farming, the persistence of hormones and other undesired chemicals, such as antibiotics, can interfere with bacterial activities and negatively affect organic matter and nitrogen mineralization rates.

15.5 Laboratory Scale Experiments

Laboratory experiments are often used to understand single processes that are almost impossible to isolate in the field from other processes. In the following paragraphs, a number of different examples are given, in which laboratory experiments have been employed to quantify the leaching behaviour of different nitrogen species and the kinetic of biogeochemical reactions.

15.5.1 Batch Experiments

In alluvial plains, intensive farming, with conspicuous use of agrochemicals, may cause land pollution and groundwater contamination. In central Po River plain, paleochannels are important links between arable lands and the underling aquifer since the latter is often confined by clay sediments that act as a barrier against contaminants migration. Therefore, paleochannels are recharge zones of particular interest that have to be protected from pollution even because they are often used for water supply. The problem was approached via batch test leaking and column elution of alluvial sediments [76]. Results from batch experiments showed fast increase in all major cations and anions, suggesting equilibrium control of dissolution reactions, limited availability of solid phases and geochemical homogeneity of samples. In column experiments, early elution and tailing of all ion breakthrough was recorded due to preferential flow paths. Conclusions indicate that, for sediments investigated in this study, dispersion, dilution and chemical reactions can reduce fertilizers at concentrations below drinking standards in a reasonable time frame when fertilizer loading is hindered. Thus, the definition of a respect area along paleochannels is recommended to improve groundwater quality.

Batch leaching experiments were performed using the saturation soil extraction (SSE) methods described by Schuwirth and Hoffman [77] with use of tap water, representative of water quality recharging the aquifer. The experiment was performed in a temperature-controlled laboratory at 20 $\pm$ 0.5 °C. Sediments were not sterilized but air-dried at room temperature to minimize heat-driven dehydration reactions

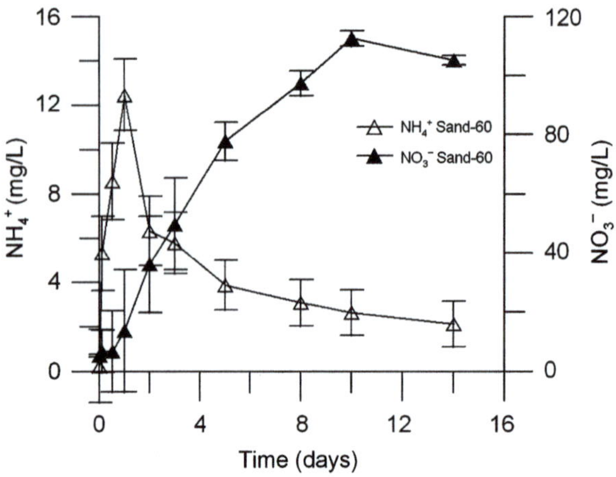

Fig. 15.3 Graphs of NH_4^+ versus NO_3^- from batches containing sediments Sand-60 (Reproduced from [76] with kind permission of © Springer-Verlag Berlin/Heidelberg)

and to avoid changes in the structure, in the ion exchange capacities and in the dissolution characteristics of clay minerals. Samples were not washed as this would preferentially remove components of the system that are associated with finer, more friable minerals, such as micas.

For a sediment sample with 60% of sand, named Sand-60, 3 batches were run with solid/liquid ratio of 1:1 (290 g of sediments and 200 ml of tap water), 1:2 (145 g and 200 ml) and 1:4 (72.5 g and 200 ml). Triplicates were prepared to derive standard deviation of dissolved species concentration. Over the 15 days duration of the experiment, 10 samples (2 ml each) were taken from each of the 18 reactor vessels at increasing time intervals. These were filtered using a 0.2-μm polypropylene filter and analysed for anion and cation by IC as described above (Sect. 15.4.1.). High concentration in NH_4^+ and NO_3^- (Fig. 15.3) found in batches are probably due to previous fertilizer loading in the area where sediment samples Sand-60 were collected. Ammonium adsorbed onto mineral surface was released in pore water in the first few hours of the experiment (Fig. 15.3). Denitrification did not start during batch experiments, as oxygen remained readily available to provide the source of electron acceptor for organic matter degradation and nitrate continued to increase via bacteria nitrification until all available ammonium was consumed.

The fast nitrification reaction found in batches suggests that in aerobic condition, ammonium is quantitatively converted to nitrate. Results highlight the role of unsaturated zone to convert urea and ammonium, usually applied as fertilizers, into nitrate. This was also confirmed by surface and groundwater sampling in the study area, where ammonium concentration was always very low (<1 mg/l) or not detectable. Thus, batch experiments were helpful to determine the time required to reach chemical equilibrium between water and solid phase.

15.5.2 Column Experiments

A column named Sand-60 filled with the same sediments used for batch experiments was constructed from 10-cm-diameter PERPEX drainpipe. Column lengths were 1 m each. A 5-cm HDPE pre and post-chamber was installed in the column to uniform inflow and outflow and to prevent washing of finer grains: chambers contained uniformly packed quartz gravel, an HDPE-pierced disc and a 50-μm NITEX mesh in contact with sediments.

Sediments were prepared identically to that used in the batch experiments. Loading took place in 15–20 increments, and each increment was lightly packed before the next one was placed on top. The packed column was completely saturated with tap water and let still for 2 weeks, until chemical equilibrium between influent tap water and sediments was reached. A constant head reservoir was applied to pump tap water in the columns and effluent tube was fixed to maintain a stable head gradient. Head gradient in Sand-60 was 1,000 times larger than the field one (average head gradient 1‰), because the extremely low natural flow pattern was not easily repeatable in experimental conditions.

The influent tube was constructed of 5 mm internal diameter to transmit input water through a steady seepage onto the pre-chamber. This guaranteed that all packed sediments obtained an equal amount of influent at the lower boundary of the column.

Effluents drained from the column via a 2-mm internal diameter pipe through a flow cell that contained permanently located O_2, pH, TDS and EC probes. Probes were calibrated using standard solutions at the beginning and end of each run. Effluent drained under gravity into conical flasks. The point of drainage was lower than the base of the columns. This preserved saturation at the base of the column and decreased the risk of air bubbles in the effluent. Samples were taken from the conical flasks at increasing time intervals for 1 month and filtered with a 0.2-μm polypropylene filter. Batch and column experiments were employed to assess changes in groundwater quality when kept in contact with sediments affected by fertilizers and to characterize mobilization and transport of major anions and cations in paleochannel sediments. Column experiments were used to identify the existence of preferential flow and stagnant zones.

The total enrichment and mineralization of water found in batches was fast diluted in columns experiment, and even if an initial high concentration of nitrate was recovered, denitrification and dispersion most probably helped to decrease concentrations at acceptable levels.

While nitrate concentration takes 0.85 days to reach tap water composition in column Sand-60, using field head gradient, nitrate disappearance will be reached only after 850 days of continuous flushing. Taking into account the rainy days in the Ferrara Province and the summer irrigation of the arable land, a total amount of 62 days of fertilizers flushing per year can be estimated. This means that nitrate disappearance in the field would take approximately 14 years for sediments similar to Sand-60. Extrapolation of laboratory findings did not aim to offer an exact quantitative calculation of effective residence time of fertilizers in the field, but gave an estimated order of magnitude of their persistence in the aquifer. Sediment properties have an important influence on concentration of chemical components and on their migration into groundwater environment. The study case indicated the high potential of arable land to act as periodic source of fertilizers, indicating that an appropriate respect zone and more restrictive standards should be imposed in delicate recharge areas as paleochannels to preserve recharge water quality and prevent aquifer contamination.

15.5.3 Tank Experiments

A long-term elution experiment to study the saturated transport of pre-accumulated fertilizer by-products was conducted within a large tank (4 × 8 × 1.4 m) equipped with 26 standard piezometers [58]. Silty sand sediments were excavated from an unconfined alluvial aquifer near Ferrara (Northern Italy) and used to fill the tank; the field site was connected to a pit lake located in a former agricultural field. To evaluate spatial heterogeneity, the tank filling material was characterized via slug tests and grain size distribution analysis. Initial tank pore water composition exhibited high concentration of nitrate (NO_3^-), sulphate (SO_4^{2-}), calcium (Ca^{2+}) and magnesium (Mg^{2+}) due to fertilizer leaching from the topsoil in the field site. The initial spatial distribution of NO_3^- and SO_4^{2-} was heterogeneous and not related to the finer grain size content (<63 μm). The tank material was flushed with purified tap water for 800 days in steady-state conditions; outflowing water was regularly sampled to monitor the migration rate of fertilizer by-products. Complete removal of NO_3^- and SO_4^{2-} took 500 and 600 days, respectively.

Clean-up times were compared with those of a previous laboratory experiment developed on columns (1 × 0.1 m) filled with the same sediments. Results emphasized organic substrates availability and spatial heterogeneities as the most important constraints to denitrification and nitrogen removal, which increase the time required to achieve remediation targets.

The experiment was performed in a large tank (4 m wide by 8 m long and 1.4 m deep) located in the Hydrogeology Laboratory at the Scientific and Technological Pole of the University of Ferrara (Fig.15.4). The tank was assembled with an internal structure of armed PVC fastened on an external structure of natural wood; the non-metallic materials were selected for future electrical resistivity tomography applications. The tank was filled with 42 m^3 of unconsolidated material (35 m^3 of natural sediments and 7 m^3 of gravel), by means of a bulldozer equipped with a 2-tons tilting shovel mounted on a 8-m-long telescopic crane.

Saturated natural sediments were poured by the tilting shovel into the tank starting from the inflow gravel wall towards the outflow wall. Once a layer of

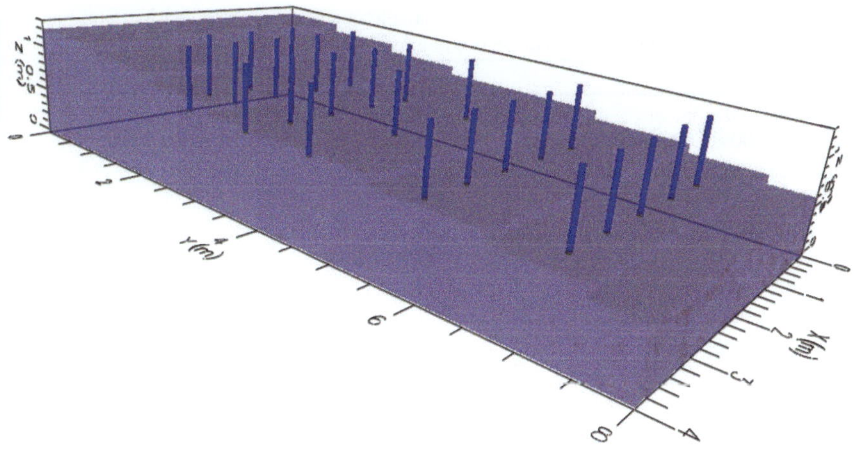

Fig. 15.4 Three-dimensional view of the tank with the saturated zone (*grey* shaded area) and piezometers (*blue lines*). The origin of the *x*, *y* and *z* axis was located to the *left bottom* corner of the inflow wall

0.2 ± 0.05 m was created, it was compacted, using the bottom of the shovel, and new layers were added. The filling procedure took 10 h. Following this, the natural compaction of sediments was monitored for four months; the average bulk compaction was found to be approximately 0.02 m.

An external reservoir (constant head) was connected to the tank via three inflow pipes (Fig. 15.4); the reservoir had a large surface area to minimize the introduction of trapped air bubbles. The constant head was introduced to create a steady-state flux with a mean head gradient of 7% in the sediments. In order to maintain a certain uniformity of the potentiometric surface, two gravel walls were built, one at the inflow and outflow of the tank. The mean head gradient in the gravel walls was calculated by the Darcy law to be 0.1 ‰.

Twenty-six piezometers were installed using a hand-driven auger, on the base of a semi-regular monitoring grid (Fig. 15.4). A detailed topographic survey was carried out using a Nikon DTM-450 total station to accurately determine the well-case position in x, y and z axis (Fig. 15.4); piezometric heads were monitored every 2 months. Seventy-eight undisturbed 1-in. cores were collected every 0.3 m by a Shelby sampler for grain size analyses. Particle size curves were obtained using a sedimentation balance for the coarse fraction and an X-ray diffraction Sedigraph 5100 Micromeritics for the finer fraction; the two regions of the particle size curve were connected using the computer code SEDIMCOL. Bulk density and total porosity were determined gravimetrically. Organic carbon content (f_{oc}) of the sediments was measured by dry combustion.

To estimate the *k* variability, 130 multilevel slug tests were performed within the saturated zone (every 0.15 m) using small inflatable straddle packers. The tests were run after the complete elution of resident water in order to avoid interference with

Table 15.1 Average grain size distribution, hydraulic conductivity, bulk density, total porosity and f_{oc} measurements for natural sediments and gravel walls

Parameter	Natural sediments	Gravel walls
Grain size (%)		
Gravel (2,000–20,000 μm)	0.0	35.0
Coarse sand (630–2,000 μm)	0.30	3.3
Medium sand (200–630 μm)	5.4	22.3
Fine sand (63–200 μm)	50.8	22.8
Silt (2–63 μm)	32.1	13.4
Clay (<2 μm)	11.4	3.2
Hydraulic conductivity (m/s)	$1.8e^{-6}$	$1.2e^{-3}$
Bulk density (kg/m^3)	1.68	1.61
Total porosity (−)	0.42	0.27
Organic carbon (−)	0.027	0

Reproduced from [57–59] with kind permission of © Springer Berlin Heidelberg

the monitoring program, due to the possibility that slug tests could modify the flow field, although transiently. Water level in every piezometer was instantaneously lowered with a pneumatic syringe. All the acquired slug test responses were analysed using the Bouwer and Rice method [78].

Once the tank elution started, groundwater sampling took place every 2 months in order to get a snap shot of the distribution of solute concentrations (Fig. 15.5). Samples were taken from every piezometer via low-flow sampling technique using Waterra inertial pumps, from the external reservoir, from the inflow gravel wall and from the outflow pipes.

A large tank experiment was implemented to evaluate the clean-up time of a shallow unconfined aquifer contaminated by fertilizer by-products. Via grain size distribution analysis and geological survey, the depositional environment was identified as a crevasse splay, characterized by a large spectrum of textures and a heterogeneous *k* field. To overcome the common problems associated with permeability testing on recovered samples, multilevel slug tests were performed. This technique allowed a precise three-dimensional reconstruction of the *k* field and its local heterogeneities within the tank.

NO_3^- degradation within the tank was limited by scarce organic substrate availability since about 40 years ago, organic fertilization ceased and labile organic load underwent to a substantial reduction and disproportion with respect to the total oxidative capacity. In agreement with results of a column experiment, using the same sediments, nitrate fate within the tank confirmed the high vulnerability of unconfined aquifers to this common pollutant and the long bioremediation time. A methodological implication of this study is that laboratory experiments can provide robust results only when the flow velocity approximates the field conditions and relatively large samples are used, because considerable heterogeneities present in sediments need to be taken into account when testing fertilizer fate and transport within the saturated zone in complex alluvial depositional environments.

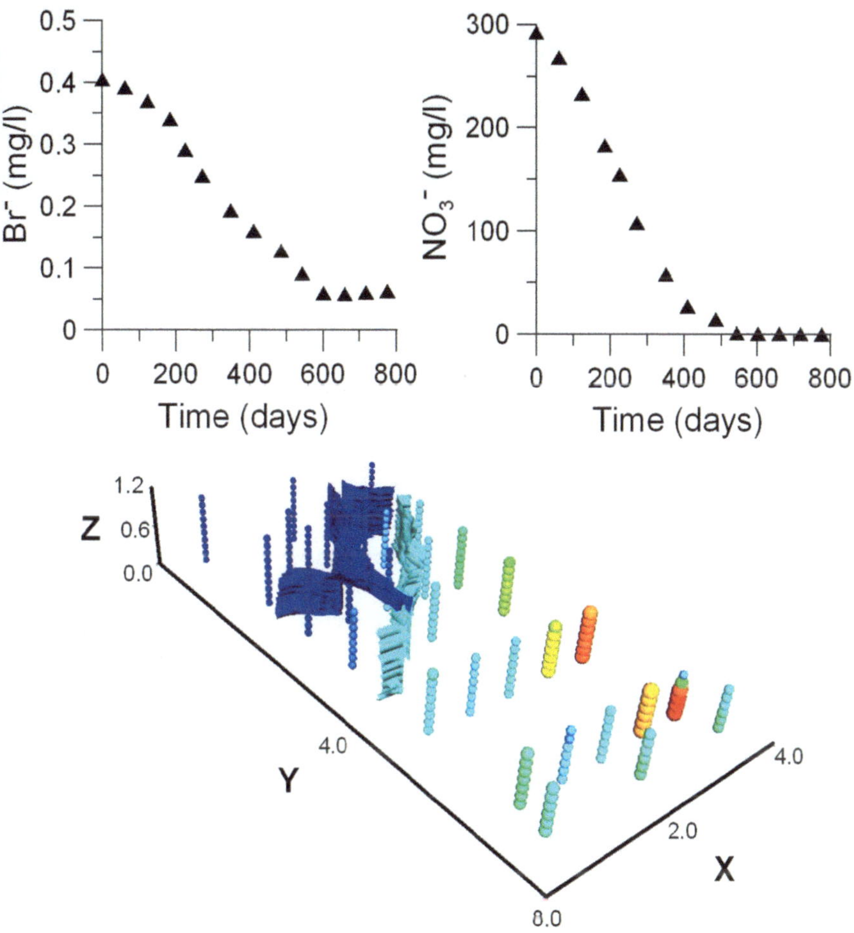

Fig. 15.5 *Upper* graphs show the time series of bromide and nitrate at the tank's outflow, while the *lower* picture shows the concentration of bromide with isosurfaces at 0.05 and 0.1 mg/l in *blue* and *light blue*, respectively. Bromide concentration is also displayed with colour bubble proportional to concentration (*red* is the maximum and *blue* is the minimum), along each piezometer after 200 days from the beginning of the experiment (Reproduced from [57–59] with kind permission of © Springer Berlin Heidelberg)

15.5.4 Reactive Modelling of Batch Experiments

The introduction of good agricultural practices in order to reduce the amount of nitrate entering the groundwater system is crucial to quantify the kinetic control on nitrate attenuation capacity. With this aim, a series of anaerobic batch experiments, consisting of loamy soils and nitrate-contaminated groundwater, were carried out using acetate as electron donor. Acetate is the main intermediate species in many biodegradation pathways of organic compounds, and hence it is a suitable carbon source for denitrification. However, the acetate denitrification efficiency is likely to be related to sorption. Sorption of acetate was determined fitting a Langmuir isotherm in both natural and artificially depleted organic matter soils. To account for the spatial variability of soil parameters, quadruplicate experiments were performed. The geochemical code PHREEQC(2) was used to simulate kinetic denitrification using Monod equation, equilibrium Langmuir sorption of acetate and equilibrium reactions of gas and mineral phases (calcite). The reactive modelling results highlighted a rapid acetate mineralization rate, suggesting that the main pathway of nitrate attenuation is through denitrification, while calcite acted as a buffer for pH [59]. Reactive modelling is thought to be an efficient and robust tool to quantify the complex biogeochemical reactions which can take place in underground environments. Again, the major issue that affects these systems is the vulnerability to water and agrochemical losses from the surface soil profile where the crops' effective rooting system subsists. This issue is also dominated by the combined effects of the physical soil fertility, the crop type and the agricultural practices [22]. Particularly, quantity and quality of organic matter play a central role in regulating denitrification, thus buffering the excess and risk of percolation to groundwater. In remediation policies and relative incentives, this point is undoubtedly central but often omitted, partly because commonly used methods for organic matter quantification do not give indications on its labile/refractory nature and therefore cannot indicate the relative role as electron donor and in denitrification pathway. On the other side, organic matter loss represents one of the most serious concerns of industrialized agriculture with widespread environmental consequences, such as loss of natural fertility that increases the demand of synthetic fertilizers, loss of water-holding capacity that increases the leaking of pollutants (like nitrate) and the shift from sink to source of carbon dioxide [32].

In a soil, the size of microbial community is largely determined by the amount of "available" carbon. Denitrifying microorganisms can compete with other heterotrophs for added C, and, hence, denitrification capacity of a soil is often highly correlated with its C content. With the exception of C deposited at the time of strata formation (geological carbon) and that produced by autotrophic bacteria, any C in the subsoil will have originated in the topsoil. This movement of soluble C therefore may be a limiting factor [33]. Compared to N, much less research has focused on quantifying soluble C leaching under field conditions and various management practices such as tillage and fertilization. The movement of soluble nutrients within the soil matrix depends on two factors: the concentration of solutes in the soil and the water flux transporting these solutes. Diffusion of solutes proceeds relatively slowly compared with the potential translocation of solutes within the soil profile by mass flow. Water infiltration and redistribution in the soil matrix are most commonly responsible for the mass flux of soluble soil constituents.

The reactive modelling (Fig. 15.6) emphasized the complete nitrate mineralization in RS-Ace microcosms, in conjunction with the concomitant alkalinity increase. These evidences, corroborated by the good model fit, indicated that the main pathway of nitrate attenuation in these sediments may occur through denitrification in presence of acetate as electron donor, while calcite acted as a buffer for pH.

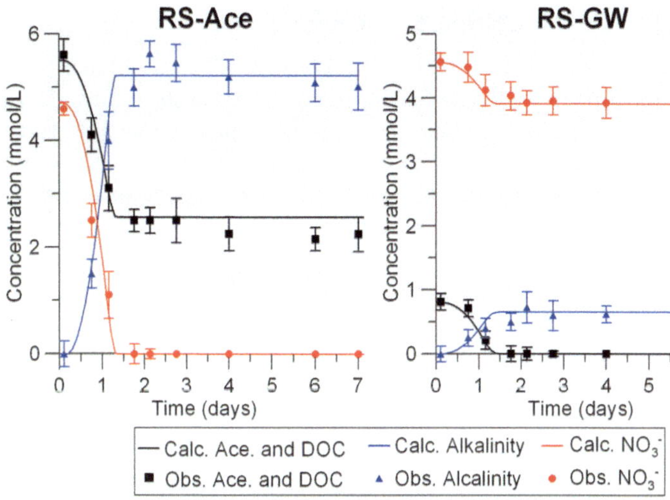

Fig. 15.6 Plots of observed and calculated values in microcosms experiments, *RS* stand for reference soil, *Ace* for acetate, *GW* for groundwater. *Symbols* denote experimental points while the curves represent model results fitted to the experimental data. *Error bars* show the standard deviation of four replicates (Reproduced from [58, 59] with kind permission of © Springer-Verlag Dordrecht)

On the contrary, when a viable carbon source is not available like in RS-GW microcosms, denitrification is likely to remain a negligible pathway of NO_3^- removal from sediments.

Overall, this study showed that in SOM-depleted sediments, NO_3^- natural attenuation could be very low compared with non-depleted sediments where a viable carbon source is present. It follows that a rationale prevention, management and mitigation of nitrate pollution should focus on agricultural practices which could augment the flux of low molecular weight organic acids towards the subsoils.

15.5.5 Field Scale Experiments

Field experiments are often used to understand multiple processes, once laboratory experiments have already cleared some specific process like the ones explained before. In the following paragraphs, a number of different examples are presented, in which field scale experiments are employed to quantify the leaching behaviour of different nitrogen species and to evaluate the vulnerability of agricultural lowlands to nitrate pollution.

15.6 Numerical Modelling at the Plot Scale

To maximize the irrigation efficiency and to protect groundwater from agrochemical pollution, two variables must be known with good accuracy: effective evapotranspiration and infiltration, especially in lowland areas were the run-off is minimal. Three different experimental plots cultivated with maize were equipped with tensiometers and TDR probes for daily monitoring of water movement in the unsaturated zone. In the study area, located in Ferrara province (Italy), at an altitude ranging from 5 to -3 m a.s.l., the three sites selected to monitor and model the water movement in the unsaturated zone were named SAP, CCR and MEZ (Table 15.2).

The surface area of each plot was 1 ha, and its slope was less than 1% (and mostly less than 0.5%). For this reason, it was assumed surface run-off to be minimal and water movement in the unsaturated zone to be dominantly vertical. The soil stratification was divided in two distinct layers: the upper one interested by tillage, roots growth and weathering and the lower undisturbed one. In SAP and MEZ sites, the upper layer was 0.65 m thick and in CCR was 0.75 m, while the lower layer extended until 2 m b.g.l. in all sites.

Other relevant parameters of the various soil layers as hydraulic conductivity and water retention curve were obtained in laboratory experiments, while boundary conditions as precipitations, temperature and root growth were obtained on-site. Inverse modelling and sensitivity analysis were performed using HYDRUS-1D to assess the degree of uncertainty on model parameters.

The finite element model HYDRUS-1D [37] was selected for simulating the one-dimensional movement of water in variably saturated media. The program numerically solves the Richards equation for saturated–unsaturated water flow:

$$\frac{\partial \theta}{\partial t} = \frac{\partial}{\partial z}\left[K(\theta)\left(\frac{\partial h}{\partial z}+1\right)\right]+S \tag{15.1}$$

where h is the water pressure head (L), θ is the volumetric water content (L3/L^3), t is time (T), z is the spatial coordinate (L) positive downward, K is the unsaturated hydraulic conductivity function of soil (L/T^1) and S is the source/sink term (L3/L^3/T^1). The Van Genuchten [81] parametric functions were selected to solve Eq. (15.1), assuming that hydraulic properties of soil can be described by

$$\theta(h) = \begin{cases} \theta_r + \frac{\theta_s-\theta_r}{[1+|\alpha h|^n]^m} & h<0 \\ \theta_s & h \geqslant 0 \end{cases} \tag{15.2}$$

$$K(h) = \begin{cases} K_s s_e^l\left[1-\left(1-s_e^{l/m}\right)^m\right]^2 & h<0 \\ K_s & h \geq 0 \end{cases} \tag{15.3}$$

$$s_e = \frac{\theta-\theta_r}{\theta_s-\theta_r} \tag{15.4}$$

Table 15.2 Soil characterization for the three field sites

	Pedological classification	Sedimentological environment	Textural classification
SAP	Calcic vertisols	Interfluvial	Silty clay
CCR	Haplic calcisols	Fluvial	Silty loam
MEZ	Thionic histosols	Marsh lagoon	Peaty

Reproduced from [79] with kind permission of © Elsevier BV

Table 15.3 Hydraulic parameters used in the numerical models for the two soil horizons of each site

	SAP	SAP	CCR	CCR	MEZ	MEZ
Parameter	Upper layer	Lower layer	Upper layer	Lower layer	Upper layer	Lower layer
K_s (cm/day)	5.0	0.3	2.0	5.3	2.0	0.6
θ_s (cm^3/cm^3)	0.46	0.44	0.41	0.38	0.66	0.56
θ_r (cm^3/cm^3)	0.11	0.08	0.05	0.06	0.029	0.05
α (1/cm)	0.0061	0.0052	0.0012	0.0051	0.0050	0.0021
n	1.31	1.50	1.45	1.53	1.20	2.22

Reproduced from [79] with kind permission of © Elsevier BV

in which θ_r and θ_s denote the residual and saturated water content (L3/L^3), respectively, K_s is the saturated hydraulic conductivity (L/T^1), α is the inverse of the air-entry value, n is a pore size distribution index and it is constrained to be >1, m is equal to $(1-1/n)$ and l is a pore connectivity parameter assumed to be 0.5. Constant pressure head tests were used to infer the average K_s of each column; θ_s was determined gravimetrically after elution of a volume corresponding to 100 pore volumes, while θ_r was measured gravimetrically in triplicates on air-dried sediments, after heating for 24 h at 105 °C (Table 15.3).

The numerical grid was discretized in 200 nodes of 0.01 m each to form a regular grid 2 m long and a surface area of 1 m^2. The grid was subdivided into two regions representing the upper and the lower soil horizons; initial water content conditions of collected soil cores (every 0.25 m) at each site were measured via gravimetric methods and linearly interpolated along the vertical axis.

At the soil surface, an atmospheric boundary condition with surface layer was selected. This condition permits water to build up on the surface. The height of the surface water layer increases due to precipitation and reduces because of infiltration and evaporation, this value was assumed to be 0.01 m in each model. As lower boundary condition variable pressure heads were specified in every model using groundwater levels.

The sink term, S, in Eq. 15.1 corresponds to the volume of water removed from a unit volume of soil per unit time due to the root water uptake and can be defined as [82]:

$$S(h) = \alpha(h)S_p \tag{15.5}$$

where $\alpha(h)$ is a prescribed dimensionless response function of the soil water pressure head ($0 \leq \alpha \leq 1$) accounting for the effects of water stress on root water uptake and S_p is the potential root water uptake rate (d^{-1}).

The average observed soil water salinity level integrated along the first soil metre was 650 mg/l at CCR, 320 mg/l at SAP and 1,260 mg/l at MEZ. Particularly in MEZ site down to 0.70 m b.g.l. salinity was approximately 430 mg/l and rapidly increased up to 3,200 mg/l from 0.70 to 1 m b.g.l. These values suggested that stress reduction was mainly attributable to water stress instead of salinity stress. Thus, salinity stress was not accounted in all the simulations.

Potential evapotranspiration was calculated in HYDRUS-1D using the FAO recommended Penman–Monteith equation for evapotranspiration (ET_0) [83].

The potential evapotranspiration was also evaluated using the much simpler Hargreaves formula [84]. Finally, another simple method, the Turk method [85], was applied for a further comparison. This method allows the calculation of the daily potential evapotranspiration as a function of air temperature, relative humidity and solar radiation.

The effect on the groundwater flux towards the saturated zone, namely, the recharge, is exemplified by Fig. 15.7, which compares the bottom flux of the three different approaches employed to calculate PET for each field site. Here, a negative flux means that groundwater leaves the bottom model boundary, while a positive flux indicates an upflow towards the ground level. Upflows can take place only when the water table of the underlying unconfined aquifer rises up or when evapotranspiration provokes groundwater movement by capillary forces, otherwise gravitational force induces water to percolate downward. From Fig. 15.7, it can be seen that large discrepancies are produced in the calculated cumulative bottom fluxes by using the three different approaches. This is especially true for the CCR and SAP models, where the water table laid often below the bottom boundaries. While, for the MEZ model, a closer behaviour between the cumulative bottom fluxes (calculated with the three different PET methods) is apparent from Fig. 15.7. This was mainly due to the water table constantly maintained at −1.5 m b.g.l. by an irrigation canal located approximately 30 m from the site. This forced situation produced a net upflow during the simulated period, while in the other two simulations, the net flow was towards the aquifer for all the three different approaches.

Numerical models of unsaturated flow in cultivated maize field with different soils were successfully calibrated using laboratory-derived parameters, PET calculated with Penman–Monteith formula and the S-shaped stress reduction formula. A comparative analysis led to exclude simple approaches to calculate PET such as Hargreaves or Turk ones, due to the worsening of model fit for every field site investigated. In addition, large differences in groundwater recharge were obtained with the different methods used to estimate PET.

It was also proved that water stress reduction was important in all field sites and that the S-shaped stress reduction formula was preferable to the Feddes one, since it leads to the same model fit for all sites, but a single parameter was needed instead of the 7 needed for the Feddes formula.

To summarize and conclude, results showed a good model fit of water content and head pressure at various depths, in each site, using Penman–Monteith formula for daily potential evapotranspiration calculation, but poor fit applying the Hargreaves and Turk formulas. Best performance of model fit was observed for S-Shaped equation employed to simulate the root water uptake reduction with respect to Feddes equation.

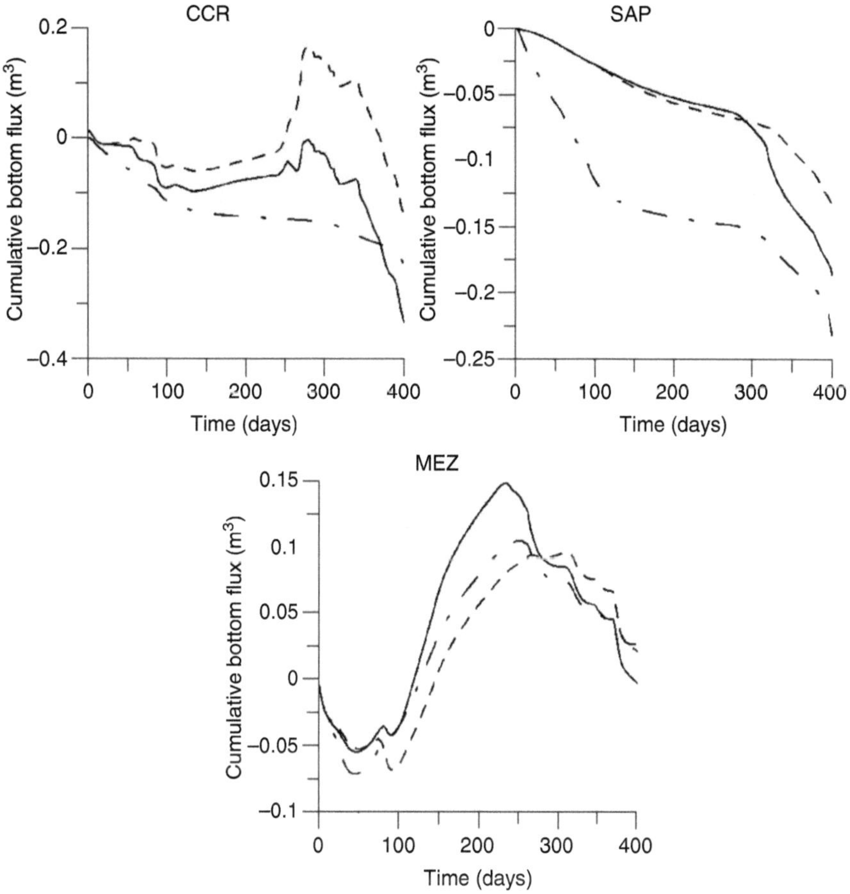

Fig. 15.7 Calculated cumulative bottom flux for the Penman–Monteith models (*black lines*), the Hargreaves models (*dashed lines*) and the Turk models (*dashed-dotted lines*) in each field site (Note the different *y* scales)

The soil parameters uncertainty was limited and within analytical errors; thus, a robust estimation of cumulative infiltration and evapotranspiration was derived [79, 80]. This study points out that evapotranspiration is the most important variable in defining groundwater recharge for maize crops in lowlands, thus in these conditions is essential to have a good estimation of evapotraspiration to assess the nitrogen leaching risk.

15.7 Numerical Flow and Transport Model in an Unconfined Aquifer

A portion of an unconfined alluvial aquifer located in the Padana Plain (Italy) was characterized following an integrated hydrogeophysical approach. Initially, an electrical resistivity tomography (ERT) survey was employed to localize the

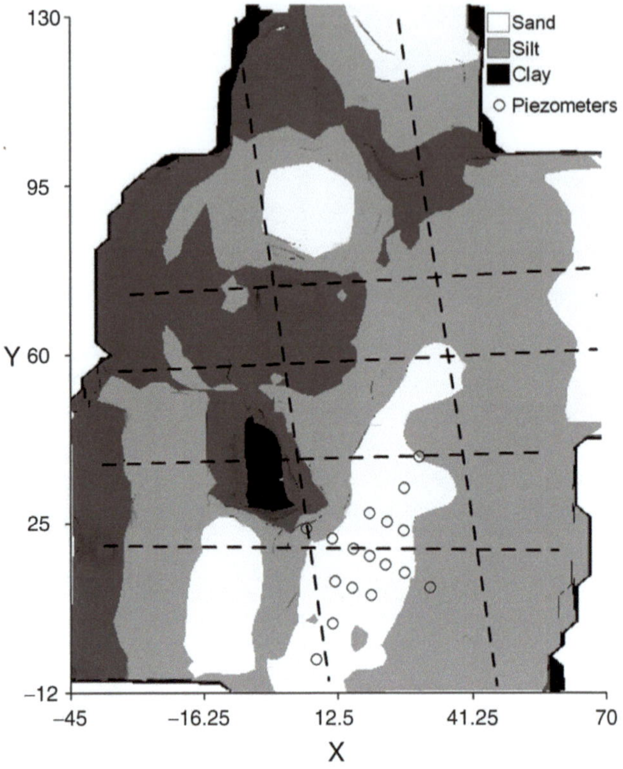

Fig. 15.8 Paleochannel reconstruction from geomorphologic and ERT surveys (*dashed lines*) and 17 stratigraphical information

boundaries of a modest paleochannel body and to design the installation of a groundwater monitoring network. Multilevel slug tests were performed to estimate the aquifer saturated hydraulic conductivities. Determined permeability values together with electrical resistivity data were correlated. The correlation resulted in a site-specific bi-logarithmic linear relationship. Based on this relationship, punctually determined hydraulic conductivities were spatially extended over the studied flow domain. Continuously measured piezometric heads were used to calibrate a 3D flow model. The integration of the ERT survey results with the classical hydrogeological tests can be conveniently applied to constrain the permeability field during flow model calibration [80] (Fig. 15.8).

Although the applicability of the determined relationship is site specific, the followed procedure is useful especially when there is a need to optimize the available resources and in case of small-scale pilot studies. After the geophysical and hydrogeological characterization was carried out, a monitoring and modelling study on fertilizers by-products fate and transport was performed in the unconfined shallow aquifer equipped with a grid of 17 piezometers [57–59].

The field site was located in a former agricultural field overlying a paleochannel near Ferrara (Northern Italy) and cultivated with cereals rotation until 2004.

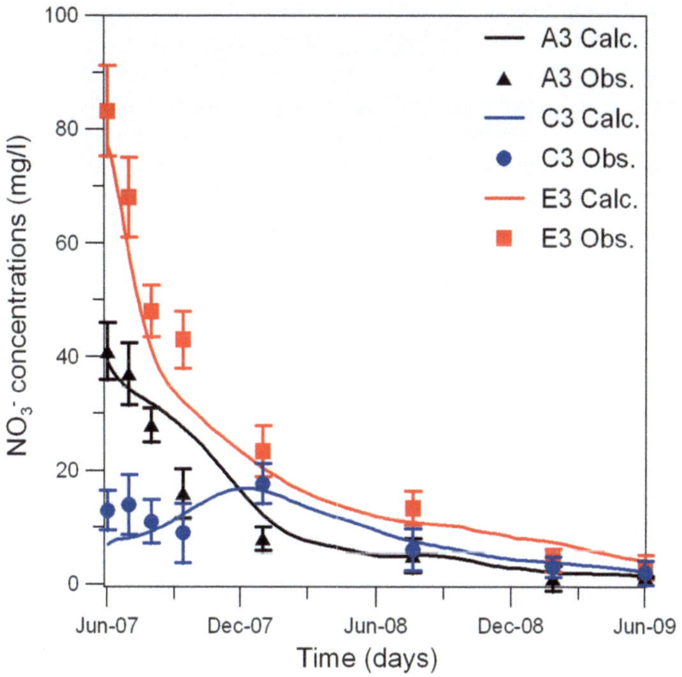

Fig. 15.9 Time series graph comparing observed and computed NO_3^- concentrations in piezometers located along a groundwater flow line (Reproduced from [57, 58] with kind permission of © Springer-Verlag Dordrecht)

To evaluate the temporal and spatial variation of groundwater heads, piezometers were monitored via pressure transducer data loggers, while an on-site meteorological station provided the data for monthly recharge rate calculations. The initial groundwater composition exhibited elevated nitrate (NO_3^-) and chloride (Cl^-) concentrations, due to fertilizer leaching from the topsoil. The initial spatial distribution of NO_3^- and Cl^- was heterogeneous, and it decreased during the monitoring period, with NO_3^- removal (below 10 mg/l) after 650 days. A transient groundwater flow and transport model was calibrated versus observed heads, NO_3^- and Cl^- concentrations. Cl^- was used as environmental tracer to quantify groundwater flow velocity and was simulated as conservative species. NO_3^- was treated as a reactive species, and denitrification was simulated by a first-order degradation rate (Fig. 15.9).

A low denitrification rate was found ($2.5e^{-3}$ 1/day) after model calibration, likely because of prevailing oxic conditions and low concentration of dissolved organic carbon. Scenario modelling (implemented with average and variable flow boundary conditions) showed that transient piezometric conditions did not exert a strong control on NO_3^- clean-up time, while transient recharge rate did. A small field site experiment was implemented to evaluate the clean-up time of a shallow unconfined aquifer contaminated by NO_3^-. The site represents a case of vulnerable area (paleochannel) where the agricultural practices were stopped to quantify the NO_3^- clean-up time. NO_3^- presence at the field confirmed the high vulnerability of unconfined aquifers to this common pollutant and the long bioremediation time required. NO_3^- degradation within the monitoring grid was limited by scarce organic substrate availability, and NO_3^- removal was mainly driven by dilution transport processes. The three-dimensional flow and transport modelling provided a useful tool to assess the NO_3^- degradation rate in the field, especially because the degradation rate was very low. Scenario modelling provided the evidence that, in a flat area, recharge is one of the key parameters in NO_3^- remediation, while variable flow field is not dramatically affecting NO_3^- clean-up time. This is because the head gradient in these areas is small (usually 1‰), and seasonal head variations do not produce large variations in the flow field. Thus, an accurate estimation of recharge rate should be taken into account when evaluating clean-up times.

15.8 GIS Approach at the Watershed Scale

The most used method to determine water and agrochemicals losses from agricultural land is the use of process-based mathematical models, describing water movement and the transport and transformations of dissolved species that were explained in the preceding paragraphs. More simple models like classification indices (vulnerability or risk indices) use fewer and more accessible data such as climatic data, topography and general soil physical properties, and they can be employed easily in large-scale application using GIS tools. The most popular vulnerability indices to describe the intrinsic vulnerability of the groundwater to contamination, by using weights and ratings, are DRASTIC index [42], SINTACS index [43] and MERLIN [46]. To avoid the complexity of the above-mentioned process-based models and to avoid the subjectivity of the indices using weights and ratings, the LOS method has been developed [86]. In this method, the loss of water and nitrogen indices are calibrated by regression analysis, via Gleams v3.0 model, using soil properties, topography and climatic conditions of a reference field crop. The results of calibration are then used as observed values. Consequently the ranking of LOS indices has a physical meaning, using units for the amounts and concentrations of water and nitrogen losses. Thus, LOS indices results are very easy to introduce into GIS software to obtain vulnerability maps of agricultural lands to nitrogen percolation. These indices are the basis for the integrated water resources management in agricultural systems. In order to calibrate the indices using multiple regression analysis, the results of GLEAMS v3.0 [88] model for combinations of different soil properties, topography and climatic conditions of a reference field crop are used as "observed" values. A case study for the application of the indices is also presented in this study.

For the application of a set of indices (LOS) to classify the intrinsic vulnerability of agricultural land to water and nitrogen losses, we considered the same case study of the Province of Ferrara. For the Ferrara Province, the mean annual water was 314 ± 37.2 mm. LOS indices application for the case showed their ability to

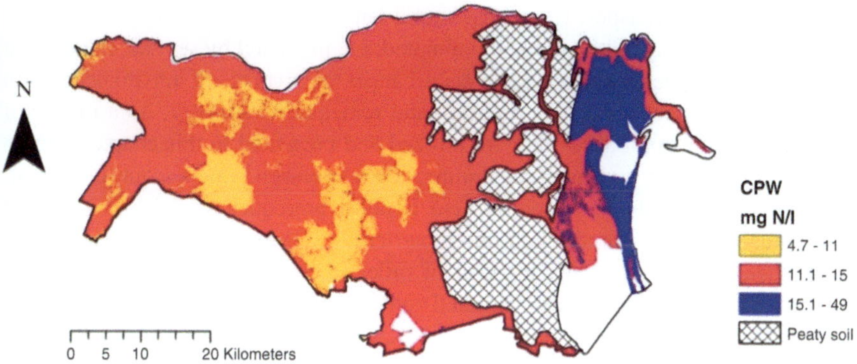

Fig. 15.10 Nitrogen concentrations in the percolated water (CPW) that will reach the unconfined aquifer. The areas with peaty soils are excluded by the calculation of the index

represent adequately the intrinsic vulnerability in lowland regions. Moreover, LOS indices showed their advantage, with respect to other indices, since their objective physical and statistical background was used in assessing the relative pollution potential of surface and groundwater, unbinding the results from site specificity and therefore giving the possibility to compare results among different regions. Finally, this case showed that nitrogen losses were proportional to water losses. This leads to consider that a good water balance is a prerequisite to understand nitrate fate in the subsurface and therefore that it is crucial to take into account seasonality as key factor (Fig. 15.10).

According to the application of the LOS indices in Ferrara Province, two regions have been clearly identified to be more vulnerable to water and nitrogen losses: the coastal region and the areas close to the riverbanks and the paleochannels of the main rivers.

The application of LOS indices in the case study responded satisfactorily to represent lowland regions, giving an integrated aspect of the intrinsic vulnerability of agricultural land to water and nitrogen losses. The assessment of groundwater pollution potential using the LOS indices showed their strong advantage to give comparable results among different regions. Although this approach is suitable to assess the vulnerability to nitrogen, new researches are currently under development to assess the risk of nitrate contamination taking into account actual crops present in the domain [87], actual irrigation rates and microbial nitrogen transformations, particularly denitrification, with respect to physical soil properties and soil organic matter biodegradability.

15.9 Conclusion

Nitrogen fertilizers used in agriculture often cause nitrate leaching towards shallow groundwater, especially in lowland areas where the flat topography minimizes the surface run-off. These features render these areas simplified case studies for nitrogen fate and transport from agricultural land to groundwater. However, these areas are intensively cultivated and populated, i.e. impacted by several, concurrent and often overlapped, anthropogenic disturbances that may result into misleading in setting up a proper monitoring program and remediation strategy. As a matter of fact, most of the approaches that we utilized and presented in this chapter remain complex, expensive and time consuming. Thus, we argue that accomplishing all experimental procedures presented here is not a prerequisite to achieve a good conceptual model of the nitrogen fate in agricultural lands. Instead, in our experience, there are several steps to fulfil in order to understand the complexity of nitrogen fate in the vadose and saturated zone. First, water fluxes must be entailed and characterized. Nitrogen species are preferentially dissolved and transported in the pore water, and, only after that water budget has been reliably characterized, nitrogen mass balance can be studied. Second, before starting any experimental activity, a general but fundamental principle is to clear up targets to avoid as much as possible (1) to waste time and energy in the jungle of unnecessary data acquisition and, even worse, (2) to omit some determinations which would turn out later as irrecoverable lack in the characterization. This principle is particularly true in the unsaturated soil.

Once collected, data set may be validated via deterministic modelling. This task is usually complex and needs a multidisciplinary approach. In addition, complexity increases due to the presence of several models created for the needs of different disciplines. Nowadays reactive and predictive numerical models are not accessible to most of the researchers dealing with nitrogen balance in agricultural lands because a multidisciplinary expertise is required to fully handle the hydrological, numerical, agronomical and biogeochemical terms. Moreover, these models require wide data set to be calibrated and validated. Nowadays, the most promising model applications are the GIS integrated models, although more research is needed to create easy but robust tools suitable for use by non-specialists.

In conclusion, in order to evaluate the sustainability of agricultural practices in a specific region, it is not strictly a matter of what model to use but of operating an accurate model calibration and validation by using a representative data set. Our experience and the most recent scientific results indicate that site specificity remains the key factor which governs nitrogen fate in subsoil.

References

1. Gruber N, Galloway JN (2008) An Earth-system perspective of the global nitrogen cycle. Nature 451:293–296
2. Mulholland PJ, Helton AM, Poole GC, Hall RO (2008) Stream denitrification across biomes and its response to anthropogenic nitrate loading. Nature 452:202–205
3. Yang J, He Z, Yang Y, Stoffella P, Yang X, Banks DM, Mishra S (2007) Use of amendments to reduce leaching loss of phosphorous and other nutrients from a sandy soil in Florida. Environ Sci Pollut Res 14(4):266–269

4. Wakida FT, Lerner DN (2005) Non-agricultural sources of groundwater nitrate: a review and case study. Water Res 39(1):3–16
5. Galloway JN, Townsend AR, Erisman JW, Bekunda M, Cai Z, Freney JR, Martinelli LA, Seitzinger SP, Sutton MA (2008) Transformation of the nitrogen cycle: recent trends, questions, and potential solutions. Science 320:889–892
6. Rivett MO, Buss SR, Morgan P, Smith JWN, Bemment CD (2008) Nitrate attenuation in groundwater: a review of biogeochemical controlling processes. Water Res 42:421–4232
7. Nixon S (1995) Coastal marine eutrophication: a definition, social causes and future concerns. Ophelia 41:199–219
8. Rabalais NN (2002) Nitrogen in aquatic ecosystems. Ambio 31:102–112
9. Nolan BT, Hitt KJ, Ruddy BC (2002) Probability of NO3- contamination of recently recharged groundwaters in the conterminous United States. Environ Sci Technol 36(10):2138–2145
10. Foster SSD (2000) Assessing and controlling the impacts of agriculture on groundwater – from barley barons to beef bans. Q J Eng Geol Hydrogeol 33(4):263–280
11. Höring H, Chapman D (2004) NO3-s and nitrites in drinking water. In: World Health Organization drinking water series. IWA Publishing, London
12. Fan AM, Steinberg VE (1996) Health implications of nitrate and nitrite in drink water: an update on methemoglobinemia occurrence and reproductive and developmental toxicity. Regul Toxicol Pharmacol 23:35–43
13. Dorsch MM, Scragg RKR, McMichael AJ, Baghurst PA, Dyer KF (1984) Congenital malformations and maternal drinking water supply in rural South Australia – a case-control study. Am J Epidemiol 119:474–486
14. Ward MH, Mark SD, Cantor KP, Weisenburger DD, Correa-Villasenor A, Zahm SH (1996) Drinking water nitrate and the risk of non-Hodgkin's lymphoma. Epidemiology 7(5):465–471
15. Thompson RB, Martinez-Gaitan C, Gallardo M, Gimenez C, Fernandez MD (2007) Identification of irrigation and N management practices that contribute to nitrate leaching loss from an intensive vegetable production system by use of a comprehensive survey. Agric Water Manage 89:261–274
16. Angelopoulos K, Spiliopoulos IC, Mandoulaki A, Theodorakopoulou A, Kouvelas A (2009) Groundwater nitrate pollution in northern part of Achaia Prefecture. Desalination 248:852–858
17. Möller A, Altfelder S, Moeller HW, Darwish T, Abdelgawad G (2003) A guide to sustainable nitrogen management in agricultural practices, vol 8. ACSAD, BGR and CNRS/L. 90 p
18. Mahvi AH, Nouri J, Babaei A, Nabizadeh R (2005) Agricultural activities impact on groundwater nitrate pollution. Int J Environ Sci Technol 2(1):41–47
19. Ju T, Kou L, Zhang S, Christie P (2006) Nitrogen balance and groundwater nitrate contamination: comparison among three intensive cropping systems on the North China Plain. Environ Pollut 143(1):117–125
20. Halwani J, Ouddane B, Baoudi M, Wartel M (1999) Contamination par les nitrates des eaux souterraines de la plaine d'Akkar au Liban du Nord. Cahiers Santé 9:219–223
21. Ray C (2001) Managing nitrate problems for domestic wells in irrigated alluvial aquifers. J Irrig Drain Eng 127(1):49–53
22. Kay P, Edwards AC, Foulger M (2009) A review of the efficacy of contemporary agricultural stewardship measures for ameliorating water pollution problems of key concern to the UK water industry. Agr Syst 99(2–3):67–75
23. Böhlke JK, Wanty R, Tuttle M, Delin G, Landon M (2002) Denitrification in the recharge area and discharge area of a transient agricultural nitrate plume in a glacial outwash sand aquifer, Minnesota. Water Resour Res 38(7):10.1–10.26
24. Wriedt G, Rode M (2006) Modelling nitrate transport and turnover in a lowland catchment system. J Hydrol 328:157–176
25. Thayalakumaran T, Bristow KL, Charlesworth PB, Fass T (2008) Geochemical conditions in groundwater systems: implications for the attenuation of agricultural nitrate. Agric Water Manage 95:103–115
26. Tesoriero AJ, Liebscher H, Cox SE (2000) Mechanism and rate of denitrification in an agricultural watershed: electron and mass balance along groundwater flow paths. Water Resour Res 36(6):1545–1559

27. Almasri MN, Kaluarachchi JJ (2007) Modeling nitrate contamination of groundwater in agricultural watersheds. J Hydrol 343:211–229
28. Coyne MS (2008) Biological denitrification. In: Schepers JS, Raun W (eds) Nitrogen in agricultural systems. ASA-CSSSA-SSSA Agronomy monograph, vol 49. Madison, WI, pp 197–249
29. Schipper LA, Robertson WD, Gold AJ, Jaynes DB, Cameron SC (2010) Denitrifying bioreactors – an approach for reducing nitrate loads to receiving waters. Ecol Eng. doi:10.1016/j.ecoleng.2010.04.008
30. Seitzinger S, Harrison JA, Böhlke JK, Bouwman AF, Lowrance R, Peterson B, Tobias C, Van Drecht G (2006) Denitrification across landscapes and waterscapes: a synthesis. Ecol Appl 16(6):2064–2090
31. Barnes RT, Raymond PA (2010) Land-use controls on sources and processing of nitrate in small watersheds: insights from dual isotopic analysis. Ecol Appl 20(7):1961–1978
32. Matson PA, Parton WJ, Power AG, Swift MJ (1997) Agricultural intensification and ecosystem properties. Science 277:504–509
33. Brye KR, Norman JM, Bundy LG, Gower ST (2001) Nitrogen and carbon leaching in agroecosystems and their role in denitrification potential. J Environ Qual 30:58–70
34. Jarvis NJ (1994) MACRO, a model of water movement and solute transport in macroporous soils. Swedish University of Agricultural Sciences, Uppsala
35. RZWQM Development Team: Hanson JD, Ahuja LR, Shaffer MD, Rojas KW, De Coursey DG, Farahani H, Johnson K (1998) RZWQM: simulating the effects of management on water quality and crop production. Agric Syst 57:161–195
36. Jasson PE, Karlberg L (2004) Coupled heat and mass transfer model for soil-plant-atmosphere systems. COUP manual, 453 pp. Web site: http://www.lwr.kth.se/vara%20datorprogram/CoupModel/index.htm
37. Šimunek J, Šejna M, Saito H, Sakai M, van Genuchten MTh (2008) The HYDRUS-1D software package for simulating the movement of water, heat, and multiple solutes in variably saturated media, version 4.0, HYDRUS Software Series 3, Department of Environmental Sciences, University of California Riverside, Riverside, California, USA, p 315
38. Gogu RC, Dassargues A (2000) Current trends and future challenges in groundwater vulnerability assessment using overlay and index methods. Environ Geol 39(6):549–559
39. Manos B, Papathanasiou J, Bournaris T, Voudouris K (2010) A DSS for sustainable development and environmental protection of agricultural regions. Environ Monit Assess 164:43–52
40. Manos B, Papathanasiou J, Bournaris T, Voudouris K (2010) A multicriteria model for planning agricultural regions within a context of groundwater rational management. J Environ Manage 91:1593–1600
41. Voudouris K, Polemio M, Kazakis N, Sifaleras A (2010) An agricultural decision support system for optimal land use regarding groundwater vulnerability. Int J Info Syst Soc Change 1 (4):66–79
42. Aller L, Bennett T, Lehr JH, Petty RJ (1985) Drastic: a standardized system for evaluating groundwater pollution potential using hydrogeologic settings. EPA/600/2-85/018. U.S. Environmental Protection Agency, Robert S. Kerr Environmental Research Laboratory, Office of Research and Development
43. Civita M, De Maio M (1997) SINTACS – Un sistema parametrico per la valutazione e la cartografia della vulnerabilità degli acquiferi all"nquinamento. Quaderni di tecniche di protezione ambientale, n. 60. Pitagora Editrice, Bologna
44. Petelte-Giraude E, Dorfliger N, Crochet P (2000) RISKE: Methode d'evaluation multicritere de la cartographie de la vulnerabilite des aquiferes karstiques. Applications aux systemes des Fontanilles et Cent-Fonts (Herault, France). Hydrogeologie 4:71–88
45. Zwahlen F (2003) Vulnerability and risk mapping for the protection of carbonate (karst) aquifers. Eur Comm Cost Act 620:42
46. Aveline A, Rousseau ML, Guichard L, Laurent M, Bockstaller C (2009) Evaluating an environmental indicator: case study of MERLIN, a method for assessing the risk of nitrate leaching. Agric Syst 100:22–30

47. Carey MA, Lloyd JW (1985) Modelling non-point sources of nitrate pollution of groundwater in the Great Ouse Chalk, UK. J Hydrol 78:83–106
48. Pierce FJ, Shaffer MJ, Brodhal MK (1991) Spatial distribution of NO3 leaching risk to water supplies under cropland using NLEAP and GRASS GIS. American Society of Agronomy, Madison, p 297
49. Shaffer MJ, Wylie BK, Hall MD (1995) Identification and mitigation of nitrate leaching hot spots using NLEAP–GIS technology. J Contam Hydrol 20:253–263
50. Birkinshaw SJ, Ewen J (2000) Nitrogen transformation component for SHETRAN catchment nitrate transport modelling. J Hydrol 230:1–17
51. Lasserre F, Razack M, Banton O (1999) A GIS-linked model for the assessment of nitrate contamination in groundwater. J Hydrol 224(3–4):81–90
52. Refsgaard JC, Thorsen M, Jensen JB, Kleeschulte S, Hansen S (1999) Large scale modeling of groundwater contamination from nitrate leaching. J Hydrol 221:117–140
53. De Paz JM, Ramos C (2002) Linkage of a geographical information system with the GLEAMS model to assess nitrate leaching in agricultural areas. Environ Pollut 118:249–258
54. De Paz JM, Delgado JA, Ramos C, Shaffer MJ, Barbarick KK (2009) Use of a new GIS nitrogen index assessment tool for evaluation of nitrate leaching across a Mediterranean region. J Hydrol 365:183–194
55. Cinnirella S, Buttafuoco G, Pirronea N (2005) Stochastic analysis to assess the spatial distribution of groundwater NO3- concentrations in the Po catchment (Italy). Environ Pollut 133:569–580
56. Onorati G, Di Meo T, Bussettini M, Fabiani C, Farrace MG, Fava A, Ferronato A, Mion F, Marchetti G, Martinelli A, Mazzoni M (2006) Groundwater quality monitoring in Italy for the implementation of the EU water framework directive. Phys Chem Earth 31:1004–1014
57. Mastrocicco M, Colombani N, Castaldelli G, Jovanovic N (2011) Monitoring and modeling nitrate persistence in a shallow aquifer. Water Air Soil Pollut 217(1–4):83–93
58. Mastrocicco M, Colombani N, Palpacelli S, Castaldelli G (2011) Large tank experiment on nitrate fate and transport: the role of permeability distribution. Environ Earth Sci 63(5):903–914
59. Mastrocicco M, Colombani N, Salemi E, Castaldelli G (2011) Reactive modeling of denitrification in soils with natural and depleted organic matter. Water Air Soil Pollut 222(1–4):205–215
60. Provini A, Crosa G, Marchetti R (1992) Nutrient export from the Po and Adige river basin over the last 20 years. Sci Total Environ 127(Suppl):291–313
61. Palmieri L, Bendoricchio G, Artioli Y (2005) Modelling nutrient emissions from river systems and loads to the coastal zone: Po River case study, Italy. Ecol Model 184:37–53
62. Gavilan P, Berengena J, Allen RG (2007) Measuring versus estimating net radiation and soil heat flux: impact on Penman–Monteith reference ET estimates in semiarid regions. Agric Water Manage 89:275–286
63. Lenka S, Singh AK, Lenka NK (2009) Water and nitrogen interaction on soil profile water extraction and ET in maize–wheat cropping system. Agric Water Manage 96:195–207
64. Drexler JZ, Snyder RL, Spano D, Paw KT (2005) A review of models and micrometeorological methods used to estimate wetland evapotranspiration. Hydrol Process 18:2071–2101
65. Sharma ML (1985) In: Hillel D (ed) Estimating evapotranspiration, vol 3, Advances in irrigation. Academic, New York, pp 213–281
66. Allen RG, Pereira LS, Raes D, Smith M (1998) Crop evapotranspiration – guidelines for computing crop water requirements. FAO Irrigation and drainage paper, 56, FAO, Rome
67. Berengena J, Gavilàn P (2005) Reference ET estimation in a highly advective semi-arid environment. J Irrig Drain Eng ASCE 131(2):147–163
68. Jensen ME, Burman RD, Allen RG (1990) Evapotranspiration and irrigation water requirements. In: ASCE manual no. 70. ASCE, New York
69. Passioura JB (2002) Environmental biology and crop improvement Funet. Plant Biol 29:537–546
70. Duchemin B, Hadria R, Erraki S, Boulet G, Maisongrande P, Chehbouni A, Escadafal R, Ezzahar J, Hoedies JCB, Kharrou MH, Khabba S, Mougenot B, Olioso A, Rodriguez JC, Simonneaux V (2006) Monitoring wheat phenology and irrigation in Central Morocco: on the

use of relationships between evapotranspiration, crops coefficients, leaf area index and remotely-sensed vegetation indices. Agric Water Manage 79:1–27
71. Klute A (1986) Methods of soil analysis, vol 9(1), 2nd edn, Agronomy. American Society of Agronomy, Madison
72. Ryden JL, Lund L, Focht D (1978) Direct in-field measurement of nitrous oxide flux from soils. Soil Sci Soc Am J 42:731–737
73. Schjørring JK (1995) Long-term quantification of ammonia exchange between agricultural cropland and the atmosphere – I. Evaluation of a new method based on passive flux samplers in gradient configuration. Atmos Environ 29:885–893
74. Stanhill G (1969) A simple instrument for the field measurement of turbulent flux. J Appl Meteorol 8:509–513
75. Mastrocicco M, Colombani N, Salemi E, Vincenzi F, Castaldelli G (2012) The role of the unsaturated zone in determining nitrate leaching to groundwater. In: Zuber A, Maloszewski P, Witczak S, Malina G (eds) Groundwater quality sustainability. CRC Press, Leiden. ISBN 9780415698412
76. Mastrocicco M, Colombani N, Palpacelli S (2009) Fertilizers mobilization in alluvial aquifer: laboratory experiments. Environ Geol 56(7):1371–1381
77. Schuwirth N, Hofmann T (2006) Comparability of and alternatives to leaching tests for the assessment of the emission of inorganic soil contamination. J Soils Sediments 6(2):102–112
78. Bouwer H, Rice RC (1976) A slug test for determining hydraulic conductivity of unconfined aquifers with completely or partially penetrating wells. Water Resour Res 12(3):423–428
79. Mastrocicco M, Colombani N, Salemi E, Castaldelli G (2010) Numerical assessment of effective evapotranspiration from maize plots to estimate groundwater recharge in lowlands. Agric Water Manage 97(9):1389–1398
80. Mastrocicco M, Vignoli G, Colombani N, Abu Zeid N (2010) Surface electrical resistivity tomography and hydrogeological characterization to constrain groundwater flow modeling in an agricultural field site near Ferrara (Italy). Environ Earth Sci 61(2):311–322
81. van Genuchten MT (1980) A closed-form equation for predicting the hydraulic conductivity of unsaturated soils. Soil Sci Soc Am J 44:892–898
82. Feddes RA, Kowalik PJ, Zaradny H (1978) Simulation of field water use and crop yield. Pudoc, Wageningen
83. Food and Agriculture Organization of the United Nations (1990) Expert consultation on revision of FAO methodologies for crop water requirements, *ANNEX V*, FAO Penman-Monteith Formula, Rome, Italy
84. Jensen DT, Hargreaves GH, Temesgen B, Allen RG (1997) Computation of Eto under nonideal conditions. J Irrig Drain 123(5):394–400
85. Turk L (1961) Estimation of irrigation water requirements, potential evapotranspiration: a simple climatic formula evolved up to date. Ann Agron 12:13–49
86. Aschonitis VG, Mastrocicco M, Colombani N, Salemi E, Kazakis N, Voudouris K, Castaldelli G (2012) Assessment of the intrinsic vulnerability of agricultural land to water and nitrogen losses, via deterministic approach and regression analysis. Water Air Soil Pollut 223 (4):1605–1614
87. Salemi E, Colombani N, Aschonitis V, Mastrocicco M (2011) Assessment of groundwater nitrogen contamination risk using LOS indices in the Ferrara Province, Italy. In: Advances in the research of aquatic environment, vol 2. Springer, Berlin. doi:10.1007/978-3-642-24076-8
88. Knisel WG, Davis FM (2000) GLEAMS, Groundwater Loading Effects from Agricultural Management Systems V3.0. Publ. No. SEWRL-WGK/FMD-050199, U.S.D.A., Tifton, Georgia

Chapter 16
Environmental Risk Assessment of WWII Shipwreck Pollution

Justyna Rogowska and Jacek Namieśnik

16.1 Introduction: Definition of a Shipwreck

To begin with, we need to address the question as follows: what is a wreck? It is important to define precisely what is meant by a 'wreck', because wrecks are objects of legal regulation. In the popular imagination, a wreck is a ship that has sunk or is a very badly damaged vessel that has been abandoned by her crew. A definition of the concept of a wreck has long been missing from international legal parlance. The United Nations Convention on the Law of the Sea (UNCLOS) of 1982, which lays down the norms for the utilization of the sea, does not contain an unequivocal definition of a wreck. Article 303 of UNCLOS obliges signatory states to protect objects of an archaeological and historical nature found at sea and to cooperate for this purpose. The same article also prohibits the removal of such objects from the seabed without the express consent of the coastal state [1]. On the other hand, the definition of 'underwater cultural heritage' in the UNESCO Convention on the Protection of the Underwater Cultural Heritage (2001) embraces '*all traces of human existence having a cultural, historical or archaeological character which have been partially or totally under water, periodically or continuously, for at least 100 years such as vessels, aircraft, other vehicles or any part thereof, their cargo or other contents, together with their archaeological and natural context*'. A definition of a wreck was incorporated into international maritime law within the framework of the Nairobi International Convention

J. Rogowska (✉) • J. Namieśnik
Department of Analytical Chemistry, Chemical Faculty, Gdańsk University of Technology, 11/12 Narutowicza Str, 80-233 Gdańsk, Poland
e-mail: j.j.rogowska@wp.pl

S.K. Sharma and R. Sanghi (eds.), *Wastewater Reuse and Management*,
DOI 10.1007/978-94-007-4942-9_16,

on the Removal of Wrecks (Nairobi Convention) of 18th May 2007. In this convention, a wreck is defined as:

(a) A sunken or stranded ship
(b) Any part of a sunken or stranded ship (including any object that is or has been on board such a ship)
(c) Any object that is lost at sea from a ship and that is stranded, sunken or adrift at sea
(d) A ship that is about, or may reasonably be expected, to sink or to strand, where effective measures to assist the ship or any property in danger are not already being taken

In the case of this last-mentioned convention, it should be noticed that it does not apply to warships or other vessels in the possession of a state and being utilized by it for non-commercial purposes (art. 4 clause 2). Clause 3 of article 4 does allow for the convention's application to warships and governmental vessels, but only by a flag State and not by a coastal State [1]. Also, the Convention on the Protection of the Underwater Cultural Heritage excludes the application of its rules to warships and ships of state (flagships): in article 3, it makes reference to the principles of international maritime law contained, inter alia, in UNCLOS.

16.2 Shipwreck: Classification

In the same way as ships, shipwrecks can be classified according to many different categories, for example, their purpose (merchant vessel, passenger ship), type of propulsion, draught (surface and underwater) and construction. The basic system for classifying ships, and therefore their wrecks, divides them, as mentioned above, into 'warships' and 'all other vessels'. The Convention on the High Seas (Geneva, 29th April 1958) defines a warship as '*a ship belonging to the naval forces of a State and bearing the external marks distinguishing warships of its nationality, under the command of an officer duly commissioned by the government and whose name appears in the Navy List, and manned by a crew who are under regular naval discipline*'. The definition of a warship in UNCLOS is very similar.

Another, equally valid criterion of classification, is their cultural value, that is,:

- Wrecks deemed to be historical finds
- Wrecks of no cultural value

Wrecks defined as historical finds are a source of knowledge about past ship construction technologies, trade, craftsmanship and military activities. A wreck is also valuable in that it provides an opportunity to become acquainted with past cultures, habits and customs, etc.

As far as the environmental effect of wrecks is concerned, their most germane classification divides them into:

- Wrecks constituting a threat to the environment
- Wrecks not constituting a threat to the environment

Such a classification is applicable to both war- and merchant ships.

16.3 The Problem of Wrecks in the Waters of the World's Seas and Oceans

The problem of wrecks lying at the bottom of seas and oceans can be addressed from two angles: a wreck may constitute a physical obstruction to shipping, or it may cause considerable damage to the marine environment and coastal ecosystems. The presence of recent wrecks in sea and ocean waters is the consequence of the intensification of (apparently profitable) maritime transport. The growing demand for oil products has resulted in an ongoing search for new oilfields on land and at sea and for increasingly cheaper and faster methods of getting petroleum products to their markets. Accordingly, ever larger tankers to transport the oil are being designed and constructed, and the tanker traffic along international sea lanes is growing progressively heavier [2]. In 2007, the total world oil production was 85 million barrels per day (mbpd); approximately half of this volume of oil (43 mbpd) was transported along the international shipping lanes of the world's seas [3]. Every tanker emergency can lead to the pollution of both the water and the adjacent shorelines. For example, as a result of the *Amoco Cadiz* disaster on 17 March 1978 on the coast of Brittany (France), more than 230,000 t of oil flowed into the sea. Between 1939 and 1997, an estimated total of 21,486 merchant vessels were recorded as total losses (actual and constructive) comprising a gross tonnage of 76 million tonnes. Approximately one third of this total, 5,915 vessels with a gross tonnage 25,537,630 tonnes, was lost between 1939 and 1946 [4]. On the other hand, Monfils states that over 7,800 vessels, including more than 860 oil tankers, were sunk during WWII worldwide [5]. The shipwrecks of merchant vessels and warships now lying at the bottom of the seas and oceans were caused primarily by hostilities at sea during WWI and WWII. It is estimated that the largest numbers of wrecks are lying at the bottom of the north-west and north-east Atlantic and in the South Asia–Pacific region [6] as a result of the intense fighting in these areas: the Battle of the Atlantic and the Pacific War gave rise to a great many environmental time bombs. It is clearly just a matter of time before petroleum products begin to seep from wrecks corroding in sea water. The risk of seepage is governed by numerous parameters characterizing not only the wreck itself (the ship's condition at the time of sinking, the type and quantity of fuel carried, the time elapsed since the sinking, the depth of the wreck) but also the environment in which it is lying (e.g. salinity, wind and wave action). The effects of seepage depend on many factors, such as the physicochemical properties of the oil, the nature of the environment and the physical, chemical and biological processes taking place in it, for example, spreading, evaporation, dispersion, dissolution, oxidation, sedimentation and biodegradation. In every case, oil spills pose a threat to marine flora and fauna, and damage marine and coastal ecosystems. Many petroleum constituents are toxic and carcinogenic and therefore exert an adverse effect on marine plankton, fish and mammals. Oil spills can also have serious economic and social consequences. Collective or public losses are usually identified with cleaning and restoration costs [7]. The clean-up of a major oil spill involves both the use of special equipment to combat oil pollution (vessels, skimmers, etc.) and extensive personnel costs [8]. But it is the indirect costs of pollution that have the greatest economic effects. Unlike the direct costs, which are relatively easily quantified, the additional extensive indirect costs are much more difficult to pinpoint. The indirect costs are due primarily to the increased stress on local services and reduced industrial incomes [8]. For example, after the *Exxon Valdez* disaster, during the 42 million litres of Alaskan North Slope crude oil released into the sea [9], the Exxon Corporation paid immediate post-spill cost to fishermen and clean-up $2.3 billion [10]. Furthermore, the Exxon was ordered to pay $900 million more over 10 years by the Federal and State officials for restoration of the natural environment [10]. In 1994, an Alaska jury ordered ExxonMobil to pay $287 million as compensation to fishermen and $5 billion as punitive compensation. In 2006, an appeals court reduced punitive damages for the *Exxon Valdez* oil spill to $2.5 billion, half the original amount. In 2008, the U.S. Supreme Court reduced that $2.5 billion in punitive damages to just more than $507 million [39].

It is not only the petroleum products, carried as fuel or cargo, that constitute a danger to marine ecosystems but also corroded ammunition, containing chemical warfare agents (CWA) among other things, and chemicals carried as cargo. As a result of the disarmament of Germany following WWII, approximately 65,000 tonnes of stockpiled CWA munitions were ordered by the allied forces to be disposed of. A significant proportion of them were subsequently dumped at sea during the late 1940s [11]. Ships filled with chemical weapons were sunk mainly in the Baltic, the Atlantic Ocean, the North Sea and the southern Adriatic [12]. Approximately 40,000 tonnes of chemical munitions containing no more than 13,000 t of chemical warfare agents are known to have been dumped in the Baltic Sea in the period after WWII [40]. The main dumping areas are south-east of Gotland (south-west of Liepaja), east of Bornholm and south of the Little Belt [15]. From 1945 to 1948, British and American occupation forces sank one German merchant vessel in the Norwegian Sea and 41–43 such vessels at two sites in the Skagerrak. These vessels contained approximately 150,000 t of chemical weapons and conventional ammunition [13]. There were two different methods of dumping: simply throwing them overboard, which was done in the Baltic, and sinking them in old ships, which was done in the North Sea (Skagerrak) [14]. These munitions have been lying on the seabed and in the sediment of the Baltic Sea for approximately 60 years now, and the extent of corrosion of the shells, and hence the release of the toxic chemicals is raising environmental and human health concerns in the countries adjoining the Baltic. Some shells will already have seeped or be seeping their contents, whereas others may still be intact [11]. Almost all warfare agents break down at varying rates into less toxic, water-soluble substances. Some compounds, however, are very poorly soluble in water and degrade very slowly (viscous mustard gas, Clark I and II and Adamsite). It is therefore unlikely that these will occur in larger concentrations in the water, so a large-scale threat to the marine environment from dissolved chemical warfare agents can be ruled out. Nevertheless, elevated levels of sparingly soluble Clark, Adamsite or mustard gas in viscous form might occur in the sediment in the immediate vicinity of dumped munitions [15]. Examination in 2002 of four wrecks lying in the Skagerrak indicated that the wrecks themselves did not show any significant signs of corrosion. Nonetheless, some of the ammunition on board had been pierced by corrosion, so the contents will probably have seeped out. This indicates that the release of chemical ammunition into the sea will go on for a long time, but the instant release of arsenicals and sulphur mustard will not result in large toxic plumes. Owing to their low solubility and high density, these compounds will spread as pieces and eventually settle on the sea floor, where they could affect local marine organisms [16]. Only the collapse of a wreck containing tabun as a result of, say, fishery activities could affect a large volume of water in a worst-case scenario. Moreover, no acute toxic effects have been caused by the chemical warfare agents in the Skagerrak [17]. The ammunition dumped in the Skagerrak can therefore be regarded as much safer than that dumped in the Baltic. The fact that the warfare agents are inside a ship's hull has also prevented further spreading [15], whereas ammunition thrown overboard does constitute a threat because they could be picked up by fishermen or thrown up onto beaches during storms.

16.4 Information on the Strategies and Programmes for Identifying Wrecks and Preventing Pollution from Them

These threats and the new opportunities for the easier penetration of wrecks have raised awareness of this problem and encouraged efforts aiming to regularize their legal status [1]. Successful attempts at identifying wrecks and determining the hazards they may pose were undertaken in the southern Pacific within the framework of the Pacific Ocean Pollution Prevention Programme (PACPOL), concerning contamination originating from ships, which is part of the South Pacific Regional Environment Programme (SPREP). During WWII, East Asia and the Pacific were a theatre of intense hostilities between Japan and the USA. Such battles as *Pearl Harbor, Guadalcanal* or *Midway* not only had a significant effect on the progress of the war but also caused enormous human and material losses. As a result, there are a great many ships lying at the bottom of the Pacific that sank with their loads and tanks of fuel. These environmental time bombs contain fuel, greases, chemicals and unexploded artillery devices. Figure 16.1 shows the distribution of wrecks in the Pacific–East Asian region [18].

One of these vessels is the aforementioned USS *Mississinewa*. This tanker was sunk in 1944 on the Ulithi Atoll (the Federated States of Micronesia). The fully loaded 24,400-tonne tanker contains around 19 million litres of highly volatile

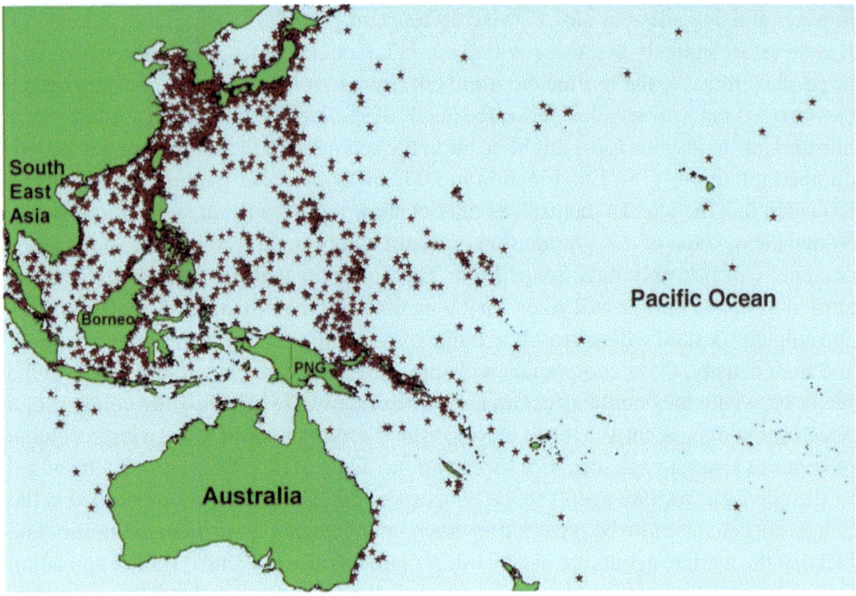

Fig. 16.1 The location of WWII shipwrecks in the Pacific–East Asian region (SPREP database)

aviation gasoline and fuel oil. In 2001, as a result of a tropical storm that passed over this area, an estimated 68,000–91,000 l of this fuel seeped into the sea in the course of some 60 days [19]. The legal upshot of this event was the development of a regional strategy for dealing with pollution originating from wrecks. The Pacific Ocean Pollution Prevention Programme of SPREP was tasked with formulating a regional strategy to address marine pollution from WWII Wrecks, which aims to:

- Prevent/minimize damage to marine and coastal environments and resources as a result of marine spills from WWII wrecks
- Ensure that any action taken will respect the sanctity of these sites, as they are often also war memorials and gravesites [18]

The strategy recommends a five-step approach to addressing the issue:

1. Data collection and analysis – to continue with the identification and analysis of wreck sites, wreck depth, access and cargoes/bunkers carried.
2. Generic risk assessment – to carry out a generic risk assessment to compare the risk levels between sites. The tool used by the US Department of Defense was identified as the most appropriate one to use. All sites will be ranked high, medium and low.
3. Agree on the interventions: high risk – direct (pump out, salvage); medium risk – manage site (contingency plan, restricted access); low risk – leave alone and monitor.
4. Site specific risk assessments – according to the level of risk, specific site assessments will need to be carried out. These would include environmental impact assessment (EIA), shoreline contamination assessment and treatment (SCAT), consultations with coastal and flag States and logistics assessments.

5. Planning implementation – drawing up of final implementation plans for interventions, including determining responsibility budgets, timing, logistical requirements, environmental and social issues [19].

The effect of implementing this strategy was to create a database containing information on more than 3,800 wrecks, of which more than 330 were tankers and oilers [19].

The creation of the Asia Pacific database inspired the formation of the Atlantic, Mediterranean and Indian Ocean (AMIO) WWII shipwreck database. This new geographic database, although still in its initial development, highlights the significant number of WWII shipwrecks globally. The AMIO database details the location and ownership of over 3,950 vessels, over 1,000 tonnes, of which 529 are oil tankers [5].

The Gulf of Mexico is another region where there is a large accumulation of wrecks of tankers, passenger and fishing vessels. During WWII, these waters were the scene of considerable German U-boat activity, which targeted the shipping lanes along which crude oil was being carried for American industry and the US Navy. An estimated 56 merchant ships were sunk in 1942 and a further 14 were damaged [20]. In the summer of 2004, a multidisciplinary team composed of archaeologists, biologists, filmmakers, oceanographers, professional surveyors and remotely operated vehicle (ROV) operators worked together on the Deep Wrecks Project to document and analyse the biological and archaeological aspects of six World War II era shipwrecks in the deep waters of the Gulf of Mexico [21]. The archaeological aspects of the project were to [22]:

1. Confirm each shipwreck identity and establish its type and date of construction, nationality, ownership (past and present), use history, mission and cargo at the time of loss
2. Determine each vessel's past and present condition, state of preservation, assess any environmental impact caused by the wreck and make observations relating to its deterioration and future research potential
3. Determine the visible extent of the debris fields surrounding each casualty
4. Analyse imagery and historical documentation to determine nomination eligibility for the National Register of Historic Places
5. Assess the impacts of biofouling communities on these shipwrecks to determine the site's stability
6. Prepare a National Register nomination form for vessels deemed eligible for nomination.

The biological part of the project aimed to [22]:

1. Characterize the environment at each site (e.g. water depth, bottom sediment type, currents).
2. Determine the biological effects of shipwreck artificial reefs at selected sites and include detailed imagery surveys of the sites at a variety of scales.
3. Determine the extent of physical and biological modification of sediments in the immediate area of wreck sites compared to sediment conditions at sites distant

from wreck areas. Sampling will include sediment coring close to and distant from wrecks to determine any 'artificial reef effect'.
4. Conduct limited sampling of fauna attached to hard substrate for taxonomic and other potential analyses such as isotope studies.
5. Analyse imagery and sample collection to examine the spatial heterogeneity of any fouling community and motile fish and invertebrate association with wrecks.

The effect of implementing this project was to assess the state of particular wrecks, their rate of deterioration and their potential threat to the marine environment of the Gulf of Mexico.

Hostilities were intense in the Baltic as well. The Danish Straits, including the Skagerrak, were a strategically important area, as they link the Baltic with the North Sea. In 2003–2007 the Scandinavian countries – Denmark, Sweden and Norway – together with the EU established cooperation within the framework of Forum Skagerrak II. This project was a continuation of the earlier Forum Skagerrak I. Its aim was to find out more about the Skagerrak and to implement concrete proposals for enhancing the attractiveness of this region. One of the areas of the programme (WP2: hazardous substances, waste and oil spills) envisaged the identification of wrecks in the Skagerrak region and an assessment of the environmental hazards they could pose. The first stage of the project, the gathering of information about ships that sank in the Skagerrak was based on the following criteria:

- Registered tonnage of at least 100 tonnes
- Sinking after 1914
- Fuelled by oil, diesel oil or similar
- Transport of dangerous cargoes, that is, crude oil, fuel oil, ammunition, chemical warfare agents, etc. (even if they did not fulfil the above criteria) [23]

This work produced a database on 261 wrecks constituting a potential threat to marine ecosystems. Most of these wrecks are from the WWII period and are badly corroded. The best known and most thoroughly investigated of these wrecks is that of the s/s *Skytteren*, now recognized as being the largest oil-seeping wreck in the Skagerrak [8]. The *Skytteren* sank during an attempt at escaping from the port of Gothenburg (Göteborg) in 1942. It is estimated that she had some 6,000 m^3 of light oil in her tanks. According to calculations performed by the Swedish Coastguard in 2005, up to 100 l of oil may be seeping into the sea in 24 h [8]. The calculations were repeated in 2006 and showed that the extent of the area affected by the seep had not changed. During flights in 2007 over the area where the s/s *Skytteren* had sunk, no signs of oil seepage were seen, which could mean that all the oil had already seeped out or that besides the light oil, there could also be heavier oil in the tanks, which would drop down to the seabed [8].

The case of the s/s *Skytteren* shows that there is an urgent need to work out long-term action plans regarding the systematic identification of wrecks and the assessment of the potential environmental threats that they may pose. Accurate identification is essential before any decision on remediation can be taken. Norway is an example of a state that possesses a detailed database on wrecks that could endanger the environment. Beginning in 1991, the Norwegian Pollution Control Authority (NPCA) (now the Norwegian Climate and Pollution Agency) implemented a programme to identify, localize and assess potential areas affected by wrecks. In time, a database was created containing information on more than 2,000 wrecks of over 100 GRT that had sunk in Norwegian territorial waters after 1914 [24]. The creation of this project was not the result of an observed seepage but of the awareness of the Norwegian authorities of the seriousness of the problem and of the possible effects of seepage. The first stage of the project was to find information on ships that had sunk in Norwegian territorial waters in domestic and foreign archives. The wrecks were categorized into one of three classes depending on the extent of the danger they posed to the environment:

C – wrecks posing a serious threat to the environment
P – wrecks posing a potential threat
N – wrecks not posing any such threat [24]

In 1993, the NPCA began a systematic investigation of the wrecks, beginning with those in class C. Fuel was pumped out of the wreck of the German cruiser *Blücher*, which had sunk in Oslofjord in 1940. During this operation some 60–80% of the fuel in the cruiser's tanks was removed.

16.5 Contamination of the Marine Environment by Substances Seeping from Wrecks: A Risk Assessment

The first step in any investigation with the objective of assessing the effect of wrecks and/or their remains on ecosystems should be the identification of the wreck, its position and the cargo it was carrying. Three situations should be taken into consideration:

1. A wreck has been found and identified.
2. An unidentified wreck has been found
3. A particular wreck is being sought on the basis of available information.

Once a wreck has been identified and its position precisely defined, one needs to gather and analyse archival data relating to the vessel and the cargo it was carrying at the time of her sinking. However, it is often the case that there is no such information for ships sunk during WWII. Moreover, in order to localize the fuel and cargo, it would be useful to have the vessel's design drawings to hand or, if these are unavailable, the drawings of a vessel of the same or a similar class. The type of ship, her design, age and purpose very often determine the amount and type of fuel or cargo. It is equally important to know the cause of sinking. The situation becomes much more complicated when an unidentified wreck has happened upon. This is most frequently the case during routine echo sounder surveys of the seabed in the search for obstructions to shipping, the compilation of bathymetric maps, during recreational diving, etc. In such a case, we need to gather every possible scrap of information that could shed light on the vessel's identity. The first step here is to collect information on the site of sinking, such as:

- The wreck's exact position
- The depth at which the wreck is lying
- The type of ship (warship, merchant ship, passenger ship, etc.)
- The wreck's dimensions

These data are acquired using different types of measuring instruments, that is, echo sounders, sonar and magnetometers. But despite these facilities and their high technical potential, the 100% certain identification of an underwater object, its condition and appearance, still sometimes requires it to be inspected visually. This can be done in a number of ways, for example, using hyperbaric methods (with human diving teams) or unmanned underwater vehicles. Unmanned underwater vehicles, most frequently of the ROV type (remotely operated vehicle), may carry a TV camera, a sonar, instruments for measuring the hydrological parameters of the water, etc. A big advantage of the ROV is its relatively high mobility and ability to work long hours at high depths, sometimes in extremely difficult conditions [25]. Once we have this basic information, we can then extend the search to archives, museums and Lloyd's Register. An example of such an accidentally discovered wreck is that of the MS *Steuben*. This was a passenger liner that was torpedoed on the Baltic as it was escaping from Pillau (since 1946 Baltiysk, Russia), becoming a tomb for 3–5 thousand people. After the wrecks of the MS *Wilhelm Gustloff* (5–10 thousand victims) [26, 27] and the MS *Goya* (6–7 thousand victims), it is the largest underwater tomb in the world. In 2004, a hydrographic vessel of the Polish Navy came across the wreck during a routine survey of the bottom of the Baltic. Some 70 km from the Polish coast, the multibeam sounder recorded a large object at a depth of 72 m. More detailed sonograms were obtained using a ROV. These, as well as photographs taken from the ROV, were compared with design drawings of the ship found in German archives. There was no doubt that the wreck was that of the MS *Steuben* [28]. The USS *Mississinewa*, mentioned earlier, was also found by accident. For 57 years, this vessel had lain at the bottom of a lagoon undiscovered until early 2001, when an adventure dive team found the ship and pinpointed her location [19].

If, however, we are looking for the wreck of a particular ship, then the procedure is carried out in reverse: first, we look up all the possible information about the ship and the possible sites of her sinking, and then we go in search of the wreck. This is the procedure adopted in the case of wrecks that have a high material value (because of the cargo they carried) or are of historical importance, both by scientists from various research institutions and by treasure-seeking amateurs or commercial ventures. Examples are the discoveries of the wreck of the RMS *Titanic* by Dr Robert Ballard's expedition in 1985 [29], or that of the British warship HMS *Victory*, which sank in the English Channel in 1744 and was localized by Odyssey Marine Exploration in 2008.

Other important factors in the search for a wreck are the course of its final voyage and attempting to establish the site of sinking. At the present time, there is hardly any object that cannot be located and explored on the seabed. Sophisticated equipment can pinpoint any anomaly on the seabed, and advanced technology enables most objects to be raised [30]. The problem here, though, concerns the extent of the search area. For example, in 2008, an expedition was organized to try to find and identify the wreck of the Polish submarine ORP *Orzeł* in the North Sea. The legendary *Orzeł* became famous in 1939 for her daring escape from internment at Tallinn (Estonia) and sailing across the Baltic through the Danish Straits without the aid of charts. After a 40-day voyage, the *Orzeł* reached her haven at Rosyth (Scotland) [31]. In 1940, the submarine disappeared during a patrol in the North Sea. Despite 3 years of preparations for this expedition and the use of specialist equipment, the wreck could not be found.

The assessment of the pollution hazards posed by wrecks is a complex problem and should be carried out on two planes:

1. An initial appraisal of the state of the environment and the wreck to assess the seriousness of possible seepage
2. The gathering of relevant information upon which to base action taken to minimize the effects of seepage

If analysis of literature data or observed leaks indicates that the wreck may pose a threat to the marine environment, the wreck needs to be inspected for damage and deformations that have resulted from:

- Lying a long time in the water
- A fire or explosion that took place before or during sinking
- The action of elevated pressure, if the wreck is lying at a great depth
- The action of strong winds and sea currents

Very often the wreck's condition is evaluated at the same time as its identification.

It should be emphasized at this juncture that seas and oceans are not environments conducive to the preservation of wrecks. The corrosion rate is influenced by dissolved oxygen, temperature, pH, salinity, current velocity, wave action and marine growth. This means that eventually some or all of the load will seep into the water. Despite the fact that corrosion processes are impeded by the formation of marine encrustation (made up of barnacles, clams, mussels and other filter-feeding marine life that leave behind a thick 'cement-like' covering of calcium carbonate on the iron or steel hulls), metal deterioration can continue owing to the presence of sulphate-reducing bacteria. These bacteria play a large part in the corrosion of metals, especially of iron and steel in salt water: as much as 60% of the corrosion of iron in salt water can be attributed to bacterial action [32]. Apart from governing biological processes, the oxygen status of water also affects the chemical degradation of metals: well-oxygenated waters lead to the rapid degradation of iron objects through the growth of hydroxides [33]. Dissolved oxygen is the only factor affecting the corrosion rate in deep waters, where the temperature is always constant and near 0 °C and the current velocity is likewise zero. Thus, the model predicting the corrosion rate in a deep sea based on field testing employs dissolved oxygen as a parameter [34]. In order to assess the state of a wreck, it may be necessary to collect samples of the wreck's skin in order to be able to determine the degree and rate of metal corrosion. Obtaining information on the corrosion processes and understanding their mechanisms is essential for working out an approximate timeline of the eventual fate of these vessels. Moreover, applying this corrosion timeline may enable us to anticipate when tanks and holds containing fuel oils or other hazardous materials are likely to fail and release their cargoes into the environment [32]. To evaluate the threat and a possible salvage operation, we also need to take samples of the fuel that the vessel's engines ran on and to analyse them for their content, possible toxicity, dispersant properties and rate of degradation. It is expected that most oil in large vessels sunk during WWII will be heavy fuel oil, along with diesel, lubrication oils and some aviation fuels and petrol. Vessels such as submarines will have been mostly diesel driven and have smaller quantities of fuel on board, whereas large carriers, battleships, destroyers and the like will probably have large quantities of heavier fuel oil [35]. It is important to sample fuel directly from a wreck, because sampling leaked fuel may yield erroneous fuel characteristics, as fuel that has leaked into the sea is subject to physical, chemical and biological transformations such as evaporation, oxidation and biological degradation. It is often very difficult to assess the quantity and type of oil a vessel was carrying. That is why, it is crucial not only to obtain documents and design drawings relating to the ship but also to associate historical and technical data in order to acquire as much valuable information as possible.

It is equally crucial, as far as environmental contamination is concerned, to analyse the immediate surroundings of the wreck. Sea water, bottom sediments and the species composition and condition of benthic organisms can provide significant information on the degree of pollution of the environment and the extent of the area affected by contaminants seeping out of the wreck.

Investigations into the state of the environment should be comprehensive and include:

- The characteristics of the environment – sea depth, water temperature, type of sediments, type of bottom, sea currents, storms, etc.
- A definition of the physical, chemical and biological changes to the bottom sediments in comparison with sediments sampled well away from the wreck
- Sampling biota in order to identify the species composition, changes in the composition of their populations, as well as changes in the individual organisms themselves

Figure 16.2 shows the oil slick that formed during the sampling of bottom sediments around the wreck of the s/s *Stuttgart*.

Investigations into the condition of the environment and identification of the type of fuel are also necessary to assess the effect of a possible oil spill on the environment. The impact of an oil spill on biota depends on a number of factors, such as:

- The rate of spread of the oil slick
- The oil composition

Fig. 16.2 The oil slick visible on the surface of the water after sampling the sediments at the site of the WWII wreck of the s/s *Stuttgart* (Lazarettschiff C) (Gulf of Gdańsk – Polish coast)

- The location of the spill
- The time or season of the accident (bird migrations)
- The properties, toxicity and stability of the petroleum substance
- The species biodiversity at the site of the oil spill
- Environmental sensitivity, that is, proximity of bird habitat, beaches, rocks, wetlands
- The number and type of habitats [2]

Three major pathways lead to long-term impacts:

1. The chronic persistence of oil and biological exposure affect populations of species that are closely associated with shallow sediments.
2. The delayed population effects of sublethal doses that may compromise health, growth and reproduction.
3. The indirect effects of the trophic interaction cascade, all of which transmit effects well beyond the acute-mortality phase [36].

Further potential effects of a spill could include:

– The loss of natural value, that is, the reduced attractiveness of a coastal area, leading to reduced income from tourism
– The temporary suspension of fisheries as a result of the pollution of the water, the contamination of fishing boats and gear, giving rise to perturbations in the fishery and fish processing sector

Table 16.1 Comparison of selected parameters affecting the cost of operations to remove oil from the wrecks of the SS *Jacob Luckenbach* and the USS *Mississinewa* [6, 38]

Vessel name	SS *Jacob Luckenbach*	USS *Mississinewa*
Year of sinking	1953	1944
Location	About 10 miles south-east of Farallon Island	Ulithi Atoll, Yap State, Federated States of Micronesia
Type of vessel	Freighter	Oiler
Type of oil	457,000 Gallons (1.7 mln l) of ship's fuel	19 Million litres of NSFO, (Navy Special Fuel Oil), gasoline and diesel fuel
Depth of wreck	175 ft (over 53 m)	Max. 130 ft (40 m)
Condition of the wreck	Massive structural damage	Good
Direct vessel access	Difficult	Straight
Sea and weather conditions	Strong currents running through a marine canyon	Weak and moderate currents
	Cold water (range 4 to 8 °C)	Warm water (app. 24 °C)
	Poor visibility under water	Good visibility under water (in the 70-foot range)

The end result of the project should be to compile a report on the state of the wreck and its effect on the environment and to assess the threat of an oil spill from it. Wrecks should thus be classified into one of three groups:

1. High risk – posing a definite environmental hazard and requires immediate remediation action
2. Medium risk – posing a potential environmental hazard, requires periodic monitoring and the drawing up of an action plan in case an oil spill occurs
3. Low risk – not posing any environmental hazard

The gathering of the maximum amount of information is crucial in the context of deciding whether to remove the fuel or just monitor the wreck. Any decision in favour of remediation must be supported by a reliable assessment of risks and costs.

At high-risk sites, the level of risk is so high that the hazard has to be eliminated or minimized through direct intervention. This will involve either the salvage of the vessel or the removal of the hazardous cargo and bunkers from the wreck [35]. The wreck's condition, its location and the oil contained will largely establish the removal techniques and tools to be used. Calculating the costs of removing fuel on the basis of such factors as the ship's operating time, crew hire, special equipment, etc., is far easier than assessing the potential costs to the environment in the case of an uncontrolled leak. The costs of remediation depend on a multitude of factors, such as the depth of the wreck, the amount and type of load, and the wreck's accessibility. For example, the operation to remove oil from the SS *Jacob Luckenbach*, which sank in 1953 with 85,000 gallons (over 320,000 l) of heavy fuel oil, costs $19 million. In contrast, the removal of 1.83 million gallons (app. 7 mln l) of oil from the *Mississinewa* in 2003 costs $5 million [37]. Table 16.1 compares the parameters affecting the cost of removing oil from the wrecks of the SS *Jacob Luckenbach* and the USS *Mississinewa*.

16.6 Legal Aspects of the Problem of WWII Shipwreck Ownership and Responsibility

On the basis of the international law of the sea, the problem of wrecks is addressed by the two conventions referred to earlier:

1. The Convention of the Protection of the Underwater Cultural Heritage
2. The Nairobi International Wreck Removal Convention

The Convention of the Protection of the Underwater Cultural Heritage applies (as already mentioned above) only to wrecks at least 100 years old. Since World War II wrecks are only 70 years old, they are not subject to this convention. On the other hand, the rules of the Nairobi Convention are applicable only to wrecks of a commercial nature and not to historic wrecks and naval wrecks, including World War II wrecks. In view of this, it is relevant to pose the questions as follows: 'Who is the owner of the World War II wrecks?' and 'Who carries responsibility for these wrecks?' And on the basis of what regulations?

During WWII, all warships and most merchant vessels were under state control and direction, regardless of asset ownership. The sovereignty of warships and merchant vessels resided in the government that had control of the vessel at the time of its sinking [18]. According to art. 95 of UNCLOS, '*Warships on the high seas have complete immunity from the jurisdiction of any State other than the flag State*'. But art. 96 concerns the immunity of ships used only on government non-commercial service. Moreover, art. 236 of UNCLOS exempts warships, naval auxiliary, other vessels or aircraft owned or operated by a State and used, for the time being, only on government non-commercial service from the obligation to protect and preserve the marine environment. The principle, formulated in UNCLOS, that warships and state ships used for non-commercial purposes are wholly immune from the jurisdiction of any other state is fully recognized in international law and cannot be challenged [1]. In the context of existing convention norms, customary norms, international practice and jurisdiction, and the predominant part of the doctrine of international law, the phrase 'flag State' is also applicable to the wrecks of the warships of that State, which retains permanent title to them, regardless of the elapse of time [1]. The practice of States confirms the well-established rule of international law that title to such vessels and aircraft is lost only by capture or surrender during battle (before sinking), by international agreement or by an express act of abandonment, gift or sale by the sovereign in accordance with relevant principles of international law and the law of the flag State governing the abandonment of government property [30]. A coastal State does not acquire the right of ownership to the wreck of a warship merely because it is lying on the seabed or in waters over which it exercises sovereignty (inland waters, territorial waters, archipelagic waters and to some extent the contiguous coastal zone) [1]. Access to such vessels and aircraft and their associated artefacts located on or embedded in the seabed of foreign archipelagic waters, territorial seas or contiguous zones, is subject to coastal State control in accordance with international law [30]. Most [coastal] States grant the requests of flag States to protect such wrecks or for permission to visit them. Access to sunken ships and aircraft located in an exclusive economic zone and on the continental shelf of a coastal State does not depend on the consent of the coastal State, although the flag State may have to request permission of the coastal State to carry out drilling operations or to erect installations, and will have to respect the latter's sovereign laws and its exercise of jurisdiction [35].

16.7 Conclusion

The problem of the pollution of the marine environment caused by wrecks is complex. The search for information about the wreck and the cargo it was carrying may come up against problems due to the lack of or restricted access to archival materials. The investigation of the state of a wreck requires the involvement of many agencies, not to mention specialist equipment; it is a very costly process. The economic aspects are often a barrier to the implementation of measures aimed at preventing this kind of pollution; as a result, the problem of wrecks is often tackled only after seepage has occurred. As the example from the Pacific and East Asian region shows, the risk of seepage from shipwrecks is a genuine problem. Consequently, defining the degree of risk is absolutely crucial in view of the very high costs incurred, either as a result of seepage or for removing toxic substances from the wreck. If we are in possession of a detailed database concerning the characteristics of wrecks and their effect on the environment, we may be in a position to minimize the risks of environmental disasters resulting from oil spills. Moreover, the drawing up of contingency plans for coping with actual spills will enable losses to marine and offshore ecosystems to be reduced and will significantly reduce costs. One should also bear in mind that the problem of wrecks is not just an environmental and economic one. There is an ethical aspect to it: wrecks are the collective graves of combatants and civilians. It has a safety aspect: numerous wrecks carry weapons and ammunition, which can endanger shipping and possibly the teams of divers investigating them. Finally, there is a legal aspect, where the rights of ownership may be in contention.

On the other hand, wrecks from which contaminating substances have been removed become underwater reefs on which marine life, particularly sessile species like mussels and clams, can flourish better than on the surrounding seabed. Wrecks have also become tourist attractions: wreck diving is a rapidly growing form of recreation.

Acknowledgements The authors would like to thank the Polish Ministry of Science and Higher Education (project no. 4221/B/T02/2009/37) for funding this investigation. This work was also supported financially by the project 'The development of interdisciplinary doctoral studies in modern technologies at the Gdansk University of Technology' (project POKL.04.01.01-00-368/09).

References

1. Symonides J (2007) The legal status of sunken warships and state vessels in the law of the sea. Int Relat 3–4:29–48 [in polish]
2. Rogowska J, Namieśnik J (2010) Environmental implications of oil spills from shipping accidents. Rev Environ Contam Toxicol 206:95–114
3. E.I.A. (2010) Energy information administration official energy statistics from the U.S. government. http://www.eia.doe.gov/cabs/World_Oil_Transit_Chokepoints/Background.html. Accessed Jan 2011
4. GESAMP (2007) Estimates of oil entering the marine environment from sea-based activities. Report and studies GESAMP no. 75. GESAMP publication, London
5. Monfils R (2005) The global risk of marine pollution from WWII shipwrecks: examples from the Seven Seas. Manuscript no. 93. http://www.seaaustralia.com/documents/The%20Global%20Risk%20of%20Marine%20Pollution%20from%20WWII%20Shipwrecks-final.pdf. Accessed Dec 2010
6. Michel J, Gilbert T, Waldron J, Blocksidge ChT, Schmidt Etkin D, Urban R (2005) Potentially polluting wrecks in marine waters, an issue paper: international oil spill conference 15–19 May 2005, Miami, USA
7. Garza-Gil MD, Suris-Regueiro JC, Varela-Lafuente MM (2006) Assessment of economic damages from the Prestige oil spill. Mar Policy 30:544–551
8. Hassellöv IM (2007) Pre-study of ship wreck assessment and remediation. The Alliance For Global Sustainability, Göteborg
9. Payne JR, Driskell WB, Short JW, Larsen ML (2008) Long term monitoring for oil in the Exxon Valdez spill region. Mar Pollut Bull 56:2067–2081
10. Paine RT, Ruesink JL, Sun A, Soulanille EL, Wonham MJ, Harley ChDG, Brumbaugh DR, Secord DL (1996) Trouble on oiled waters: lessons from the Exxon Valdez oil spill. Annu Rev Ecol Syst 27:197–235
11. Sanderson H, Fauser P, Thomsen M, Sørensen PB (2009) Human health risk screening due to consumption of fish contaminated with chemical warfare agents in the Baltic Sea. J Hazard Mater 162:416–422
12. Valkovic V, Matika D, Kollar R, J Obhodas J, Sudac D (2009) Environmental Security of the Adriatic Coastal Sea Floor. ANIMMA international conference, 7–10 June 2009, Marseille, France
13. Glasby GP (1997) Disposal of chemical weapons in the Baltic Sea. Sci Total Environ 206:267–273
14. Andrulewicz E (2007) Chemical weapons dumped in the Baltic Sea. In: Gonenc IE (ed) Assessment of the fate and effects of toxic agents on water resources. Springer, Dordrecht
15. HELCOM CHEMU (1994) Report on chemical munitions dumped in the Baltic Sea. Report to the 15th meeting of Helsinki Commission 8–11 March 1994 from the Ad Hoc Working Group on Dumped Chemical Munition, January
16. Szarejko A, Namieśnik J (2009) The Baltic Sea as a dumping site of chemical munitions and chemical warfare agents. Chem Ecol 25:13–26
17. Tørnes JA, Voie ØA, Ljønes M, Opstad AM, Bjerkeseth Leif H, Hussain F (2002) Investigation and risk assessment of ships loaded with chemical ammunition scuttled in Skagerrak, (TA-1907/2002). http://www.klif.no/publikasjoner/avfall/1907/ta1907.pdf. Accessed Nov 2010
18. Monfils R, Gilbert T, Nawadra S (2006) Sunken WWII shipwrecks of the Pacific and East Asia: the need for regional collaboration to address the potential marine pollution threat. Ocean Coast Manage 49:779–788
19. Nawadra S, Gilbert T (2002) Risk of marine spills in the Pacific Islands region and its evolving response arrangements, SPILLCON 2002, Sydney, Australia
20. Church RA, Warren DJ (2008) The 2004 deepwrecks project: analysis of World War II era shipwrecks in the gulf of Mexico. Int J Hist Archaeol 12:82–102

21. Church RA, Warren DJ, Irion JB (2009) Analysis of deepwater shipwrecks in the Gulf of Mexico artificial reef effect of six World War II shipwrecks. Oceanography 22:50–63
22. Church RA, Warren DJ, Cullimore R, Johnston L, Schroeder W, Patterson W, Shirley T, Kilgour M, Morris N, Moore J (2007) Archaeological and biological analysis of World War II shipwrecks in the Gulf of Mexico: artificial reef effect in deep water. U.S. Dept. of the Interior, Minerals Management Service, Gulf of Mexico OCS Region, New Orleans, LA. OCS Study MMS 2007-015
23. Lindström P (2006) Vrak i Skagerrak. Sammanfattning av kunskaperna kring miljöriskerna med läckande vrak i Skagerrak. Forum Skagerrak II Report [in Swedish]. http://www.forumskagerrak.com/download/533/x/Vrak%20i%20Skagerrak-web.pdf. Accessed Dec 2010
24. Idaas K (2005) Norwegian pollution control authority work on shipwrecks. http://www.iosc.org/papers/01378.pdf. Accessed Nov 2010
25. Olejnik A (2008) Visual identification of underwater objects using a ROV-type vehicle: "Graf Zeppelin" wreck investigation. Pol Marit Res 15:72–79
26. Gröner E (1988) Die Deutschen Kriegsschiffe 1815–1945. Bernard&Graefe Verlag, Koblenz, Band 5 [in german]
27. Wilhelm Gustloff (2010) http://www.wilhelmgustloff.com. Accessed Jan 2011
28. Jamkowski M (2010) Duchy z głębin Bałtyku. Strauben , Gustloff, Goya. Świat Książki, Warszawa [in polish]
29. Michel JL, Ballard RD (1994) The RMS titanic 1985 discovery expedition. IEEE oceans conference, 13–16 Sept 1994, Brest, France
30. Roach JA (1996) Sunken warships and military aircraft. Mar Policy 20:351–354
31. Orzeł Expedition (2008) www.orzeł.balex.eu. Accessed Dec 2010
32. Overfield ML (2005) Corrosion on deep gulf shipwrecks of World War II. International oil spill conference, 15–19 May 2005. Miami, USA. http://www.iosc.org/papers/IOSC%202005%20a377.pdf. Accessed Nov 2010
33. Wheeler AJ (2002) Environmental controls on shipwreck preservation: the Irish context. J Archaeol Sci 29:1149–1159
34. Kuroda T, Takai R, Kobayashi Y, Tanaka Y, Hara Sh (2008) Corrosion rate of shipwreck structural steels under the sea. OCEANS 2008 – MTS/IEEE Kobe Techno-Ocean, 8–11 April 2008, Kobe, Japan
35. SPREP (2002) Regional strategy to address marine pollution from World War II shipwrecks. Thirteenth SPREP meetings of officials (Item 7.2.2.1) Majuro, Marshall Islands 21–25 July 2002. (Pub. SPREP). http://www.sprep.org. Accessed Nov 2010
36. Peterson ChH, Rice SD, Short JW, Esler D, Bodkin JL, Ballachey BE, Irons DB (2003) Long-term ecosystem response to the Exxon Valdez oil spill. Science 302:2082–2086
37. Albertson A (2004) Potentially polluting wrecks warrant further dialog. Mar Technol Soc J 38:15–16
38. Meza M (2006) Relative challenges to oil recovery from wrecks: a comparison of the USS Mississinewa and the SS Jacob Luckenbach (Paper 52). INTERSPILL 2006, 21–23 March 2006, London, UK
39. Center for American Progress (2010) Oil spills by the numbers. The devastating consequences of Exxon Valdez and BP Gulf. http://www.americanprogress.org/issues/20S10/04/oil_numbers.html. Accessed Dec 2010
40. HELCOM www.helcom.fi

Chapter 17
Methods to Countermeasure the Intrusion of Seawater into Coastal Aquifer Systems

Andreas Kallioras, Fotios-Konstantinos Pliakas, Christoph Schuth, and Randolf Rausch

17.1 Introduction

Seawater intrusion is a widely occurring environmental problem, which concerns the majority of the coastal aquifers in the world, having large negative impacts – from an economic, developmental, environmental as well as social point of view – on the coastal local communities. The fact that almost 40% of the world's population lives within a distance of 80 km from the coastline gives rise to the significance of the aforementioned problem and emphasises the necessity for its prevention. Seawater intrusion have been reported in many European and Mediterranean coastal aquifers, including Spain [1–4], Portugal [5], Italy [6, 7], Israel [8], Turkey [9], Egypt [10], Libya [11] and Morocco [4].

Coastal aquifers are in hydraulic connection with the sea, and therefore, a mixture between fresh and brackish/saline groundwater takes place at all times. Under natural conditions (i.e. no anthropogenic intervention), these two fluids are in hydrodynamic balance separated by a transition zone with increasing salinity or salts concentration is observed. Any lowering in the piezometric (or potentiometric) surface of the

A. Kallioras (✉)
School of Mining and Metallurgical Engineering, National Technical University of Athens, Athens, Greece
e-mail: kallioras@metal.ntua.gr

F.-K. Pliakas
Department of Civil Engineering, Democritus University of Thrace, Xanthi, Greece

C. Schuth
Institute of Applied Geosciences, Darmstadt Technical University, Darmstadt, Germany

R. Rausch
Deutsche Gesellschaft fuer Internationale Zusammenarbeit (GIZ) GmbH, Riyadh Office, Riyadh, Kingdom of Saudi Arabia

S.K. Sharma and R. Sanghi (eds.), *Wastewater Reuse and Management*, DOI 10.1007/978-94-007-4942-9_17, © Springer Science+Business Media Dordrecht 2013 freshwater coastal aquifer will disrupt this hydrodynamic balance between the two fluids, and seawater wedge will encroach furthermore towards the mainland, fact which will cause a qualitative degradation to the fresh groundwater of the coastal aquifer. A matter of paramount importance for the rate of intrusion is the form and transformation of the hydraulic gradient of the aquifer. Usually, under natural conditions, the hydraulic gradient is observed towards the coast, and there is a constant discharge of freshwater into the sea. When the aforementioned hydraulic gradient has been reversed, then the term of active seawater intrusion – as described by Fetter [12] – is more applicable, referring to the rapid and extensive encroachment of the seawater wedge into the mainland.

As most of the engineering hydraulic works, seawater intrusion countermeasures depend mainly on the extent of the seawater wedge as well as on the availability of alternative freshwater resources in the vicinity of the study area. Whatever the method of remediation followed, a strong database of field data [13] is essential for the analysis of the intrusion mechanism as well as on the origin of salinity.

17.2 Management of Coastal Aquifers Subjected to Seawater Intrusion

When pumpage is initiated in a freshwater lens with a sharp interface floating on salt water, the freshwater removed from the system comes from (1) drainage from the water table, (2) drainage at the interface where freshwater is being replaced by salt water and (3) water released due to elastic storage effects [14]. Therefore, the saline wedge moves towards the mainland, and the transition zone between the two fluids of different density becomes more thick. The hydrodynamic balance between the two fluids is the most governing factor with respect to the selection of the appropriate remediation technology which will be applied. Todd and Mays [15] categorise a list a methods for the prevention or remediation of aquifers subjected to groundwater salinisation, which can be summarised as follows:

1. Modification of pumping pattern
2. Artificial recharge
3. Extraction barrier
4. Injection barrier
5. Subsurface barrier

Oude Essink [16] provides an additional list for applicable water works engineering methods to combat seawater intrusion in coastal aquifers:

1. Freshwater injection barriers through injection or (deep-well) infiltration of fresh (purified sewage) water near the shoreline.
2. Extraction of saline and brackish groundwater. However, this could regrettably result in undesirably low piezometric heads, especially in shallow coastal aquifers. Furthermore, the disposal of the extracted saline or brackish groundwater could meet with problems.

3. Modifying pumping practice through reduction of withdrawal rates or adequate relocation of extraction wells. The desired extraction rate should preferably be extracted by well-distributed shallow wells to prevent excessive upconing. Groundwater withdrawal for domestic, agricultural and industrial water supply may not be reduced during periods of droughts so that salt water intrusion tends to occur anyway.
4. Land reclamation, thus creating a foreland where a freshwater body may develop which could delay the inflow of saline groundwater.
5. Increase of (artificial) recharge in upland areas to enlarge the outflow of fresh groundwater through the coastal aquifer and thus to reduce the length of the salt water wedge.
6. Creation of physical barriers, such as sheet piles, clay trenches and injection of chemicals. This solution is only applicable in shallow aquifers and at high cost.

17.3 Methods for Seawater Intrusion Prevention

With respect to the phenomenon of seawater intrusion, one could argue that the available methods proposed for the prevention of further encroachment can be further categorised into direct and indirect ones. Direct methods are the ones which are directly applied (i.e. by physical means) to the aquifer system and provide immediate remediation of the groundwater quality (such as all types of coastal barriers as analysed in the following sections), whereas indirect are the ones which are not related to the aforementioned (such as alternative managerial proposals with respect to the pumping and abstraction practices).

17.3.1 Direct Methods

All of these methods with respect to the installations of different types of engineering water works are related to large financial costs, which in some cases are not feasible to be applied in the field. Sometimes the financial feasibility of the applied method might proved even more important than the actual engineering result itself.

17.3.1.1 Extraction Barriers

Extraction or abstraction barriers are simply a series of groundwater wells which are installed at a close distance with the coastline and continuously pump groundwater of varying quality (fresh and brackish). The installation of such wells should consider that the filter of the well is located in the vicinity of the toe of the seawater wedge and in the zone of transition. Todd and Mays [15] mention that in this way, a pumping trough is formed along the shoreline, whereas seawater flows towards the trough (inland) and freshwater flow towards the opposite direction (i.e. seawards).

Abstraction or extraction barriers have been installed and applied in different cases worldwide in order to prevent seawater intrusion at coastal aquifers where extensive pumping of groundwater has resulted in seawater encroachment. Kashef [17] argues that the main disadvantage of this method is the high financial cost which is usually involved. Most of the cases which were proposed for the application of such method were later proved financially unfeasible. Petalas et al. [18] also argue that this method could be proved unfeasible in the case of coastal aquifers which are characterised by a low groundwater resources potential, and therefore, any further extraction of groundwater could have negative impacts (e.g. significant lowering of the piezometric surface of the aquifer system).

17.3.1.2 Injection Barriers

Injection barriers are similar in design with the previous ones (extraction wells) by means that they are also related to the installation of a series of groundwater wells along the coastline of the coastal aquifer. However, the principle and therefore the hydrogeological mechanism is quite opposite in comparison with the one above. Freshwater is continuously injected into the aquifer and again in the vicinity of the shoreline; however, in this case, the filter of the injection well should be located exactly at the toe of the interface between the two fluids (fresh and saline groundwater). Todd and Mays [15] refer to the formation of a pressure ridge which is maintained as water is constantly injected into the aquifer, and this can be thought schematically as a hydraulic barrier.

This technique requires significant amounts of alternative sources of water of very high quality. The quality of the injected water should be such that it will not create qualitative degradation of the groundwater resources of the aquifer. High quality of water to be injected is mainly suggested for the prevention of well clogging and hence decrease in the productivity of injection. The alternative freshwater resources should be located in the vicinity of the study area in order to also minimise related water transfer costs. Bruington [19] suggests redevelopment of the injection wells when clogging takes place and the effectiveness of the injection barrier is significantly decreased.

The quality standards of the injected water should be according to the drinking water standards [20, 21] in order to eliminate contamination or pollution of the aquifer which is under remediation.

17.3.1.3 Underground Barriers

Another type of coastal barrier for the prevention of further seawater encroachment in a coastal aquifer is the installation of underground construction works for the immediate prevention of the seawater wedge intrusion. In principle, this technique could be thought as the design of an underground dam which directly stops the intrusion of the seawater immediately at the toe of the interface between fresh- and seawater. This method simply involves a physical impermeable vertical subsurface barrier along the coastline, with a length which depends on the extent of the seawater wedge along the coast. Todd and Mays [15] quote that the materials to be used could include sheet piling, puddled clay, emulsified asphalt, cement grout, bentonite, silica gel, calcium acrylate or plastics.

As in any case of underground construction works, the feasibility of this technique is directly related to:

- Depth of the construction
- The length of the barrier along the coastline
- The availability of construction materials
- The geochemical properties of soils

Kashef [17] argues that some aspects should be considered prior to the construction of such physical subsurface barriers, such as the ones listed below:

- If the barrier extends through the entire depth of a deep artesian aquifer, the cost is prohibitive, but the results of increasing the potentiometric heads are desirable.
- If soil grouting is used, there should be sufficient effective weight above the upper boundary of the artesian aquifer to counterbalance the increasing upward pressures produced during construction.
- If the depth of the barrier covers part of the outflow surface in artesian aquifers, the freshwater flow will emerge from underneath the barrier with very high velocities that may lead to backward erosion and ultimately to land subsidence.

17.3.1.4 Biological Barriers

Petalas et al. [18] argue that due to the fact that seawater intrusion threatens coastal groundwater resources which could be used for drinking purposes, the prevention of such a phenomenon should involve the application of pilot and state-of-the-art technologies.

James et al. [22] and Cunningham et al. [23] describe the application of biofilm barriers (through the subsurface injection of bacteria and/or nutrient solutions), which can aid the prevention of saltwater intrusion by reducing the subsurface hydraulic conductivity, thereby enhancing freshwater injection control strategies. The subsequent growth of the bacterial biofilms and production of extracellular polymeric substances (EPS) occludes pore throats in the porous matrix, providing significant reductions in the effective hydraulic conductivity (down to 99.4% decrease). The application of biofilm barriers relies on the principle of hydraulic conductivity decrease and therefore the transformation of the aquifer matrix into a physical subsurface barrier.

17.3.1.5 Artificial Recharge

Pliakas et al. [24] quote that artificial recharge of groundwater can be achieved with surface infiltration systems (basins, spreading facilities), wells, trenches, pits, shafts, stream-channel modification or augmentation, Soil Aquifer Treatment (SAT) systems, Aquifer Storage and Recovery (ASR) wells, vadose-zone wells, combination of surface and subsurface recharge, induced recharge, incidental recharge, etc. According to Todd and Mays [15] groundwater levels can be raised and maintained by artificial recharge, using surface spreading for unconfined aquifers and recharge wells for confined aquifers; however, this necessitates development of a supplemental water source.

Haddad [25] describes the Monterey County Water Recycling Project (MCWRP), a centralised tertiary-treatment facility that began operating in April, 1998, and delivers urban recycled water to growers at the mouth of the Salinas River, a site of extensive seawater intrusion to coastal aquifers.

Zacharias and Koussouris [26] propose an alternative to traditional water supply models in European small islands consisting of desalination of brackish groundwater, instead of seawater, coupled with an effective strategy for controlling saltwater intrusion that enhances the hydrologic budget through reuse of wastewater.

17.3.1.6 Seawater Intrusion Prevention at River Estuaries

River estuaries are surface water bodies which are very susceptible in changes of the hydrological and climate regime of the area. They are considered a unique case of seawater intrusion by means that they provide a different pathway for the seawater to encroach towards the mainland. Intrusion takes place in a twofold manner, both superficially (within the river flow) and subficially (either by the typical intrusion into the aquifer or through percolation from the river to the water table in the pattern of groundwater-surface water interactions).

During extended dry seasons, the river flow is usually significantly decreased, fact which negatively alters the hydrodynamic balance between the two fluids (fresh and saline water) and hence enhances the further encroachment of seawater through the river flow.

Prevention of seawater intrusion in such cases can be achieved by the following direct methods:

- Installation of superficial barriers (usually rubber dams, which can be transferred in different locations within the river bed) in order to provide a direct barrier to seawater encroachment.
- Construction of appropriate canals which provide pathway for the saline water to migrate towards selected areas of the study area.
- River interventions upstream of the estuary such as the construction of controlling dams which can store water during wet periods and release significant quantities of water during dry periods in order to sustain a constant environmental flow throughout the entire hydrologic year.

17.3.1.7 Non-conventional Countermeasures

Patel and Shah [27] describe a list of alternative seawater direct barriers which are used in different case studies worldwide, such as:

- *Slurry Walls*: The construction of a slurry wall involves the cutting of a bentonite slurry stabilised trench to prevent seawater encroachment.
- *Grout Curtains*: Grout curtains are constructed by drilling 2- to 6-in.-diameter holes along a single line or multiple parallel lines.
- *Air Injection*: Compressed air injected into the groundwater attempts to cause a piezometric rise in water level that can be used to alter groundwater gradients and flow directions.

17.3.2 Indirect Methods

Indirect methods for combating seawater intrusion refer to a series of managerial practices which can be proved effective to prevent further seawater encroachment. Although these can be generalised to some extent, they are mostly limited in a local extent. Furthermore, this category of prevention methods includes the use of computer codes as they can be effectively applied as supporting scientific tools to all aforementioned prevention methods (direct and indirect) by means that they can forecast the impacts of such a prevention method in the aquifer system.

17.3.2.1 Monitoring of Pumping

In the case of seawater intrusion, the two fluids (fresh- and seawater) are in hydrodynamic balance, which is mainly governed by the rate of groundwater abstraction within the aquifer system. The seawater wedge moves further inland as pumping practices are enhanced, while it moves further backwards the sea as the groundwater resources potential of the coastal aquifer is replenished and the piezometric surface is elevated. For the proper and effective management of such an aquifer system, a network of monitoring wells should be carefully designed in order to observe in details the hydrogeologic (quantitative and qualitative) regime of the aquifer.

17.3.2.2 Relocation of Pumping Scheme

Relocation of the pumping scheme is mainly related to the areas where certain groundwater well groups are located with respect to their distance from the shoreline. In general terms, it is advised that wells which are located inland of the coastal aquifer should be operated during dry seasons while the ones located close to the coast should be operated mainly during the wet season or during a period when the piezometric surface of the aquifer is rather elevated. This practice is directly related to the hydraulic gradient of the coastal aquifer and the conservation of a constant aquifer outflow towards the sea. Headworth and Fox [28] introduced appropriate relocating schemes to reduce the effects of saline intrusion in the Chalk of Great Britain by utilising two sets of boreholes, one inland and one coastal. Patel and Shah [27] mention that the economic feasibility of countermeasures should be investigated (for instance, it is recommended to derive the optimum position of well lines and rates of extraction or infiltration), whereas the countermeasures should be adapted and optimised in the course of the realisation of the measure, based on changes in the salinity of the subsoil. Todd and Mays [15] also argue that changing the locations of pumping wells, typically by dispersing them in inland areas, can aid in re-establishing a stronger seaward hydraulic gradient.

17.3.2.3 Modification of Pumping Scheme

Similar to the aforementioned, beneficial effects can be also achieved by modifying the entire pumping scheme of the aquifer system in a way which provides optimal extraction rates without causing any further encroachment of seawater into the mainland. As upconing of saline interface has a major impact to the quality of the extracted groundwater, all groundwater wells which are located above the seawater wedge should be as shallow as possible, closer to the water table. According to Patel and Shah [27], in most situations, groundwater withdrawal for domestic, agricultural and industrial water supply has not been reduced during periods of droughts, so that salt water intrusion tends to occur anyway. In the case where an aquifer is underlain by saline fluid, upconing can be limited by proper design and operation of wells [29] such as low abstraction rates of shallow groundwater wells.

An alternative proposed by Das Gupta [30] includes the installation of a horizontal pumping well acting as an infiltration gallery, in contrast to a heavy pumping vertical well which could produce further upconing.

17.3.2.4 Seawater Intrusion Prevention at River Estuaries

As already mentioned, coastal river estuaries may be seen as a unique study case, by means that it involves certain characteristic (combination of surface and subsurface seawater intrusion). For that reason, methods of prevention in such cases are listed separately. Indirect countermeasures in coastal river estuaries can be achieved by:

- Use of alternative freshwater resources
- Restriction of groundwater abstractions within the river basin during dry periods or periods of tidal effects and sea level rise

17.3.2.5 Abandoned Groundwater Wells

Abandoned or improperly constructed groundwater wells can provide a vertical pathway for groundwater to migrate from different saline aquifer layers to freshwater ones. This is mainly due to the change of hydraulic head within the aquifer layers which contain saline groundwater. Additionally, improper insulation of boreholes and abandoned wells provide a pathway for the saline water to move upwards. As Richter and Kreitler [31] quote that they provide an artificial hydraulic connection between non-contaminated and contaminated aquifer layers.

17.3.2.6 Computer Codes

As mentioned earlier, the application of appropriate computer codes for the analysis of groundwater flow and pollutant transport in coastal aquifers could be used as a supporting managerial tool accompanying all series of direct or indirect countermeasures. Groundwater models can be used to forecast the effectiveness of a proposed seawater intrusion method or even suggest alternative countermeasures by applying different hydrogeological condition scenarios. As obvious, this involves the development of an integrated database of field measurements, historical hydrogeological data and a monitoring network of groundwater wells for the quantitative and qualitative inspection of the investigated aquifer system.

Simulation-optimization methods, applied with adequate sensitivity tests, can provide useful quantitative guidance for controlling seawater intrusion, as demonstrated in an application to the West Coast Basin of coastal Los Angeles by Reichard and Johnson [32] who considered two management options for improving hydraulic control of seawater intrusion: increased injection into barrier wells and in lieu delivery of surface water to replace current pumpage.

Bray and Yeh [33] used a calibrated simulation model linked with two optimisation models to investigate alternatives for enhancing seawater intrusion barrier operations for the Alamitos Barrier Project in Los Angeles, and they analysed two types of management problems: the optimal scheduling problem – OSP – and the optimal well location problem. (The OSP objective was to minimise the total injected water subject to constraints on the state variables: hydraulic head and chloride concentration at target locations.)

Hallaji and Yazicigil [34] developed seven groundwater models to determine the optimal planning and operating policies of a coastal aquifer in Southern Turkey threatened by seawater intrusion. They developed optimal pumpage policies under three management objectives that maximised agricultural water withdrawals and minimised drawdowns and pumping costs, subject to constraints related to the system's response equations; demand requirements; drawdown limitation in saltwater intrusion control locations and pumping wells; and discharge bounds.

Aliewi et al. [35] investigated the feasibility of applying skimming and scavenger pumping as a means to exploit the freshwater lenses and to control saline water upconing in the aquifers, and they used two simulation models that coupled density-dependent fluid flow and solute transport to simulate and predict the movement of saltwater under different hydrogeological and operational conditions of skimming and scavenger wells in the two aquifers.

Mercurio et al. [36] used MODFLOW in order to construct groundwater flow and geochemical models, delineate wellhead protection area and develop a comprehensive management program.

Shammas and Jacks [37] used the codes MODFLOW and MT3DMS for solute transport to determine the movement of the freshwater/saltwater interface. The study proposed the protection of the groundwater in Salalah plain aquifer in Oman, from further encroachment, by artificial recharge with reclaimed water, preferably along the Salalah coastal agricultural strip.

Kallioras et al. [38] used MODFLOW in order to simulate the deterioration of the groundwater resources potential of a coastal aquifer (N. Greece) subjected to overexploitation.

17.4 Conclusions

Seawater intrusion is a widespread environmental problem especially for the Mediterranean countries, where local communities totally depend on the groundwater resources of coastal management in order to sustain their agricultural activities. Global changes, with special reference to the increase in global population, have led to the increase in water resources demand to be used either for drinking purposes or for agricultural activities. This resulted in overexploitation of coastal aquifers in such regions, i.e. quantitative deterioration of the groundwater resources potential, which in turn resulted in qualitative degradation of the underlying aquifer layers due to seawater intrusion. This environmental problem has nowadays multifold dimensions, which are expanded to a social, economic as well as developmental level.

Due to the above, a number of different approaches have been followed in order to combat further seawater intrusion in coastal aquifers, all applying single or a combination of different engineering water work methodologies. These generally fall in two main categories such as the direct and indirect ones, the former referring to the application of specific engineering water works (such as hydraulic barriers, application of artificial recharge and others) while the latter refers to managerial practices which can be followed in order to combat further encroachment of the seawater wedge. The selection criteria are mainly dependent on the availability of two specific sources: financial and freshwater.

One could argue that the category of indirect countermeasures, including alteration of pumping conditions such as modification of pumping scheme, relocation of groundwater together with a constant and up-to-date quantitative as well as qualitative monitoring of the aquifer's groundwater resources would result in long-term positive effects. This would certainly contribute to achieve rational management of the groundwater resources of the contaminated aquifer system at a local scale.

References

1. Calvache ML, Pulido-Bosch A (1997) Effects of geology and human activity on the dynamics of seawater intrusion in three coastal aquifers in southern Spain. Environ Geol 30(3/4): 215–223
2. Gimenez E, Morell I (1997) Hydrogeochemical analysis of salinisation processes in the coastal aquifer of Oropesa (Castellon, Spain). Environ Geol 29(1–2):118–131
3. Iribar V, Carrera J, Custodio E, Medina A (1997) Inverse modeling of seawater intrusion in the Llobregat delta deep aquifer. J Hydrol 198(1–4):226–244
4. Pulido-Bosch A, Tahiri A, Vallejos A (1999) Hydrogeochemical characteristics of processes in the Temara aquifer in northwestern Morocco. Water Air Soil Pollut 114(3–4):323–337
5. Stigter TY, van Ooijen SPJ, Post VEA, Appelo CAJ, Carvalho Dill AMM (1998) A hydrogeo logical and hydrochemical explanation of the groundwater composition under irrigated land in a Mediterranean environment, Algavre, Portugal. J Hydrol 208(3–4):262–279
6. Barazzuoli P, Bouzelboudjen M, Cucini S, Kiraly L, Menicory P, Salleolini M (1999) Olocenic alluvial aquifer of the River Cornia coastal plain (southern Tuscany, Italy)-database design for groundwater management. Environ Geol 39(2):123–143
7. Chiocchini U, Gissoti G, Macioce A, Manna F, Bolasco A, Lucarini C, Patrizi GM (1997) Environmental geology problems in the Tyrrhenian coastal area of Santa Marinella, province of Rome, central Italy. Environ Geol 32(1):1–8
8. Yakirevich A, Melloul A, Sorek S, Shaath S, Borishov V (1998) Simulation of seawater intrusion into the Khan Yunis area of the Gaza Strip coastal aquifer. Hydrogeol J 6(4):549–559
9. Karahanoglou N (1997) Assessment of seawater intrusion in a coastal aquifer by using correlation, principal component, and factor analyses. Water Environ Res 69(3):331–342
10. El-Bihery MA, Lachmar TE (1994) Groundwater quality degradation as a result of overpumping in the Delta Wadi El-Arish Area, Sinai Peninsula, Egypt. Environ Geol 24(4): 293–305
11. El Asswad RM (1995) Agricultural prospects and water resources in Libya. Ambio 24(6): 324–327
12. Fetter CW (2001) Applied hydrogeology, 4th edn. Prentice-Hall, Inc, Upper Saddle River, 598 p
13. Kallioras A, Pliakas F, Diamantis I, Kallergis G (2010) SWOT analysis in groundwater resources management of coastal aquifers: a case study from Greece. Water Int 35(4):425–441
14. Essaid HI (1986) A comparison of the coupled fresh water-salt water flow and the Ghyben-Herzberg sharp interface approaches to modeling of transient behavior in coastal aquifer systems. J Hydrol 86:169–193
15. Todd DK, Mays LW (2005) Groundwater hydrology, 3rd edn. Wiley, New York, 636 p
16. Oude Essink GHP (2001) Improving fresh groundwater supply – problems and solutions. Ocean Coast Manag 44:429–449
17. Kashef AI (1977) Management and control of salt-water intrusion in coastal aquifers. Crit Rev Environ Control 7:217–275
18. Petalas C, Pliakas F, Kallioras A, Diamantis I (2002) Engineering, managerial and economical elements of recent trends for the prevention of seawater intrusion in coastal aquifers. In: Proceedings of 6th hydrogeological conference of the Greek Hydrogeological Committee of the Geological Society of Greece, Xanthi, Greece, pp 23–34, [in Greek]

19. Bruington AE (1968) The amelioration of prevention of salt-water intrusion in aquifers-experience in Los Angeles County, California. In: Salt-water encroachment into aquifers: proceedings of the limited professional symposium held at Louisiana State University, Baton Rouge, Louisiana, 4–5 May 1967. Water resources Research Institute, Baton Rouge, Louisiana, Bulletin 3, Oct 1968
20. Bloetscher F (1999) Comments on the May 17, 1999. Draft of the class V underground injection control study: saltwater intrusion barrier wells information summary. Florida Governmental Utility Authority. 16 June 1999
21. Crook J, Asano T, Nellor M (1991) Ground water recharge with reclaimed water in California. Municipal wastewater reuse: selected readings on water reuse. US Environmental Protection Agency. E.P.A. 430/09-91-022, pp 67–74
22. James GA, Warwood BK, Hiebert R (2000) In-situ biological barriers to the spread of pollution. In: Valdez JJ (ed) Bioremediation. Kluwer Academic Publishers, Boston, pp 1–13
23. Cunningham A, Warwood B, Sturman P, Horrigan K, James G, Costerton JW, Hiebert R (1997) Biofilm processes in porous media – practical applications. In: Amy PS, Haldeman DL (eds) The microbiology of the terrestrial deep subsurface. Lewis Publishers, New York, pp 325–344
24. Pliakas F, Petalas C, Diamantis I, Kallioras A (2005) Modeling of groundwater artificial recharge by reactivating an old stream bed. Water Res Manag 19:279–294
25. Haddad BM (2002) Monterey county water recycling project: institutional study. J Water Res Plann Manag (ASCE) 128(4):280–287
26. Zacharias I, Koussouris T (2000) Sustainable water management in the European Islands. Phys Chem Earth (B) 25(3):233–236
27. Patel AS, Shah DL (2008) Water management – conservation, harvesting and artificial recharge. New Age International (P) Ltd. Publishers, New Delhi, 331 p
28. Headworth HG, Fox GB (1986) The South Downs Chalk aquifer: its development and management. J Inst Water Eng Sci 40:345–361
29. Bowen R (1986) Groundwater, 2nd edn. Elsevier Applied Science Publishers, London, 427 pp
30. Das Gupta A (1983) Steady interface upcoming beneath a coastal infiltration gallery. Groundwater 21(4):465–474
31. Richter BC, Kreitler CW (1993) Geochemical techniques for identifying sources of groundwater salinization, Library of Congress, Cataloging-in-Publication data, CRC Press, Inc., p 257
32. Reichard EG, Johnson TA (2005) Assessment of regional management strategies for controlling seawater intrusion. J Water Res Plann Manag (ASCE) 131(4):280–291
33. Bray BS, Yeh WW-G (2008) Improving seawater barrier operation with simulation optimization in southern California. J Water Res Plann Manag (ASCE) 134(2):171–180
34. Hallaji K, Yazicigil H (1996) Optimal management of coastal aquifer in southern Turkey. J Water Res Plann Manag (ASCE) 122(4):233–244
35. Aliewi AS, Mackay R, Jayyousi A, Nesereddin K, Mushtaha A, Yaqubi A (2001) Numerical simulation of the movement of saltwater under skimming and scavenger pumping in the Pleistocene aquifer of Gaza and Jericho Areas, Palestine. Transp Porous Media 43:195–212
36. Mercurio JW, Beljin MS, Barry J, Maynard J (1999) Groundwater models and wellfield management: a case study. Environ Eng Policy 1(3):155–164
37. Shammas MI, Jacks G (2007) Seawater intrusion in the Salalah plain aquifer, Oman. Environ Geol 53:575–587
38. Kallioras A, Pliakas F, Diamantis I (2010) Simulation of groundwater flow in a sedimentary aquifer system subjected to overexploitation. Water Air Soil Pollut 211:177–201

ERRATUM TO

Chapter 8
Hospital Wastewaters: Quali-Quantitative Characterization and for Strategies for Their Treatment and Disposal

P. Verlicchi, A. Galletti, and M. Al Aukidy

DOI 10.1007/978-94-007-4942-9_18

List of corrections not included in the final version already published:

Chapter title the right version is:
Hospital Wastewaters: Quali-Quantitative Characterization and Strategies for Their Management and Treatment

Abstract (it is not present in the book but in corrected proofs it is). The mistakes refer to three measure of unit:
replace μg L#1 with μg L^{-1}.

Keywords: (they are not present). I suggested:
Hospital effluent, Urban effluent, Pharmaceuticals, Management, Proper treatment, Co-treatment

Page 233: There are four reference citations in the below paragraph, 51 was missing:
Pharmaceutical Compounds: A great number of PhCs are used in hospitals for treatment, diagnosis and research purposes. Table 8.3 reports the most widely investigated compounds, grouped according to their specific therapeutic use. In this table, the third and the fourth columns report the range of concentrations of each of the compounds cited, measured in HWws and, for comparison, in UWws [5, 39, 51, 52].

Page 233: Reference number added to the citation:
In 2003, Kummerer and Henninger (67), . . .

S.K. Sharma and R. Sanghi (eds.), *Wastewater Reuse and Management*,
DOI 10.1007/978-94-007-4942-9_8,

Index

S.K. Sharma and R. Sanghi (eds.), *Wastewater Reuse and Management*,
DOI 10.1007/978-94-007-4942-9, © Springer Science+Business Media Dordrecht 2013

MIX
Papier aus verantwortungsvollen Quellen
Paper from responsible sources
FSC® C105338

If you have any concerns about our products,
you can contact us on
ProductSafety@springernature.com

In case Publisher is established outside the EU,
the EU authorized representative is:
Springer Nature Customer Service Center GmbH
Europaplatz 3, 69115 Heidelberg, Germany

Printed by Libri Plureos GmbH
in Hamburg, Germany